VOLUME FOUR HUNDRED AND FORTY-ONE

METHODS IN
ENZYMOLOGY

Nitric Oxide, Part G
Oxidative and Nitrosative
Stress in Redox Regulation
of Cell Signaling

METHODS IN ENZYMOLOGY

Editors-in-Chief

JOHN N. ABELSON AND MELVIN I. SIMON

Division of Biology
California Institute of Technology
Pasadena, California

Founding Editors

SIDNEY P. COLOWICK AND NATHAN O. KAPLAN

VOLUME FOUR HUNDRED AND FORTY-ONE

METHODS IN ENZYMOLOGY

Nitric Oxide, Part G Oxidative and Nitrosative Stress in Redox Regulation of Cell Signaling

EDITED BY

ENRIQUE CADENAS

Professor and Chairman
Molecular Pharmacology and Toxicology
School of Pharmacy
University of Southern California
Los Angeles, CA

LESTER PACKER

Department of Molecular Pharmacology and Toxicology
School of Pharmacy
University of Southern California
Los Angeles, CA

AMSTERDAM • BOSTON • HEIDELBERG • LONDON
NEW YORK • OXFORD • PARIS • SAN DIEGO
SAN FRANCISCO • SINGAPORE • SYDNEY • TOKYO
Academic Press is an imprint of Elsevier

ELSEVIER

Academic Press is an imprint of Elsevier
525 B Street, Suite 1900, San Diego, California 92101-4495, USA
84 Theobald's Road, London WC1X 8RR, UK

This book is printed on acid-free paper. ∞

For information on all Elsevier Academic Press publications
visit our Web site at www.elsevierdirect.com

ISBN-13: 978-0-12-374309-1

PRINTED IN THE UNITED STATES OF AMERICA
08   09   10   11   9   8   7   6   5   4   3   2   1

CONTENTS

22. Microscopic Technique for the Detection of Nitric Oxide-Dependent Angiogenesis in an Animal Model   393

Seung Namkoong, Byoung-Hee Chung, Kwon-Soo Ha, Hansoo Lee, Young-Guen Kwon, and Young-Myeong Kim

Contributors

Lisong Ai
Department of Biomedical Engineering and Division of Cardiovascular Medicine, School of Engineering & School of Medicine, University of Southern California, Los Angeles, California

Christopher Asmus
Department of Pharmaceutical Chemistry, University of Kansas, Lawrence, Kansas

Rui M. Barbosa
Faculty of Pharmacy and Center for Neurosciences and Cell Biology, University of Coimbra, Coimbra, Portugal

Silvina Bartesaghi
Department of Histology and Embryology and Department of Biochemistry and Center for Free Radical and Biomedical Research, Facultad de Medicina, Universidad de la República, Montevideo, Uruguay

Carlos Batthyány
Institut Pasteur de Montevideo and Center for Free Radical and Biomedical Research, Facultad de Medicina, Montevideo, Uruguay

Farideh Beigi
Division of Cardiology and Interdisciplinary Stem Cell Institute, Miller School of Medicine, University of Miami, Miami, Florida

Nigel Benjamin
Peninsula Medical School, Universities of Exeter and Plymouth, St. Luke's Campus, Exeter, United Kingdom

Timothy R. Billiar
Department of Surgery, Medical School, University of Pittsburgh, Pittsburgh, Pennsylvania

Charles A. Bosworth
Department of Physiology and Biophysics, University of Alabama at Birmingham, Alabama

Catherine Bregere
Department of Pharmacology and Pharmaceutical Sciences, University of Southern California, Los Angeles, California

Vittorio Calabrese
Department of Chemistry, Clinical Biochemistry and Clinical Molecular Biology
Chair, University of Catania, Italy

Orazio Cantoni
Istituto di Farmacologia e Farmacognosia, Università degli Studi di Urbino
"Carlo Bo," Urbino, Italy

Laura Castro
Department of Biochemistry and Center for Free Radical and Biomedical Research,
Facultad de Medicina, Universidad de la República, Montevideo, Uruguay

Adriana María Cassina
Department of Biochemistry and Center for Free Radical and Biomedical
Research, Facultad de Medicina, Universidad de la República, Montevideo,
Uruguay

Liana Cerioni
Istituto di Farmacologia e Farmacognosia, Università degli Studi di Urbino
"Carlo Bo," Urbino, Italy

Byung-Min Choi
Medicinal Resources Research Institute, Wonkwang University, Iksan, Korea

Carolin Cornelius
Department of Chemistry, Clinical Biochemistry and Clinical Molecular Biology
Chair, University of Catania, Italy

Hun-Taeg Chung
Medicinal Resources Research Institute, Wonkwang University, Iksan, Korea

Byoung-Hee Chung
Vascular System Research Center, Kangwon National University, Chunchon,
KoreaHa

Jack H. Crawford
Department of Pathology, University of Alabama at Birmingham, Alabama

Claire A. Davies
Genzyme Corporation, Framingham, Massachusetts

Ruba S. Deeb
Department of Pathology and Center for Vascular Biology, Weill Cornell Medical
College, New York, New York

Albena Dinkova-Kostova
Biomedical Research Centre, Ninewells Hospital and Medical School, University
of Dundee, Scotland, UK and Department of Pharmacology and Molecular
Sciences and Department of Medicine, Johns Hopkins University, Baltimore,
MD, USA

Jeannette E. Doeller
Center for Free Radical Biology and Department of Environmental Health Sciences, University of Alabama at Birmingham, Alabama

Elena S. Dremina
Department of Pharmaceutical Chemistry, University of Kansas, Lawrence, Kansas

Paul Eggleton
Peninsula Medical School, Universities of Exeter and Plymouth, St. Luke's Campus, Exeter, United Kingdom

Mariana Ferrari
Center for Free Radical and Biomedical Research and Department of Immunology, Facultades de Ciencias y Química, Universidad de la República, Uruguay

Ana María Ferreira
Center for Free Radical and Biomedical Research and Department of Immunology, Facultades de Ciencias y Química, Universidad de la República, Uruguay

Gerardo Ferrer-Sueta
Instituto de Química Biológica, Facultad de Ciencias and Center for Free Radical and Biomedical Research, Facultad de Medicina, Universidad de la República, Montevideo, Uruguay

Denise C. Fernandes
Vascular Biology Laboratory, Heart Institute (InCor), University of São Paulo School of Medicine, São Paulo, Brazil

Timothy K. Gallaher
Department of Pharmacology and Pharmaceutical Sciences, University of Southern California, Los Angeles, California

Greg A. Gerhardt
Department of Anatomy and Neurobiology, Center for Microelectrode Technology, University of Kentucky, Lexington, Kentucky

Putrika Gharini
Department of Medicine, Division of Cardiology, Pulmology and Vascular Medicine, CardioBioTech Research Group, University Hospital Aachen, Aachen, Germany

Gregory I. Giles
Department of Pharmacology and Toxicology, Otago School of Medical Sciences, University of Otago, Dunedin, New Zealand

Mark T. Gladwin
Critical Care Medicine Department, Clinical Center and Pulmonary and Vascular Medicine Branch, National Heart, Lung and Blood Institute, National Institutes of Health, Bethesda, Maryland

Steven S. Gross
Department of Pharmacology and Center for Vascular Biology, Weill Cornell Medical College, New York, New York

Marijke Grau
Department of Medicine, Division of Cardiology, Pulmology and Vascular Medicine, CardioBioTech Research Group, University Hospital Aachen, Aachen, Germany

Kwon-Soo Ha
Vascular System Research Center, Kangwon National University, Chunchon, Korea

Richard Haigh
Department of Rheumatology, Princess Elizabeth Orthopaedic Centre, Royal Devon and Exeter NHS Foundation Trust (Wonford), Exeter, United Kingdom and Peninsula Medical School, Universities of Exeter and Plymouth, St. Luke's Campus, Exeter, United Kingdom

David P. Hajjar
Department of Pathology and Center for Vascular Biology, Weill Cornell Medical College, New York, New York

Joshua M. Hare
Division of Cardiology and Interdisciplinary Stem Cell Institute, Miller School of Medicine, University of Miami, Miami, Florida

Ulrike Hendgen-Cotta
Department of Medicine, Division of Cardiology, Pulmology and Vascular Medicine, CardioBioTech Research Group, University Hospital Aachen, Aachen, Germany

Neil Hogg
Department of Biophysics and Free Radical Research Center, Medical College of Wisconsin, Milwaukee, Wisconsin

Sung Jung Hong
Department of Pharmaceutical Chemistry, University of Kansas, Lawrence, Kansas

Tzung Hsiai
Department of Biomedical Engineering and Division of Cardiovascular Medicine, School of Engineering & School of Medicine, University of Southern California, Los Angeles, California

Peter Huettl
Department of Anatomy and Neurobiology, Center for Microelectrode Technology, University of Kentucky, Lexington, Kentucky

T. Scott Isbell
Center for Free Radical Biology and Department of Pathology, University of Alabama at Birmingham, Alabama

Joy Joseph
Department of Biophysics and Free Radical Research Center, Medical College of Wisconsin, Milwaukee, Wisconsin

Balaraman Kalyanaraman
Department of Biophysics and Free Radical Research Center, Medical College of Wisconsin, Milwaukee, Wisconsin

Nicholas J. Kettenhofen
Department of Biophysics and Free Radical Research Center, Medical College of Wisconsin, Milwaukee, Wisconsin

Malte Kelm
Department of Medicine, Division of Cardiology, Pulmology and Vascular Medicine, CardioBioTech Research Group, University Hospital Aachen, Aachen, Germany

Jacque Killmer
Department of Pharmaceutical Chemistry, University of Kansas, Lawrence, Kansas

Young-Myeong Kim
Vascular System Research Center, Kangwon National University, Chunchon, Korea

Petra Kleinbongard
Department of Medicine, Division of Cardiology, Pulmology and Vascular Medicine, CardioBioTech Research Group, University Hospital Aachen, Aachen, Germany

Iona A. Knight
Peninsula Medical School, Universities of Exeter and Plymouth, St. Luke's Campus, Exeter, United Kingdom

Jeffrey R. Koenitzer
Department of Biology, University of Alabama at Birmingham, Alabama

David W. Kraus
Center for Free Radical Biology and Department of Biology, University of Alabama at Birmingham, Alabama

Young-Guen Kwon
Department of Biochemistry, College of Science, Yonsei University, Seoul, Korea

Jack R. Lancaster
Center for Free Radical Biology, Department of Physiology and Biophysics and Department of Anesthesiology, University of Alabama at Birmingham, Alabama

Francisco R. M. Laurindo
Vascular Biology Laboratory, Heart Institute (InCor), University of São Paulo School of Medicine, São Paulo, Brazil

Xiaobao Li
Department of Pharmaceutical Chemistry, University of Kansas, Lawrence, Kansas

Hansoo Lee
Vascular System Research Center, Kangwon National University, Chunchon, Korea

Cátia F. Lourenço
Faculty of Pharmacy and Center for Neurosciences and Cell Biology, University of Coimbra, Coimbra, Portugal

João Laranjinha
Faculty of Pharmacy and Center for Neurosciences and Cell Biology, University of Coimbra, Coimbra, Portugal

Cesare Mancuso
Institute of Pharmacology, Catholic University School of Medicine, Roma, Italy

Hee-Jun Na
Vascular System Research Center, Kangwon National University, Chunchon, Korea

Seung Namkoong
Vascular System Research Center, Kangwon National University, Chunchon, Korea

Tal Nuriel
Department of Pharmacology, Weill Cornell Medical College, New York, New York

Rakesh P. Patel
Center for Free Radical Biology and Department of Pathology, University of Alabama at Birmingham, Alabama

Gonzalo Peluffo
Department of Biochemistry and Center for Free Radical and Biomedical Research, Facultad de Medicina, Universidad de la República, Montevideo, Uruguay

Justin Pennington
Department of Pharmaceutical Chemistry, University of Kansas, Lawrence, Kansas

Francois Pomerleau
Department of Anatomy and Neurobiology, Center for Microelectrode Technology, University of Kentucky, Lexington, Kentucky

Rafael Radi
Department of Biochemistry and Center for Free Radical and Biomedical Research, Facultad de Medicina, Universidad de la República, Montevideo, Uruguay

Nicolo' Ragusa
Department of Chemistry, Clinical Biochemistry and Clinical Molecular Biology
Chair, University of Catania, Italy

Tienush Rassaf
Department of Medicine, Division of Cardiology, Pulmology and Vascular Medicine,
CardioBioTech Research Group, University Hospital Aachen, Aachen, Germany

Igor Rebrin
Department of Pharmacology and Pharmaceutical Sciences, University of Southern
California, Los Angeles, California

Sophie A. Rocks
Microsystems and Nanotechnology Centre, Department of Materials, School of
Applied Sciences, Cranfield University, Cranfield, United Kingdom

Mahsa Rouhanizadeh
Department of Biomedical Engineering and Division of Cardiovascular Medicine,
School of Engineering & School of Medicine, University of Southern California,
Los Angeles, California

Homero Rubbo
Center for Free Radical and Biomedical Research, Department of Biochemistry,
Facultad de Medicina, Universidad de la República, Uruguay

Ricardo M. Santos
Faculty of Pharmacy and Center for Neurosciences and Cell Biology, University of
Coimbra, Coimbra, Portugal

Célio X. C. Santos
Vascular Biology Laboratory, Heart Institute (InCor), University of São Paulo
School of Medicine, São Paulo, Brazil

Giovanni Scapagnini
Department of Health Sciences, University of Molise, Campobasso, Italy

Christian Schöneich
Department of Pharmaceutical Chemistry, University of Kansas, Lawrence, Kansas

Victor S. Sharov
Department of Pharmaceutical Chemistry, University of Kansas, Lawrence, Kansas

Frances L. Shaw
Peninsula Medical School, Universities of Exeter and Plymouth, St. Luke's Campus,
Exeter, United Kingdom

Anna Signorile
Department of Biochemistry, University of Bari, Italy

Rajindar S. Sohal
Department of Pharmacology and Pharmaceutical Sciences, University of Southern California, Los Angeles, California

José M. Souza
Department of Biochemistry and Center for Free Radical and Biomedical Research, Facultad de Medicina, Universidad de la República, Uruguay

John F. Stobaugh
Department of Pharmaceutical Chemistry, University of Kansas, Lawrence, Kansas

Wakako Takabe
Department of Biomedical Engineering and Division of Cardiovascular Medicine, School of Engineering & School of Medicine, University of Southern California, Los Angeles, California

Xinjun Teng
Department of Pathology, University of Alabama at Birmingham, Alabama

Maria Thorson
Department of Pharmaceutical Chemistry, University of Kansas, Lawrence, Kansas

Ilaria Tommasini
Istituto di Farmacologia e Farmacognosia, Università degli Studi di Urbino "Carlo Bo," Urbino, Italy

Andrés Trostchansky
Center for Free Radical and Biomedical Research, Department of Biochemistry, Facultad de Medicina, Universidad de la República, Uruguay

Madia Trujillo
Department of Biochemistry and Center for Free Radical and Biomedical Research, Facultad de Medicina, Universidad de la República, Montevideo, Uruguay

Konstantinos Tziomalos
Division of Cardiology and Interdisciplinary Stem Cell Institute, Miller School of Medicine, University of Miami, Miami, Florida

Bernardo Ventimiglia
Department of Science of Senescence, Urology and Neuro-Urology, University of Catania, Italy

Xunde Wang
Pulmonary and Vascular Medicine Branch, National Heart, Lung and Blood Institute, National Institutes of Health, Bethesda, Maryland

Peter Wardman
University of Oxford, Gray Cancer Institute, Mount Vernon Hospital, Northwood, Middlesex, United Kingdom

Matthew Whiteman
Peninsula Medical School, Universities of Exeter and Plymouth, St. Luke's Campus, Exeter, United Kingdom

Paul G. Winyard
Peninsula Medical School, Universities of Exeter and Plymouth, St. Luke's Campus, Exeter, United Kingdom

Hongyu Yu
Department of Biomedical Engineering and Division of Cardiovascular Medicine, School of Engineering & School of Medicine, University of Southern California, Los Angeles, California

Hao Zhang
Department of Biophysics and Free Radical Research Center, Medical College of Wisconsin, Milwaukee, Wisconsin

METHODS IN ENZYMOLOGY

Protein 3-Nitrotyrosine in Complex Biological Samples: Quantification by High-Pressure Liquid Chromatography/Electrochemical Detection and Emergence of Proteomic Approaches for Unbiased Identification of Modification Sites

Tal Nuriel,[*,‡] Ruba S. Deeb,[†,‡] David P. Hajjar,[†] and Steven S. Gross[*]

Contents

[*] Department of Pharmacology, Weill Cornell Medical College, New York, New York
[†] Department of Pathology and Center for Vascular Biology, Weill Cornell Medical College, New York, New York
[‡] These authors contributed equally to this work

Methods in Enzymology, Volume 441   © 2008 Elsevier Inc.
ISSN 0076-6879, DOI: 10.1016/S0076-6879(08)01201-9   All rights reserved.

Abstract

Nitration of tyrosine residues by nitric oxide (NO)-derived species results in the accumulation of 3-nitrotyrosine in proteins, a hallmark of nitrosative stress in cells and tissues. Tyrosine nitration is recognized as one of the multiple signaling modalities used by NO-derived species for the regulation of protein structure and function in health and disease. Various methods have been described for the quantification of protein 3-nitrotyrosine residues, and several strategies have been presented toward the goal of proteome-wide identification of protein tyrosine modification sites. This chapter details a useful protocol for the quantification of 3-nitrotyrosine in cells and tissues using high-pressure liquid chromatography with electrochemical detection. Additionally, this chapter describes a novel biotin-tagging strategy for specific enrichment of 3-nitrotyrosine-containing peptides. Application of this strategy, in conjunction with high-throughput MS/ MS-based peptide sequencing, is anticipated to fuel efforts in developing comprehensive inventories of nitrosative stress-induced protein-tyrosine modification sites in cells and tissues.

1. INTRODUCTION

A large and growing body of evidence has associated the accumulation of 3-nitrotyrosine (3-NT) in proteins with major neurological (Alzheimer's, Parkinson's, multiple sclerosis, and stroke) and cardiovascular (atherosclerosis, myocardial infarction, coronary artery disease, hypertension, and diabetic vasculopathy) diseases that share inflammation as a contributor to pathogenesis. While inflammation-associated protein-3-NT forms predominantly via the reaction of proteinaceous tyrosine residues with peroxynitrite ($ONOO^-$), the latter arising from the near diffusion-limited reaction of nitric oxide ($^\bullet NO$) with superoxide ($O_2^{\bullet-}$) (Beckmann *et al.*, 1994; Pacher *et al.*, 2007), other nitration mechanisms have also been implicated (Eiserich *et al.*, 1998). Regardless of the mechanism of formation *in vivo*, protein 3-NT levels provide a useful measure of the severity of tissue exposure to reactive nitrogen species and a telling indicator of inflammatory disease severity and progression Brodsky *et al.*, 2004; Halejcio-Delophont *et al.*, 2001; Skinner *et al.*, 1997; Upmacis *et al.*, 2007). Semiquantitative immunological methods (Ischiropoulos, 1998), as well as quantitative high-pressure liquid chromatography (HPLC)-based methods that utilize ultraviolet-visible (UV/VIS) absorption, electrochemistry, and mass spectrometry (MS) for detection (Frost *et al.*, 2000; Maruyama *et al.*, 1996; Schwedhelm *et al.*, 1999; Shigenaga *et al.*, 1997; Yi *et al.*, 2000), have all been employed for the quantification of protein 3-NT residues; each method offers its own particular strengths and limitations. Importantly, these alternative 3-NT detection methods differ widely with regard to their relative sensitivity, specificity, throughput, and accessibility (because of differing requirements for expensive and/or specialized instrumentation). The reader is referred to an excellent survey of proteomic and MS-based 3-NT assays by Kanski and Schoneich (2005), as well as chapters in this volume containing complementary information.

The goal of this chapter is threefold: (1) detail a useful protocol for assay of protein 3-NT in cells and tissues, (2) consider approaches for unbiased identification of protein 3-NT modification sites, and (3) describe a novel biotin-tagging approach to selectively enrich 3-NT-containing peptides for more comprehensive 3-NT site identification by liquid chromatography-mass spectrometry/mass spectrometry (LC-MS/MS).

2. QUANTIFICATION OF 3-NT IN PROTEINS USING HPLC SEPARATION AND ELECTROCHEMICAL (EC) DETECTION

This section provides procedural details for the quantification of protein-bound 3-NT using isocratic reverse-phased HPLC paired with EC detection. This method is essentially as described previously by Maruyama and colleagues (Maruyama et al., 1996; Skinner et al., 1997), but with some recommendations and implemented modifications. Strengths of this method are its relatively low cost and sufficient sensitivity for reliable measurement of protein 3-NT at the relatively low basal levels found in most "healthy" tissues. It is notable that the use of EC detection provides ≈100 times greater sensitivity than that which can be obtained using UV/VIS absorption. Protein hydrolysis for the release of free 3-NT from proteins is performed under neutral conditions, minimizing the potential for artifactual nitration that may occur in acidified nitrite-containing solutions (Oldreive et al., 1998). When performed as described, the efficiency of protein 3-NT retrieval from crude extracts of cells and tissues routinely exceeds 85%.

3. PROTOCOL FOR QUANTIFICATION OF 3-NT IN HYDROLYZED PROTEINS USING HPLC-EC

3.1. Materials

3-Nitrotyrosine, sodium octanesulfonate, acetonitrile, proteinase K, sodium hydrosulfite, and all other reagents are purchased from Sigma-Aldrich in the best available grade (minimum 95% purity; HPLC grade where available). An isocratic HPLC system with autosampler, pump, tubings, and fittings, as well as a multichannel electrochemical CoulArray detector and EC cell, are from ESA, Inc. HPLC mobile phase and all buffers and standard solutions are prepared using 18 MΩ resistance water, either purchased or prepared using a Milli-Q water purification system (Waters Inc.) or equivalent. The HPLC mobile phase is vacuum degassed and filtered through a 0.2-μm nylon membrane (Whatman) to reduce background electrochemical noise, prolonging the electrochemical cell lifetime and enhancing the 3-NT detection of sensitivity. 3-NT detection sensitivity can be enhanced progressively by continuous recirculation of the mobile phase through the EC cell to attenuate background oxidizable species. Centrifugal molecular weight cutoff filters (Microcon Ultracel YM-10; 10 kDa) are from Millipore Corporation. Polypropylene microcentrifuge tubes (2 ml) are from Sorenson Biosciences, ultracentrifuge tubes (1.5 ml) are from Beckman Instruments, and autosampler vials are from Fisher Scientific. Any reversed-phase C18 column may be used for 3-NT analysis; however, columns with smaller particle sizes (3 μm, or less) and longer column lengths (>100 mm) will enhance resolution, except at the expense of analysis time. We often use a relatively inexpensive Microsorb-MV 100 mm C18 (5-μm particle size) HPLC column from Varian Instruments. Also required for this protocol is a handheld homogenizer (Branson Scientific), microultracentrifuge (e.g., Beckman Optima TLX and TLA100.3 rotor), and a Speed-Vac concentrator (Savant).

3.2. Sample preparation

Tissues and cells to be analyzed (wet weight of $\approx$50–100 mg) are rinsed of blood or media, respectively, by multiple washes in iced phosphate-buffered saline (PBS) buffer (pH 7.2–7.4). Tissue samples are placed in a weigh boat containing a volume of up to 500 μl buffer A, comprising 50 mM Tris-HCl, 150 mM NaCl, 0.1 mM EDTA, and 20 mM CHAPS (pH 7.4), and minced with a fine scissors. The minced tissue is transferred to a 1.5-ml ultracentrifugation tube and disrupted on ice using three 10-s bursts of a small handheld homogenizer. Cells in culture are scrape harvested (after washing in iced PBS), transferred quantitatively to a 15-ml conical tube, pelleted at 800 g, transferred in up to 500 μl of buffer A to an ultracentrifugation tube, and subjected to three 10-s bursts of a handheld homogenizer. Tissue and cell homogenates are ultracentrifuged at 100,000 g for 60 min in a Beckman TLA100.3 rotor, and supernatants are retained for isolation of 3-NT. A 20-μl aliquot of the supernatant should be preserved for the analysis of total protein content using the Bio-Rad DC assay (or other comparable protein assay) using bovine serum albumin as a standard. Note that assay of total protein in each sample is essential, as this determines the amount of protease to be added to the supernatant for complete protein digestion and additionally provides the normalization factor for final determination of protein 3-NT (i.e., protein

3-NT content is typically expressed on a pmol/mg protein basis). If a pure protein is the subject of 3-NT quantification, then 50 μg is recommended as a convenient starting quantity of protein. For analysis of 3-NT in pure proteins, the homogenization and ultracentrifugation steps can be bypassed.

Supernatants obtained after homogenization and ultracentrifugation (or known quantities of pure proteins) are treated with a 3:1 v/v of ice-cold acid precipitation buffer (0.1 M phosphoric acid, 0.23 M TCA). The pellet is resuspended in Buffer A and subjected to complete proteolytic digestion with proteinase K (10 U/mg protein) in a total volume of up to 500 μl. The digestion mixture should be allowed to incubate for 8 h at 55°C in 2-ml capacity polypropylene microcentrifuge tubes (note that these tubes accommodate the subsequently added volume of acetonitrile used for 3-NT extraction, described later). It was shown previously that proteinase K is more efficient than either pronase or trypsin for complete digestion of both plasma and tissue samples (Skinner et al., 1997). Using these digestion conditions, 30% of the proteinase K is self-hydrolyzed (determined by UV/VIS spectroscopy) and thus contributes to the total pool of amino acids analyzed by HPLC-EC. Nevertheless, proteinase K autolysis poses no interference for 3-NT detection in samples. Indeed, Fig. 1.1 (chromatogram B) affirms that HPLC-EC signals derived from the proteinase K autolysis are well separated from the 3-NT signal in a typical chromatogram and would not result in false 3-NT quantification. In addition, Fig. 1.1 (chromatogram A) demonstrates that proteinase K itself does not contribute a detectable level of 3-NT to a study sample.

Digested protein solutions are cooled to room temperature, followed by extraction in the same tubes with ice-cold acetonitrile (3:1 v/v; acetonitrile: sample). Following the addition of acetonitrile, samples are vortexed, incubated on ice for 5 min, and then centrifuged at 12,000 g for 15 min at 4°C. Note that during the extraction step, precipitation is anticipated; however,

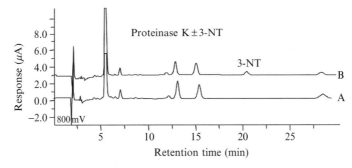

Figure 1.1 HPLC-EC analysis of 10 units proteinase K in the presence (chromatogram B) and absence (chromatogram A) of (200 pmol) 3-NT. Proteinase K ($\pm$ 3-NT) was treated to the full procedure of heating, extraction, and isolation as for analysis of 3-NT in an unknown sample.

the supernatant (acetonitrile extract) retains the hydrolyzed amino acids, including 3-NT. Next, 500-μl volumes of the supernatants are transferred to new microcentrifuge tubes and concentrated to dryness at room temperature in a Speed-Vac (Savant Instruments; note that limiting the supernatant volume to 500 μl/tube expedites the drying process). Depending on protein concentration and sample volume, this step typically takes $\approx$4 to 6 h. After drying, samples may be stored indefinitely at $-80°$ prior to analysis for 3-NT content.

For analysis of 3-NT, dried samples are resuspended in HPLC mobile phase and pooled (when desired) to give a final combined volume of 250 μl. Prior to HPLC injection, the resuspended samples are then filtered through a 10-kDa cutoff centrifugal concentrator to remove residual high molecular weight species. The HPLC mobile phase buffer is as described previously (Crabtree *et al.*, 2003), containing 90 mM sodium acetate, 35 mM citric acid, 130 μM EDTA, and 460 μM sodium octane sulfonate (pH 4.35) in Milli-Q water (vacuum filtered as described earlier).

3.3. Reverse-phase HPLC-EC analysis of 3-NT in samples

HPLC analysis is performed isocratically using a C18 reversed-phase column and mobile phase (see earlier), composed as described previously. Using a flow rate of 0.75 ml/min and on an optimally prepared C18 column, the retention time of 3-NT may range from 10 to 20 min, depending on the choice of column length, particle size, length and bore of tubing connections, and operating temperature. Confirmation of the authenticity of a putative 3-NT signal can be performed as described later. The optimal potential for complete oxidation of 3-NT and hence electrochemical detection is 800 mV. To eliminate interfering signals that arise from species that are oxidized more readily than 3-NT, and additionally affirm complete oxidation of 3-NT at 800 mV, the multichannel EC detector is additionally set to flanking voltages of 700 and 900 mV, respectively (Crabtree *et al.*, 2003).

The reproducibility of HPLC-EC chromatogram signal intensity and the retention time of 3-NT are determined by the quality of the C18 column used and the proper maintenance of the electrochemical cell. Optimal reproducibility is ensured by (1) routine washing of the column between uses, with methanol (with the electrochemical cell disconnected), (2) safeguarding the electrochemical cell by ensuring that the potentials are not applied when the mobile phase is not flowing or when the system is being flushed with organic solvent (>10%), and (3) periodic reconditioning of the electrode when a performance loss is observed (briefly, this involves the repeated application of alternating potentials between 1000 and -450 mV, during perfusion with the mobile phase).

3.4. Interpretation of HPLC chromatograms for quantification of 3-NT

Quantification of 3-NT is performed by comparing the peak area of an unknown sample to the peak area of an external standard comprising a known concentration of pure 3-NT. External standards may range from 1 to 1000 pmol, although the expected range for biological samples is typically 1 to 400 pmol per mg protein. The limit of 3-NT detection by HPLC-EC with a pristine EC cell is ≈500 fmol, rising to ≈4 pmol with a 4-year-old extensively used but otherwise well-maintained EC cell.

The identity of a peak in a complex chromatogram as arising from 3-NT is evidenced by the fulfillment of multiple expectations: (1) retention time of the peak is indistinguishable from authentic 3-NT; (2) electrochemical properties of the peak are identical to authentic 3-NT, that is, minimal oxidation at 700 mV and complete oxidation at 800 mV; (3) the peak coelutes with authentic 3-NT in a 3-NT-"spiked" sample; and (4) treatment of the sample with a final concentration of 10 mM sodium hydrosulfite (dithionite) prior to HPLC-EC analysis reduces 3-nitrotyrosine to 3-aminotyrosine (Riordan and Sokolovsky, 1971), silencing the electrochemical signal of bona fide 3-NT (see note later on the dithionite reaction). Generally, it is desirable to employ more than one approach to confirm the identity of the 3-NT peak—nonetheless, sample availability may limit validation. Along with the retention time and electrochemical properties of the 3-NT peak (approaches 1 and 2 given earlier), chemical reduction of the 3-NT signal provides a convincing method for 3-NT confirmation (approach 5). It is notable that the addition of sodium dithionite typically results in a cloudy precipitate immediately after addition to test samples. Therefore, it is necessary to filter dithionite-treated samples with a syringe filter (0.22-μm Millex GV filters; Millipore Corp.) prior to sample injection.

Figure 1.2 shows a chromatogram (trace A) for ONOO⁻-treated bovine serum albumin (BSA), revealing multiple peaks and a complex array of preexisting oxidizable species and ONOO⁻-derived oxidation products. Trace B in Figure 1.2 demonstrates that the 3-NT peak in ONOO⁻-treated BSA can be identified by its retention time, relative to a 3-NT standard, and its disappearance following exposure of the sample to 10 mM dithionite.

4. BEYOND QUANTIFICATION: SPECIFICATION OF 3-NT SITES IN PROTEINS

While quantification of total protein 3-NT levels is extremely useful for identifying changing levels of exposure to nitrosative stress in cells and tissues and can be accomplished effectively using the HPLC-EC method

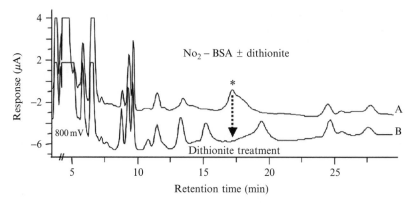

Figure 1.2 (A) HPLC-EC chromatogram of 3-NT in bovine serum albumin, following reaction of BSA with 500 μM peroxynitrite (ONOO$^-$). The retention time for 3-NT is denoted by the asterisk. (B) HPLC-EC chromatogram for ONOO$^-$-treated BSA is as in chromatogram A, but following chemical reduction of 3-NT using sodium dithionite. As highlighted by the arrow, dithionite treatment silences the 3-NT signal by reducing the 3-nitro moiety to 3-amino, providing confirmation of the identity of the presumed 3-NT peak.

detailed earlier, mechanistic questions regarding *nitrosative signaling* are not answered by knowledge of the total protein 3-NT content. For example, in neurodegenerative illnesses such as Alzheimer's disease, it is established that 3-NT levels increase in the brain, especially in disease-associated plaques (Smith *et al.*, 1997; Tohgi *et al.*, 1999), yet it is uncertain whether protein nitration is a cause or effect of the disease process. To understand how tyrosine nitration may contribute to Alzheimer's disease pathogenesis, one would need to identify the particular proteins that become nitrated, specify their sites of nitration, and elucidate the functional consequences of this modification. Further, one would want to know the degree of nitration at these sites (i.e., quantity of nitrated vs non-nitrated tyrosines) to infer the extent of nitration-evoked change in the function of a given protein. Unfortunately, methods that enable an unbiased identification and quantification of protein nitration sites in biologically complex protein mixtures have been elusive. To date, discovery of the majority of currently recognized biologically relevant protein nitration sites have been arduously identified on a protein-by-protein basis, using an iterative approach involving mass spectrometry-based peptide sequencing, Tyr mutagenesis, and protein functional analysis. Notwithstanding, various strategies have been described in the literature for unbiased protein tyrosine-nitration site identification, although none can be applied comprehensively to discover low endogenous modification levels, physiologically present in protein extracts from tissues. Furthermore, no proteomic strategy has yet been devised to enable the simultaneous, high-throughput and unbiased quantification of changes in 3-NT content in what is likely to be thousands of individual 3-NT-containing proteins present in tissues following exposure to acute or chronic nitrosative stress.

Among the strategies that have been presented for unbiased identification of protein nitration sites, bottom-up mass spectrometry (MS; i.e., analysis of protein digests) seems to be uniquely suited for 3-NT site specification (Kanski et al., 2005). However, given the enormous complexity of protein biological tissues, coupled with a relatively low abundance of protein tyrosine nitration, peptide prefractionation and enrichment steps are likely to be crucial for broad specification of 3-NT modification sites. A recent step toward success in this endeavor is described in a report by Smith and colleagues, who were able to identify 29 nitrated proteins from healthy mouse brain extracts using first-dimension HPLC prefractionation, followed by nano-LC-MS/MS analysis (Sacksteder et al., 2006). It is notable that this relative success was achieved using extensive sample prefractionation and MS instrument time, sufficient to enable the identification of 7792 proteins in total, culminating in the discovery of a mere 29 nitration sites. Although this represents a significant advance, it is reasonable to assume that a far larger number of nitrated proteins and their cognate sites went undiscovered.

Because most researchers who seek to identify endogenously nitrated proteins do not have access to the powerful mass spectrometry resources employed by Smith and co-workers for the study just described (Sacksteder et al., 2006), other investigators have attempted enrichment of nitrated proteins prior to analysis by MS (Kanski and Schoneich, 2005). For example, some investigators have used anti-nitrotyrosine antibodies to immunoprecipitate nitrated proteins prior to further purification and analysis by MS or MS/MS (Kanski et al., 2005). Although immunoprecipitation can be used for effective enrichment of 3-NT-containing proteins, it is notable that a significant number of nonnitrated proteins coimmunoprecipitate. This apparent lack of specificity may arise from undesired interactions of anti-3-NT antibodies with 3-NT-free proteins (Aulak et al., 2004), such as those which contain 5-hydroxynitrotryptophan (Rebrin et al., 2007), and the predicted pull down of proteins that associate with 3-NT-containing proteins.

Perhaps the most commonly used unbiased approach for identifying 3-NT-containing proteins has been two-dimensional (2D) gel electrophoresis, wherein protein spots that are specifically recognized by Western blotting with an anti-3-NT antibody are cored for subsequent in-gel trypsinolysis and protein identification by MS or MS/MS (Castegna et al., 2003; Turko et al., 2003). Although this method has been used successfully to identify novel nitrated proteins, it suffers from several shortcomings (Kanski and Schoneich, 2005). Principal among these shortcomings is that, only in the odd case where a specific site of nitration can be assigned by MS/MS to a peptide ion (e.g., by the observed addition of 45 Da as nitrate, or loss of 181 Da as a hallmark immonium product ion), the identity of a putative 3-NT-containing protein can only be considered as tentative. An additional weakness of the approach is that the labor involved in individually extracting and processing proteins from the 2D gel limits the number of proteins that can be analyzed practically by MS. Furthermore, the resolution of proteins on 2D gels is biased against hydrophobic proteins and proteins at the extremes of the pI and molecular mass ranges. Moreover, low-abundance 3-NT-containing proteins will be challenging to detect by Western blotting and hence, will tend to not be selected for MS analysis. Finally, the 2D gel approach can result in numerous false positives, resulting from either the unintentional extraction of overlapping proteins or the identification of proteins that were detected nonspecifically by the 3-NT antibody.

The relatively low abundance of 3-NT compared to non-3-NT-containing proteins in cells and tissues, combined with the limitations of 2D gel electrophoresis and Western blotting for recognition of 3-NT-containing proteins, evidences the need for a more effective approach to enrich endogenously nitrated proteins and/or peptides for high-throughput sequence analysis by MS. One solution would be to develop an affinity-tagging strategy for the selective purification of 3-NT-containing peptides in digests of biologically complex protein mixtures. To address an analogous methodological need, our laboratory previously developed the *SNOSID* method for isolation and unbiased specification of *S*-nitrosylation sites on cysteine residues in complex protein mixtures (Hao *et al.*, 2006). The *SNO-SID* method builds on the biotin-switch technique (Jaffrey *et al.*, 2001) for introduction of a chemically cleavable biotin-containing "tag" at sites of protein *S*-nitrosylation. The biotin tag enables highly selective capture of peptides on immobilized avidin and allows facile peptide release upon cleavage of the tag. Thus, just as the *SNOSID* method enabled the identification of endogenously *S*-nitrosylated proteins, we hypothesize that a similar biotin-based affinity purification strategy may be applied to broadly identify sites of endogenous nitration on peptides in complex mixtures and specify their proteins of origin.

A review of the protein 3-NT literature finds that two such affinity-purification strategies have previously been proposed, both of which take significant steps toward the goal of unbiased identification of modification sites in endogenously nitrated proteins. In 2003, Nikov *et al.* proposed a strategy that takes advantage of the ability of sodium dithionite to selectively convert 3-NT to 3-aminotyrosine, thus providing a chemical means to introduce a new amino group into a peptide that can be targeted for modification with an amine-reactive affinity tag. To test their approach, the authors nitrated human serum albumin and then treated it with a 200-fold molar excess of sodium dithionite in 50 mM ammonium bicarbonate buffer. After removal of the dithionite on a desalting column, the protein was reconstituted in 50 mM sodium acetate buffer, pH 5.0, in 0.1% sodium dodecyl sulfate and treated for 30 min with a 2-fold molar excess of an amine-reactive biotin reagent, sulfo-NHS-SS-biotin. The resulting biotinylated proteins were affinity captured on immobilized streptavidin, digested with trypsin, and then analyzed by MS-based sequencing—details of this approach are schematized in Fig. 1.3.

The Nikov *et al.* (2003) strategy represents an important conceptual advance and enabled the investigators to identify an *in vitro* nitrated protein that had been spiked into the complex milieu of mouse plasma. However, the authors were unable to accomplish the practical goal of discovering endogenously nitrated proteins and specifying their modification sites. This shortcoming can be partly explained by the authors' reliance on pH to guide

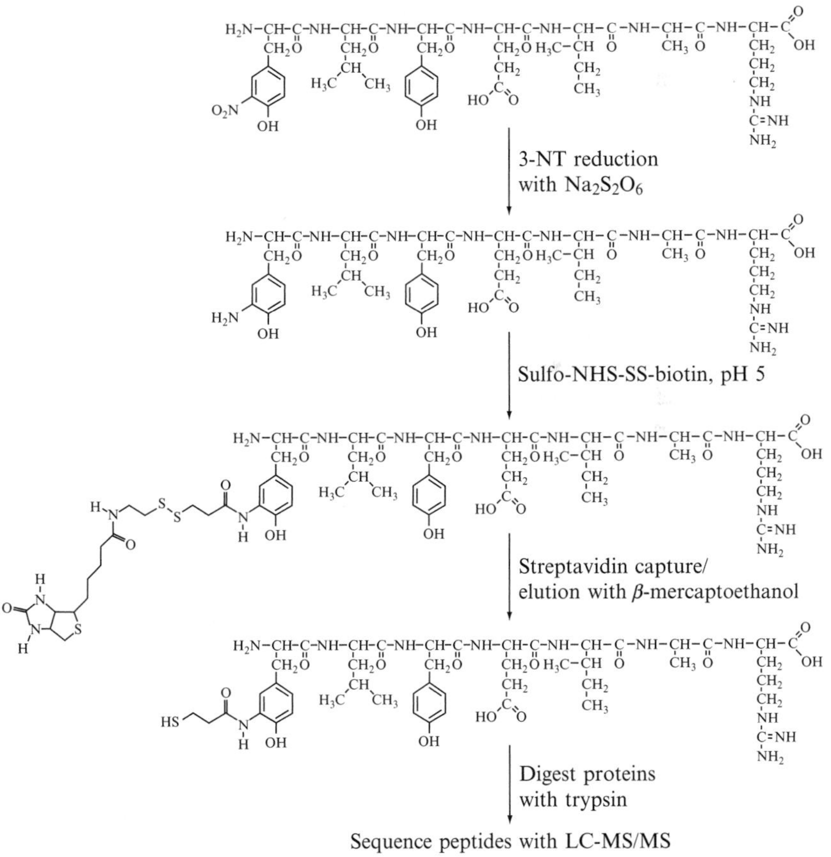

Figure 1.3 A strategy described by Tannenbaum and colleagues (Nikov *et al.*, 2003) for immunoprecipitation of 3-NT-containing proteins and enrichment of 3-NT peptides for potential unbiased identification (modified from Nikov *et al.*, 2003).

the specificity of the sulfo-NHS-SS-biotin reagent. Because sulfo-NHS-SS-biotin reacts only with nonprotonated amines and the amino group of 3-aminotyrosine has a pK_a of ≈ 5 (more than three orders of magnitude less than the pK_a of ε-lysine and N-terminal amino groups; 10.5 and 8.5, respectively), the authors reasoned that aminotyrosines would be selectively biotinylated at pH 5.0. However, this selectivity was apparently not achieved, likely a consequence of the far greater number of ε-lysine and N-terminal amines in the samples (relative to aminotyrosine) and the progressive reaction of sulfo-NHS-SS-biotin with all amine groups over time.

This limitation was clearly appreciated by Zhang *et al.* (2007), who modified the Nikov *et al.* strategy and thereby enabled the identification of a number of protein nitration sites in a peroxynitrite-treated brain extract. The Zhang *et al.* (2007) protocol involves protein digestion with trypsin as a first step, followed by reduction and alkylation of all free cysteine thiols with 5 mM dithiothreitol and 20 mM iodoacetamide. In this procedure, all free amines are then blocked by acetylation, using a 200-fold molar excess of acetic anhydride in ammonium bicarbonate buffer, pH 8.5. (*Note:* undesired acetylation of hydroxyl residues is reversed by treatment with 20 mM hydroxylamine.) After removal of excess acetic anhydride with a desalting column, peptide 3-NT residues are reduced to 3-aminotyrosine with sodium dithionite treatment. However, instead of biotinylating the newly formed 3-aminotyrosine groups with an amine-reactive biotinylating reagent, as in the Nikov *et al.* (2003) strategy, the peptides are treated with a 50-fold molar excess of *N*-succinimidyl *S*-acetylthioacetate (SATA), followed by treatment with 0.5 M hydroxylamine. This SATA/hydroxylamine treatment results in replacement of the amino group on 3-aminotyrosine with a sulfhydryl group on a chemical linker. The authors then employ a thiol-enrichment technique, pioneered in their laboratory (Liu *et al.*, 2004), and the resulting thiol-enriched peptides are subjected to MS/MS-based amino acid sequence analysis—procedural details are schematized in Fig. 1.4.

Using this strategy, the authors confidently identified 150 nitration sites on 102 unique proteins in a mouse brain homogenate that had been nitrated previously *in vitro*. While this is recognized as a significant advance over the Nikov *et al.* (2003) strategy, which was only used to identify a single nitrated protein spiked into plasma, it fell short of the goal of identifying endogenously nitrated proteins. Furthermore, the 150 protein nitration sites identified by Zhang *et al.* (2007) represented only 35% of the sequenced spectra—the other 65% of peptides that were "enriched" by the affinity protocol were nonnitrated. The relative inefficiency of the Zhang *et al.* (2007) strategy for the enrichment of *in vitro* nitrated peptides is likely to preclude the effective use of this approach to capture and sequence peptides in protein digests containing low endogenous levels of tyrosine nitration, estimated at <0.1% of total tyrosine residues. We predict that the high false-positive peptide capture rate with the Zhang *et al.* (2007) method lies in the

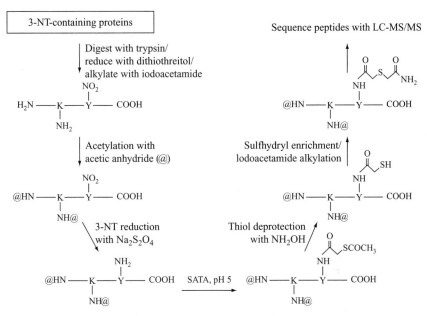

Figure 1.4   A strategy described by Smith and colleagues (Zhang *et al.*, 2007) for unbiased enrichment of nitrated peptides and sequencing by MS/MS (modified from Zhang *et al.*, 2007). Peptide enrichment is performed by the reduction of peptidyl-3-NT to aminotyrosine (using dithionite), which is then reacted with SATA to introduce a sulfhydryl group for selective peptide capture and release on a thiol affinity matrix.

choice of affinity tag. Notably, the efficiency and selectivity of thiol-peptide capture are likely to be many orders of magnitude less than that which can be achieved by taking advantage of the femtomolar K_d interaction of biotin for avidin, one of the strongest interactions known to occur in biological systems (Green, 1963). Incomplete blocking of cysteine thiols may have also contributed to the significant capture of nonnitrated peptides on thiol-Sepharose beads by Zhang *et al.* (2007).

To overcome some of the perceived shortcomings of the Zhang *et al.* (2007) protocol, moving it closer to the goal of unbiased identification of endogenous protein nitration sites, we suggest several key modifications. While it is essential that an approach is used that results in virtually complete blocking of all free amine groups (ε-lysine amino and N-terminal amino) prior to reducing peptide 3-NT to aminotyrosine residues, it is important that a positive charge is retained. Notably, a survey of the nitrated peptides discovered by Zhang *et al.* (2007) reveals that 109 of the 150 identified peptides contain a C-terminal arginine residue, despite the prediction of an approximately equal number of C-terminal lysine- and C-terminal arginine-containing tryptic peptides. This discrepancy is reconciled by the fact that acetic anhydride-mediated acetylation of peptides with a C-terminal lysine will neutralize the positive charge on this amine (in addition to neutralizing the positive charge on N termini), limiting detection by positive ion monitoring MS. In contrast, C-terminal arginine residues will not be acetylated and therefore will predictably retain a single positive charge, allowing them to be detected by positive ion monitoring MS (although sensitivity is likely to be diminished for not being doubly charged). Thus, the choice of acetic anhydride as an amino–blocking agent will result in an unacceptable loss of peptide ion signals by MS.

To overcome the problem of signal loss on MS, we are employing reductive dimethylation to block free amines on tryptic peptides, while retaining their positive charge (see Fig. 1.5). As has been shown previously, reductive dimethylation is a rapid, specific, irreversible, and, most importantly for the present application, allows the modified amines to retain their positive charge at acidic pH (Hsu *et al.*, 2007). We found that 10 μg of a tryptic digest of bovine serum albumin is >99% dimethylated following treatment with a combination of 0.48% formaldehyde and 50 mM sodium cyanoborohydride for 1 h at 37° in 50 mM potassium phosphate buffer (pH 7) and results in peptide ions that can be detected readily in positive ion mode by MS/MS. Another significant advantage of reductive dimethylation is that excess formaldehyde can then be consumed efficiently by the addition of 100 mM glycine preventing the need for a desalting step and attendant peptide losses.

For reduction of the 3-NT residues to 3-aminotyrosine, we are employing dithionite treatment (5 mM for < 1 min.). However, dithionite treatment has been shown to result in a small percentage of side-products. Therefore, we are currently investigating alternative means for reduction. After reduction, 3-Aminotyrosine residues can then be biotinylated using the thiol-cleavable amine-reactive reagent such as that employed by Nikov *et al.* (2003). Thus, by treating the dimethylated peptides with excess sulfo-NHS-SS-biotin in sodium acetate buffer (pH 5.0), biotinylation of 3-aminotyrosine becomes specific, as ε-lysine amines and N-terminal amines are unavailable for reaction. Moreover, even if dimethylation was incomplete, a low abundance of unblocked lysine and N-terminal amines would predictably be tolerated inasmuch as the biotinylation of these amines should be relatively inefficient under the acidic conditions employed. After removing the excess biotin reagent, the biotinylated proteins are affinity purified on streptavidin-Sepharose and eluted using β-mercaptoethanol. These eluted peptides are biotin free, but retain a thiol vestige on the aminotyrosine residue that can be utilized by MS analysis to verify their authenticity as formerly nitrated peptides.

The combination of amine blocking and biotin tagging is anticipated to greatly improve the purification efficiency of 3-NT-containing peptides, relative to the two previously reported strategies discussed earlier. Although the procedure awaits optimization, we expect a version of this strategy to

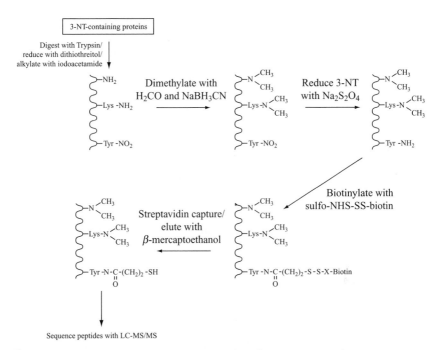

Figure 1.5 A proposed proteomic strategy for efficient capture of 3-NT-containing peptides in tryptic digests of complex protein mixtures to enable broad proteomic identification of 3-NT modification sites using MS/MS (Nuriel and Gross, unpublished). The schematized approach is exemplified for a representative dipeptide, NO₂-Tyr-Lys. The overall strategy is an extension of the biotin-switch method (Jaffrey and Snyder, 2001) adapted for high-throughput identification of S-nitrosylation sites on protein cysteine residues (Hao *et al.*, 2006). After complete blocking of free amino groups (by dimethylation), peptides are treated with dithionite to chemically reduce NO₂-Tyr to NH₂-Tyr. This newly formed amine is then modified by reaction with sulfo-NHS-SS-biotin, a thiol-cleavable and amine-selective biotinylating agent (note that an acid-cleavable biotinylating agent can similarly be employed). The biotinylated peptides are affinity captured on streptavidin-Sepharose and washed extensively, and nonbiotinylated peptides are eluted by treatment with β-mercaptoethanol for LC-MS/MS-based amino acid sequence analysis. This approach is expected to provide a more confident and comprehensive identification of protein 3-NT-modification sites than obtained previously.

significantly advance the goal of unbiased and comprehensive identification of endogenously nitrated proteins. Furthermore, because N termini of nitrated peptides are dimethylated during the course of discovery, it should be a simple matter to modify this protocol for isotopically encoded modifications (i.e., with H2/C13 vs H1/C12-labeled formaldehyde) to allow quantification of changes in endogenously nitrated protein expression with diseases and treatments.

ACKNOWLEDGMENT

This work was supported by NIH Grants HL46403, HL50656, and HL80702.

REFERENCES

Aulak, K. S., Koeck, T., Crabb, J. W., and Stuehr, D. J. (2004). Proteomic method for identification of tyrosine-nitrated proteins. *Methods Mol. Biol.* **279,** 151–165.

Beckmann, J. S., Ye, Y. Z., Anderson, P. G., Chen, J., Accavitti, M. A., Tarpey, M. M., and White, C. R. (1994). Extensive nitration of protein tyrosines in human atherosclerosis detected by immunohistochemistry. *Biol. Chem. Hoppe Seyler* **375,** 81–88.

Brodsky, S. V., Gealekman, O., Chen, J:, Zhang, F., Togashi, N., Crabtree, M., Gross, S. S., Nasjletti, A., and Goligorsky, M. S. (2004). Prevention and reversal of premature endothelial cell senescence and vasculopathy in obesity-induced diabetes by ebselen. *Circ. Res.* **94,** 377–384.

Castegna, A., Thongboonkerd, V., Klein, J. B., Lynn, B., Markesbery, W. R., and Butterfield, D. A. (2003). Proteomic identification of nitrated proteins in Alzheimer's disease brain. *J. Neurochem.* **85,** 1394–13401.

Crabtree, M., Hao, G., and Gross, S. S. (2003). Detection of cysteine S-nitrosylation and tyrosine 3-nitration in kidney proteins. *Methods Mol. Med.* **86,** 373–384.

Eiserich, J. P., Hristova, M., Cross, C. E., Jones, A. D., Freeman, B. A., Halliwell, B., and van der Vliet, A. (1998). Formation of nitric oxide-derived inflammatory oxidants by myeloperoxidase in neutrophils. *Nature* **391,** 393–397.

Frost, M. T., Halliwell, B., and Moore, K. P. (2000). Analysis of free and protein-bound nitrotyrosine in human plasma by a gas chromatography/mass spectrometry method that avoids nitration artifacts. *Biochem. J.* **345,** 453–458.

Green, N. M. (1963). Avidin. 1. The use of (14-C)biotin for kinetic studies and for assay. *Biochem. J.* **89,** 585–591.

Halejcio-Delophont, P., Hoshiai, K., Fukuyama, N., and Nakazawa, H. (2001). No evidence of NO-induced damage in potential donor organs after brain death. *J. Heart Lung Transplant.* **20,** 71–79.

Hao, G., Derakhshan, B., Shi, L., Campagne, F., and Gross, S. S. (2006). SNOSID, a proteomic method for identification of cysteine S-nitrosylation sites in complex protein mixtures. *Proc. Natl. Acad. Sci. USA* **103,** 1012–1017.

Hsu, J. L., Chen, S. H., Li, D. T., and Shi, F. K. (2007). Enhanced a1 fragmentation for dimethylated proteins and its applications for N-terminal identification and comparative protein quantitation. *J. Proteome Res.* **6,** 2376–2383.

Ischiropoulos, H. (1998). Biological tyrosine nitration: A pathophysiological function of nitric oxide and reactive oxygen species. *Arch. Biochem. Biophys.* **356,** 1–11.

Jaffrey, S. R., Erdjument-Bromage, H., Ferris, C. D., Tempst, P., and Snyder, S. H. (2001). Protein s-nitrosylation: A physiological signal for neuronal nitric oxide. *Nat. Cell Biol.* **3,** 193–197.

Jaffrey, S. R., and Snyder, S. H. (2001). The biotin switch method for the detection of S-nitrosylated proteins. *Sci. STKE* **2001,** PL1.

Kanski, J., Behring, A., Pelling, J., and Schoneich, C. (2005). Proteomic identification of 3-nitrotyrosine-containing rat cardiac proteins: Effects of biological aging. *Am. J. Physiol. Heart Circ. Physiol.* **288,** H371–H381.

Kanski, J., and Schoneich, C. (2005). Protein nitration in biological aging: Proteomic and tandem mass spectrometric characterization of nitrated sites. *Methods Enzymol.* **396,** 160–171.

Liu, T., Qian, W. J., Strittmatter, E. F., Camp, D. G., 2nd, Anderson, G. A., Thrall, B. D., and Smith, R. D. (2004). High-throughput comparative proteome analysis using a quantitative cysteinyl-peptide enrichment technology. *Anal. Chem.* **76**, 5345–5353.

Maruyama, W., Hashizume, Y., Matsubara, K., and Naoi, M. (1996). Identification of 3-nitro-L-tyrosine, a product of nitric oxide and superoxide, as an indicator of oxidative stress in the human brain. *J. Chromatogr. B Biomed. Appl.* **676**, 153–158.

Nikov, G., Bhat, V., Wishnok, J. S., and Tannenbaum, S. R. (2003). Analysis of nitrated proteins by nitrotyrosine-specific affinity probes and mass spectrometry. *Anal. Biochem.* **320**, 214–222.

Oldreive, C., Zhao, K., Paganga, G., Halliwell, B., and Rice-Evans, C. (1998). Inhibition of nitrous acid-dependent tyrosine nitration and DNA base deamination by flavonoids and other phenolic compounds. *Chem. Res. Toxicol.* **11**, 1574–1579.

Pacher, P., Beckman, J. S., and Liaudet, L. (2007). Nitric oxide and peroxynitrite in health and disease. *Physiol. Rev.* **87**, 315–424.

Rebrin, I., Bregere, C., Kamzalov, S., Gallaher, T. K., and Sohal, R. S. (2007). Nitration of tryptophan 372 in succinyl-CoA:3-ketoacid CoA transferase during aging in rat heart mitochondria. *Biochemistry* **46**, 10130–10144.

Riordan, J. F., and Sokolovsky, M. (1971). Reduction of nitrotyrosyl residues in proteins. *Biochim Biophys. Acta* **236**, 161–163.

Sacksteder, C. A., Qian, W. J., Knyushko, T. V., Wang, H., Chin, M. H., Lacan, G., Melega, W. P., Camp, D. G., 2nd, Smith, R. D., Smith, D. J., Squier, T. C., and Bigelow, D. J. (2006). Endogenously nitrated proteins in mouse brain: Links to neurodegenerative disease. *Biochemistry.* **45**, 8009–8022.

Schwedhelm, E., Tsikas, D., Gutzki, F. M., and Frolich, J. C. (1999). Gas chromatographic-tandem mass spectrometric quantification of free 3-nitrotyrosine in human plasma at the basal state. *Anal. Biochem.* **276**, 195–203.

Shigenaga, M. K., Lee, H. H., Blount, B. C., Christen, S., Shigeno, E. T., Yip, H., and Ames, B. N. (1997). Inflammation and NOx-induced nitration: Assay for 3-nitrotyrosine by HPLC with electrochemical detection. *Proc. Natl. Acad. Sci. USA* **94**, 3211–3216.

Skinner, K. A., Crow, J. P., Skinner, H. B., Chandler, R. T., Thompson, J. A., and Parks, D. A. (1997). Free and protein-associated nitrotyrosine formation following rat liver preservation and transplantation. *Arch. Biochem. Biophys.* **342**, 282–288.

Smith, M. A., Richey Harris, P. L., Sayre, L. M., Beckman, J. S., and Perry, G. (1997). Widespread peroxynitrite-mediated damage in Alzheimer's disease. *J. Neurosci.* **17**, 2653–2657.

Tohgi, H., Abe, T., Yamazaki, K., Murata, T., Ishizaki, E., and Isobe, C. (1999). Alterations of 3-nitrotyrosine concentration in the cerebrospinal fluid during aging and in patients with Alzheimer's disease. *Neurosci. Lett.* **269**, 52–54.

Turko, I. V., Li, L., Aulak, K. S., Stuehr, D. J., Chang, J. Y., and Murad, F. (2003). Protein tyrosine nitration in the mitochondria from diabetic mouse heart: Implications to dysfunctional mitochondria in diabetes. *J. Biol. Chem.* **278**, 33972–33977.

Upmacis, R. K., Crabtree, M. J., Deeb, R. S., Shen, H., Lane, P. B., Benguigui, L. E., Maeda, N., Hajjar, D. P., and Gross, S. S. (2007). Profound biopterin oxidation and protein tyrosine nitration in tissues of ApoE-null mice on an atherogenic diet: Contribution of inducible nitric oxide synhase. *Am. J. Phys. Heart Circ. Phys.* **293**, H2878–H2887.

Yi, D. H., Ingelse, B. A., Duncan, M. W., and Smythe, G. A. (2000). Quantification of 3-nitrotyrosine in biological tissues and fluids: Generating valid results by eliminating artifactual formation. *J. Am. Soc. Mass Spectrom.* **11**, 578–586.

Zhang, Q., Qian, W. J., Knyushko, T. V., Clauss, T. R., Purvine, S. O., Moore, R. J., Sacksteder, C. A., Chin, M. H., Smith, D. J., Camp, D. G., 2nd, Bigelow, D. J., and Smith, R. D. (2007). A method for selective enrichment and analysis of nitrotyrosine-containing peptides in complex proteome samples. *J. Proteome Res.* **6**, 2257–2268.

CHAPTER TWO

Selective Fluorogenic Derivatization of 3-Nitrotyrosine and 3,4-Dihydroxyphenylalanine in Peptides: A Method Designed for Quantitative Proteomic Analysis

Victor S. Sharov, Elena S. Dremina, Justin Pennington, Jacque Killmer, Christopher Asmus, Maria Thorson, Sung Jung Hong, Xiaobao Li, John F. Stobaugh, *and* Christian Schöneich

Contents

Abstract

There is a need for the selective derivatization and enrichment of posttranslational protein modifications from tissue samples. This chapter describes a method for the selective derivatization of 3-nitrotyrosine (after reduction to 3-amino-tyrosine) and 3,4-dihydroxyphenylalanine with benzylamine derivatives to yield 6-amino- and 6-benzylamine-substituted benzoxazoles, which display characteristic

Department of Pharmaceutical Chemistry, University of Kansas, Lawrence, Kansas

Methods in Enzymology, Volume 441
ISSN 0076-6879, DOI: 10.1016/S0076-6879(08)01202-0

fluorescence properties. The methodology can be expanded to other substituted benzylamines, which carry functional groups for affinity enrichment.

1. INTRODUCTION

The accumulation of oxidative posttranslational protein modifications represents a hallmark of oxidative stress, accompanying numerous pathologies and biological aging (Dalle-Donne *et al.*, 2005, 2006). In the past, special attention has been given to the modification of protein tyrosine residues: here nitration to 3-nitrotyrosine (3-NY) represents an indicator for the enhanced generation of nitric oxide and oxidative metabolites of nitric oxide (Greenacre and Ischiropoulos, 2001; Ischiropoulos and Beckman, 2003; Radi, 2004; Turko and Murad, 2002), and hydroxylation to 3,4-dihydroxyphenylalanine (DOPA) represents a general marker for an increased abundance of reactive oxygen species (Fu *et al.*, 1998a,b). The extent to which these posttranslational modifications exert functional changes within cells or cellular compartments can only be evaluated through the identification of protein targets and quantification of these modifications. Proteomic studies have been designed to achieve this goal and have yielded important information regarding the nature of modified proteins and the location of posttranslational modifications within the protein sequences specifically for 3-NY (Aulak *et al.*, 2001; Casoni *et al.*, 2005; Kanski *et al.*, 2003, 2005a,b; Koeck *et al.*, 2004; Sacksteder *et al.*, 2006; Turko *et al.*, 2003; Zhan and Desiderio, 2004, 2006). In contrast, such proteomic compilations are largely absent for DOPA. Nevertheless, these studies have by far not covered the entire proteome, merely due to the abundance of potential target sites: in a first approximation the number of protein tyrosine residues available for nitration and/or oxidation can be calculated to approximately 510,000. This estimate is based on the existence of approximately 30,000 genes, an average molecular mass of proteins of about 50 kDa, corresponding to approximately 500 amino acids, and an average abundance of Tyr of 3.4% (Brooks and Fresco, 2002). These numbers suggest that even sophisticated global proteomic approaches may not be able to cover the entire "3-nitroproteome." In an effort to reduce the sample size for analysis, researchers have designed several enrichment strategies for 3-nitrotyrosine-containing peptides. Generally, these depend on the reduction of 3-nitrotyrosine to 3-aminotyrosine, followed by electrophilic derivatization of the aromatic amine. The first approach was reported by Nikov *et al.* (2003), where 3-aminotyrosine-containing peptides and proteins were reacted with sulfosuccinimidyl-2-(biotinamido)ethyl-1,3-dithiopropionate at pH 5. This reagent represents a multifunctional compound containing an *N*-hydroxysuccinimide ester for reaction with 3-aminotyrosine, a reducible disulfide linker, and a biotin affinity tag. After enrichment of the derivatized peptides on immobilized streptavidin, the biotin tag can be cleaved through reduction, and after alkylation of the resulting thiol function, the peptides can be analyzed by mass spectrometry. An inherent problem of this method is the side reaction of the hydroxysuccinimide ester with N-terminal amino groups and/or Lys residues of nonnitrated peptides. Taking the observed ratio of [Tyr]:[3-NY] $\approx 10^6$ (Tsikas and Caidahl, 2005) in the plasma of healthy humans and an average abundance of Tyr of 3.4%, we estimate that even at pH 5 the ratio of nonnitrated and non-Tyr-containing peptides with a deprotonated N terminus to 3-aminotyrosine-containing peptides (after reduction of 3-NY) would be on the order of $>10^4$. Hence, without efficient alkylation/ acetylation of free amines prior to the reduction of 3-NY, coupling with N-hydroxysuccinimide esters may not provide the desired selectivity for enrichment. Therefore, an acetylation step was introduced by Zhang *et al.* (2007) prior to derivatization of 3-aminotyrosine residues with N-succinimidyl S-acetylthioacetate (SATA). Following derivatization with SATA and reaction with hydroxylamine, the resulting thiol-containing peptides were enriched through a solid-phase enrichment method specific for thiols and subsequently analyzed by capillary liquid chromatography/mass spectrometry/mass spectrometry. This method yielded 150 unique nitrated peptides recovered from mouse brain homoge-nate. Two disadvantages of this method are (i) the absence of a suitable functional group or linker, which would allow for relative quantitation of separate samples through differential isotopic labeling with either H- or D-containing reagents, and (ii) the absence of a chromophore/fluorophore, which would allow for absolute quantification by a method independent on mass spectrometry. Therefore, we designed an alternative fluorescence tagging method, which is described in more detail here.

2. DESIGN OF A FLUOROGENIC DERIVATIZATION METHOD

Our objective was the design of a method that would (i) show selectivity toward 3-NY and DOPA-containing peptides and proteins, (ii) yield a fluorescent reaction product with excitation and emission properties far from those of native proteins and/or common oxidation products (e.g., bityrosine, N-formylkynurenin), (iii) be suitable for isotopic labeling for relative quantitation, and (iv) be suitable for the introduction of additional ligands for the improvement of chromatographic behavior and/or affinity enrichment. Based on previous reports that the reaction of catechols with diphenylethylenediamine yielded a fluorescent benzoxazole-type product

(Nohta *et al.*, 1992), we focused our efforts toward the reaction of catechols and aminophenols with benzylamine and benzylamine derivatives. These reactions utilize both aromatic substituents (the two hydroxyl groups in the case of DOPA, and the hydroxyl and amino group in the case of 3-aminotyrosine) for the generation of a fluorescent benzoxazole, while no fluorescent benzoxazole product can be generated if one of these substituents would be absent, that is, in the case of Tyr. Moreover, benzylamine can be added in such excess to these reactions that competitive reactions with deprotonated N-terminal amines or Lys residues from peptides will be kinetically insignificant. Finally, the aromatic moiety of benzylamine may be labeled isotopically (with D or ^{13}C) for relative quantitation or appended with *para* substituents, which enhance polarity or allow for affinity enrichment. This chapter summarizes some model studies with 4-methylcatechol (a model for DOPA) and 2-aminocresol (a model for 3-aminotyrosine), which led to the correct structural identification of the reaction product with benzylamine (different from that originally proposed), a reaction mechanism, the stoichiometry of isotopic labeling, and the optimization of some reaction conditions. Subsequently, we extend these studies to a 3-nitrotyrosine-containing model peptide, tagging with 4-aminobenzylbenzene sulfonic acid, and the use of an ammonia-containing buffer vs phosphate buffer.

3. MODEL STUDIES WITH 4-METHYLCATECHOL AND 2-AMINOCRESOL

The derivatization of 4-methylcatechol and 2-aminocresol with benzylamine (BA) led to the structural characterization of a novel common fluorescent reaction product **1** (*N*-benzyl-5-methyl-2-phenylbenzo[d]oxazol-6-amine) for both reactants, displayed in Scheme 2.1, with an excitation maximum of $\lambda_{exc} = 344$ nm and an emission maximum of $\lambda_{em} = 467$ nm (Pennington *et al.*, 2007). Here, 4-methylcatechol serves as a model for DOPA, and 2-aminocresol serves as a model for 3-aminotyrosine (i.e., the reduction product of 3-nitrotyrosine). Mechanistically, the formation of **1** is rationalized by a multistep process (reactions 1–7). In our initial studies, all reactions were performed in mixtures of methanol and water (9:1, v/v), to which were added in the following order: analyte (final concentration of 5–100 nM), $K_3Fe(CN)_6$ (final concentration of 5 mM) and BA (final concentration of up to 0.1 M). Reactions have to be carried out at alkaline pH (≈ 9.0), which is maintained through the high concentrations of BA (alternative procedures are described later in which the alkaline pH is maintained through the addition of suitable buffers). Maximal yields of product 1 were obtained after rather short reaction times of 10 min, when BA was present in at least 10^5-fold excess over the analyte (alternative reaction

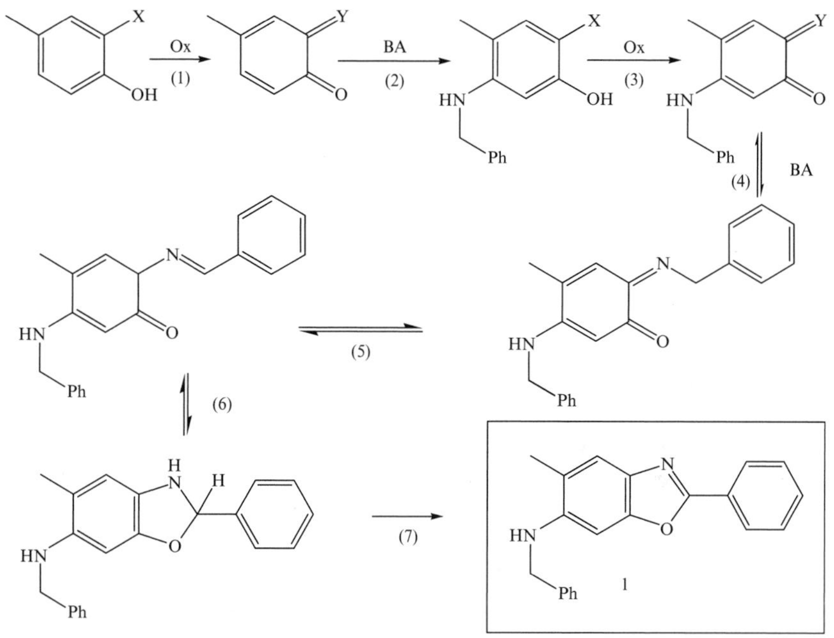

Scheme 2.1   Formation of 6-benzylamino-substituted benzoxazole 1 from the reaction of 4-methylcatechol and 2-aminocresol with benzylamine. For 4-methylcatechol, $X = OH$ and $Y = O$; for 2-aminocresol, $X = NH_2$ and $Y = NH$.

conditions for peptides are described later). The incorporation of two BA moieties into product 1 offers an advantage for relative quantification through isotopic labeling, where derivatization with $C_6D_5CH_2NH_2$ instead of $C_6H_5CH_2NH_2$ will ultimately lead to a mass difference of 10 amu. This was demonstrated not only for 4-methylcatechol and 2-aminocresol, but also for DOPA-containing model peptides derived from the hydroxylation of Tyr-Gly-Gly and angiotensin II_{3-8} (Pennington *et al.*, 2007).

4. Derivatization of a 3-Nitrotyrosine-Containing Model Peptide with Benzylamine and 4-Aminomethylbenzene Sulfonic Acid

4.1. Fluorescence spectroscopy

To evaluate the derivatization of a 3-nitrotyrosine-containing peptide, we synthesized FSAY(3-NO2)LER from Fmoc-protected amino acids using an ACT 90 peptide synthesizer (Advanced ChemTech, Louisville, KY).

This nitrated sequence originates from mouse glycogen phosphorylase (F[545]SAYLER[551]) and will serve in future studies as an internal standard to quantitate the *in vivo* nitration of the nearly homologous rat sequence F[546]AAYLER[552] in a model for biological aging. After complete reduction of FSAY(3-NO_2)LER to the 3-aminotyrosine-containing peptide, FSAY (3-NH_2)LER, using $NaBH_4$ in the presence of Ni/Pt (Killmer, 2007), the following derivatization conditions are employed: (i) to 40 μM FSAY (3-NH_2)LER in 0.1 M sodium phosphate buffer, adjusted to pH 9, are added a final concentration of 2 mM BA or 4-aminomethylbenzene sulfonic acid (ABS), respectively, and the reaction is run for 30 min at room temperature; (ii) to 40 μM FSAY(3-NH_2)LER in 0.2 M NH_4HCO_3 buffer, adjusted to pH 9, is added a final concentration of 0.5 mM $K_3Fe(CN)_6$ and the mixture is incubated for 120 min; then a final concentration of 2 mM BA or ABS, respectively, is added and the reaction is run for 30 min at room temperature. Figure 2.1A displays fluorescence spectra recorded after derivatization of FSAY(3-NH_2)LER with 2 mM BA or 2 mM ABS in phosphate buffer. Derivatization with BA yields a product with an emission maximum around $\lambda_{em} = 460$ nm ($\lambda_{exc} = 340$ nm), which is expected based on previous model studies on the reaction of BA with 2-aminocresol (see earlier discussion; Pennington *et al.*, 2007). However, derivatization with ABS yields a product with a red-shifted emission maximum around $\lambda_{em} = 490$ nm ($\lambda_{exc} = 360$ nm). This red shift must originate from introduction of the sulfonate group into the final product. Importantly, the fluorescence intensity of the reaction product formed from ABS is significantly higher compared to that formed from BA. Products with rather identical fluorescence properties are generated when the derivatization is carried out in NH_4HCO_3, that is,

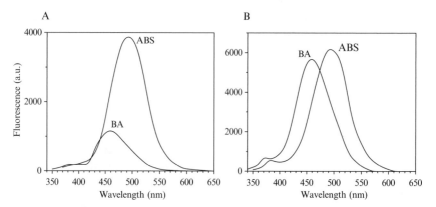

Figure 2.1 Fluorescence spectra recorded after derivatization 40 μM FSAY(3-NH_2) LER with 2 mM BA or ABS in the presence of 0.5 mM $K_3Fe(CN)_6$ at pH 9 in (A) 0.1 M sodium phosphate buffer or (B) 0.2 M NH_4HCO_3 buffer. Excitation wavelengths were 340 and 360 nm for BA and ABS derivatization, respectively, in phosphate buffer, and 330 and 340 nm for BA and ABS, respectively, in NH_4HCO_3 buffer.

$\lambda_{em} = 460$ nm for reaction with BA ($\lambda_{exc} = 330$ nm) and $\lambda_{em} = 490$ nm for reaction with ABS ($\lambda_{exc} = 340$ nm). The corresponding spectra are displayed in Fig. 2.1B. However, no difference in the fluorescence intensities was observed in this buffer.

4.2. Analysis by HPLC coupled to fluorescence detection

When the reaction products formed in phosphate and NH_4HCO_3 buffer were analyzed by HPLC coupled to fluorescence detection, important differences were noted. Derivatization of FSAY(3-NH$_2$)LER with BA in phosphate buffer yielded a product eluting as a rather broad peak with $t_R \approx 34$ to 35 min (Fig. 2.2A), while derivatization of FSAY(3-NH$_2$)LER with BA in NH_4HCO_3 yielded a product eluting with $t_R = 22$ min (Fig. 2.2B). Derivatization of FSAY(3-NH$_2$)LER with ABS in phosphate buffer yielded a product eluting with $t_R = 16$ min (Fig. 2.2A), while derivatization of FSAY(3-NH$_2$)LER with ABS in NH_4HCO_3 yielded a product eluting with $t_R = 12$ to 13 min (Fig. 2.2B). All chromatographic separations were carried out on a Vydac C18 column (4.6 mm ID × 250 mm). Samples were injected onto columns equilibrated at 10% ACN/90% H$_2$O/0.1%TFA (v/v/v), and peptides were separated by a linear increase of ACN in the mobile phase (1%/min) followed by a washing step for 10 min at 80% (v/v) (ACN = acetonitrile).

4.3. Analysis by HPLC-MS/MS

Important differences were then realized upon mass spectrometry analysis of the various reaction products (for mass spectrometry conditions, see later). Derivatization of FSAY(3-NH$_2$)LER with ABS and BA in phosphate buffer

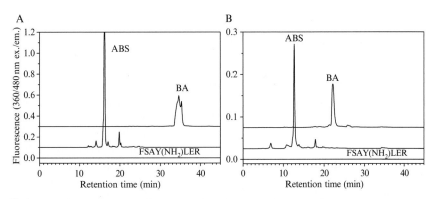

Figure 2.2 HPLC–fluorescence analysis after derivatization of 40 μM FSAY(3-NH$_2$) LER with 2 mM BA or ABS in the presence of 0.5 mM K$_3$Fe(CN)$_6$ at pH 9 in (A) 0.1 M sodium phosphate buffer or (B) 0.2 M NH$_4$HCO$_3$ buffer. The injection volume was 200 μl; for separation conditions, see text.

(Fig. 2.3) yielded the expected products with molecular masses of 1251.5 amu ([M+H]$^+$ for ABS; Fig. 2.3A) and 1091.6 ([M+H]$^+$ for BA; Fig. 2.3B), indicating the incorporation of two ABS or BA moieties into the peptide, respectively, analogous to the reaction of BA with 2-aminocresol shown in Scheme 2.1. The corresponding structures for the peptide are shown in Scheme 2.2. The respective MS/MS spectra confirm the nature of structures **2** and **3** in Scheme 2.2. However, compared to phosphate buffer, molecular masses for the derivatization products obtained in NH$_4$HCO$_3$ buffer are 170 amu lower for ABS (m/z 1081.5; [M+H]$^+$) and 90 amu lower for BA (m/z 1001.6; [M+H]$^+$) (Figs. 2.4A and 2.4B). These data suggest that the reaction in NH$_4$HCO$_3$ buffer leads to the incorporation of one NH$_2$ group and one benzylamine derivative, as shown in structures **4** and **5** in Scheme 2.3, and this is confirmed by the respective MS-MS data shown in Figs. 2.4C and 2.4D. Importantly, incorporation of NH$_2$ vs benzylamine has no significant consequences for the fluorescence properties of the products, that is, fluorescence spectra of **2** and **4** are identical, as are those of products **3** and **5**. However, specifically for derivatization with BA, the substitution of one BA by one NH$_2$ group significantly shortens the retention time during reversed-phase chromatography. The latter property may be useful for the analysis of larger, more hydrophobic peptides.

Our mass spectrometry data suggest a modified reaction mechanism for the formation of 6-substituted benzoxazoles **4** and **5** in NH$_4$HCO$_3$ buffer, displayed in Scheme 2.4. It appears that the Michael addition of the small NH$_3$ at the 6 position is kinetically favored over the addition of BA or ABS at the low (2 mM) concentrations of the benzylamines used for peptide derivatization. Hence, by the choice of buffer, or merely through the addition of an ammonium salt to a suitable buffer, products **4** and **5** may be generated, which have different chromatographic properties compared to **2** and **3**.

4.4. Mass spectrometry conditions

Peptide samples (5 μl) are submitted to nano-HPLC-nanoelectrospray ionization-tandem mass spectrometry analysis on either a ThermoElectron LCQ Duo or a ThermoElectron Classic (San Jose, CA), equipped with a nanoelectrospray source (ThermoElectron). Separation of peptides is achieved online prior to MS/MS analysis on in-house-packed BioBasic C18 stationary phase (Thermo Electron) nanoflow columns (300 Å, 10 cm × 75 μm, 15-μm tip size) (New Objective, Woburn, MA) with the following chromatographic conditions. Mobile phase A: 0.1% formic acid in water, mobile phase B: 0.1% formic acid in ACN. The flow rate is 0.5 μl/min, delivered by a MicroTech Scientific Ultra Plus II pump (after 1:20 split) or by a MicroTech Xtreme Simple nanoflow pump (direct flow). The following gradient profile is used to increase mobile phase B linearly to

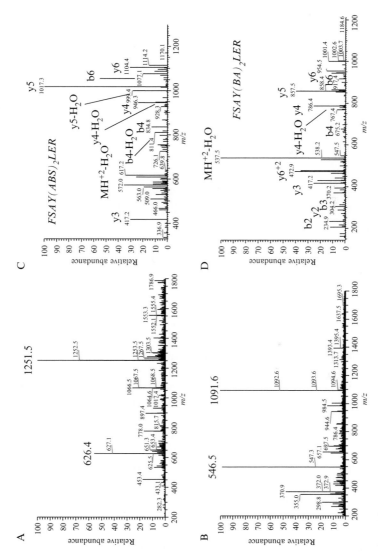

Figure 2.3 MS1 and MS/MS analysis after derivatization of 40 μM FSAY(NH$_2$)LER with 2 mM BA or ABS in the presence of 0.5 mM K$_3$Fe(CN)$_6$ in 0.1 M sodium phosphate buffer, adjusted to pH 9.0. For conditions of MS analysis, see text.

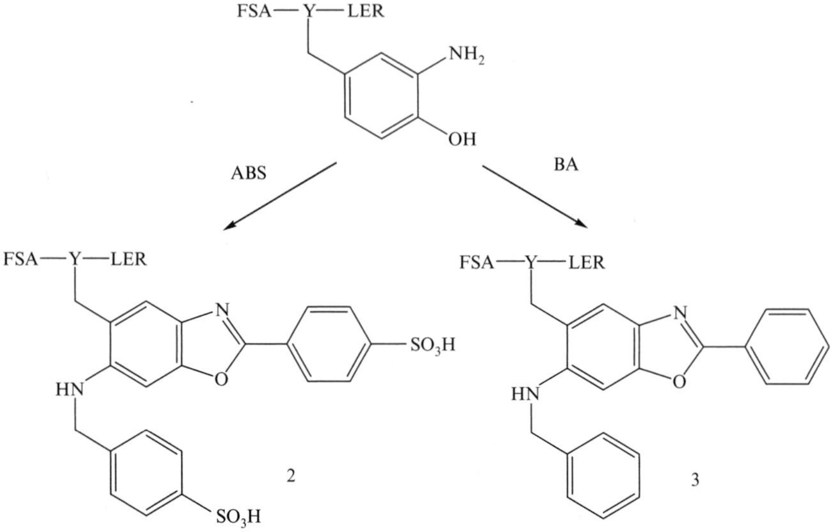

Scheme 2.2 Structures of 6-benzylamino-substituted benzoxazoles 2 and 3 from the derivatization of FSAY(3-NH₂)LER in sodium phosphate buffer with ABS and BA, respectively.

the following fractions: from a 0- to 5-min gradient held at 10% B, increased to 60% B within 105 min, and continued at 60% B for an additional 5 min. After each run, the column is washed by a short gradient (0–60% B for 20 min) and allowed to reequilibrate to the initial conditions for 15 min. The following instrumental conditions are used for mass spectrometric analysis: number of microscans is 3, length of microscans is 200 ms, capillary temperature is 160°, spray voltage is 1.9 kV, capillary voltage is 35 V, and tube lens offset is −14 V. The mass spectrometer is tuned using the static nanospray setup with a 5 μM solution of angiotensin I (MW 1296.5) infused by a picotip emitter (New Objective). Data acquisition is performed in a data-dependent fashion, that is, an MS scan followed by three or four MS/MS scans of the three or four most intense peaks with the normalized collision energy for MS/MS set at 35% and an isolation width of 2.0 m/z. A minimal signal for MS/MS acquisition is set to 2×10^6. Additionally, the dynamic exclusion option is enabled and set with the following parameters: repeat count is 3, repeat duration is 5 min, exclusion list size is 25, exclusion duration is 5, and exclusion mass width is 3. Peptide sequence analysis is achieved with the ThermoElectron Bioworks 3.1 software package with the most current nonredundant NCBI protein database downloaded from ftp.ncbi. nlm.nih.gov/blast/db. Additionally, MS/MS spectra of interest are examined

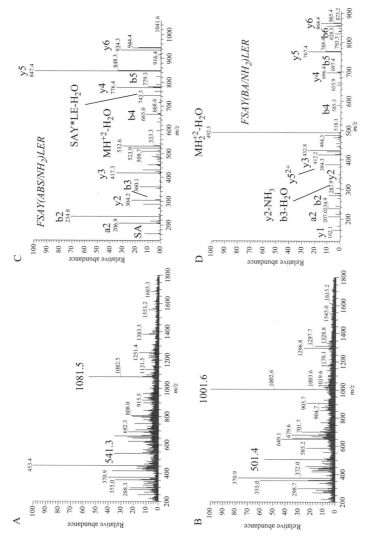

Figure 2.4 MS1 and MS/MS analysis after derivatization of 40 μM FSAY(NH$_2$)LER with 2 mM BA or ABS in the presence of 0.5 mM K$_3$Fe(CN)$_6$ in 0.2 MNH$_4$HCO$_3$ buffer, adjusted to pH 9.0. For conditions of MS analysis, see text.

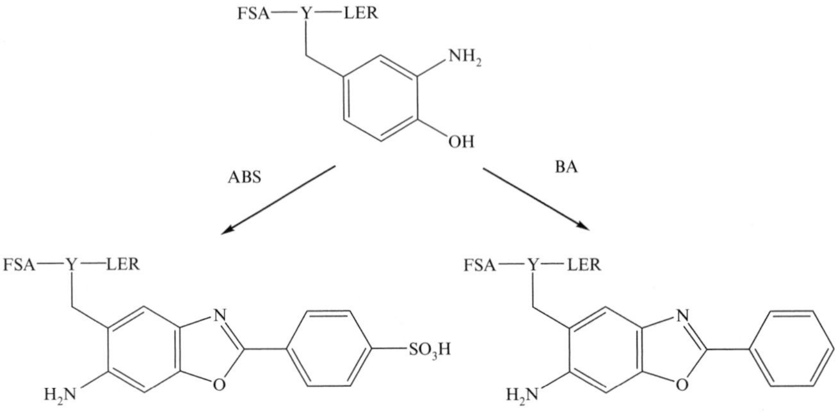

Scheme 2.3 Structures of 6-amino-substituted benzoxazoles 4 and 5 from the derivatization of FSAY(3-NH$_2$)LER in NH$_4$HCO$_3$ buffer with ABS and BA, respectively.

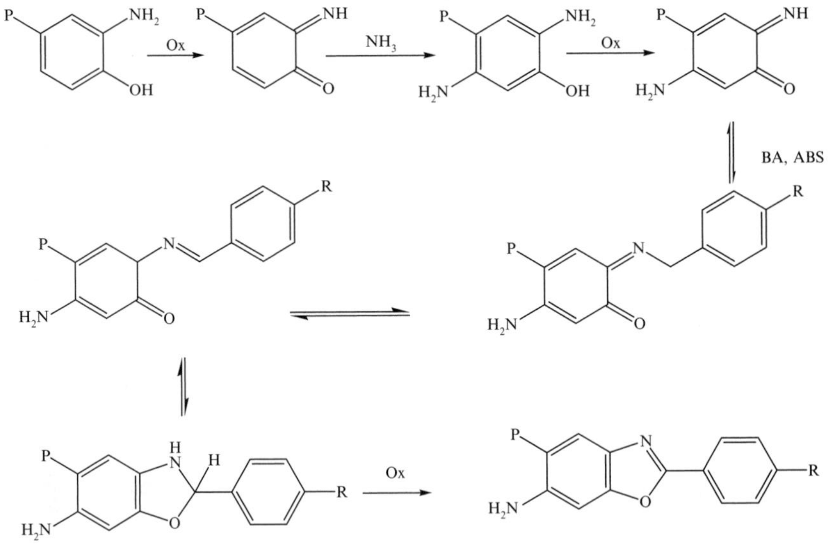

Scheme 2.4 Formation of 6-amino-substituted benzoxazoles from the reaction of FSAY(3-NH$_2$)LER in NH$_4$HCO$_3$ buffer with ABS (R = SO$_3$H) and BA (R = H), respectively. The substituent P denotes residual peptide moiety.

manually for the presence of neutral losses and internal fragments. Analysis of MS/MS spectra is based on a search for the major sequence-indicating ions resulting from cleavage of the parent ion at specific locations relative to the peptide bond.

5. Conclusions

The derivatization of DOPA and 3-aminotyrosine with benzylamine yields 6-amino or 6-benzylamino-substituted benzoxazole derivatives with characteristic fluorescent properties, which can be used for the quantitative analysis of these posttranslational modifications in protein samples. Our laboratory is currently focusing on the incorporation of additional substituents into the 4-position of aminomethylbenzene, which would allow for affinity enrichment of the derivatized peptides.

ACKNOWLEDGMENT

This research was supported by grants from the NIH (AG23551, AG25350).

REFERENCES

Aulak, K. S., Miyagi, M., Yan, L., West, K. A., Massilon, D., Crabb, J. W., and Stuehr, D. J. (2001). Proteomic method identifies proteins nitrated *in vivo* during inflammatory challenge. *Proc. Natl. Acad. Sci. USA* **98,** 12056–12061.

Brooks, D. J., and Fresco, J. R. (2002). Increased frequency of cysteine, tyrosine, and phenylalanine residues since the last universal ancestor. *Mol. Cell. Proteomics.* **1,** 125–131.

Casoni, F., Basso, M., Massignan, T., Gianazza, E., Cheroni, C., Salmona, M., Bendotti, C., and Bonetto, V. (2005). Protein nitration in a mouse model of familial amyotrophic lateral sclerosis: Possible multifunctional role in the pathogenesis. *J. Biol. Chem.* **280,** 16295–16304.

Dalle-Donne, I., Rossi, R., Colombo, R., Giustarini, D., and Milzani, A. (2006). Biomarkers of oxidative damage in human disease. *Clin. Chem.* **52,** 601–623.

Dalle-Donne, I., Scaioni, A., Giustarini, D., Cavarra, E., Tell, G., Lungarella, G., Colombo, R., Rossi, R., and Milzani, A. (2005). Proteins as biomarkers of oxidative/ nitrative stress in diseases: The contribution of redox proteomics. *Mass Spectr. Rev.* **24,** 55–99.

Fu, S., Dean, R. T., Southan, M., and Truscott, R. (1998a). The hydroxyl radical in lens nuclear cataractogenesis. *J. Biol. Chem.* **273,** 28603–28609.

Fu, S., Fu, M., Baynes, J. W., Thorpe, S. R., and Dean, R. T. (1998b). Presence of DOPA and amino acid hydroperoxides in proteins modified with advanced glycation end products (AGEs): Amino acid oxidation products as a possible source of oxidative stress induced by AGE proteins. *Biochem. J.* **330,** 233–239.

Greenacre, S. A. B., and Ischiropoulos, H. (2001). Tyrosine nitration: Localization, quantification, consequences for protein function and signal transduction. *Free Radic. Res.* **34,** 541–581.

Ischiropoulos, H., and Beckman, J. S. (2003). Oxidative stress and nitration in neurodegeneration: Cause, effect, or association. *J. Clin. Invest.* **111,** 163–169.

Kanski, J., Alterman, M., and Schöneich, Ch. (2003). Proteomic identification of age-dependent protein nitration in rat skeletal muscle. *Free Radic. Biol. Med.* **35,** 1229–1239.

Kanski, J., Behring, A., Pelling, J., and Schöneich, Ch. (2005a). Proteomic identification of 3-nitrotyrosine-containing rat cardiac proteins: Effect of biological aging. *Am. J. Physiol. Heart Circ. Physiol.* **288,** H371–H381.

Kanski, J., Hong, S. J., and Schöneich, Ch. (2005b). Proteomic analysis of protein nitration in aging skeletal muscle and identification of nitrotyrosine-containing sequences *in vivo* by nanoelectrospray ionization tandem mass spectrometry. *J. Biol. Chem.* **280,** 24261–24266.

Killmer, J. (2007). "Development of Mass Labels for the Comparative Quantitative Analysis of 3-Nitrotyrosyl Proteins." Ph.D. Thesis, Department of Pharmaceutical Chemistry, University of Kansas.

Koeck, T., Fu, X., Hazen, S. L., Crabb, J. W., Stuehr, D. J., and Aulak, K. S. (2004). Rapid and selective oxygen-regulated protein tyrosine denitration and nitration in mitochondria. *J. Biol. Chem.* **279,** 27257–27262.

Nikov, G., Bhat, V., Wishnok, J. S., and Tannenbaum, S. R. (2003). Analysis of nitrated proteins by nitrotyrosine-specific affinity probes and mass spectrometry. *Anal. Biochem.* **320,** 214–222.

Nohta, H., Lee, M. K., and Ohkura, Y. (1992). Fluorescent products of the reaction for the determination of catecholamines with 1,2-diphenylethylenediamine. *Anal. Chim. Acta* **267,** 137–139.

Pennington, J. P., Schöneich, Ch, and Stobaugh, J. F. (2007). Selective fluorogenic derivatization with isotope coding of catechols and 2-aminophenols with benzylamine: A chemical basis for the relative determination of 3-hydroxytyrosine and 3-nitrotyrosine peptides. *Chromatographia,* **66,** 649–659.

Radi, R. (2004). Nitric oxide, oxidants, and protein tyrosine nitration. *Proc. Natl. Acad. Sci. USA* **101,** 4003–4008.

Sacksteder, C. A., Qian, W. -J., Knyushko, T. V., Wang, H., Chin, M. H., Lacan, G., Melega, W. P., Camp, D. G., Smith, R. D., Smith, D. J., Squier, T. C., and Bigelow, D. J. (2006). Endogenously nitrated proteins in mouse brain: Links to neurodegenerative disease. *Biochemistry.* **45,** 8009–8022.

Tsikas, D., and Caidahl, K. (2005). Recent methodological advances in the mass spectrometric analysis of free and protein-associated 3-nitrotyrosine in human plasma. *J. Chromatogr. B Analyt. Technol. Biomed. Life Sci.* **814,** 1–9.

Turko, I. V., Li, L., Aulak, K. S., Stuehr, D. J., Chang, J.-Y., and Murad, F. (2003). Protein tyrosine nitration in the mitochondria from diabetic mouse heart. *J. Biol. Chem.* **278,** 33972–33977.

Turko, I. V., and Murad, F. (2002). Protein nitration in cardiovascular diseases. *Pharmacol. Rev.* **54,** 619–634.

Zhan, X., and Desiderio, D. M. (2004). The human pituitary nitroproteome: Detection of nitrotyrosyl-proteins with two-dimensional Western blotting, and amino acid sequence determination with mass spectrometry. *Biochem. Biophys. Res. Commun.* **325,** 1180–1186.

Zhan, X., and Desiderio, D. M. (2006). Nitroproteins from a human pituitary adenoma tissue discovered with a nitrotyrosine affinity column and tandem mass spectrometry. *Anal. Biochem.* **354,** 279–289.

Zhang, Q., Qian, W., Knyushko, T. V., Clauss, T. R. W., Purvine, S. O., Moore, R. J., Sacksteder, C. A., Chin, M. H., Smith, D. J., Camp, D. G., II, Bigelow, D. J., and Smith, R. D. (2007). A method for selective enrichment and analysis of nitrotyrosine-containing peptides in complex proteome samples. *J. Proteome Res.* **6,** 2257–2268.

CHAPTER THREE

NITROALKENES: SYNTHESIS, CHARACTERIZATION, AND EFFECTS ON MACROPHAGE ACTIVATION

Ana María Ferreira,[*,†] Andrés Trostchansky,[†] Mariana Ferrari,[*,†] José M. Souza,[†] *and* Homero Rubbo[†]

Contents

* Department of Immunology, Facultadas de Ciencias y Química, Universidad de la República, Uruguay
† Center for Free Radical and Biomedical Research, Department of Biochemistry, Facultad de Medicina, Universidad de la República, Uruguay

Methods in Enzymology, Volume 441
ISSN 0076-6879, DOI: 10.1016/S0076-6879(08)01203-2

Abstract

Nitroalkenes derivatives of free as well as esterified unsaturated fatty acids are present in human plasma and tissue, representing novel pluripotent cell signaling mediators. Lipid nitration occurs in response to pro-inflammatory stimuli as an adaptive mechanism to downregulate inflammatory responses. This chapter first discusses the generation of nitroalkenes during macrophage activation following chemical and biological characterization. In particular, it describes procedures for (a) synthesizing and characterizing esterified (cholesteryl-nitrolinoleate, $CLNO_2$) as well as free (nitroarachidonate, $AANO_2$) nitroalkenes, (b) determining nitration of cholesteryl linoleic acid during macrophage activation by inflammatory stimuli, (c) examining the modulatory effects of nitroalkenes on the expression of inducible enzymes by activated macrophages, and (d) discussing the signaling pathways involved in nitroalkene-mediated anti-inflammatory actions.

1. INTRODUCTION

Macrophages are key cells in the onset of inflammatory responses, playing a role in the first line of defense against invading pathogens. They possess a repertoire of receptors, including those called Toll-like receptors (TLR), that distinguish between a variety of motifs present in pathogens (ranging from viral nucleic acids to bacterial cell wall components)(Kawai and Akira, 2005; Lien and Ingalls, 2002). Ligand binding to these receptors activates signaling pathways linked to the production of inflammatory mediators (i.e., cytokines and biolipids) and highly reactive oxygen and nitrogen species (Dobrovolskaia and Vogel, 2002; Kaisho and Akira, 2006). In particular, nitric oxide (NO·) is one of the most important free radicals synthesized during macrophage activation by TLR ligands (Thoma-Uszynski *et al.*, 2001; Werling *et al.*, 2004). It is formed by the oxidation of L-arginine to L-citrulline by the inducible nitric oxide synthase (NOS2), and its toxic potential is mainly associated with its ability to react with the superoxide (O_2-) anion to yield peroxynitrite (Radi *et al.*, 2000; Rubbo *et al.*, 1994), as well as with O_2 forming nitrogen dioxide ($NO_2·$) and other oxidant and nitrosating species (Moller *et al.*, 2007; Wink *et al.*, 1993). These NO·-derived reactive species, usually termed reactive nitrogen species (RNS), constitute potent toxic molecules for intracellular invading pathogens because they are capable of reacting with cell constituents, modifying many classes of biomolecules in an irreversible way (Alvarez *et al.*, 1999; Eiserich *et al.*, 1998; Tien *et al.*, 1999). However, RNS overproduction could cause damage, not only to the invading pathogen but paradoxically to macrophage themselves and surrounding cells. Thus, RNS formation during macrophage activation is subjected to a tight regulation for eradicating pathogens, avoiding tissue injury.

Several mechanisms are likely involved in the fine-tuning of RNS formation. Overproduction of RNS may be considered as a signal that triggers mechanisms for negative regulation of NOS2 expression. In fact, accumulation of peroxynitrite has been associated with enhanced induction of heme-oxygenase 1 (HO-1)(Srisook and Cha, 2005; Srisook et al., 2005a), a cytoprotective enzyme whose expression was found to correlate with a decrease in NO• generation (Srisook et al., 2005b). HO-1 effects on NO• production involve control mechanisms at the level of NOS activity and expression. First, HO-1 catalyzes oxidative degradation of the free heme group, releasing, among other products, carbon monoxide (CO), which binds avidly to the heme iron of NOS2, interfering with electron transfer reactions required for NO• production (Srisook et al., 2006). Second, an enhanced induction of HO-1 correlates with a decrease in NF-κB-dependent NOS2 expression; direct and/or indirect involvement of CO in HO-1-mediated inhibition of NF-κB activation may be involved (Lee et al., 2003; Srisook et al., 2006). Furthermore, recent studies suggest that NO• production may be downregulated by nitroalkenes derived from the nitration of unsaturated fatty acids. These products are capable of transducing signaling pathways, leading to the induction of HO-1 (Cui et al., 2006) and downregulation of NOS2 expression in macrophages (Trostchansky et al., 2007). Thus, overproduction of NO• could drive lipid nitration, and the generated nitroalkenes may act as endogenous, secondary mediators capable of attenuating NOS2 expression and limiting the proinflammatory actions of NO•-derived reactive species. This hypothesis, even if very attractive, still needs further studies because nitroalkene concentrations in resting and activated macrophages have not been determined to ensure that the inhibitory effects observed in vitro are of significance in physiological conditions. Current data regarding nitroalkene levels in plasma and macrophages, as well as the mechanisms used by them to modulate macrophage activation, are discussed briefly here.

Nitroalkene derivatives of unsaturated fatty acids, mainly nitrolinoleic acid (9-, 10-, 12-, and 13-nitro-9,12-cis-octadecadienoic acid, LNO$_2$), nitrooleic acid (9- and 10-nitro-9-cis-octadecenoic acid, OA-NO$_2$), and the cholesteryl ester of linoleic acid (CLNO$_2$), have been synthesized and structurally characterized (Baker et al., 2004, 2005; Lima et al., 2002, 2003; Alexander et al., 2006). To date, most quantitative analyses have focused mainly on nitroalkene derivatives of free fatty acid on blood and urine of healthy humans, with their total levels (free plus esterified) being around 500 nM (Baker et al., 2004; Lima et al., 2003). Because their structural diversity and electrophilic reactivity result in the formation of protein-nitroalkene derivatives (Batthyany et al., 2006), this value may underestimate nitroalkene concentrations. Concentrations of nitrated free fatty acid derivatives in an inflammatory milieu remain also to be fully defined.

Nitration of cholesteryl linoleate during lipopolysaccharide (LPS) plus inter-feron (IFN)γ-driven macrophage activation has been detected (Ferreira *et al.*, unpublished); it is likely that oxidative and nitrative inflammatory conditions would induce greater extents of nitration of fatty acids and other biomole-cules (Schopfer *et al.*, 2003), yielding nitroalkenes in the micromolar range. Moreover, exogenous addition of nitroalkenes to macrophages inhibited NOS2 expression and cytokine secretion in response to LPS, supporting the concept that lipid nitration derivatives are endogenous mediators that serve to modulate the inflammatory response of macrophages. The capacity of nitroalkenes either to activate the peroxisome proliferator–activated receptor γ (Baker *et al.*, 2005; Schopfer *et al.*, 2005b) or to release NO$\cdot$ during decay in aqueous milieu via a Nef reaction (Lima *et al.*, 2005; Schopfer *et al.*, 2005a) cannot explain their anti–inflammatory effects; they are likely consequences of the electrophilic nature of nitroalkenes that allows them to mediate reversible nitroalkylation reactions with the transcriptional factors Keap1 and NF-κB, leading to the activation of Keap1/Nrf2/ARE (Villacorta *et al.*, 2007) and the inhibition of NF-κB pathways (Cui *et al.*, 2006), respectively.

This chapter describes protocols intended for evaluating the generation of nitroalkenes during macrophage activation and the effects that nitroalk-enes could exert on macrophage activation and differentiation. In particular, this chapter describes procedures for (a) synthesizing and characterizing ester-ified (cholesteryl-nitrolinoleate, CLNO$_2$), as well as free (nitroarachidonate, AANO$_2$) nitroalkenes, (b) determining nitration of cholesteryl linoleate dur-ing macrophage activation by inflammatory stimuli, (c) examining the effects of exogenous nitroalkene addition on NOS2 expression by activated macro-phages, and (d) exploring the ability of nitroalkenes to induce HO-1 and interfere with NF-κB activation.

2. Synthesis and Characterization of Cholesteryl-Nitrolinoleate

Esterified nitroalkenes can be synthesized successfully using protocols for nitroselenylation of alkenes (Baker *et al.*, 2004; Coles *et al.*, 2002). However, a simple nitration procedure based on the reaction of the lipid precursor with nitrite (NO$_2^-$) at acid pH can also be utilized (Lima *et al.*, 2002). For preparation of CLNO$_2$, first prepare a mixture of hexane and 1% sulfuric acid [1:1 (v/v), 4 ml final volume] in a stoppered round-bottom flask; degass the solution under a stream of N$_2$ for 3 to 5 min (do not degass for longer periods to avoid hexane volatilization). Then, using a Hamilton syringe, add 1.5 mmol of cholesteryl linoleate (CL, Nu-Check Prep) in chloroform to the mixture and immediately proceed to the addition of 3.75 mmol of NaNO$_2$, twice with 15-min intervals. The biphasic reaction system should be stirred vigorously for 30 min at 25 °C. Stop the reaction by organic extraction of the lipid products; separate and wash the organic layer twice with 0.15 M NaCl and dry under a stream of N_2 (if the solution looks cloudy, add a spatula tip of NaCl until the solution clarifies). Then dilute the reaction mixture with methylene chloride and separate the lipids by thin-layer chromatography (TLC) on silica gel HF TLC plates (Uniplate, Analtech) using a mixture of hexane:diethyl ether (80:20; v/v) as the solvent. Detect the separated lipid components by ultraviolet (UV) absorption and by charring the following reaction with concentrated sulfuric acid. In this condition, $CLNO_2$ looks like a double band that migrates less than CL (termed $CLNO_2$ top and bottom isomers, in reference to chromatographic R_f properties). Regions of silica containing $CLNO_2$ can be extracted using conventional protocols (Bligh and Dyer, 1959), dried under vacuum, dissolved in isopropanol, and stored at $-20\,^\circ$C. $CLNO_2$ quantitation can be performed by HPLC using a nitrogen detector, using caffeine as the standard to obtain calibration curves. In parallel, and in comparison with nitrogen detection, gravimetry is a good option for quantitating $CLNO_2$ when the amount of the purified product is more than 10 mg.

Perform mass spectrometric analysis of $CLNO_2$ by ESI- MS and ESI-MS/MS in a QTrap 2000 (Applied Biosystems/MDS Sciex). Inject $CLNO_2$ and CL in isopropanol containing 4 mM ammonium acetate and analyze in the positive ion mode following formation of the ammonium adduct ($[M+NH_4]^+$) of the nitroalkene (Lima *et al.*, 2003). The ammonium adduct of $CLNO_2$ (both top and bottom fractions) displays an m/z of 711 (Fig. 3.1A). Analysis of MS/MS of the species with m/z 711 is necessary to confirm the identity of the product, as well as to determine where nitration occurred. Both $CLNO_2$ isomers yield daughter ions of m/z 369 (cholesterol-OH)$^+$ and m/z 326 (LNO_2); the demonstration that nitration occurred on the fatty acid carbon chain of CL can be followed by the presence of the ion with m/z 279 that corresponds to the neutral loss of the nitro group [(LNO_2 - HNO_2)$^+$, Fig. 3.1B] of the ion of m/z 326.

Nuclear magnetic resonance (NMR) analysis of the two bands confirmed that nitration occurred on the acyl moiety of CL and corresponded to nitroalkene isomers: $CLNO_2$ bottom data correspond to cholesteryl-10-nitrolinoleate and its conjugated isomer, whereas $CLNO_2$ top data correspond to a mixture of cholesteryl-9-nitrolinoleate and cholesteryl-10-nitrolinoleate.

3. SYNTHESIS AND IDENTIFICATION OF NITROALKENES ISOMERS: MASS SPECTROMETRY ANALYSIS OF NITROARACHIDONATE

Arachidonic acid (AA) is a 20 carbon fatty acid with four double bonds whose nitration yields many positional nitroalkenes isomers. Thus, it is an adequate model to characterize and identify the position of the nitro group at the carbon chain. Nitration of AA can be performed with acidic

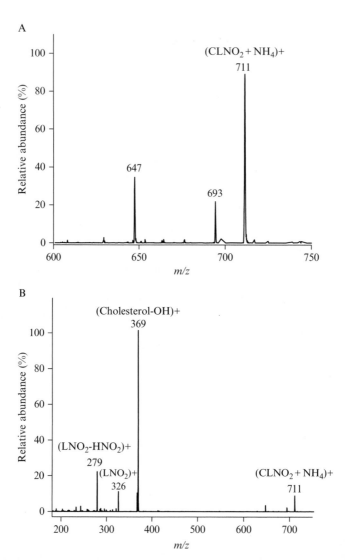

Figure 3.1   MS characterization of $CLNO_2$. $CLNO_2$ is separated by TLC showing two main nitrated products, which are extracted from the silica, dried under vacuum, and analyzed by MS (positive mode, direct injection, isopropanol 4 mM ammonium acetate). (A) MS spectra of the ammonium adducts of $CLNO_2$ (m/z 711). (B) MS/MS of parent molecule showing LNO_2 as a daughter ion.

NO_2^- (Balazy *et al.*, 2001; Napolitano *et al.*, 2002, 2004; Panzella *et al.*, 2003; Trostchansky *et al.*, 2007). In hexane, treat AA for 1 h at 25 °C with $NaNO_2$, at a 1:1 molar ratio, under continuous stirring; initiate the reaction by adjusting the pH to 3.0. Stop the reaction by solvent extraction with ethyl acetate/NaCl (1:2, v/v); collect the organic phase, remove the remaining H_2O with ammonium sulfate anhydride, and then dry under a stream of N_2. Dissolve the resultant lipid products in methanol (MeOH). To separate, identify, and characterize $AANO_2$ positional isomers, separate the reaction mixture by reversed-phase HPLC using a 250 × 2-mm C18 Phenomenex Luna column (3-μm particle size). Elute lipids using a solvent system consisting of A (H_2O containing 0.05% formic acid) and B ($CNCH_3$ containing 0.05% formic acid) under the following gradient conditions: 35% B (0–10 min); 35–100% B (10–35 min), and 100–35% B (35–45 min). Follow elution of the oxidized and nitrated lipid products by UV detection at 215 and 274 nm (Baker *et al.*, 2005; Balazy and Poff, 2004; Balazy *et al.*, 2001; Trostchansky *et al.*, 2007); collect fractions, carry out a lipid extraction using the method described by Bligh and Dyer and store at −20 °C in MeOH. Under these experimental conditions three fractions eluted before AA exhibit MS and MS/MS spectra characteristic of $AANO_2$ (*m/z* 348). NMR and IR analysis confirmed that these fractions are mononitrated *cis*-nitroalkenes derived from AA (Trostchansky *et al.*, 2007).

Because AA has four double bonds, up to eight positional isomers of $AANO_2$ nitroalkenes can be formed. To determine which isomers are formed in the acidic nitration of AA, analyze the purified fractions by MS in the positive ion mode in the presence of lithium acetate (Trostchansky *et al.*, 2007). The presence of lithium locks the charge at the carboxy moiety and allows the cleavage directly across the nitroalkene double bond, depending on the position of the NO_2 group, generating two specific fragments. The fragments carrying the charge, allowing its detection by MS, correspond to those having the carbon closer to the carboxyl group in the nitroalkene double bond. Specific patterns of collision-induced dissociation fragments for each theoretical isomer were obtained (Table 3.1). MS/MS spectra showed the presence of the $[M+Li]^+$ ion (*m/z* 356), as well as the one that suffered the neutral loss of the NO_2 group (*m/z* 309, Fig. 3.2). From MS/MS spectra, four major isomers (Fig. 3.2, labeled with an asterisk) were identified: 12- and 15-nitroarachidonate (*m/z* 203 and *m/z* 243, respectively) in fraction 1, the most polar one; 9-nitroarachidonate (*m/z* 163) in fraction 2, medium polarity; and 14-nitroarachidonate (*m/z* 256) in fraction 3, the less polar one.

4. DETERMINATION OF LIPID NITRATION DURING MACROPHAGE ACTIVATION

4.1. Cell culture

The murine macrophage-like cell line J774.1 is from American Type Culture Collection (ATCC, number TIB-67). Maintain cells by passage in Dulbecco's modified Eagle medium (DMEM, Sigma) containing 4 mM

Table 3.1 Theoretical fragmentation patterns of AANO$_2$–lithium adducts

	Isomer	m/z of [M-H]	m/z of fragment (II) lithium derived	m/z of fragment (I) lithium derived
	5-nitroAA	356	136	—
	6-nitroAA	356	—	123
	8-nitroAA	356	176	—
	9-nitroAA	356	—	163
	11-nitroAA	356	216	—
	12-nitroAA	356	—	203
	14-nitroAA	356	256	—
	15-nitroAA	356	—	243

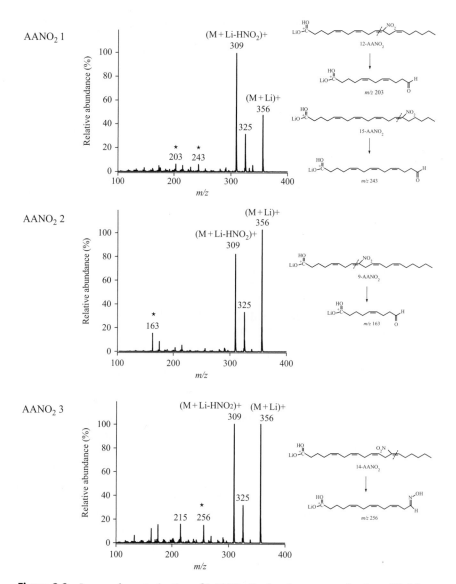

Figure 3.2 Isomer characterization of AANO$_2$. Product ion spectra of each purified fraction of AANO$_2$ in the presence of lithium are shown. Structure of the detected ions (marked with an asterix) is also shown. Reproduced with permission from Trostchansky *et al.* (2007).

L-glutamine, 100 U/ml penicillin, 100 μg/ml streptomycin, and 10 mM HEPES and supplement with 10% heat-inactivated fetal calf serum (FCS, Gibco/Invitrogen) at 37 °C in humidified air containing 5% CO$_2$. Prepare subcultures by dislodging cells from the flask with a cell scraper, using a subcultivation ratio between 1:3 and 1:6 as recommended for ATCC.

4.2. Preparation of lipid extracts from activated macrophages

Harvest J774A.1 cells by scraping and seeding in a six–well flat bottom tissue culture plate. When a confluent cell monolayer is obtained, activate cells by adding 1 μg/ml LPS (*Escherichia coli* serotype O127:B8) plus 400 U/ml interferon (IFN)γ (Sigma). Prepare a control by culturing cells in the absence of stimulus. Follow the generation of $NO\cdot$ during cell activation by measuring $NO_2\cdot$ formation in the supernatant using a spectrophotometric assay based on the Griess reaction (Green *et al.*, 1982); is expected that $NO_2\cdot$ levels increase around ninefold after 24 h of stimulation. Because phenol red interferes with the colorimetric detection of $NO_2\cdot$, DMEM free of phenol red is recommended for these assays. At different times poststimulation, wash cells with cold 0.01 M phosphate-buffered saline, pH 7.4 (PBS), containing 5 mM EDTA and 20 μM butyl-hidroxitolueno, harvest by scraping, and centrifuge at 100g for 10 min. Extract lipids using the Bligh and Dyer method. During the monophase stage of lipid extraction, add [^{13}C]LNO$_2$ as an internal standard for normalization purposes.

4.3. Identification and quantitation of CLNO$_2$ in lipid extracts obtained from macrophages

Determine the presence of $CLNO_2$ in lipid extracts derived from macrophages by HPLC ESI MS/MS using a 150 × 2.1-mm C18 GraceVydac columm (5-μm particle size) coupled to the Q Trap. Inject $CLNO_2$ standard, [^{13}C]LNO$_2$, and lipid extracts from macrophages in MeOH and elute using an isocratic solvent system consisting of isopropanol/acetonitrile/ammonium acetate (60/40/4 mM) (Figs. 3.3A and 3.3B). For quantitative analysis, monitor two multiple reaction monitoring (MRM) transitions: m/z 711 → 369 for the ammonium adduct of $CLNO_2$ and m/z 342 → 295 for [^{13}C]LNO$_2$ in the positive and negative mode, respectively (Baker *et al.*, 2005; Lima *et al.*, 2003). Confirm the identity of $CLNO_2$ by monitoring fragmentation of $CLNO_2$ to LNO_2 (m/z 711 → 326) and enhance product ion spectra (EPI) of the m/z 711 product (Fig. 3.3). Results are expressed as the ratio of the peak areas of $CLNO_2$ relative to the internal standard.

5. EFFECTS OF NITROALKENES ON NOS2 INDUCTION BY ACTIVATED MACROPHAGES

5.1. Macrophage activation and treatment with nitroalkenes

The induction of NOS2 in macrophages occurs in response to TLR ligands via NF-κB signaling and is amplified by IFNγ via several mechanisms, including upregulation of TLRs and coreceptor/accessory molecules and the increase of

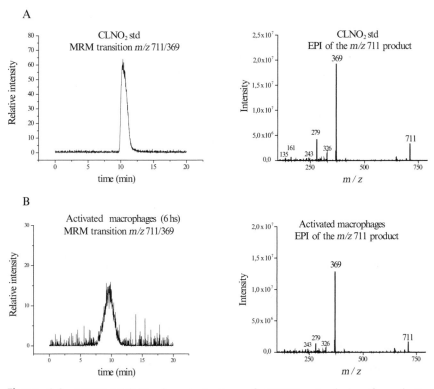

Figure 3.3 HPLC-MS/MS characterization of CLNO₂. Analysis of synthetic CLNO₂(A) and of a lipid extract obtained from activated macrophages (B) showing a representative HPLC elution profile using MRM scan mode for the m/z 711/369 transition (left panels) and the EPI of the m/z 711 product (right panels).

promoter activation through synergy between TLR and IFNγ-induced transcription factors (NF-κB and STAT1, respectively)(Lowenstein *et al.*, 1993; Schroder *et al.*, 2006.). Thus, macrophage activation with LPS alone or plus IFNγ has been widely used to study the mechanisms involved in NOS2 induction. For analyzing the effect of nitroalkenes (particularly AANO₂, LNO₂ and CLNO₂), LPS alone (0.1 and 1 μg/ml) is recommended as the stimulus, as it provides a more sensitive assay; the magnitude of nitroalkene-mediated-inhibition is reduced significantly in the presence of IFNγ (Fig. 3.4A), which suggests that the interference occurs mainly at the level of NF-κB activation. Murine macrophage cell lines J774A.1 and RAW 264.7 (both from ATCC) can be used for these assays; monocytes or human macrophages obtained from THP-1 monocyte-like cells (ATCC) are not recommended because LPS-mediated NOS2 induction is not easily detectable by conventional methods.

For NOS2 induction, expose confluent monolayers of cells to 0.1–1 μg/ ml LPS in the presence of nitroalkene (1 to 20 μM), non-nitrated lipids (1 to

A

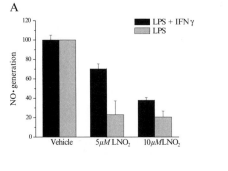

B

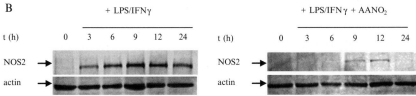

Figure 3.4 Nitroalkene inhibited ·NO generation by activated macrophages. (A) Macrophages were stimulated with LPS or LPS/IFNγ in the presence of LNO₂ or vehicle as a control. After 5 h stimulation, NO· generation was measured by fluorimetry. Results are expressed as NO· generation relative to the control. (B) Activation of J774.1 macrophages was performed in the presence or absence of AANO₂ (5 μM) as indicated. At different time points post stimulation, cells were lysed and analysed for NOS2 expression by Western blot. Results show a representative profile of NOS2 expression along macrophage activation; actin levels were also determined for controlling the amount of sample analysed.

20 μM), or vehicle as control. Perform cell activation in DMEM containing 10% FCS (DMEM-FCS); the final concentration of the vehicle must be lower than 0.5% (v/v). Dilutions of lipids in DMEM-FCS must be done in a sterile glass tube using a Hamilton syringe. In the case of esterified nitrolipids, do not add a small volume of the sample in the total volume of medium because it could be difficult to dissolve it. A better solubilization can be obtained by adding the required volume of DMEM-FCS to the nitroalkene solution in various steps and mixing vigorously with a vortex between each addition (start with a small volume, e.g., fivefold the volume of nitroalkene solution). Finally, in order to optimize the assay, the effect of administration of the nitroalkene prior to or concurrently with LPS should be studied. After incubation (37 °C, 5% CO_2), analyze cells for NOS2 induction by determining NO· production by fluorimetry or detecting the enzyme by Western blot (Fig. 3.4) and/or immunohistochemistry. Cell viability is also examined as described later.

5.2. Determination of NO· generation by fluorimetry

NO· is detected in LPS-stimulated macrophages using NO· fluorescent indicators such as 4,5-diaminofluorescein diacetate (DAF-2 diacetato) and 4-amino-5-methylamino-2′,7′-difluorofluorescein (DAF-FM diacetato);

these membrane-permeable poorly fluorescent indicators react with NO· to yield highly fluorescent benzotriazole products (Chatton and Broillet, 2002). These dyes are trapped in the cell as a result of intracellular esterases, which hydrolyze their ester moieties to yield the negatively charged, membrane-impermeable form. Stock solutions of these dyes should be prepared in anhydrous dimethyl sulfoxide (DMSO) at 5 to 10 mM and kept in small aliquots at −20 °C in the dark.

The time course of NO· production is monitored in activated macrophages for at least 90 min; this can be performed after 3 to 4 h of activation, but maximal levels are usually achieved after 8 h. For this assay, cell activation steps can be done using confluent cell monolayers prepared in 96-well plates, which allows examining a high number of activation conditions. After incubation, discard cell supernatants and wash cells with PBS and incubate (37 °C) in DMEM (free of red phenol) containing 5 μM DAF-2DA (Alexis) and 1 mM L-arginine. Record the fluorescence (λ_{ex} 495 nm, λ_{em} 515 nm) for at least 1 h using a FLUOstar-OPTIMA fluorimeter (BMG Labtechnologies GmbH, Germany). Plot fluorescence vs time and calculate NO· generation as the slope of this curve in its lineal range (usually between 30 and 90 min; the initial 30 min are required for the loading and equilibration of the dye into cells).

5.3. Immunocytochemical detection of NOS2 expression

For these assays, plate cells on chamber slides and stimulate as described earlier. After a 6-h activation, fix cells with 4% paraformaldehyde in PBS at room temperature for 30 min. Incubate slides for 2 h at room temperature with anti-NOS2 (1:100, Sigma) followed by AlexaFluo 488-conjugated secondary IgG antibody (1:100, Invitrogen-Molecular Probes). Stain the nucleus by incubation with 1 μg/ml 4′,6-diamidino-2-phenylindole (Invitrogen-Molecular Probes) for 10 min. Sections can be analyzed using an Olympus microscope and images captured using a digital video camera and Adobe Photoshop (Adobe Systems, CA). Using this protocol we observed that nitroalkenes, but not their non-nitrated lipid precursors, were capable of downregulating NOS2 expression in J774A.1 macrophages stimulated with LPS or LPS plus IFNγ.

5.4. Western blot for NOS2 expression

For this analysis, experiments should be done with around 2 × 10^6 cells; a practical way is to prepare confluent monolayers of cells in six-well flat-bottom tissue culture plates to perform activation steps. NOS2 is detected after 3 to 4 h, but maximal expression requires longer times (between 8 and 12 h). After activation, wash cells once with ice-cold PBS and then scrape in the presence of ice-cold lysis buffer containing 10 mM HEPES, pH 7.5, 10 mM KCl, 1.5 mM MgCl$_2$, 2 mM EDTA, 1 mM dithiothreitol (DTT), and a protease inhibitor mixture [5 mM benzamidine, 1 mM phenylmethyl-sulfonyl fluoride (PMSF), 1 mM E-64, and 1 μM pepstatin A]. Sonicate the cell lysate on ice (three 30-s bursts at maximal amplitude with 10-s intervals) using an Ultrasonic homogenizer 4710 series (Cole Parmer Instruments). Remove insoluble material by centrifugation (4 °C, 5000g, 5 min). Collect supernatants and determine protein concentration by the BCA protein assay (Calbiochem/Novabiochem).

Mix equal amounts of proteins obtained from cells (25–30 μg protein) with sample buffer [0.2 M Tris pH 8, 2% sodium dodecyl sulfate (SDS), 8 M urea, 6 mg/ml DTT, 0.01% bromophenol blue], heat for 5 min at 100 °C, and subject to electrophoresis using 10% SDS-polyacrylamide gels (SDS-PAGE). A commercial protein standard should also be used for indication of molecular weights. Perform electrophoresis for about 90 min at 100 V in Tris-glycine electrophoresis buffer (192 mM glycine, 25 mM Tris, 0.1% SDS). Transfer separated proteins onto nitrocellulose membranes (Milli-pore) by overnight 35-V electroblotting in transfer buffer (50 mM Tris, 380 mM glycine, 0.1% SDS, 20% methanol) at 4 °C. Polyvinylidene fluoride (PVDF) membranes are also a good choice for blotting, but protocols for PVDF and nitrocellulose membranes have a few differences. PVDF is a hydrophobic membrane and requires prewetting in methanol (15 s) before it is used (the membrane should uniformly change from opaque to semi-transparent). Then, place it carefully in ultrapure water, soak for 2 min, and equilibrate for at least 5 min in transfer buffer. Another change to note is that PVDF membranes are much more sensitive to SDS levels and that proteins already bound to the membrane could slip off; for this reason, SDS levels in buffers should never exceed 0.05% when using PVDF membranes.

For detecting NOS2, block membranes (1 h at room temperature or overnight at 4 °C) with PBS containing 0.1% (v/v) Tween 20 and 1% bovine serum albumin (BSA). Incubate the membranes at room tempera-ture for 2 h with rabbit anti-NOS2 (Sigma) or rabbit antiactin (Sigma) diluted 1:5000 and 1:2000 in PBS containing 0.05% Tween 20 and 0.05% BSA (PBS-TBSA), respectively. After washing [three times with excess PBS containing 0.05% Tween 20 (PBS-T) and shaking], incubate membranes with goat antirabbit IgG conjugated to peroxidase (Calbiochem/Novabio-chem) diluted 1:10,000 in PBS-TBSA. After washing, develop specific binding using the SuperSignal chemiluminescent substrate (Pierce); acqui-sition and analysis of images can be performed using a ChemiDoc XRS imaging station and Quantity One software (Bio-Rad).

A time–course study of NOS2 expression by J774A.1 macrophages activated with LPS in the presence of LNO_2 (10 μM) or vehicle (control) is shown in Fig. 3.5; this demonstrates that LNO_2 is capable of down-regulating LPS-mediated NOS2 induction.

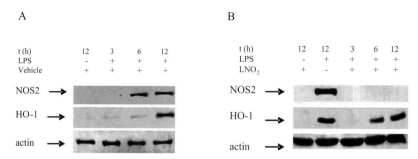

Figure 3.5 Induction of HO-1 expression in macrophages by nitroalkenes. Mouse J774.1 cells at confluence were stimulated with LPS (1 μg/ml) in the presence or absence of LNO$_2$ (10 μM). At different time poststimulation (3, 6, 12 h), the presence of NOS2, HO-1, and actin was examined by Western blot.

5.5. Analysis of cell viability

Assess cell viability by measuring the mitochondrial-dependent reduction of 3-[4,5-dimethylthiazol-2yl]-2,5-diphenyltetrazolium bromide (MTT, Sigma) to formazan. Prepare a 0.2 mg/ml solution of MTT in DMEM free of red phenol. Add the MTT solution to cells and incubate at 37 °C for 4 h; use 500 μl of MTT solution for 5 × 10^5 cells. After removing media, add 0.1 M glycin, pH 10.5 (50 μl) and dissolve formazan crystals in DMSO (500 μl). Read absorbance at 570 nm using a microplate spectrophotometer. Results are expressed as the percentage respect to control values (stimulated cells in the presence of the vehicle).

6. ANALYSIS OF HO-1 INDUCTION AND NF-κB ACTIVATION

6.1. Detection of HO-1 by Western blot

Carry out analysis of the capacity of nitroalkenes to induce HO-1 expression by exposing confluent monolayers of J774A.1 cells to nitrated or their corresponding non-nitrated lipids (1 to 10 μM), CL (5 to 50 μM), or vehicle as control. Prepare lipid dilutions in DMEM containing 10% FCS following the recommendations described earlier (see details under Section 5.1). Incubate cells (37 °C, 5% CO$_2$) for different times (e.g. 3, 6, 9, 12, and 24 h) and evaluate the presence of HO-1 by Western blot in cell lysates following the same procedures described for the detection of NOS2 (see details under Section 5.4), but using a rabbit antimouse HO-1 (Stressgen Biotechnologies Inc.) diluted 1:2000 in PBS-TBSA. Compare the time course of NOS2 and HO-1 expression during macrophage activation to analyse whether HO-1 is involved in to the nitrolipid-mediated-NOS2 modulation (Fig. 3.5).

6.2. Analysis of NF-κB activation

Nuclear factor κB is an inducible eukaryotic transcription factor that has been shown to regulate the expression of numerous inflammatory genes (Siebenlist *et al.*, 1994). Active DNA-binding forms of NF-κB are dimeric complexes composed of various combinations of members that constitute the Rel family. Of these, the p50–p65 heterodimeric complex is the most abundant of all dimeric NF-κB complexes, being expressed in almost all cell types. In resting macrophages, NF-κB is sequestered in the cytoplasm by association with an inhibitory protein, IκB. Proinflammatory stimuli, such as LPS, activate an IκB kinase, resulting in IκB phosphorylation, which in turn leads to IκB ubiquitination and degradation by the proteasome. Upon proteolytic degradation of IκB, the active, free NF-κB p50–p65 dimer translocates from the cytoplasm to the nucleus to interact with the promoter–enhancer region of target genes. Thus, a simple strategy to determine NF-κB activation is to evaluate the p65 and/or p50 translocation to the nucleus by determining the presence of these proteins in nuclear extracts.

Stimulate confluent J774.1 cells with LPS (0.5 μg/ml) for 60 min at 37 °C and 5% CO_2 in the presence and absence of $CLNO_2$, CL, or vehicle as control. The stimulation step could be shorter because NF-κB translocation in LPS-activated macrophages can be detected after 15 min (Perera *et al.*, 1996). Preparation of nuclear extracts is carried out as follows. Wash cells once with ice-cold PBS, scrape in the presence of the same buffer, transfer to Eppendorf tubes, and centrifuge at 4 °C and 100g for 5 min. Discard the supernatant and quickly freeze cell pellets in a dry ice/ethanol bath (2–5 min). Resuspend the pellet in ice-cold 10 mM HEPES, pH 7.5, 10 mM KCl, 1.5 mM MgCl$_2$, 2 mM EDTA, 1 mM DTT, and a protease inhibitor mixture (5 mM benzamidine, 1 mM PMSF, 1 mM E-64, and 1 μM pepstatin A), incubate for 10 min, and centrifuge at 4 °C and 1200g for 10 min. Resuspend the pellet (nuclei) in 20 mM HEPES, pH 7.5, 400 mM NaCl, 1.5 mM MgCl$_2$, 2 mM EDTA, and 1 mM DTT containing the protease inhibitor mixture and incubate for 30 min. After centrifugation at 4 °C and 15,000g, collect the supernatant (nuclear extract) and store at −70 °C until use. Determine the protein concentration using the BCA protein assay (Calbiochem/Novabiochem).

Carry out detection of NF-κB p65 by Western blotting following the same protocol described for the detection of NOS2 (see details under Section 5.4), but using a rabbit polyclonal anti-NF-κB (Calbiochem/Novabiochem, diluted 1:1000 in PBS-TBSA).

ACKNOWLEDGMENTS

We thank Carlos Batthyány, Paul Baker, and Francisco Schopfer for participation in parts of the experimental work presented here and Valerie O'Donnell and Bruce Freeman for helpful discussions and support. This work was supported by grants from Fogarty-NIH,

Guggengheim Foundation, Wellcome Trust (HR), Programa para el Desarrollo de las Ciencias Básicas (PEDECIBA, Uruguay) and Programa de Desarrollo Tecnológico and Fondo Clemente Estable, Uruguay (AF, AT, JS, and HR).

REFERENCES

Alexander, R. L., Bates, D. J., Wright, M. W., King, S. B., and Morrow, J. (2006). Multidrug Resistance Protein 1 (MRP-1): Glutathione Conjugation and MRP-1-Mediated Efflux Inhibit Nitrolinoleic Acid-Induced PPAR gamma-Dependent Transcription Activation. *Biochemistry* **45**, 7889–7896.

Alvarez, B., Ferrer-Sueta, G., Freeman, B. A., and Radi, R. (1999). Kinetics of peroxynitrite reaction with amino acids and human serum albumin. *J. Biol. Chem.* **274**, 842–848.

Baker, P. R., Lin, Y., Schopfer, F. J., Woodcock, S. R., Groeger, A. L., Batthyany, C., Sweeney, S., Long, M. H., Iles, K. E., Baker, L. M., Branchaud, B. P., Chen, Y. E., et al. (2005). Fatty acid transduction of nitric oxide signaling: Multiple nitrated unsaturated fatty acid derivatives exist in human blood and urine and serve as endogenous peroxisome proliferator-activated receptor ligands. *J. Biol. Chem.* **280**, 42464–42475.

Baker, P. R., Schopfer, F. J., Sweeney, S., and Freeman, B. A. (2004). Red cell membrane and plasma linoleic acid nitration products: Synthesis, clinical identification, and quantitation. *Proc. Natl. Acad. Sci. USA* **101**, 11577–11582.

Balazy, M., Iesaki, T., Park, J. L., Jiang, H., Kaminski, P. M., and Wolin, M. S. (2001). Vicinal nitrohydroxyeicosatrienoic acids: Vasodilator lipids formed by reaction of nitrogen dioxide with arachidonic acid. *J. Pharmacol. Exp. Ther.* **299**, 611–619.

Balazy, M., and Poff, C. D. (2004). Biological nitration of arachidonic acid. *Curr. Vasc. Pharmacol.* **2**, 81–93.

Batthyany, C., Schopfer, F. J., Baker, P. R., Duran, R., Baker, L. M., Huang, Y., Cervenansky, C., Braunchaud, B. P., and Freeman, B. A. (2006). Reversible posttranslational modification of proteins by nitrated fatty acids in vivo. *J. Biol. Chem.* **281**, 20450–20463.

Bligh, E. G., and Dyer, W. L. (1959). A rapid method of total lipid extraction and purification. *Can. J. Biochem. Pysiol.* **37**, 911–917.

Chatton, J. Y., and Broillet, M. C. (2002). Detection of nitric oxide production by fluorescent indicators. *Methods Enzymol.* **359**, 134–148.

Coles, B., Bloodsworth, A., Eiserich, J. P., Coffey, M. J., McLoughlin, R. M., Giddings, J. C., Lewis, M. J., Haslam, R. J., Freeman, B. A., and O'Donnell, V. B. (2002). Nitrolinoleate inhibits platelet activation by attenuating calcium mobilization and inducing phosphorylation of vasodilator-stimulated phosphoprotein through elevation of cAMP. *J. Biol. Chem.* **277**, 5832–5840.

Cui, T., Schopfer, F. J., Zhang, J., Chen, K., Ichikawa, T., Baker, P. R., Batthyany, C., Chacko, B. K., Feng, X., Patel, R. P., Agarwal, A., Freeman, B. A., and Chen, Y. E. (2006). Nitrated fatty acids: Endogenous anti-inflammatory signaling mediators. *J. Biol. Chem.* **281**, 35686–35698.

Dobrovolskaia, M. A., and Vogel, S. N. (2002). Toll receptors, CD14, and macrophage activation and deactivation by LPS. *Microbes Infect.* **4**, 903–914.

Eiserich, J. P., Hristova, M., Cross, C. E., Jones, A. D., Freeman, B. A., Halliwell, B., and van der Vliet, A. (1998). Formation of nitric oxide-derived inflammatory oxidants by myeloperoxidase in neutrophils. *Nature* **391**, 393–397.

Green, L. C., Wagner, D. A., Glogowski, J., Skipper, P. L., Wishnok, J. S., and Tannenbaum, S. R. (1982). Analysis of nitrate, nitrite, and [15N]nitrate in biological fluids. *Anal. Biochem.* **126**, 131–138.

Kaisho, T., and Akira, S. (2006). Toll-like receptor function and signaling. *J. Allergy Clin. Immunol.* **117**, 979–987.

Kawai, T., and Akira, S. (2005). Pathogen recognition with Toll-like receptors. *Curr. Opin. Immunol.* **17**, 338–344.

Lee, T. S., Tsai, H. L., and Chau, L. Y. (2003). Induction of heme oxygenase-1 expression in murine macrophages is essential for the anti-inflammatory effect of low dose 15-deoxy-Delta 12,14-prostaglandin J2. *J. Biol. Chem.* **278**, 19325–19330.

Lien, E., and Ingalls, R. R. (2002). Toll-like receptors. *Crit. Care Med.* **30**, S1–S11.

Lima, E. S., Bonini, M. G., Augusto, O., Barbeiro, H. V., Souza, H. P., and Abdalla, D. S. (2005). Nitrated lipids decompose to nitric oxide and lipid radicals and cause vasorelaxation. *Free Radic. Biol. Med.* **39**, 532–539.

Lima, E. S., Di Mascio, P., and Abdalla, D. S. (2003). Cholesteryl nitrolinoleate, a nitrated lipid present in human blood plasma and lipoproteins. *J. Lipid Res.* **44**, 1660–1666.

Lima, E. S., Di Mascio, P., Rubbo, H., and Abdalla, D. S. (2002). Characterization of linoleic acid nitration in human blood plasma by mass spectrometry. *Biochemistry.* **41**, 10717–10722.

Lowenstein, C. J., Alley, E. W., Raval, P., Snowman, A. M., Snyder, S. H., Russell, S. W., and Murphy, W. J. (1993). Macrophage nitric oxide synthase gene: Two upstream regions mediate induction by interferon gamma and lipopolysaccharide. *Proc. Natl. Acad. Sci. USA* **90**, 9730–9734.

Moller, M. N., Li, Q., Lancaster, J. R., Jr., and Denicola, A. (2007). Acceleration of nitric oxide autoxidation and nitrosation by membranes. *IUBMB Life.* **59**, 243–248.

Napolitano, A., Camera, E., Picardo, M., and d'Ishida, M. (2002). Reactions of hydro(pero)xy derivatives of polyunsaturated fatty acids/esters with nitrite ions under acidic conditions: Unusual nitrosative breakdown of methyl 13-hydro(pero)xyoctadeca-9,11-dienoate to a novel 4-nitro-2-oximinoalk-3-enal product. *J. Org. Chem.* **67**, 1125–1132.

Napolitano, A., Panzella, L., Savarese, M., Sacchi, R., Giudicianni, I., Paolillo, L., and d'Ischia, M. (2004). Acid-induced structural modifications of unsaturated fatty acids and phenolic olive oil constituents by nitrite ions: A chemical assessment. *Chem. Res. Toxicol.* **17**, 1329–1337.

Panzella, L., Manini, P., Crescenzi, O., Napolitano, A., and d'Ischia, M. (2003). *In* Nitrite-induced nitration pathways of retinoic acid, 5,6-epoxyretinoic acid, and their esters under mildly acidic conditions: Toward a reappraisal of retinoids as scavengers of reactive nitrogen species. *Chem. Res. Toxicol.* **16**, 502–511.

Perera, P. Y., Qureshi, N., and Vogel, S. N. (1996). Paclitaxel (Taxol)-induced NF-kappaB translocation in murine macrophages. *Infect. Immun.* **64**, 878–884.

Radi, R, Denicola, A, Alvarez, B, Ferrer-Sueta, G, and Rubbo, H (2000). "Nitric Oxide, Biology and Pathobiology" (L. Ignarro, ed.), Academic Press, San Diego57–82.

Rubbo, H., Radi, R., Trujillo, M., Telleri, R., Kalyanaraman, B., Barnes, S., Kirk, M., and Freeman, B. A. (1994). Nitric oxide regulation of superoxide and peroxynitrite-dependent lipid peroxidation: Formation of novel nitrogen-containing oxidized lipid derivatives. *J. Biol. Chem.* **269**, 26066–26075.

Schopfer, F. J., Baker, P. R., and Freeman, B. A. (2003). NO-dependent protein nitration: A cell signaling event or an oxidative inflammatory response? *Trends Biochem. Sci.* **28**, 646–654.

Schopfer, F. J., Baker, P. R., Giles, G., Chumley, P., Batthyany, C., Crawford, J., Patel, R. P., Hogg, N., Branchaud, B. P., Lancaster, J. R., Jr., and Freeman, B. A. (2005a). Fatty acid transduction of nitric oxide signaling: Nitrolinoleic acid is a hydrophobically stabilized nitric oxide donor. *J. Biol. Chem.* **280**, 19289–19297.

Schopfer, F. J., Lin, Y., Baker, P. R., Cui, T., Garcia-Barrio, M., Zhang, J., Chen, K., Chen, Y. E., and Freeman, B. A. (2005b). Nitrolinoleic acid: An endogenous peroxisome proliferator-activated receptor gamma ligand. *Proc. Natl. Acad. Sci. USA* **102**, 2340–2345.

Schroder, K., Sweet, M. J., and Hume, D. A. (2006). Signal integration between IFNgamma and TLR signalling pathways in macrophages. *Immunobiology* **211**, 511–524.

Siebenlist, U., Franzoso, G., and Brown, K. (1994). Structure, regulation and function of NF-kappa B. *Annu. Rev. Cell Biol.* **10**, 405–455.

Srisook, K., and Cha, Y. N. (2005). Super-induction of HO-1 in macrophages stimulated with lipopolysaccharide by prior depletion of glutathione decreases iNOS expression and NO production. *Nitric Oxide.* **12**, 70–79.

Srisook, K., Han, S. S., Choi, H. S., Li, M. H., Ueda, H., Kim, C., and Cha, Y. N. (2006). CO from enhanced HO activity or from CORM-2 inhibits both O2- and NO production and downregulates HO-1 expression in LPS-stimulated macrophages. *Biochem. Pharmacol.* **71**, 307–318.

Srisook, K., Kim, C., and Cha, Y. N. (2005a). Cytotoxic and cytoprotective actions of O2- and NO (ONOO-) are determined both by cellular GSH level and HO activity in macrophages. *Methods Enzymol.* **396**, 414–424.

Srisook, K., Kim, C., and Cha, Y. N. (2005b). Role of NO in enhancing the expression of HO-1 in LPS-stimulated macrophages. *Methods Enzymol.* **396**, 368–377.

Thoma-Uszynski, S., Stenger, S., Takeuchi, O., Ochoa, M. T., Engele, M., Sieling, P. A., Barnes, P. F., Rollinghoff, M., Bolcskei, P. L., Wagner, M., Akira, S., and Norgard, M. V., *et al.* (2001). Induction of direct antimicrobial activity through mammalian toll-like receptors. *Science.* **291**, 1544–1547.

Tien, M., Berlett, B. S., Levine, R. L., Chock, P. B., and Stadtman, E. R. (1999). Peroxynitrite-mediated modification of proteins at physiological carbon dioxide concentration: pH dependence of carbonyl formation, tyrosine nitration, and methionine oxidation. *Proc. Natl. Acad. Sci. USA* **96**, 7809–7814.

Trostchansky, A., Souza, J. M., Ferreira, A., Ferrari, M., Blanco, F., Trujillo, M., Castro, D., Cerecetto, H., Baker, P. R., O'Donnell, V. B., and Rubbo, H. (2007). Synthesis, isomer characterization, and anti-inflammatory properties of nitroarachidonate. *Biochemistry.* **46**, 4645–4653.

Villacorta, L., Zhang, J., Garcia-Barrio, M. T., Chen, X. L., Freeman, B. A., Chen, Y. E., and Cui, T. (2007). Nitro-linoleic acid inhibits vascular smooth muscle cell proliferation via the Keap1/Nrf2 signaling pathway. *Am. J. Physiol. Heart Circ. Physiol.* **293**, H770–H776.

Werling, D., Hope, J. C., Howard, C. J., and Jungi, T. W. (2004). Differential production of cytokines, reactive oxygen and nitrogen by bovine macrophages and dendritic cells stimulated with Toll-like receptor agonists. *Immunology* **111**, 41–52.

Wink, D. A., Darbyshire, J. F., Nims, R. W., Saavedra, J. E., and Ford, P. C. (1993). Reactions of the bioregulatory agent nitric oxide in oxygenated aqueous media: Determination of the kinetics for oxidation and nitrosation by intermediates generated in the NO/O2 reaction. *Chem. Res. Toxicol.* **6**, 23–27.

CHAPTER FOUR

IN-GEL DETECTION OF S-NITROSATED PROTEINS USING FLUORESCENCE METHODS

Nicholas J. Kettenhofen,* Xunde Wang,[†] Mark T. Gladwin,[†,‡] and Neil Hogg*

Contents

* Department of Biophysics and Free Radical Research Center, Medical College of Wisconsin, Milwaukee, Wisconsin
† Pulmonary and Vascular Medicine Branch, National Heart, Lung and Blood Institute, National Institutes of Health, Bethesda, Maryland
‡ Critical Care Medicine Department, Clinical Center, National Institutes of Health, Bethesda, Maryland

Methods in Enzymology, Volume 441
ISSN 0076-6879, DOI: 10.1016/S0076-6879(08)01204-4

Abstract

Gel-based detection of protein *S*-nitrosothiols has relied on the biotin-switch method. This method attempts to replace the nitroso group with a biotin label to allow detection and isolation of *S*-nitrosated proteins and has been used extensively in the literature. This chapter describes a modification of this method that differs from the original in two major ways. First, it uses a combination of copper ions and ascorbate to achieve selective reduction of the *S*-nitrosothiol. Second, it replaces the biotin label with fluorescent cyanine dyes in order to directly observe the modified proteins in-gel and perform comparative studies using difference gel electrophoresis analysis in two dimensions.

1. INTRODUCTION

While *S*-nitrosation has become increasingly recognized as an important posttranslational modification of cysteinyl residues, the detection of such modifications has been somewhat troublesome (Gladwin *et al.*, 2006). It is relatively straightforward to detect the total *S*-nitrosothiol level, as has been described in detail elsewhere (Bryan and Grisham, 2007; Fang *et al.*, 1998 Samouilov and Zweier, 1998; Yang *et al.*, 2003); however, the detection of *S*-nitrosation on specific proteins (PSNO) relies on indirect methodology (Jaffrey *et al.*, 2001). The reasons for this are that (i) the nitroso group is not appropriate for radiolabeling and (ii) there is no reliable antibody for the direct detection or immunoprecipitation of *S*-nitrosated proteins. In addition, no strategies for modifying the S-NO group directly and stably have been devised. Consequently, the only strategy so far described involves complete destruction of the *S*-nitroso functional group and its replacement by a label. This was first exemplified by the biotin-switch assay in which the *S*-nitroso group is replaced by a biotin label (Jaffrey *et al.*, 2001). The three steps that comprise this method (Fig. 4.1) involve the blockade of all free thiol groups, the ascorbate–mediated reduction of *S*-nitrosothiols to thiols, and the labeling of the nascent thiols with a biotinylating agent. While this technique has been used for many studies, it remains somewhat controversial due to (i) the possibilities of false positives because of incomplete thiol blocking, (ii) the slow kinetics of *S*-nitrosothiol reduction by ascorbate, (iii) the possibility of disulfide reduction by ascorbate, and (iv) the widespread (although anecdotal) inconsistency experienced by many investigators.

This chapter describes the modifications made to the original method to allow sensitive detection of protein *S*-nitrosothiols using both traditional biotin–based methods and novel fluorescence methods.

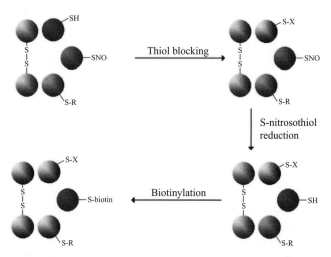

Figure 4.1 The biotin-switch method. The biotin-switch assay involves selectively repla-cing the nitroso group with a biotin label. This is accomplished in three steps. The first step involves blockade of all thiol groups. The second step involves the selective reduction of *S*-nitrosothiols to their parent thiols, avoiding reduction protein–protein disulfides or protein-mixed disulfides. The third step involves labeling of the nascent thiols with a thiol-specific biotinylation agent, thus tagging only the *S*-nitrosated proteins. (See color insert.)

 ## 2. Role of Trace Metal Ions in Ascorbate-Mediated Reduction of *S*-Nitrosothiols

We have previously investigated the biotin-switch method in detail and have reported that ascorbate is a very poor reducer of *S*-nitrosothiols. In order to get significant biotin labeling, much higher levels of ascorbate are required than originally proposed (Zhang *et al.*, 2005). This has been confirmed by others (Forrester *et al.*, 2007) and raises the question of how other investigators achieved positive results in the biotin–switch assay using the original low ascorbate conditions. We examined this issue and observed that the presence of contaminating metal ions can dramatically enhance the ability of ascorbate to reduce *S*-nitrosothiols and that the addition of copper ions and the removal of metal ion chelators allow the biotin-switch assay to work at the originally proposed ascorbate levels (Wang *et al.*, 2008). It is therefore not ascorbate that is reducing the *S*-nitrosothiol but contaminat-ing transition metal ions. In the case of copper ions, the role of ascorbate is to reduce the cupric ions back to the cuprous state, which can then rapidly reduce the *S*-nitrosothiol to form the parent thiol, nitric oxide and cupric ion [Eqs. (4.1) and (4.2)] (Askew *et al.*, 1995; Singh *et al.*, 1996):

$$\text{Ascorbate} + \text{Cu}^{2+} \rightarrow \text{Ascorbyl} + \text{Cu}^{+} \qquad (4.1)$$

$$\text{RSNO} + \text{Cu}^{+} \rightarrow \text{RSH} + \text{NO} + \text{Cu}^{2+} \qquad (4.2)$$

It is likely that other metal ions may also catalyze ascorbate-dependent *S*-nitrosothiol reduction by a similar mechanism to copper. Consequently, in the following protocols we have removed metal ion chelators from the buffers used in the reduction/labeling step, and supplemented these buffers with copper ions.

3. DETECTION OF PROTEIN *S*-NITROSATION USING FLUORESCENT LABELING METHODS

While the replacement of protein RSNO with a biotin tag provides a handle for affinity-precipitation experiments that have been utilized successfully for the identification of modified proteins (Greco *et al.*, 2006; Hao *et al.*, 2006), it has the disadvantage that Western analysis is required for in-gel visualization of the modified proteins. We have modified the original biotin-switch method to develop a direct in-gel fluorescence method for the determination of *S*-nitrosated proteins using cyanine dye technology and difference gel electrophoresis (DIGE) (Alban *et al.*, 2003). This allows direct in-gel visualization of the labeled proteins using fluorescence scanning. The availability of two spectrally resolvable versions of maleimide CyDyes (Maeda *et al.*, 2004) allows us to exploit DIGE technology for comparative proteomic studies of protein *S*-nitrosation. In addition to the advantage of directly visualizing the labeled proteins in-gel, this technology allows direct comparisons of samples to be made within the same gel, therefore eliminating gel-to-gel variations that commonly plague two-dimensional gel proteomic studies. The labeling scheme is shown in Fig. 4.2. Protein thiols exist in various states of modification, and the basic approach of the DIGE method is to examine the difference in modification before and after exposure to an experimental condition (e.g., exposure to a nitric oxide donor or to an inflammatory stimulus). Both control samples and experimental samples are treated in an identical manner with the exception that the control and experimental samples are labeled with different cyanine dyes. In Fig. 4.2, control samples are labeled with the green fluorescence dye Cy3 and the experimental sample is labeled with the red fluorescence dye Cy5. However, it is useful to also run the opposite labeling scheme to ensure that any observations are not dye specific. In an identical manner to the biotin-switch assay (Fig. 4.1), the first step is to block thiol groups. Any thiol groups that escape this blockade will be labeled in both the control and the experimental samples. The next step is to reduce the modified thiols to the

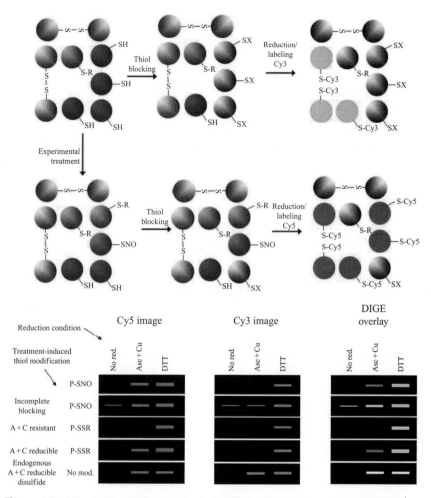

Figure 4.2 The CyDye-switch method. (Top) Protein cysteine residues can exist in protein disulfides as mixed disulfides and as free thiols. The experimental treatment (e.g., addition of NO donor or activation of NOS) will convert some of these thiols to S-nitrosothiols and some to disulfides. In the CyDye-switch method, the control sample is treated with a thiol blocker, and the resulting proteins are reduced with ascorbate/Cu and labeled with green Cy3. Proteins that show up as green fluorescence will therefore represent unblocked thiols and endogenous disulfides that were reduced (e.g., by the ascorbate/Cu treatment or by DTT). The experimental sample is treated in the same way, but labeled with red Cy5. In this case, red fluorescence will be associated with the same proteins that are labeled in the control experiment, but in addition, both S-nitrosothiols, and potentially some mixed disulfides that were formed from the experimental treatment, will also be labeled in red. The labeled protein from both control and experiment are then pooled at equal protein concentration and run on a gel. (Bottom) Typical patterns that may be expected in various scenarios. (See color insert.)

parent thiol and label them with the appropriate dye. Various reduction strategies can be employed. For example, ascorbate alone, ascorbate with copper ions, dithiothreitol (DTT), and other reducing agents, such as triphenylphosphine, can all be used to reduce different pools of modified thiol. For the specific detection of *S*-nitrosothiols we employ an ascorbate/Cu(II) mixture as described later. As shown in Fig. 4.2, any background thiol modifications present in the control sample that are reduced to thiols in the reduction step will be labeled green. For the treated sample, these same proteins will also be labeled red, but, in addition, any thiol modifications that occur as a result of treatment, which are also reduced by the specific reduction strategy employed, will also be labeled red. Consequently, the difference between red fluorescence and green fluorescence will report the treatment-dependent modifications. After labeling, both control and experimental samples are pooled at equal protein load and co-separated in the same lane of a 1-dimensional gel or on the same 2-dimensional gel. The gel is then scanned twice—once for red and once for green fluorescence. The difference between red and green fluorescence scans will then reveal the treatment-dependent modifications. As mentioned earlier and illustrated later, it is advisable to do an identical experiment with reverse labeling to make sure that the differences observed are not a function of the labeling order.

The bottom part of Fig. 4.2 shows some possible outcomes for this experimental paradigm when using an ascorbate/Cu(II) mixture or DTT as the reducing agent. The top line represents a protein that is *S*-nitrosated upon treatment and shows red in ascorbate/Cu(II) but red and green in DTT. The overlay of the two scans is highly informative, as in this case the ascorbate/Cu(II) treatment will show red and the DTT yellow (but may still have a higher intensity of red than green, depending upon the ratio of *S*-nitrosothiol and disulfide in that particular protein). The second line illustrates an *S*-nitrosated protein with incomplete thiol blocking (on either the same or a different thiol on the protein that contains the *S*-nitrosothiol). In this case, a yellow band is observed in the absence of a reducing agent as a result of equal red and green labeling. This yellowish color in the overlay image will also be present in all reduction schemes due to the green background fluorescence in the control sample. However, it may still be possible to distinguish this modification from the difference between the intensity of the red and green fluorescence. Nitrosative/oxidative treatment of cellular proteins can cause disulfide formation as well as *S*-nitrosation. If the disulfide is not reduced by ascorbate/Cu(II), then it will only show up upon DTT reduction (line 3 in the bottom portion of Fig. 4.2); however, if it is reduced by ascorbate/Cu(II), then it will be indistinguishable from an *S*-nitrosothiol (line 4 in bottom part of Fig. 4.2). Finally, any basal disulfides that are also reducible by ascorbate/Cu(II) will show up yellow on the overlay and therefore will not show a false positive. In summary, this illustration shows that only disulfides formed as a result of the experimental treatment, which are then reducible by ascorbate/Cu(II), will show up as "false positive" S-nitrosothiols. Consequently, before a positive identification of protein S-nitrosation can be made, it is important to rule out this possibility. This is not an easy task, but can be done by attempting to S-thiolate cellular proteins with an oxidant stress or diamide/thiol mixtures or preferably (if the protein is known and is available in a pure form) specifically synthesizing and testing S-thiolated forms of the protein.

4. Current Protocol for CyDye-Switch Method

Our recommended protocol for the CyDye-Switch method is as follows.

4.1. Sample preparation and free thiol blocking

Throughout the entire CyDye-switch protocol, as well as any other RSNO analysis, samples should be kept in the dark whenever possible. A vast excess of N-Ethylmaleimide (NEM; 50 mM) or any other thiol-reactive molecule of choice should be present in the lysis buffer (HDN: 250 mM HEPES with 1 mM DTPA and 100 μM neocuproine, pH 7.7, containing protease inhibitors) for cellular or tissue studies or added immediately following the protein treatment for *in vitro* studies. Immediate thiol blocking is recommended to prevent any additional PSNO formation during sample processing as well as quench the low molecular weight thiol (GSH) pool that may react with PSNO to "reverse" the modification. Complete blocking is accomplished by denaturing proteins with SDS [2.5% (w/v) final] and incubating samples at 50° for 30 to 60 min with frequent vortexing. Complete blocking is essential to prevent nonspecific protein labeling, which may create a high background.

4.2. Separating sample proteins from excess blocking agent

The excess thiol-blocking agent (NEM) must be removed in order to prevent any competition for the labeling reaction. This can be done by various techniques of protein precipitation (TCA, EtOH, acetone, etc.) or size-exclusion column chromatography (also possibly dialysis). For precipitation, add 2 volumes of 15% TCA and spin down protein pellet (2000 g for 10 min). Aspirate all supernatant and wash the pellet two times with 15% TCA with repeated spins to ensure complete NEM removal. Resuspend the pellet in HS (25 mM HEPES with 1% SDS), as no metal chelators should be present for any samples using copper-mediated reductions.

4.3. PSNO reduction/CyDye labeling

Specific reduction of PSNO is mediated by the addition of ascorbate (1 mM final) and copper sulfate (1 μM final) in the presence of a thiol-reactive maleimide CyDye in order to label the resulting free thiols at cysteine residues that were previously S-nitrosated. Cu(I) salts (e.g., CuCl) can be used in place of Cu(II) and may be preferable as the metal ion is already reduced. However, solutions of Cu(I) salts are generally more unstable. CyDye (Cy3 or Cy5) is added to a level of 100 pmol dye/μg of protein (as determined by BCA assay). The reduction/labeling reaction is allowed to proceed for 15 to 60 min at room temperature. Samples for one-dimensional analysis are mixed 1:1 with Laemmli buffer containing β-mercaptoethanol.

4.4. Pool matched sample pairs prior to protein separation for DIGE analysis

Mix equal amounts of Cy3-labeled control sample and Cy5-labeled treatment sample (or vice versa) to allow for direct in-gel comparisons. Load pooled sample on a single lane or single IPG strip for one- or two-dimensional analysis, respectively. Samples can then be coseparated by electrophoresis using gels that are cast in low-fluorescence glass plates. The band/spot intensities for Cy3 and Cy5 fluorescence can be compared following gel fluorescence scanning using a Typhoon Trio Imager or a similar instrument. Alternatively, a single CyDye can be used for side-by-side comparison of samples, as is done with the classic biotin switch.

4.5. CyDye Switch Protocol for 2D Analysis

Two-dimensional gel electrophoresis (2D-E) is a standard technique for separating complex protein mixtures in proteomic studies. As stated earlier, the comparative DIGE analysis that is possible using multi-fluorophore CyDye labeling is especially advantageous under these conditions. However, sample preparation procedures must be slightly modified in order to make the CyDye switch labeling protocol compatible with 2D-E. The presence of the anionic detergent SDS during several steps of the typical labeling protocol interferes with isoelectric focusing (IEF) employed in 2D experiments. With this in mind, IEF-compatible buffers must be substituted when samples are prepared for 2D analysis. This is accomplished by switching to Urea/CHAPS-based buffers that retain the thiol-blocking (denaturant) and resuspension (protein solubilization) characteristics of the SDS-based buffers, while providing the additional IEF compatibility.

For 2D-DIGE CyDye switch experiments, cell/tissue samples are lysed in TUC buffer (30 mM Tris, 8 M Urea, 4% w:v CHAPS, pH 7.4) containing 1 mM DTPA, 50 mM NEM and 1% protease inhibitor cocktail. Thiol-blocking and protein precipitation steps proceed as described for typical 1D analysis. However, protein pellets must be resuspended in TUC buffer (with no supplements) after which PSNOs can be labeled as described previously. Additionally, the reduction/labeling reactions should be quenched by the addition of GSH or cysteine in molar excess of maleimde CyDye. Preparation of stock solutions of GSH or cysteine are dissolved in 1 m*M* DTPA so that the quench step will also chelate the copper ions added for PSNO reduction. This step is included in order to prevent subsequent labeling of protein disulfides that are reduced by the DTT present in IEF buffer. Sample concentration via Microcon centrifugation is recommended prior to pooling experimentally matched samples (i.e. control-Cy3 + treated-Cy5) with IEF buffer and appropriate ampholytes. The samples are then loaded together on a single IPG strip for co-separation in two dimensions (IEF and SDS–PAGE) and DIGE analysis can proceed following imaging of the gels.

5. DIFFERENCE GEL ELECTROPHORESIS ANALYSIS OF A MODEL PROTEIN MIXTURE

In order to test these protocols we have selected a model mixture of four proteins with differing thiol/disulfide content and molecular weight. The four proteins used were aldolase, which has a high thiol/disulfide ratio (8 total cys, 3.6 free thiols as determined by the DTNB assay), catalase (4 total cys, 0.8 free thiols), bovine serum albumin (BSA; 35 total cys, 0.4 free thiols), and lactoferrin (34 total cys, 0.2 free thiols). The thiol levels were determined using Elman's reagent. In each case we generated *S*-nitrosated versions of these proteins, pooled them, and subjected them to the CyDye-switch method. This experiment used both senses of labeling and only showed the overlay of the two fluorescent images. As shown in Fig. 4.3, in the absence of reducing agents, no bands were visible. This indicates that thiol blocking was complete. In the presence of ascorbate alone, a faint band was seen for aldolase, but only in the treated samples. This indicates that the *S*-nitrosothiol of aldolase is particularly sensitive to reduction and illustrates that ascorbate alone will emphasize such proteins. Reduction by ascorbate/Cu(II) shows very clear labeling of aldolase, catalase, and BSA, but not lactoferrin. In all cases the labeling is only present for the *S*-nitrosated protein and not for the control protein, indicating that in all cases this treatment did not reduce endogenous disulfides. Finally, reduction using DTT shows that aldolase and, to some extent, catalase still present a clear distinction between treated and control samples, suggesting that disulfide reduction and labeling are not big issues with these proteins. In contrast, lactoferrin and BSA show indistinguishable labeling between control and *S*-nitrosated samples (Fig. 4.3, yellow in the overlay). This illustrates nicely that the ascorbate/Cu(II) mixture is still able to detect *S*-nitrosation in proteins that contain many disulfides, such as BSA.

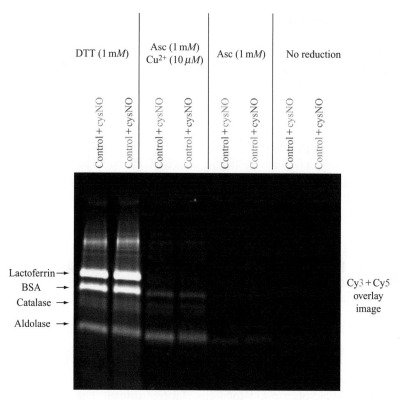

Figure 4.3 CyDye-switch analysis of a simple protein mixture. An equimolar mix (1 mg/ml total protein) of aldolase, catalase, BSA, and lactoferrin was treated with either 100 μM *S*-nitrosocysteine (CysNO) or buffer alone (control) for 30 min. Samples were then labeled with Cy3 (green) or Cy5 (red) by CyDye switch using various reduction conditions as indicated. Equal amounts of experimentally paired samples were then pooled and separated by SDS-PAGE. An overlay image of Cy3 and Cy5 fluorescence is presented. (See color insert.)

 ## 6. Sensitivity of Detection of the CyDye Label

In order to examine the sensitivity of the CyDye method we prepared *S*-nitroso HSASNO and diluted it with untreated HSA to varying degrees so that the total protein content remained equal. These mixtures were subjected to the CyDye-switch protocol using only Cy3 as the labeling agent. Figure 4.4A shows both the fluorescence image and the Scion-Image densitometric band intensity as a function of the calculated amount of initial protein *S*-nitrosothiol loaded onto the gel. These data are recapitulated in Fig. 4.4B with a conventional evenly spaced x axis. This illustrates that the CyDye-switch method can detect down to about 40 fmol of *S*-nitrosated protein and that the response is relatively linear with the exception of the

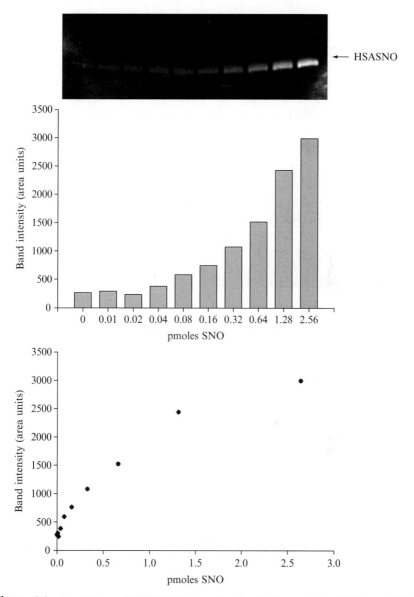

Figure 4.4 Sensitivity of HSASNO detection by CyDye switch. *S*-Nitroso-HSA samples (constant total protein, varied SNO) were Cy3 labeled by CyDye switch with ascorbate (1 m*M*)/Cu(II) (0.1 *μM*) reduction. One microgram of protein containing varying amounts of SNO (indicated below graphs) was resolved in each lane by SDS–PAGE. The gel was imaged for Cy3 fluorescence (top), and band intensities were quantified using Scion Image and plotted versus the amount of SNO loaded (bar and scatter plots).

highest level examined where the response tails off. This illustrates that fluorescence imaging is at least as sensitive as the biotin–switch method. Of course, these results are specific for the particular reduction condition (i.e., copper concentration and time) used, as well as the sensitivity setting on the fluorescence scanner. These data using HSASNO represent a conservative estimate of sensitivity given the low Cu(II) concentration ($0.1\ \mu M$) and short reduction/labeling time (15 min). Additionally, as seen in data from our simple protein mixture, sensitivity will most likely be dependent on the individual protein being analyzed.

7. FLUORESCENCE DETECTION OF *S*-NITROSATED PROTEINS IN PLASMA

We have employed the CyDye-switch method to examine *S*-nitrosation of plasma proteins after the treatment of plasma with *S*-nitrosocystiene. Figure 4.5A illustrates the level of total RSNO observed in plasma after treatment with *S*-nitrosocystiene, as measured by tri-iodide-based chemiluminescence. Plasma proteins were then subjected to CyDye-switch labeling; under the conditions employed, only a band representing *S*-nitroso-HSA was observed, which is perhaps unsurprising as this is the major thiol-containing protein in plasma. Again as low as 40 fmol of protein *S*-nitrosothiol could be clearly detected above background levels. This also is a conservative estimate, as identical reduction/labeling conditions were used for the plasma samples and the pure HSASNO samples presented earlier. Similar results were obtained for the Cy5 label (not shown).

8. TWO-DIMENSIONAL DIFFERENCE GEL ELECTROPHORESIS DETECTION OF *S*-NITROSATED PROTEINS

The DIGE technology works best when run in conjunction with two-dimensional gels. Figure 4.6 illustrates a two-dimensional gel of normal human bronchial epithelial (NHBE) cell lysate proteins that were *S*-nitrosocysteine-treated *ex vivo*. The total protein *S*-nitrosothiol content for this sample was an extremely high and nonphysiological level (>10 nmol/mg protein). For this experiment we used the ascorbate/Cu(II) reduction scheme for labeling. As can be seen, ascorbate/Cu(II) reduction revealed a multiplicity of spots that were labeled in the treated sample (Cy5, red) but not present in the control (Cy3, green). The relative sparseness of labeling in the control sample is very encouraging as it suggests that thiol blocking is complete and endogenous disulfides are not reduced by the ascorbate/Cu(II) treatment.

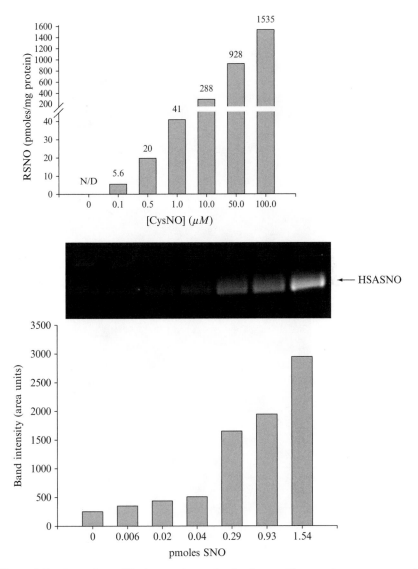

Figure 4.5 Detection of S-nitrosated proteins in plasma. Human plasma was treated with various concentrations of CysNO for 60 min. Total levels of SNO were determined by tri-iodide-based chemiluminescence and normalized to protein content (top). Samples were labeled with Cy3 using CyDye switch with ascorbate $(1 \, mM)/Cu(II)$ $(0.1 \, \mu M)$ reduction. Proteins were then resolved by SDS-PAGE, and Cy3 gel fluorescence was imaged (middle). The gel image was analyzed using Scion Image, and band intensities for HSASNO were plotted versus the total amount of SNO loaded per lane (bottom).

pH 3   pH 10
  −   +

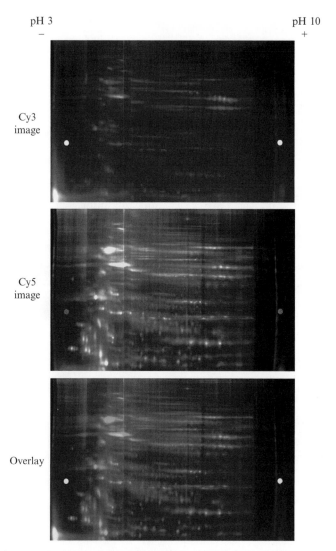

Figure 4.6 Two-dimensional DIGE detection of *S*-nitrosated proteins in a complex mixture. NHBE cells were lysed, and cytosolic proteins were treated with either 100 μM *S*-nitrosocysteine (CysNO) or buffer alone (control) for 30 min. Samples were then labeled using the CyDye-switch protocol with ascorbate (1 mM)/Cu(II) (10 μM) reduction. The control sample was labeled with Cy3 (green) while the treated sample was labeled with Cy5 (red). Equal amounts of protein from each sample were then pooled and coseparated by two-dimensional electrophoresis. The gel was imaged for both Cy3 and Cy5 fluorescence. Cy3 (top), Cy5 (middle), and overlay (bottom) images are presented. (See color insert.)

9. MODIFICATIONS OF THE ORIGINAL BIOTIN-SWITCH METHOD

The original biotin switch method described by Jaffrey *et al.* (2001) consisted of the following steps (see Fig. 4.1). First, sample proteins were incubated with the thiol-blocking agent *S*-methyl methanethiosulfonate (MMTS). The efficiency of this step dictates the specificity and sensitivity of this method, as any unblocked thiols will show up as false positives. Using a simple calculation, it can be seen that if 1% of protein thiols are nitrosated, the blocking efficiency needs to be 99% for a signal:background ratio of 1:1. *S*-Nitrosated proteins are generally much lower in abundance than this, and the signal:background needs to be much higher than 1:1, suggesting that the blocking efficiency needs to be much higher than 99%. Although blocking efficiency is crucial, it is relatively easy to test and account for if appropriate control samples are available. Our experience is that alternative blocking agents, such as NEM or iodoacetamide, work as well as MMTS. In fact, MMTS can be problematic, as it blocks thiols through formation of a disulfide, which is generally a more easily reducible modification than a thioether. The second step is the reduction of *S*-nitrosothiols by incubation with 1 mM ascorbate for 1 h in the presence of EDTA and neocuproine in HEPES buffer. The goal of this treatment is to selectively reduce the *S*-nitrosothiols but not reduce disulfides so that all thiol groups generated by this reduction process derive from protein RSNO. This step has generated the most problems with this procedure. At first glance this step appears completely untenable as the rate constants for the reduction of RSNO by ascorbate have been established and predict that 1 mM ascorbate for 1 h would reduce only a small fraction of the total *S*-nitrosothiol (Holmes and Williams, 2000; Kashiba-Iwatsuki *et al.*, 1996). For example, the reaction between GSNO and ascorbate has a rate constant of about 12 $M^{-1}s^{-1}$ at pH 7.3 (Holmes and Williams, 2000), giving a pseudo–first-order half-time for GSNO decay of about 16 h with 1 mM ascorbate. *S*-Nitroso bovine serum album appears to be reduced even more slowly than GSNO (Zhang *et al.*, 2005). In the original studies, the efficiency of this reduction step was never examined, but studies have demonstrated that this level of ascorbate hardly makes a dent in the total RSNO level in a complex mixture of cellular *S*-nitrosothiols (Kettenhofen *et al.*, 2007). This was consistent with our experience with the biotin-switch assay, which only gave positive signals at ascorbate levels of 30 mM and above (Zhang *et al.*, 2005), and others have clearly experienced the same problems (Forrester *et al.*, 2007). Interestingly, in previous studies, we consistently used DTPA in place of EDTA in the buffers, as DTPA is far better at diminishing the oxidative effects of trace iron and copper than EDTA. We closely examined the effects of iron chelators on the ability of ascorbate to promote RSNO decay and showed that the "dirtier" the experiment, the better the result. In other words, if buffers were chelex treated or if sufficient quantities of freshly prepared metal chelators were used, the degree of biotin labeling was diminished greatly (Wang *et al.*, 2008). However, if copper ions were purposely added to the incubation mixture, the biotinylation signals were enhanced greatly. This led us to propose that the reduction/labeling step of the biotin-switch assay should be performed in the absence of metal chelators and with the addition of copper ions to the buffers (see later). The final step is labeling of the newly formed thiols. (*N*-(6-(Biotinamido)hexyl)-3′-(2′-pyridyldithio)-propiona-mide (biotin-HPDP) was originally proposed as the biotinylation agent as it forms a disulfide with the protein thiol, which can aid in the release of the protein from agarose beads after a precipitation experiment. However, as with MMTS, the disulfide is a weaker link than a thioether, and the attachment of biotin through a maleimido or an iodoacetamide linking agent are other possibilities.

10. CURRENT PROCEDURE FOR THE BIOTIN-SWITCH ASSAY

Our recommended procedure for the biotinylation of protein *S*-nitrosothiols is as follows.

1. Protein thiols should be blocked as soon as possible after cessation of the experiment. Protein mixtures should not be acidified or precipitated before all thiols are blocked. The blocking buffer contains 250 mM HEPES, pH 7.7, containing 1 mM EDTA (or DTPA), 0.1 mM neocuproin, 2.5% SDS, and 50 mM MMTS (or NEM). Amber tubes should be used for all steps of the reaction, and direct light should be avoided to prevent photolytic decomposition of *S*-nitrosothiols. While *S*-nitrosothiols are not acutely light sensitive, prolonged exposure to a light source will result in some photolysis of the S–N bond. In addition, problems with direct sunlight have been reported in the ascorbate reduction step (Forrester *et al.*, 2007). This mixture is incubated for 30 min at 50° with occasional vortexing.

2. Excess blocking agent can be removed by either passage down a G25 microspin column (three times) or protein precipitation with ethanol or acetone. Both methods allow buffer exchange to a metal chelator-free buffer consisting of HEPES (25 mM, pH 7.7) and 1% SDS. Ascorbate (1 mM), biotin-HPDP (2 mM, or alternative biotinylation agent), and copper ions are added to this mixture and incubated for 1 h. Our experience suggests that the oxidation state of the copper salt is not particularly crucial, and we have achieved good results with both Cu(I)

chloride and Cu(II) sulfate. This is because ascorbate reduces the Cu(II) to Cu(I), which effectively reduces the S–NO bond. The amount of copper required may be significantly system dependent, and it is essential that the ability of the Cu/ascorbate solution to reduce *S*-nitrosothiols is checked by an independent method (such as tri-iodide chemiluminescence) in each system. With pure proteins, an amount as low as 10 nM of copper was able to reduce protein RSNO, whereas up to 1 μM may be required in complex mixtures. It should be noted that copper ions are catalytic and are undergoing a Cu(I) to Cu(II) redox cycle during ascorbate-dependent *S*-nitrosothiol decay. Although we have not examined this specifically, it is likely that too much copper may become problematic, as Cu(II) can oxidize thiols and potentially lead to a reduction in labeling. Also, the addition of high levels of Cu(I) may also result in superoxide and hydrogen peroxide generation through the reduction of oxygen, thus affecting the integrity of the proteins.

3. After labeling, proteins are detected on Western blots using either a peroxidase conjugated antibiotin antibody or peroxidase-conjugated streptavidin. For quantitative determination of the level of biotinylation, biotinylated cytochrome *c* can be used as an internal standard as described previously (Landar *et al.*, 2006). It should be stressed that use of a biotinylated standard does not quantify the amount of *S*-nitrosothiol, but allows determination of the amount of biotinylated proteins. Other controls (discussed later) and the use of standard *S*-nitrosated proteins are required to fully understand the relationship between the level of biotinylation and the original level of *S*-nitrosothiol. Instead of Western analysis, labeled proteins can be isolated using streptavidin–agarose for direct mass analysis.

4. It is essential for every experiment that a number of controls are included. First, a control needs to be performed in the absence of ascorbate/Cu in order to determine the success of the blocking step. Second, the experimental sample should be fully reduced with DTT before the blocking step to ensure that labeling is specific. Third, it is advisable to run a positive control consisting of a DTT-reduced sample in which the blocking step is omitted. This will give an idea of the total amount of a particular thiol-containing protein and can be used not only to test if the methods are working, but also to ascertain the fraction of a particular protein that has been modified.

While these modifications of the method allow reproducible biotinylation of *S*-nitrosated proteins, the method still contains an intrinsic problem regarding possible false-positive signals from disulfides that may be reduced by the ascorbate/Cu step. While a thiol blockade is generally very efficient, suggesting that the reduction of endogenous disulfides is not a major issue, the detection of mixed disulfides generated during NO/RSNO exposure cannot be fully ruled out. While we see no reduction of preformed mixed disulfides on human serum albumin using this technique (Wang *et al.*, 2008), this needs to be tested on a protein-by-protein basis, and positive identification of an *S*-nitrosothiol can only be confirmed if the possible formation of mixed disulfides is ruled out. Despite this limitation, it should be emphasized that mixed disulfide formation is also an important posttranslational NO-dependent thiol modification, and so even if the type of modification is not 100% clarified, this method still provides valuable information.

11. Conclusion

This chapter described current protocols for both the original biotin-switch method and a novel CyDye-switch method to detect *S*-nitrosated proteins. The major modification to the original method is the use of ascorbate/Cu(II) mixtures to facilitate the *S*-nitrosothiol reduction. Under these conditions, reduction appears facile and relatively specific. Although we cannot rule out signals from other modifications that may be generated from the treatment of cells with oxidants, such as mixed disulfides, as least in the case of serum albumin, such modifications do not give false positives (Wang *et al.*, 2008). The CyDye-switch method allows the comparative assessment of two states of cellular proteins directly in-gel without the necessity of a Western blot. It is highly sensitive and specific (with the same caveat mentioned earlier for the conventional biotin-switch method). The fact that proteins can be detected directly in-gel removes the necessity of spot registration and allows for the automated processing of positive spots for protein identification. These fluorescence-based methods should provide an additional and in many ways superior method for the detection and identification of *S*-nitrosated proteins.

ACKNOWLEDGMENT

This work was supported by National Institutes of General Medicine grant GM55792.

REFERENCES

Alban, A., David, S. O., Bjorkesten, L., Andersson, C., Sloge, E., Lewis, S., and Currie, I. (2003). A novel experimental design for comparative two-dimensional gel analysis: Two-dimensional difference gel electrophoresis incorporating a pooled internal standard. *Proteomics* **3**, 36–44.

Askew, S. C., Barnett, D. J., McAninly, J., and Williams, D. L. H. (1995). Catalysis by Cu^{2+} of nitric oxide release from *S*-nitrosothiols (RSNO). *J. Chem. Soc. Perkin Trans.* **2,** 741–745.

Bryan, N.S, and Grisham, M.B (2007). Methods to detect nitric oxide and its metabolites in biological samples. *Free Radic. Biol. Med.* **43,** 645–657.

Fang, K., Ragsdale, N. V., Carey, R. M., Macdonald, T., and Gaston, B. (1998). Reductive assays for S-nitrosothiols: Implications for measurements in biological systems. *Biochem. Biophys. Res. Commun.* **252,** 535–540.

Forrester, M. T., Foster, M. W., and Stamler, J. S. (2007). Assessment and application of the biotin switch technique for examining protein *S*-nitrosylation under conditions of pharmacologically induced oxidative stress. *J. Biol. Chem.* **282,** 13977–13983.

Gladwin, M. T., Wang, X., and Hogg, N. (2006). Methodological vexation about thiol oxidation versus S-nitrosation: A commentary on "An ascorbate-dependent artifact that interferes with the interpretation of the biotin-switch assay." *Free Radic Biol. Med.* **41,** 557–561.

Greco, T. M., Hodara, R., Parastatidis, I., Heijnen, H. F. G., Dennehy, M. K., Liebler, D. C., and Ischiropoulos, H. (2006). Identification of *S*-nitrosylation motifs by site-specific mapping of the *S*-nitrosocysteine proteome in human vascular smooth muscle cells. *Proc. Natl. Acad. Sci. USA.* **103,** 7420–7425.

Hao, G., Derakhshan, B., Shi, L., Campagne, F., and Gross, S. S. (2006). SNOSID, a proteomic method for identification of cysteine *S*-nitrosylation sites in complex protein mixtures. *Proc. Natl. Acad. Sci. USA* **103,** 1012–1017.

Holmes, A. J., and Williams, D. L. H. (2000). Reaction of ascorbic acid with *S*-nitrosothiols: Clear evidence for two distinct reaction pathways. *J. Chem. Soc. Perkin Trans.* **2,** 1639–1644.

Jaffrey, S. R., Erdjument-Bromage, H., Ferris, C. D., Tempst, P., and Snyder, S. H. (2001). Protein S-nitrosylation: A physiological signal for neuronal nitric oxide. *Nat. Cell Biol.* **3,** 193–197.

Kashiba-Iwatsuki, M., Yamaguchi, M., and Inoue, M. (1996). Role of ascorbic acid in the metabolism of S-nitroso-glutathione. *FEBS Lett.* **389,** 149–152.

Kettenhofen, N. J., Broniowska, K. A., Keszler, A., Zhang, Y., and Hogg, N. (2007). Proteomic methods for analysis of *S*-nitrosation. *J. Chromatogr. B* **851,** 152–159.

Landar, A., Oh, J. Y., Giles, N. M., Isom, A., Kirk, M., Barnes, S., and Darley-Usmar, V. M. (2006). A sensitive method for the quantitative measurement of protein thiol modification in response to oxidative stress. *Free Radic. Biol. Med.* **40,** 459–468.

Maeda, K., Finnie, C., and Svensson, B. (2004). Cy5 maleimide labelling for sensitive detection of free thiols in native protein extracts: Identification of seed proteins targeted by barley thioredoxin h isoforms. *Biochem. J.* **378,** 497–507.

Samouilov, A., and Zweier, J. L. (1998). Development of chemiluminescence-based methods for specific quantitation of nitrosylated thiols. *Anal. Biochem.* **258,** 322–330.

Singh, R. J., Hogg, N., Joseph, J., and Kalyanaraman, B. (1996). Mechanism of nitric oxide release from S-nitrosothiols. *J. Biol. Chem.* **271,** 18596–18603.

Wang, X., Kettenhofen, N. J., Shiva, S., Hogg, N., and Gladwin, M. T. (2008). *Free Radic Biol. Med.* **44**(7), 1362–1372.

Yang, B. K., Vivas, E. X., Reiter, C. D., and Gladwin, M. T. (2003). Methodologies for the sensitive and specific measurement of *S*-nitrosothiols, iron-nitrosyls, and nitrite in biological samples. *Free Radic. Res.* **37,** 1–10.

Zhang, Y., Keszler, A., Broniowska, K. A., and Hogg, N. (2005). Characterization and application of the biotin-switch assay for the identification of *S*-nitrosated proteins. *Free Radic. Biol. Med.* **38,** 874–881.

CHAPTER FIVE

THE ARACHIDONATE-DEPENDENT SURVIVAL SIGNALING PREVENTING TOXICITY IN MONOCYTES/MACROPHAGES EXPOSED TO PEROXYNITRITE

Orazio Cantoni, Ilaria Tommasini, *and* Liana Cerioni

Contents

Abstract

Cells belonging to the monocyte/macrophage lineage are in general highly resistant to peroxynitrite. Resistance is not dependent on the scavenging of peroxynitrite itself, or of other secondary reactive species, but is rather associated with the prompt activation of a survival signaling leading to the prevention of toxicity in cells otherwise committed to mitochondrial permeability transition (MPT)-dependent necrosis. The signaling pathway is triggered by cytosolic phospholipase A_2-released arachidonic acid, leading to the sequential activation of 5-lipoxygenase (5-LO) and protein kinase $C\alpha$, an event associated with the cytosolic accumulation of Bad. Hence, inhibition of 5-LO (or that of any of the aforementioned enzymes involved in the signaling cascade) was associated with the mitochondrial accumulation of Bad and Bax and with a rapid

Istituto di Farmacologia e Farmacognosia, Università degli Studi di Urbino "Carlo Bo," Urbino, Italy

Methods in Enzymology, Volume 441
ISSN 0076-6879, DOI: 10.1016/S0076-6879(08)01205-6

MPT-dependent toxicity. These results contribute to the definition of the mechanism(s) whereby monocytes/macrophages survive to peroxynitrite in inflamed tissues and provide insights for the development of novel anti-inflammatory therapies based on the suppression of inflammatory cell survival.

1. INTRODUCTION

In the context of the inflammatory response, various cell types, including neutrophils, monocytes, and macrophages, produce an array of toxic molecules, resulting in tissue damage. Peroxynitrite, the coupling product of nitric oxide and superoxide, is one of these species that, because of its high reactivity, is likely to generate extensive damage in the same cell in which it is being produced. How these cells cope with their own peroxynitrite and, more generally, how they can survive in an environment in which other cell types die remain unanswered questions. Our laboratory has been actively involved in studies investigating the mechanism(s) whereby cells belonging to the monocyte/macrophage lineage survive to peroxynitrite. We have found that their intrinsic resistance is not associated with a remarkable scavenging capacity and/or efficient antioxidant system. Rather, these cells appeared to be as susceptible as other cell types to the damaging effects of peroxynitrite, but survived because they are able to respond to specific stimuli with a signaling preventing toxicity in cells otherwise committed to death. Hence, resistance of these cells was lost under conditions in which the survival signaling was inhibited, thereby leading to a very rapid necrosis.

Using U937 cells as a monocyte cellular system, but reproducing critical findings in human monocytes and macrophages, we found that nontoxic concentrations of peroxynitrite nevertheless commit these cells to mitochondrial permeability transition (MPT), however prevented by a parallel signaling triggered by arachidonic acid (ARA; Tommasini *et al.*, 2002). The lipid messenger, largely available at the inflammatory sites since actively released by inflammatory cells, penetrates the plasma membranes easily and may thus reach target cells to promote survival signaling. Furthermore, peroxynitrite itself is a potent stimulus for cytosolic phospholipase A$_2$ (cPLA$_2$) (Tommasini *et al.*, 2002, 2004), thereby directly fueling the downstream target of the survival signaling, 5-lipoxygenase (5-LO; Tommasini *et al.*, 2006). The remaining downstream events leading to survival are represented by the mitochondrial translocation of protein kinase Cα (PKCα) (Cerioni *et al.*, 2006; Guidarelli *et al.*, 2005a) and by the ensuing phosphorylation of Bad (Cerioni *et al.*, 2006; Guidarelli *et al.*, 2005b). As discussed later, phosphorylation leads to the cytosolic accumulation of Bad, a condition associated with the prevention of

MPT in damaged mitochondria and thus in the survival of cells committed to MPT-dependent toxicity.

Monocytes/macrophages therefore adopt an ingenious strategy to cope with peroxynitrite: they use a most common inflammatory product, ARA, to trigger a survival signaling based on events associated with the cytosolic accumulation of Bad.

2. Materials and Methods

2.1. Cell culture and treatments

U937 human myeloid leukemia cells are cultured in suspension in RPMI 1640 medium (Sigma-Aldrich, Milan, Italy) supplemented with 10% fetal bovine serum (HyClone Laboratories, Logan, UT), penicillin (100 U/ml), and streptomycin (100 μg/ml) (HyClone), at 37 °C in T-75 tissue culture flasks (Sarstedt, Nümbrecht, Germany) gassed with an atmosphere of 95% air–5% CO_2. Peroxynitrite is synthesized by the reaction of nitrite with acidified H_2O_2, as described previously (Radi *et al.*, 1991), with minor modifications (Tommasini *et al.*, 2002). Treatments are performed in pre-warmed saline A (8.182 g/liter NaCl, 0.372 g/liter KCl, 0.336 g/liter $NaHCO_3$, and 0.9 g/liter glucose, pH 7.4, at 37 °C) containing 2.5×10^5 cells/ml. The cell suspension is inoculated into 15-ml plastic tubes (2 ml for cytotoxicity experiments) or into 50-ml plastic tubes (20 ml for Western blot experiments) before the addition of peroxynitrite. Peroxynitrite is added on the wall of these tubes and immediately mixed to equilibrate its concentration on the cell suspension. To avoid changes in pH due to the high alkalinity of the peroxynitrite stock solution, an appropriate amount of 1.5 *N* HCl is also added to the wall of the tubes prior to peroxynitrite. Note that this procedure has to be fast, as peroxynitrite decomposes rapidly at neutral pH.

2.2. Cytotoxicity assay

Cytotoxicity is determined with the trypan blue exclusion assay immediately after treatments. Briefly, an aliquot (50 μl) of the cell suspension is diluted 1:1 (v/v) with 0.4% trypan blue and viable cells are counted with a hemocytometer. Note that the toxic treatments employed in this study reduce the number of viable cells without a parallel increase in the number of trypan blue positive cells, as cell lysis is very fast. Indeed, these cells remain viable for at least 30 to 40 min and then undergo an about a 3- to -5-min process in which they first swell and then rapidly lose their membrane integrity and lyse.

2.3. Cell fractionation

Isolation of cytosolic and mitochondria-enriched fractions requires at least 5×10^6 U937 cells. After treatments (10 min), cells are collected by centrifugation at 1300g for 3 min at 4 °C. Cold extraction buffer [20 mM HEPES-KCl (pH 7.4), 10 mM KCl, 250 mM sucrose, 1.5 mM MgCl$_2$, 1 mM sodium EDTA, 1 mM sodium EGTA], completed immediately before use with 1 mM dithiothreitol (DTT), 1 mM phenylmethylsulfonyl fluoride (PMSF), 10 μg/ml leupeptin, 10 μg/ml pepstatin, and 10 μg/ml aprotinin, is added to the final cell pellet (35 μl extraction buffer/1 $\times$ 10^6 cells). Cells are kept for 10 min at 4 °C, the temperature of all subsequent steps, and are then homogenized by 40 passages through a 26-gauge needle. Note that the number of passages necessary to lyse the cells may differ significantly among different cell types. The status of the cells can be monitored conveniently after the addition of trypan blue under a microscope. Homogenates are next centrifuged at 1000g for 5 min to remove nuclei, unbroken cells, and cell debris. The resulting supernatant is centrifuged at 12,000g for 30 min to obtain the cytosolic fraction, further purified by an additional centrifugation at 10,000g for 1 h. The cytosolic fraction is maintained at 4 °C prior to electrophoresis. The pellet of the first centrifugation, representing the enriched mitochondrial fraction, is then suspended in 55 μl of cold lysis buffer [20 mM Tris (pH 7.5), 150 mM NaCl, 1% Nonidet P-40, 1 mM DTT, 1 mM PMSF, 10 mM sodium orthovanadate, 10 mM sodium fluoride, 10 μg/ml leupeptin, 10 μg/ml pepstatin, and 10 μg/ml aprotinin], incubated on ice for 20 min, and finally centrifuged at 15,000g for 5 min to obtain the solubilized enriched mitochondrial fraction.

The just described fractionation procedure of U937 cells normally yields about 1.7 mg/ml of proteins for the cytosolic fraction and 1.2 mg/ml of proteins for the mitochondrial fraction. The protein concentration is determined with the Bio-Rad dye-binding protein assay (Bio-Rad Laboratories, Hercules, CA).

2.4. Western blot analysis

Equal amounts (20 μg) of the mitochondrial and cytosolic fractions are diluted in sodium dodecyl sulfate (SDS) sample buffer [1.25 mM Tris (pH 6.9), 2% SDS, 10% glycerol, 0.002% bromophenol blue] and incubated for 10 min at 95 °C. Samples are electrophoresed in parallel with prestained molecular weight markers (Bio-Rad) on a denaturing SDS-polyacrylamide gel using a 4% stacking gel and a 12% (Bad/Bax) or 15% (cytochrome c) resolving gel. Gels are run at room temperature (25 °C) at constant current (25 mA for Bad and Bax; 60 mA for cytochrome c). Electrophoresed gels are soaked (30 s) in transfer buffer (192 mM glycine in 25 mM Tris, pH 8.3) containing 20% methanol and then electroblotted (Bad or Bax, 36 V for 75 min; cytochrome c, 100 V for 60 min) to polyvinyldiene difluoride (PVDF) membranes in a full immersion electroblotting system (Bio-Rad). PVDF membranes are wetted prior to use in 100% methanol for 1 min, washed in transfer buffer for 1 to 2 min, and finally equilibrated in the same buffer for at least 30 min. Nonspecific protein binding is prevented by a 1-h incubation at room temperature with 5% (w/v) nonfat milk protein in 50 mM Tris (pH 7.2), 140 mM NaCl, and 0.1% Tween 20 (TTBS). Membranes are then incubated overnight (4 °C) with a 1:700 dilution of monoclonal antibody in blocking solution. The antibody against Bad is obtained from BD Transduction Laboratories (Lexington, KY), whereas antibodies against Bax or cytochrome c are from Santa Cruz (Santa Cruz, CA). In order to optimize the use of antibodies, volumes of incubations are kept very low and solutions are stored for up to 3 weeks at −20 °C and used in two to three separate experiments. After overnight incubation, membranes are washed (1 × quick rinse, 1 × 15 min, 2 × 10 min) in TTBS at room temperature and incubated for 90 min (25 °C) with horseradish peroxidase (HRP)–conjugated goat antimouse secondary antibodies (Santa Cruz) at 1:2000 dilution in blocking solution. Membranes are finally washed (as indicated for the primary antibodies, see earlier discussion), and immunoreactive bands are detected by enhanced chemiluminescence of HRP substrates. Antibodies against actin (Sigma-Aldrich) and HSP-60 (Santa Cruz) are used to assess equal loading of the lanes and the purity of the mitochondrial fractions.

2.5. Statistical analysis

Statistical analysis of data for multiple comparisons is performed by ANOVA followed by a Dunnett's test.

3. RESULTS AND DISCUSSION

Bad is an important member of the Bcl-2 family critically involved in the regulation of MPT through protein–protein interactions. In particular, under conditions in which it translocates to the mitochondria, Bad may heterodimerize with Bcl-2 and/or Bcl-X$_L$ (Adams and Cory, 1998; Zamzami et al., 1996), thereby abolishing their anti-MPT activity (Letai et al., 2002; Zha et al., 1996). It follows that the cytosolic localization of Bad, therein complexed to the 14-3-3 protein (Adams and Cory 1998; Zha et al., 1996), creates optimal conditions for the anti-MPT functions of Bcl-2/Bcl-X$_L$ and, not surprisingly, various survival pathways converge in promoting this response. The accumulation of Bad in the cytosol is tightly regulated by hosphorylation in two serine residues, Ser[136], mediated by phosphatidylinositol 3-kinase and Akt/PKB (Datta *et al.*, 1997; Del Peso *et al.*, 1997), and Ser[112], mediated by either PKA- or PKC/p90[RSK] (Bertolotto *et al.*, 2000; Tan *et al.*, 1999).

As indicated in Section 1, our work has documented the involvement of a cPLA$_2$/5-LO/PKCα pathway leading to the cytosolic accumulation of Bad in cells belonging to the monocyte/macrophage lineage exposed to peroxynitrite. It should be noted that U937 cells, a promonocytic cell line widely utilized in our studies, display high levels of mitochondria-associated Bad under normal conditions and that a nontoxic dose (100 μM) of peroxynitrite indeed promotes events leading to the mitochondria to cytosol translocation of the protein (Fig. 5.1A). This event is not mediated by PKA or PKB but, rather, by PKCα (Cerioni *et al.*, 2006; Guidarelli *et al.*, 2005b). In particular, the upstream signaling triggered by peroxynitrite caused the mitochondrial translocation of PKCα sensitive to pharmacological inhibitors of cPLA$_2$, 5-LO, or PKCα or to knockdown of these proteins (Cerioni

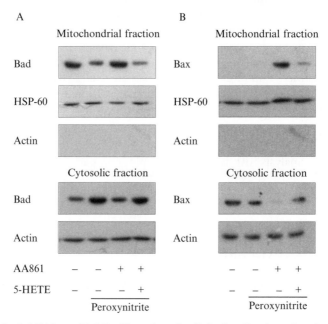

Figure 5.1 Inhibition of 5-LO affects the subcellular localization of Bad and Bax in cells exposed to an otherwise nontoxic concentration of peroxynitrite. Cells were exposed for 3 min to peroxynitrite (100 μM) and for an additional 7 min to AA861 (1 μM) alone or associated with 5-HETE (0.3 μM). Cells were then processed to obtain the mitochondrial and cytosolic fractions for Western blot analysis using antibodies against Bad (A) and Bax (B). Blots were then washed and reprobed for actin or HSP-60.

et al., 2006; Guidarelli *et al.*, 2005b). Figure 5.1A provides an example in this direction by showing that a 5-LO inhibitor, AA861 (Ashida *et al.*, 1983), causes loss of cytosolic Bad associated with the mitochondrial accumulation of the protein. This response is, however, abolished by nanomolar levels of the 5-LO product 5-hydroxyeicosatetraenoic acid (5-HETE).

Bax is an additional member of the Bcl-2 family (Oltvai *et al.*, 1993), also localized in the cytosol of healthy cells, that translocates to membrane sites, including mitochondria, in response to death signals (Gross *et al.*, 1998; Hsu *et al.*, 1997). The cytosolic localization of Bad prevents Bax translocation to the mitochondria (Tafani *et al.*, 2001), whereas the mitochondrial translocation of Bad prevents the interaction of Bax with Bcl-2/Bcl-X_L and allows the formation of Bax oligomers, causing permeabilization of the outer mitochondrial membrane (Gross *et al.*, 1998).

The results illustrated in Fig. 5.1B document the cytosolic localization of Bax in untreated cells or in cells exposed to the nontoxic dose of peroxynitrite. As observed with Bad, however, inhibition of 5-LO promoted the mitochondrial translocation of Bax via a mechanism sensitive to exogenous 5-HETE.

Thus, it appears that Bad is largely, but not entirely, localized in the cytosol of healthy cells and that the survival signaling triggered by peroxynitrite is based on activation of the cPLA$_2$/5-LO/PKCα pathway to further implement the cytosolic pool of Bad. This event is associated with prevention of the mitochondrial translocation of Bax and of the ensuing MPT-dependent toxicity.

The results illustrated in Fig. 5.2A indeed indicate that cells survive treatment with 100 μM peroxynitrite and that inhibition of 5-LO signaling promotes a prompt lethal response sensitive to either 5-HETE or cyclosporin A (CsA, 0.5 μM). The latter is an inhibitor of MPT (Halestrap *et al.*, 1997), and the observed cytoprotection emphasizes the notion that toxicity takes place via a MPT-dependent mechanism. Consistently, a significant mitochondrial release of cytochrome *c*, sensitive to either 5-HETE or CsA, was elicited by 5-LO inhibition in cells exposed to the otherwise ineffective dose of peroxynitrite (Fig. 5.2B).

In conclusion, our work has identified an ingenious strategy adopted by cells belonging to the monocyte/macrophage lineage to cope with peroxynitrite. These cells use a product largely available at the inflammatory sites to enforce a signaling leading to the cytosolic accumulation of Bad, thereby optimizing the anti-MPT functions of Bcl-2/Bcl-X_L. Under these conditions, Bax is retained in the cytosol and MPT is prevented. The observation that selective inhibition of the survival signaling causes toxicity in cells exposed to otherwise nontoxic levels of peroxynitrite provides ground for the development of therapies leading to the demise of inflammatory cells.

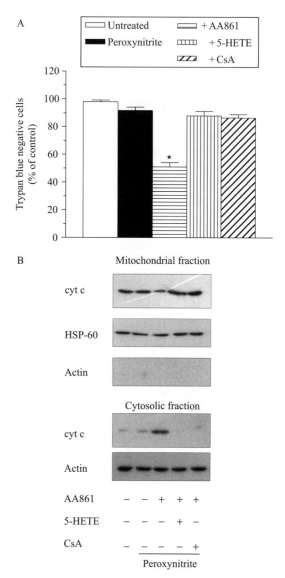

Figure 5.2 Inhibition of survival signaling mediated by an otherwise nontoxic concentration of peroxynitrite induces cell death preceded by the loss of mitochondrial cytochrome *c*. (A) Cells were exposed for 3 min to peroxynitrite (100 μM) and for a further 57 min to AA861 (1 μM) alone or associated with either 5-HETE (0.3 μM) or CsA (0.5 μM). Cytotoxicity was then determined using the trypan blue exclusion assay. (B) Cells were exposed for 3 min to peroxynitrite and for a further 7 min to AA861, 5-HETE, and CsA under the same conditions indicated in A. Cells were then processed to obtain the mitochondrial and cytosolic fractions for Western blot analysis using the antibody against cytochrome *c*. Blots were then washed and reprobed for actin or HSP-60. Results illustrated in A represent the means $\pm$ SEM calculated from four separate experiments. $\star P < 0.01$ compared with untreated cells (one-way ANOVA followed by Dunnett's test).

ACKNOWLEDGMENTS

This work was supported by a grant from the Associazione Italiana per la Ricerca sul Cancro and from Ministero dell'Università e della Ricerca Scientifica e Tecnologica, Progetti di Interesse Nazionale (OC).

REFERENCES

Adams, J., and Cory, S. (1998). The Bcl-2 protein family: Arbiters of cell survival. *Science.* **281,** 1322–1326.

Ashida, Y., Saijo, T., Kuriki, H., Makino, H., Terao, S., and Maki, Y. (1983). Pharmacological profile of AA861, a 5-lipoxygenase inhibitor. *Prostaglandins.* **26,** 955–972.

Bertolotto, C., Maulon, L., Filippa, N., Baier, G., and Auberger, P. (2000). Protein kinase C theta and epsilon promote T-cell survival by a rsk-dependent phosphorylation and inactivation of BAD. *J. Biol. Chem.* **275,** 37246–37250.

Cerioni, L., Palomba, L., Brüne, B., and Cantoni, O. (2006). Peroxynitrite-induced mitochondrial translocation of PKCα causes U937 cell survival. *Biochem. Biophys. Res. Commun.* **339,** 126–131.

Datta, S. R., Dudek, H., Tao, X., Masters, S., Fu, H., Gotoh, Y., and Greenberg, M. E. (1997). Akt phosphorylation of BAD couples survival signals to the cell-intrinsic death machinery. *Cell.* **91,** 231–241.

del Peso, L., Gonzalez-Garcia, M., Page, C., Herrera, R., and Nunez, G. (1997). Interleukin-3-induced phosphorylation of BAD through the protein kinase Akt. *Science.* **278,** 687–689.

Gross, A., Jockel, J., Wei, M. C., and Korsmeyer, S. J. (1998). Enforced dimerization of BAX results in its translocation, mitochondrial dysfunction and apoptosis. *EMBO J.* **17,** 3878–3885.

Guidarelli, A., Cerioni, L., Tommasini, I., Brüne, B., and Cantoni, O. (2005a). A downstream role for protein kinase Calpha in the cytosolic phospholipase A$_2$-dependent protective signaling mediated by peroxynitrite in U937 cells. *Biochem. Pharmacol.* **69,** 1275–1286.

Guidarelli, A., Cerioni, L., Tommasini, I., Fiorani, M., Brüne, B., and Cantoni, O. (2005b). Role of Bcl-2 in the arachidonate-mediated survival signaling preventing mitochondrial permeability transition-dependent U937 cell necrosis induced by peroxynitrite. *Free Radic. Biol. Med.* **39,** 1638–1649.

Halestrap, A. P., Connern, C. P., Griffiths, E. J., and Kerr, P. M. (1997). Cyclosporin A binding to mitochondrial cyclophilin inhibits the permeability transition pore and protects hearts from ischaemia/reperfusion injury. *Mol. Cell. Biochem.* **174,** 167–172.

Hsu, Y. T., Wolter, K. G., and Youle, R. J. (1997). Cytosol-to-membrane redistribution of Bax and Bcl-X(L) during apoptosis. *Proc. Natl. Acad. Sci. USA.* **94,** 3668–3672.

Letai, A., Bassik, M., Walensky, L., Sorcinelli, M., Weiler, S., and Korsmeyer, S. J. (2002). Distinct BH3 domains either sensitize or activate mitochondrial apoptosis, serving as prototype cancer therapeutics. *Cancer Cell.* **2,** 183–192.

Oltvai, Z., Milliman, C., and Korsmeyer, S. J. (1993). Bcl-2 heterodimerizes *in vivo* with a conserved homolog, Bax, that accelerates programmed cell death. *Cell.* **74,** 609–619.

Radi, R., Beckman, J. S., Bush, K. M., and Freeman, B. A. (1991). Peroxynitrite oxidation of sulfhydryls: The cytotoxic potential of superoxide and nitric oxide. *J. Biol. Chem.* **266,** 4244–4250.

Tafani, M., Minchenko, D. A., Serroni, A., and Farber, J. (2001). Induction of the mitochondrial permeability transition mediates the killing of HeLa cells by staurosporine. *Cancer Res.* **61,** 2459–2466.

Tan, Y., Ruan, H., Demeter, M. R., and Comb, M. J. (1999). p90(RSK) blocks Bad-mediated cell death via a protein kinase C-dependent pathway. *J. Biol. Chem.* **274,** 34859–34867.

Tommasini, I., Guidarelli, A., and Cantoni, O. (2004). Non-toxic concentrations of peroxynitrite commit U937 cells to mitochondrial permeability transition-dependent necrosis that is however prevented by endogenous arachidonic acid. *Biochem. Pharmacol.* **67,** 1077–1087.

Tommasini, I., Palomba, L., Guidarelli, A., Cerioni, L., and Cantoni, O. (2006). 5-Hydroxyeicosatetraenoic acid is a key intermediate of the arachidonate-dependent protective signaling in monocytes/macrophages exposed to peroxynitrite. *J. Leukocyte Biol.* **80,** 929–938.

Tommasini, I., Sestili, P., Guidarelli, A., and Cantoni, O. (2002). Peroxynitrite stimulates the activity of cytosolic phospholipase A_2 in U937 cells: The extent of arachidonic acid formation regulates the balance between cell survival or death. *Cell Death Differ.* **9,** 1368–1376.

Zamzami, B. N., Susin, S. A., Marchetti, P., Hirsch, T., Gómez-Monterrey, I., Castedo, M., and Kroemer, G. (1996). Mitochondrial control of nuclear apoptosis. *J. Exp. Med.* **183,** 1533–1544.

Zha, J., Harada, H., Yang, E., Jockel, J., and Korsmeyer, S. J. (1996). Serine phosphorylation of death agonist BAD in response to survival factor results in binding to 14-3-3 not BCL-X(L). *Cell.* **87,** 619–628.

CHAPTER SIX

Practical Approaches to Investigate Redox Regulation of Heat Shock Protein Expression and Intracellular Glutathione Redox State

Vittorio Calabrese,* Anna Signorile,[†] Carolin Cornelius,* Cesare Mancuso,[‡] Giovanni Scapagnini,[§] Bernardo Ventimiglia,[¶] Nicolo' Ragusa,* and Albena Dinkova-Kostova[∥]

Contents

* Department of Chemistry, Clinical Biochemistry and Clinical Molecular Biology Chair, University of Catania, Italy
[†] Department of Biochemistry, University of Bari, Italy
[‡] Institute of Pharmacology, Catholic University School of Medicine, Roma, Italy
[§] Department of Health Sciences, University of Molise, Campobasso, Italy
[¶] Department of Science of Senescence, Urology and Neuro-Urology, University of Catania, Italy
[∥] Biomedical Research Centre, Ninewells Hospital and Medical School, University of Dundee, Scotland, UK and Department of Pharmacology and Molecular Sciences and Department of Medicine, Johns Hopkins University, Baltimore, MD, USA.

Methods in Enzymology, Volume 441

ISSN 0076-6879, DOI: 10.1016/S0076-6879(08)01206-8

Abstract

The products of vitagenes such as heat shock protein 32 (Hsp32, heme oxygenase 1) and Hsp70, the family of inducible cytoprotective proteins regulated by the Keap1/Nrf2/ARE pathway, and small molecule antioxidants such as glutathione provide the cell with powerful means to counteract and survive various conditions of stress. Among these protective systems, the heat shock proteins represent a highly conserved and robust way for preservation of correct protein conformation, recovery of damaged proteins, and cell survival. Their regulation is dependent on the redox status of the cell, thus redox regulation is rapidly evolving as an important metabolic modulator of cellular functions, and is being increasingly implicated in many chronic inflammatory and degenerative diseases. Protein thiols play a key role in redox sensing, and regulation of cellular redox state is crucial mediator of multiple metabolic, signalling and transcriptional processes in the brain. Nitric oxide, and reactive nitrogen species induce the transcription of vitagenes and Keap1/Nrf2/ARE-dependent genes whose functional products protect against a wide array of subsequent challenges. Emerging interest is now focusing on exogenous small molecules that are capable of activating these systems as a novel target to minimize deleterious consequences associated with free radical-induced cell damage, such as during neurodegeneration. This chapter describes methods that can be used to assess the expression of heat shock proteins and the cellular glutathione redox status and discusses their relevance to mechanisms modulating the onset and progression of neurodegenerative diseases.

1. INTRODUCTION

The importance of oxidative stress in the pathogenesis of several diseases, such as neurodegenerative disorders, diabetes, and atherosclerosis, has been well established (Kaneto *et al.*, 2007; Madamanchi and Runge, 2007; Mancuso *et al.*, 2007). Reactive oxygen species (ROS), which are generated as a result of a number of physiological and pathological processes (Bellomo *et al.*, 2006), include superoxide anion ($O_2^{-\bullet}$), hydroxyl radical ($OH^\bullet$), singlet oxygen ($^1O_2$), and hydrogen peroxide (H_2O_2) (Bergamini *et al.*, 2004). Each of these ROS is highly reactive and unstable because of an unpaired electron in their outer electron shell (Bergamini *et al.*, 2004). This increased reactivity promotes the ability of ROS to interact rapidly with cellular macromolecules such as proteins, lipids, and nucleic acids

(Bergamini *et al.*, 2004). If there is an increased level of ROS, as in the case of inflammation, or the intracellular level of antioxidant molecules is low, oxidative stress damage occurs. As a result of oxidative stress, protein, lipid, and DNA oxidation are common features. Oxidative stress can ultimately induce neuronal damage, modulate intracellular signaling, and lead to neuronal death by apoptosis or necrosis (Loh *et al.*, 2006). Interestingly, the propensity or sensitivity of cells to oxidative stress appears to be cell type specific, with cells exhibiting dramatic differences with regards to their sensitivity to accumulate oxidized molecules and cope with toxicity during periods of high ROS exposure. The basis for this cell type specificity is poorly understood but is clearly an important topic for aging, hepatic, cardiovascular, cancer, and neuroscience research. There are several mechanisms by which ROS may be generated, including aerobic respiration, nitric oxide synthesis, and NADPH oxidase pathways during inflammation. In aerobic respiration, the mitochondrial respiratory chain produces ROS as it transfers electrons during the reduction of molecular oxygen to water. During this process, some electrons escape the electron transport chain and interact with oxygen to generate superoxide, hydrogen peroxide, or hydroxyl radical (Calabrese *et al.*, 2007a; Mancuso *et al.*, 2007; Papa *et al.*, 2006a). Activated neutrophils can also release the enzyme myeloperoxidase, which produces the highly active oxidant hypochlorous acid (HOCl) from hydrogen peroxide and chloride ions (Winterbourn *et al.*, 2006). In addition to these biological mechanisms of ROS generation, there are also exogenous sources of free radicals, including pollutants, environmental toxins, cigarette smoke (Churg, 2003).

The nitric oxide synthase (NOS) enzymes produce nitric oxide (NO) via oxidation of the terminal guanidine nitrogen of L-arginine to L-citrulline (Calabrese *et al.*, 2007b). NO is not highly reactive per se but can interact with other intermediates such as oxygen, superoxide, and transition metal generating products that affect the functionality of macromolecules. The term "nitrosative stress" has been used to indicate the cellular damage elicited by nitric oxide and its congeners peroxynitrite, N_2O_3, nitroxyl anion, and nitrosonium (all can be indicated as reactive nitrogen species or RNS) (Mancuso *et al.*, 2006; Ridnour *et al.*, 2004; Kroncke, 2003).

From a molecular point of view, the cell is able to fight against oxidant stress using many resources, including vitamins (A, C, and E), bioactive molecules (glutathione and flavonoids), enzymes [heat shock protein (HSP)-32, superoxide dismutase, catalase, glutathione peroxidases, thioredoxin reductase], and redox-sensitive protein transcriptional factors (AP-1, NFkB, Nrf2, HSF) (Calabrese *et al.*, 2004a, 2007a; Mancuso *et al.*, 2007) (Fig. 6.1). Heat shock proteins (Hsps) are one of the most studied defense system active against cellular damage. Consistently, integrated responses exist in the cells to detect and control diverse forms of stress. This is accomplished by a complex network of the so-called longevity assurance processes, which

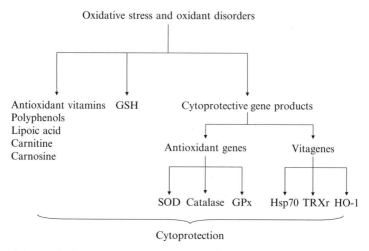

Figure 6.1 Multiple intracellular pathways activated in the fight against oxidative stress. Under free radical attack, significant changes in the intracellular *milieu* occur and several intracellular pathways are activated. Among these, it is noteworthy to mention antioxidant vitamins (A, C and E), glutathione (GSH) and enzymes such as superoxide dismutase (SOD), catalase and glutathione peroxidase (GPx). A main role is also played by the vitagenes heat shock protein 70 (Hsp70) and heme oxygenase-1 (HO-1). All these intracellular systems, by acting in concert to modulate the pro-oxidant/antioxidant balance confer cytoprotection.

are composed of several genes termed *vitagenes* (Calabrese *et al.*, 2004a,c, 2006a, 2007a; Mancuso *et al.*, 2007). Among these, heat shock proteins form a highly conserved system responsible for the preservation and repair of the correct protein conformation. Paradoxically, damage to cells can engage one of two opposing responses: apoptosis, a form of cell death that removes damaged cells to prevent inflammation, and the heat shock or stress response that prevents damage or facilitates recovery to maintain cell survival. Studies have shown that the heat shock response contributes to establishing a cyto-protective state in a wide variety of human diseases, including inflammation, cancer, aging, and neurodegenerative disorders. Given the broad cytopro-tective properties of the heat shock response, there is now strong interest in discovering and developing pharmacological agents capable of inducing the heat shock response (Calabrese *et al.*, 2004a, 2007a; Mancuso *et al.*, 2007). In this paper we describe recent methods employed to measure cellular stress response in a neuronal cell line, as well as the key role played by the heat shock response, particularly the heme oxygenase-1 (also referred to Hsp32) and Hsp70 pathways in brain stress tolerance. Increasing evidence underscores the high potential of the Hsp system as a target for new neuroprotective strategies, especially those aimed at minimizing deleterious consequences associated with oxidative stress, such as in neurodegenerative disorders and brain aging. This chapter also reviews evidence for the emerging role of redox-dependent mechanisms in the regulation of vitagenes in the brain as a potential mechanism to potentiate brain stress tolerance.

2. NITRIC OXIDE AND CELLULAR STRESS RESPONSE: ROLE OF VITAGENES

2.1. The vitagene system

The term vitagenes refers to a group of genes that are strictly involved in preserving cellular homeostasis during stressful conditions. The vitagene family is actually composed of the heat shock proteins Hsp32 and Hsp70 (Calabrese *et al.*, 2004a, 2006a, 2007a; Mancuso *et al.*, 2007) (Fig. 6.2). Among these genes, heme oxygenase-1 (HO-1), is receiving considerable attention because of its major role in counteracting both oxidative and nitrosative stress. In fact, HO-1 induction is one of the earlier events in the cell response to stress. Heme oxygenase-1 exerts a protective role by degrading the intracellular levels of the prooxidant heme and by producing

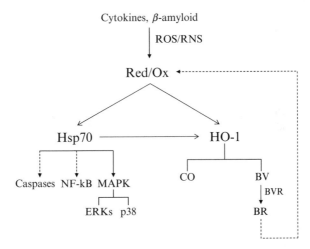

Figure 6.2 The adaptive stress response to oxidative/nitrosative stress and the role of vitagenes. In response to pro-oxidant stimuli, such as cytokines or β-amyloid, free radicals (FR) which include reactive oxygen species (ROS), nitric oxide (NO) and reactive nitrogen species (RNS), induce the expression and/or activity of both heat shock protein 70 (Hsp70) and heme oxygenase-1 (HO-1). Hsp70 inhibits caspases and NFkB, while activating members of the MAPK family, such as ERKs and p38. Conversely, HO-1 converts heme into carbon monoxide (CO) and biliverdin (BV) which is then reduced by biliverdin reductase (BVR) into bilirubin (BR). This latter is a well known scavenger for FR thus allowing a negative feedback to prevent uncontrolled formation of ROS, NO and RNS. Notably, Hsp70 induces HO-1, thus potentiating cytoprotective action. Straight arrows, activation; dashed arrows, inhibition/scavenging.

biliverdin, the precursor of bilirubin; the latter is an endogenous molecule with powerful antioxidant and antinitrosative functions (Calabrese *et al.*, 2007a; Mancuso *et al.*, 2003, 2004, 2006a,b, 2007).

Heat shock protein 70 (Hsp70) is induced in the nervous system following a variety of oxidative injuries, including cerebral ischemia or neurodegenerative disorders (Calabrese *et al.*, 2004a; Giffard *et al.*, 2004). The intracellular mechanisms by which Hsp70 exerts neuroprotection are related to the inhibition of the NF-κB signaling pathway, as well as to the inhibition of both extrinsic and intrinsic apoptotic pathways (Chan *et al.*, 2004; Feinstein *et al.*, 1996; Matsumori *et al.*, 2006). Furthermore, Hsp70 induction has been associated with activation of the ERK1/2 and p38 members of the MAPK system in the rat cerebellum and hippocampus (Maroni *et al.*, 2003) (Fig. 6.2).

2.2. Heme oxygenase: Regulation by nitrosative stress

Heme oxygenase exists in two main isoforms: the inducible one (HO-1) and the constitutive one (HO-2). Heme oxygenase-1 is induced by several stimuli, including oxidative and nitrosative stress, heat shock, lipopolysaccharide, cytokines, inorganic metals, phenolic compounds (i.e., curcumin and ferulic acid) and the neuroprotective agent neotrofin (Fig. 6.2). In contrast, the constitutive HO-2 is responsible for physiologic heme metabolism and responds only to glucocorticoids. Interestingly, evidence suggests a novel role for HO-2 as an endogenous sensor for gaseous signaling molecules such as oxygen CO and NO (Calabrese *et al.*, 2006a; Maines 2005, 1997; Mancuso 2004). The central nervous system is endowed with high HO activity under basal conditions, mostly accounted for by HO-2, with the latter being expressed in neuronal populations in forebrain, hippocampus, hypothalamus, midbrain, basal ganglia, thalamus, cerebellum, and brain stem. The inducible isoform is instead present in small amounts and is localized in sparse groups of neurons, including ventromedial and paraventricular nuclei of the hypothalamus (Maines, 1997; Mancuso, 2004). Heme oxygenase-1 is also found within cells of glial lineage, where its gene expression can be induced by oxidative stress (Mancuso, 2004). In 1997, Maines and her group described a third HO isoform called HO-3 (McCoubrey *et al.*, 1997). In the brain, this isoform is constitutively expressed in astrocytes of the hippocampus, cerebellum, and cortex (Scapagnini *et al.*, 2002). The regulation of HO-3 gene expression and its synthesis is not well understood and its possible role in physiology and pathology remains to be further clarified. With regard to modulation of HO by NO it is important to distinguish between the two HO isoforms and the tissues where this interaction occurs.

Nitric oxide and RNS induce the HO-1 gene and protein in different conditions by a mechanism not fully understood (Motterlini *et al.*, 2002). However, taking into consideration the strong prooxidant activity of NO and RNS, it is plausible to conclude that HO-1 induction has to be considered as a mechanism by which cells can react to stressful conditions. In fact, HO-1 induction by NO is important in selected cells, such as macrophages, for two reasons: first, because HO-1 activity depletes heme from cells; heme is toxic if in excess. Second, the production of bilirubin (BR) and CO through HO activity ensures an efficient scavenging of ROS and RNS and a further inhibition of NADPH-oxidase and inducible NOS (iNOS), thus contributing to the resolution of oxidative conditions (Srisook *et al.*, 2005). In addition, peroxynitrite and nitroxyl anion have been shown to increase, in a dose-dependent manner, HO-1 expression in endothelial cells and human colorectal adenocarcinoma cells (Foresti *et al.*, 1999; Hara *et al.*, 1999; Motterlini *et al.*, 2000; Naughton *et al.*, 2002). In brain cells, NO has been shown to induce HO-1 expression in rat astrocytes and microglia (Kitamura *et al.*, 1998b; Son *et al.*, 2005), as well as in rat hippocampus (Kitamura *et al.*, 1998a). Nitric oxide inhibits or stimulates HO activity, and this differential modulation depends on the tissue or cell line. In particular, studies carried out on endothelial or smooth muscle cells have shown that NO is able to increase HO activity (Durante *et al.*, 1997; Motterlini *et al.*, 1996; Sammut *et al.*, 1998), whereas Willis *et al.* (1995, 1996) demonstrated that NO (released by sodium nitroprusside) reduced HO activity in rat brain and spleen homogenates. The reason for this dual effect of NO on HO activity was clearly explained by Maines (1997) on the basis of the chemical structure of NO: because of its free radical nature, NO can reduce HO activity either by inactivating proteins, in particular those rich in thiol groups such as HO-2, or by forming nitrosyl-heme, which prevents oxygen binding to HO, which is mandatory for its activation. By virtue of these actions, NO can reduce HO activity; this effect is particularly relevant in brain because of the abundance of neuronal HO-2. Meanwhile, the free radical nature of NO can induce HO-1 protein and HO activity, and this biochemical event is important in those cells (endothelial and smooth muscle cells) in which HO-1 is predominant. Furthermore, NO can regulate HO activity by modulating the activity of δ-aminolevulinic acid synthase, the rate-limiting enzyme in heme synthesis, or ferritin, the iron storage protein (Maines, 1997). Moreover, peroxynitrite and nitroxyl anion share with NO the dual effect on HO activity because the first has been shown to decrease HO activity in rat brain or spleen microsomal preparations (Kinobe and Nakatsu, 2004) and increase oxygenase activity in endothelial cells (Foresti *et al.*, 1999), whereas the second increased HO activity in vascular cells (Naughton *et al.*, 2002). Taken together, these data demonstrated that the role of NO and RNS in regulating HO activity strictly depends on the cell type and HO isoform. An interesting corollary emerged by these studies: it has been demonstrated that BR is able to interact with NO and RNS and, as a result of this interaction, formation of a *N*-nitrosated product of BR or the oxidized product BV occurs (Mancuso *et al.*, 2003, 2006a,b). The ability of BR to scavenge NO and RNS is quite important in the brain. In fact, the brain lacks BR conjugating enzymes, allowing the bile pigment to accumulate from neuronal HO-2 activity

(Ewing and Maines, 1992; Snyder and Barañano, 2001). This finding becomes more intriguing in light of the evidence that neurons have relatively low concentrations of glutathione (GSH) (Raps *et al.*, 1989; Slivka *et al.*, 1987; Sun *et al.*, 2006), a tripeptide involved in the detoxification of ROS that is very abundant in almost all mammalian tissues; this suggests a possible role of BR as an alternative endogenous antioxidant molecule in neurons (Baranano and Snyder, 2001). Biliverdin shares with BR this scavenging effect, even if the biological importance of the BV–NO interaction is limited because of the rapid transformation of BV in BR by biliverdin reductase (BVR). Furthermore, even CO, the gaseous product of HO activity, inhibits NO-mediated vasodilation in the adult rat cerebral microcirculation, and this effect is probably a consequence of the photoreversible gas binding to the prosthetic heme of NOS (Ishikawa *et al.*, 2005). Therefore, it is possible to hypothesize a negative feedback between HO products and NO/RNS: in this frame, CO, BV, and BR could act in concert to reduce an unnecessary stimulation of HO by NO (Fig. 6.2).

2.3. Hsp70 and nitrosative stress

The 70-kDa family of stress proteins is one of the most extensively studied. To this family belong the constitutive form or heat shock cognate (Hsc70), the inducible Hsp70 (also referred to as Hsp72), and a constitutively expressed glucose-regulated protein found in the endoplasmic reticulum (GRP75) (Calabrese *et al.*, 2006a). After a variety of central nervous system (CNS) insults, Hsp70 is synthesized at high levels and is present in the cytosol, nucleus, and endoplasmic reticulum (Yenari *et al.*, 1999). A high expression of Hsc70 was observed in neuronal populations in the dentate gyrus, CA1 and CA2 regions of the hippocampus, and Purkinje neurons of the cerebellum (Belay and Brown, 2006). Whether or not stress proteins are neuroprotective has been the subject of much debate, as it has been speculated that these proteins might be merely an epiphenomenon unrelated to cell survival. Only recently, however, with the availability of transgenic animals and gene transfer, has it become possible to overexpress the gene encoding Hsp70 to test directly the hypothesis that stress proteins protect cells from injury, and it has been demonstrated that overproduction of Hsp70 leads to protection in several different models of nervous system injury (Fink *et al.*, 1997; Kelly *et al.*, 2002; Narasimhan *et al.*, 1996). A large body of evidence now suggests a correlation between mechanisms of oxidative and/or nitrosative stress and Hsp induction (Calabrese *et al.*, 2002; Lai *et al.*, 2005; Sultana *et al.*, 2005). Current opinion also holds the possibility that the heat shock response can exert its protective effects through inhibition of the NF-κB signaling pathway (Heneka *et al.*, 2000, 2001) (Fig. 6.2). We have demonstrated that cytokine-induced nitrosative stress in astroglial cell cultures is associated with an increased synthesis of Hsp70 stress proteins (Calabrese *et al.*, 2000, 2005a). The molecular mechanisms regulating the NO-induced activation of the heat shock signal seems to involve the cellular oxidant/antioxidant balance, mainly represented by the glutathione status and the antioxidant enzymes (Calabrese *et al.*, 2000, 2004b). Induction of Hsp70 under stress conditions is often accompanied by the induction of other Hsps, which act in concert to protect neuronal cells from oxidative damage (Fig. 6.2). This paradigm has been recently confirmed in a recent study from our laboratory (Sultana *et al.*, 2005) demonstrating that ferulic acid ethyl esther protects cortical neurons from β-amyloid toxicity by acting at three different levels: (i) inducing HO-1 and Hsp70 proteins; (ii) decreasing the neuronal 3-nitrotyrosine levels and, therefore, iNOS activity; and (iii) by the well-known direct free radical quenching activity (Kanski *et al.*, 2002; Sultana *et al.*, 2005). These data provide consistent evidence that a profound interplay between Hsps exists and further sustain the importance of Hsps in mediating neuroprotective effects.

A growing body of literature has unraveled the antioxidant and anticarcinogenic activities of polyphenolic compounds, such as curcumin. Curcumin, a well-known spice used commonly in India to make foods colored and flavored, is also used in traditional medicine to treat mild or moderate human diseases. Based on the ability of this compound to regulate a number of cellular signal transduction pathways, it is emerging as a potential therapeutic drug for the treatment of neurodegenerative disorders. Particularly interesting is the interaction of curcumin with the vitagene system. In particular, curcumin increased the expression of HO-1 in human cardiac myoblasts, hepatocytes, monocytes, and endothelial cells (Abuarqoub *et al.*, 2007; Jeong *et al.*, 2006; McNally *et al.*, 2006; Rushworth *et al.*, 2006), rat neurons, and astrocytes (Scapagnini *et al.*, 2006), as well as porcine endothelial cells (Balogun *et al.*, 2003). In several rodents and human cells, the curcumin-induced HO-1 overexpression was correlated with the production of mitochondrial ROS, activation of transcription factors Nrf 2 and NF-κB, induction of MAPK p38, and inhibition of phosphatase activity (Andreadi *et al.*, 2006; McNally *et al.*, 2007; Rushworth *et al.*, 2006). Moreover, curcumin upregulated Hsp70 in human colorectal carcinoma cells, proximal tubule cells (Chen *et al.*, 1996, 2001; Rashmi *et al.*, 2003; Sood *et al.*, 2001), and rat glioma cells (Kato *et al.*, 1998).

2.4. Induction of vitagenes by small molecules via the Keap1/Nrf2/ARE pathway

It is noteworthy that the levels of glutathione and HO-1 can be upregulated by small molecules (inducers) via the Keap1/Nrf2/ARE system which is now a very well recognized stress-response pathway that leads to induction of a battery of cytoprotective proteins (Prestera *et al.*, 1995; Motohashi and Yamamoto, 2004; Kobayashi and Yamamoto, 2006; Kensler *et al.*, 2007). One of the recently discovered inducers is nitric oxide (Buckley *et al.*, 2008;

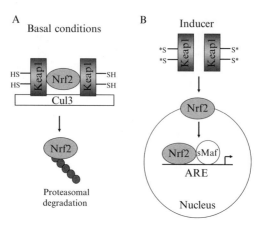

Figure 6.3 Mechanism of induction of cytoprotective proteins. Nuclear factor-erythroid2-related factor 2 (Nrf 2) is a transcription factor responsible for the induction of several phase-2 genes related to the cellular stress response, including HO-1 and GC-ligase. (A) Under basal conditions Keap1 binds and targets transcription factor Nrf2 for ubiquitination and proteasomal degradation via association with Cullin 3 (Cul3)-based E3 ubiquitin ligase. (B) Inducers react and chemically modify specific reactive cysteine residues of Keap1 which consequently loses its ability to repress Nrf 2. This leads to increased stabilization of Nrf 2, its nuclear translocation, binding to the ARE (in combination with small Maf), and ultimately transcriptional activation of cytoprotective (phase 2) genes.

Calabrese *et al.* 2007b). Inducers belong to 10 structurally distinct chemical classes and have a single common property: the ability to react with sulfhydryl groups (Prestera *et al*, 1993; Dinkova-Kostova *et al.*, 2001; 2005). Upon entry into the cell they modify highly reactive cysteine residues of the protein sensor Keap1 (Dinkova-Kostova *et al.*, 2002), which then loses its ability to target transcription factor Nrf2 for ubiquitination and proteasomal degradation (McMahon *et al.*, 2003). Consequently, Nrf 2 undergoes nuclear translocation where it binds in heterodimeric combinations with a small Maf protein to the antioxidant response elements (AREs) that are present in the promoter regions of phase 2 proteins and activates their transcription (Figure 6.3). In addition to HO-1, g-glutamatecysteine ligase, the enzyme catalyzing the rate-limiting step in the glutathione biosynthesis, glutathione reductase, and thioredoxin reductase, enzymes involved in maintaining glutathione in it reduced state, are all inducible through the Keap1/Nrf2/ARE pathway and such induction correlates with protection against various cytotoxic insults (Li *et al.*, 2007; Dinkova-Kostova and Talalay, 2008).

Because both heme oxygenase 1 and the thioredoxin/thioredoxin reductase system can be upregulated in an Nrf 2/ARE-dependent manner, the questions arise whether: (i) the third member of the vitagene family,

Hsp70, is also inducible by other inducers, and (ii) there could be a common regulatory mechanism. Indeed, many inducers of Nrf2-dependent genes have been shown to increase the protein levels of Hsp70. Among them are the cyclopentenone prostaglandin 15-deoxy-^{12,14}-prostaglandin J_2 (15dPGJ$_2$), and the vicinal dithiol reagent phenylarsine oxide, (Zhang et al., 2004; Rokutan et al., 2000). Importantly, all of these compounds react with sulfhydryl groups and the transcriptional activation of both Nrf2 and heat shock factor 1 (HSF1), the major activator of Hsp70 gene expression, depend on cysteine modification either within the Nrf2 regulator Keap1 (Wakabayashi et al., 2004; Dinkova-Kostova et al., 2002), or within HSF1 itself (Liu *et al.*, 1996).

3. MATERIAL AND METHODS

3.1. Chemicals

5,5'-Dithiobis-(2-nitrobenzoic acid) (DTNB), 1,1,3,3-tetraethoxypropane, purified bovine blood SOD, NADH, reduced GSH, oxidized glutathione disulfide (GSSG), β-NADPH (type 1, tetrasodium salt), glutathione reductase (type II from baker's yeast), N^G-monomethyl-L-arginine (a nonisoform-specific NOS inhibitor), and glucose oxidase (which generates hydrogen peroxide in the culture medium) are from Sigma Chemicals Co. (St. Louis, MO). Zinc protoporphyrin IX, a specific inhibitor of HO activity, is from Porphyrin Product (Logan, UT). Acetylcarnitine (99.99% pure) was a generous gift from Sigma Tau Co. (Pomezia, Italy). All other chemicals are from Merck (Germany) and of the highest grade available.

3.2. Cell cultures and treatment

Human neuroblastoma SH-SY5Y cells are seeded at 1×10^4 cells per 1-cm^2 plastic wells with cultured Dulbecco's modified Eagle's medium (DMEM/F-12 Ham) supplemented with 10% fetal calf serum, 1.0 mM glutamine, and 100 units/ml penicillin plus 10 μg/ml streptomycin and incubated in 5% CO$_2$ humidified at 37 °C. The medium is replaced every 3 days. Undifferentiated confluent cells are treated with 0.5 to 3 mM 3-morpholinosydno-nimine (SIN-1) for 7 to 24 h. After treatments, cells are harvested in phosphate-buffered saline (154 mmol/liter of NaCl, 0.61 mmol/liter of Na$_2$HPO$_4$, and 0.38 mmol/liter of KH$_2$PO$_4$, pH 7.4) containing 0.04% (wt/vol) EDTA, washed three times, and stored as pellets at −70 °C until use. The cellular pellet is resuspended in 0.32 M sucrose, 1 mM EDTA, 10 mM Tris (pH 7.4), and 0.5 mM phenylmethylsulfonyl fluoride and homogenized.

3.3. Western blot analysis of heme oxygenase-1, Hsp70, protein carbonyls, and 4-hydroxy-nonennals

Samples of human neuroblastoma SH-SY5Y cells are analyzed for HO-1 and Hsp70 protein expression, as well as protein carbonyl (DPNH) and 4-hydroxynonenal (4-HNE) levels, by using a Western immunoblot technique. Briefly, an equal amount of proteins (30 μg) for each sample is separated by sodium dodecyl sulfate–polyacrylamide gel electrophoresis (SDS-PAGE) and transferred overnight to nitrocellulose membranes, and the nonspecific binding of antibodies is blocked with 3% nonfat dried milk in phosphate-buffered saline (PBS). Immunodetection of HO-1 and Hsp70 protein expression is performed with a polyclonal rabbit anti-HO-1 antibody (SPA-895; Stressgen Biotechnologies, Glanford, Victoria, Canada) and a monoclonal mouse anti-Hsp70 antibody (SPA-810; Stressgen Biotechnologies), respectively. When probed for 4-HNE levels and for DPNH content, membranes are incubated with a rabbit anti-4-hydroxynonenal (HNE11-S, Lost Lane, San Antonio, TX) and rabbit anti-2,4-dinitrophenol antibodies (V0401, Dako, Denmark). A goat polyclonal antibody specific for β-actin is used as a loading control (sc-1615, Santa Cruz Biotechnology). Blots are then visualized with either an amplified horseradish peroxidase-conjugated antirabbit immunoglobulin G (sc-2030, Santa Cruz Biotechnology) when probing for HO-1, 4-HNE, and DPNH, a goat antimouse IgG when probing for Hsp70 (SAB-100, Stressgen Biotechnologies), and donkey antigoat IgG in the case of β-actin (sc-2020, Santa Cruz Biotechnology). Immunoreactive bands are scanned by a laser densitometer (LKB Ultroscan XL). Molecular weights of the proteins detected are determined by using a standard curve obtained with proteins of known molecular weight.

3.4. Preparation of nuclear extract and Western blot for Nrf 2

Samples of human neuroblastoma SH-SY5Y cells are washed twice with PBS, harvested in 1 ml PBS, and centrifuged at 3000 rpm for 3 min at 4 °C. The cell pellet is carefully resuspended in 200 μl of cold buffer A, consisting of 10 mM HEPES (pH 7.9), 10 mM KCl, 0.1 mM EDTA, 0.1 mM EGTA, 1 M dithiothreitol (DTT), and complete protease inhibitor cocktail (Roche, Mannheim, Germany). The pellet is then incubated on ice for 15 min to allow cells to swell. After this time, 15 μl of 10% NP-40 is added and the tube is vortexed for 10 s. The homogenate is then centrifuged at 3000 rpm for 3 min at 4 °C. The resulting nuclear pellet is resuspended in 30 μl of cold buffer B consisting of 20 mM HEPES (pH 7.9), 0.4 M NaCl, 1 mM EDTA, 1 mM EGTA, 1 μM DTT, and protease inhibitors. The pellet is then incubated on ice for 15 min and vortexed for 10 to 15 s every 2 min. The nuclear extract is finally centrifuged at 13,000 rpm for 5 min at 4 °C. The supernatant containing the nuclear proteins is loaded on a SDS-polyacrylamide gel, and Western blot analysis using Nrf 2 antibodies (1:1000 dilution) is performed as described earlier.

3.5. Real-time quantitative polymerase chain reaction

Total RNA from cell cultures is extracted using Trizol (Sigma, St. Louis, MO) and is treated with RNase-free DNase to remove any residual genomic DNA. Single-stranded cDNAs are synthesized by incubating total RNA (1 μg) with SuperScript II RNase H reverse transcriptase (200 U), oligo $(dT)_{12-18}$ primer (100 nM), dNTPs (1 mM), and RNase inhibitor (40 U) at 42 °C for 1 h in a final volume of 20 μl. The reaction is terminated by incubating at 70 °C for 10 min. Forward (F) and reverse (R) primers used to amplify HO-1 were, respectively, HO-1-F: TCTCTTGGCTGG-CTTCCTTA and HO-1-R: ATTGCCTGGATGTGCTTTTC (*GenBank accession no.* NM_002133), and the expected amplification products for HO-1 was 132 bp. To control the integrity of RNA and for differences attributable to errors in experimental manipulation from tube to tube, primers for rat phosphoglycerate kinase 1 (PGK 1) a housekeeping gene that is consistently expressed in brain cells, were used in separate PCR reactions (PGK-F: AGGTGCTCAACAACATGGAG, PGK-R: TACCAGAGGCCACAG-TAGCT, *GenBank accession no.* M31788), and the expected amplification products for this gene was 183 bp, or HO-2 (Figure 6.8) (HO2-F: CCCTTTCTACGCTGCTGAAC, and HO-2-R: TGCTGTCAGAC-GAGGCTCTA (*GenBank accession no.* NM_002134) and the expected amplification products for HO-1 was 317 bp. Aliquots of cDNA (0.1 μg) and known amounts of external standard (purified PCR product, 10^2 to 10^8 copies) are amplified in parallel reactions using the forward and reverse primers. Each PCR reaction (final volume, 20 μl) contains 0.5 μM of primers, 2.5 mM Mg^{2+}, and 1× Light Cycler DNA Master SYBR Green (Roche Diagnostics, Indianapolis, IN). PCR amplifications are performed with a Light Cycler (Roche Molecular Biochemicals) using the following four cycle programs: (i) denaturation of cDNA (one cycle: 95 °C for 10 min); (ii) amplification (40 cycles: 95 °C for 0 s, 58 °C for 5 s, 72 °C for 10 s); (iii) melting curve analysis (one cycle: 95 °C for 0 s, 70 °C for 10 s, 95 °C for 0 s); and (iv) cooling (one cycle: 40 °C for 3 min). The temperature transition rate is 20 °C/s except for the third segment of the melting curve analysis, where it is 0.2 °C/s. The fluorimeter gain value is 6. Real-time detection of fluorimetric intensity of SYBR Green I, indicating the amount of PCR product formed, is measured at the end of each elongation phase. Quantification is performed by comparing the fluorescence of PCR products of unknown concentration with the fluorescence of the external standards. For this analysis, fluorescence values measured in the log-linear phase of amplification are considered using the second derivative maximum method of the Light Cycler Data Analysis software (Roche Molecular Biochemicals). Specificity of PCR products obtained is characterized by melting curve analysis followed by gel electrophoresis, visualized by ethidium bromide staining, and DNA sequencing.

3.6. Total reduced and oxidized glutathione assay

Total (i.e., cytosolic and mitochondrial) reduced GSH and total GSSG are measured by the NADPH–dependent GSSG reductase method, modified as follows. Cells are homogenized on ice for 10 s in 100 mM potassium phosphate, pH 7.5, which contains 12 mM disodium EDTA. For reduced glutathione, aliquots (0.1 ml) of homogenates are immediately added to 0.1 ml of a cold solution containing 10 mM DTNB and 5 mM EDTA in 100 mM potassium phosphate, pH 7.5. The samples are mixed by tilting and are centrifuged at 12,000g for 2 min at 4 °C. An aliquot (50 μl) of the supernatant is added to a cuvette containing 0.5 U of GSSG reductase in 100 mM potassium phosphate and 5 mM EDTA, pH 7.5 (buffer 1). After 1 min of equilibration, the reaction is initiated with 220 nmol of NADPH in buffer 1 for a final reaction volume of 1 ml. Formation of a GSH–DTNB conjugate is then measured at 412 nm. The reference cuvette contains equal concentrations of DTNB, NADPH, and enzyme, but not sample. For assay of total GSSG, aliquots (0.5 ml) of homogenate are immediately added to 0.5 ml of a solution containing 10 mM N-ethylmaleimide (NEM) and 5 mM EDTA in 100 mM potassium phosphate, pH 7.5. The sample is mixed by tilting and is centrifuged at 12,000g for 2 min at 4 °C. An aliquot (500 μl) of the supernatant is passed at one drop per second through a Sep-Pak C18 column (Waters, Framingham, MA) that has been washed with methanol followed by water. The column is then washed with 1 ml of buffer 1. Aliquots (865 μl) of the combined eluates are added to a cuvette with 250 nmol of DTNB and 0.5 U of GSSG reductase. The assay then proceeds as in the measurement of total GSH. GSH and GSSG standards in the ranges between 0 and 10 nmol and 0.010 and 10 nmol, respectively, added to control samples are used to obtain the relative standard curves, and results are expressed in nanomoles of GSH or GSSG, respectively, per milligram of protein.

3.7. Determination of reduced glutathione and oxidized glutathione in the cytosol

Cells are harvested from petri dishes with 0.05% trypsin, 0.02% EDTA and washed in phosphate-buffered saline, pH 7.4, with 5% fetal bovine serum. Cells are collected by centrifugation at 500g for 3 min at room temperature, and the pellet ($\approx$5 × 106 cells) is resuspended in 0.5 ml of 100 mM potassium phosphate buffer, pH 7.5, containing 5 mM EDTA (buffer 1). The cell suspension is homogenized by five strokes; the temperature of the suspension is maintained at 4 °C during homogenization and all subsequent steps. The homogenate is divided into two aliquots; for the total glutathione (GSH + GSSG) assay, 0.25 ml of homogenate is added to an equal volume of 100 mM potassium phosphate buffer, pH 7.5, containing 17.5 mM EDTA and 10 mM DTNB (sample SS1). For the oxidized glutathione (GSSG) assay, 0.25 ml of homogenate is added to 100 mM potassium phosphate buffer, pH 6.5, containing 17.5 mM EDTA and 10 mM NEM (sample SS2). The samples are centrifuged at 800g for 20 min, and the supernatant fractions are then centrifuged at 10,000g for 30 min. The supernatants of SS1 and SS2 represent the cytosolic fractions and are used for the spectrophotometric assay of total or oxidized glutathione. Before spectrophotometric determination, a 0.25-ml aliquot of the SS2 sample is passed through a C18 Sep-Pak cartridge (Waters, Watford, UK) to remove excess NEM and washed with 0.5 ml of buffer (100 mM potassium phosphate buffer, pH 7.5, containing 5 mM EDTA). The spectrophotometric assay of glutathione is performed by adding the samples to a cuvette containing 0.5 unit of glutathione reductase, 0.2 mM DTNB in a final volume of 1 ml of 100 mM potassium phosphate buffer, pH 7.5, 5 mM EDTA and the reaction is initiated by adding NADPH (220 nmol). The change in absorbance at 412 nm is recorded over a period of 5 min for the SS1 sample or 10 min for the SS2 sample using a reference cuvette containing equal concentrations of NADPH, DTNB, and enzyme. The GSH and GSSG content, expressed as nanomoles per milligram of protein, is determined by comparison with a standard curve obtained with GSH and GSSG solution.

3.8. Determination of total glutathione (GSH + GSSG) and oxidized glutathione (GSSG) in mitochondria

The pellet of SS1 and SS2 samples obtained after centrifugation at 10,000g is utilized for the determination of total and oxidized glutathione in mitochondria. The SS1 pellet is resuspended in 0.3 ml of 100 mM potassium phosphate buffer, pH 7.5, 10 mM EDTA, 5 mM DTNB. The SS2 pellet is resuspended in 0.5 ml of 100 mM potassium phosphate buffer, pH 6.5, containing 10 mM EDTA and 5 mM NEM. Mitochondria are mixed, sonicated, and centrifuged at 10,000g for 11 min, and the supernatants are used for the spectrophotometric determination assay. Before spectrophotometric determination, 0.25 ml of the SS2 sample is passed through a C18 Sep-Pak cartridge to remove excess NEM and is washed with 0.5 ml of buffer 100 mM potassium phosphate buffer, pH 7.5, containing 5 mM EDTA.

Spectrophotometric assay of glutathione is performed by adding the samples to a cuvette containing 0.5 unit of glutathione reductase, 0.2 mM DTNB in a final volume of 1 ml of 100 mM potassium phosphate buffer, pH 7.5, 5 mM EDTA, and the reaction is initiated by adding NADPH

(220 n*M*). The change in absorbance at 412 nm is recorded over a period of 5 min for the SS1 sample or 10 min for the SS2 sample using a reference cuvette containing equal concentrations of NADPH, DTNB, and enzyme. Protein concentration is determined in the samples with NEM, according to the Bradford method, using bovine serum albumin as standard. The GSH and GSSG content, expressed as nanomoles per milligram of protein, is determined by comparison with a standard curve obtained with known concentrations of GSH and GSSG solution.

3.9. Determination of protein

Proteins are estimated by the bicinchoninic acid reagent.

3.10. Statistical examination

Results are expressed as means ± SEM of at least eight separate experiments. Statistical analyses are performed using the software package SYSTAT (Systat Inc., Evanston, IL). The significance of the differences, evaluated by two-way ANOVA, followed by Duncan's new multiple range test, is considered significant at $P < 0.05$. Correlation analysis is considered statistically significant if the coefficient of determination R is ≥ 0.8.

4. RESULTS

As mentioned previously, tissues, particularly brain, counteract oxidative stress by several mechanisms, the most important of which are antioxidant molecules, such as GSH, and antioxidant enzymes, including HO-1 and Hsp70. In a human neuroblastoma cell line (SH-SY5Y), nitrosative stress induced by the administration of SIN-1 (1 m*M*), which produces both NO and superoxide, thus resulting in increased peroxynitrite formation, resulted in a significant decrease in the total content of GSH associated with significantly increased GSSG levels (Fig. 6.4). A further dissection of this effect demonstrated that SIN-1 affected the GSH/GSSG balance, particularly in the cytosol, whereas mitochondria were not a target for this compound (Fig. 6.4). Furthermore, SIN-1 (0.5–2 m*M* for 7 or 24 h) increased other indices of oxidative stress, such as protein carbonyls (DNPH) and lipid peroxidation products (4-HNE) (Figs. 6.5 and 6.6). This documented prooxidant effect of SIN-1 was paralleled by upregulation of both HO-1 and Hsp70 protein levels in SH-SY5Y cells (Fig. 6.7). The increased protein expression of HO-1 in SH-SY5Y cells exposed to SIN-1 was accompanied by increased message expression (Fig. 6.8). Results shown in Fig. 6.8 demonstrate that at the concentration used, SIN-1 is sufficient to increase the expression of HO-1 at

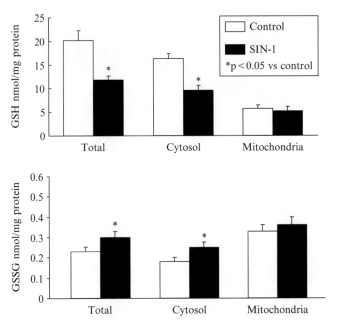

Figure 6.4 Effect of SIN-1 treatment on GSH and GSSG content in SH-SY5Y cells. Human neuroblastoma cells (SH-SY5Y) were treated with SIN-1 (1 mM) for 7 h. At the end of incubation, cells were collected and lysed, and the amount of total, cytosolic, and mitochondrial GSH and GSSG was analyzed as described in the text. Data are expressed as mean ± SEM of eight independent experiments.

the mRNA (measured at 3 h from the beginning of the treatment, Figs. 6.8D and 6.8F) and protein levels and is sufficient to increase the activity of HO-1 (data not shown). All this thus enforces the hypothesis that cells upregulate vitagenes to effectively counteract oxidative and nitrosative stress.

5. DISCUSSION

Mitochondrial dysfunction is characteristic of several neurodegenerative disorders, and evidence for mitochondria being a site of damage in neurodegenerative disorders is partially based on decreases in respiratory chain complex activities in such diseases (Calabrese *et al.*, 2001; Mancuso *et al.*, 2007; Papa *et al.*, 2006; Scacco *et al.*, 2006). Such defects in respiratory complex activities, possibly associated with oxidant/antioxidant balance perturbation, are thought to underlie defects in energy metabolism and induce cellular degeneration. Evidence that mitochondrial dysfunction

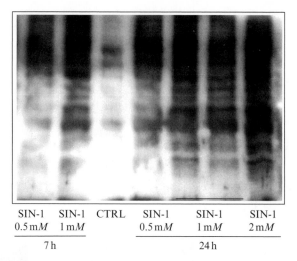

SIN-1   SIN-1   CTRL   SIN-1   SIN-1   SIN-1
0.5 m*M*   1 m*M*   0.5 m*M*   1 m*M*   2 m*M*

   7 h   24 h

Figure 6.5   SIN-1 increases protein oxidation in neuronal cells. Human neuroblastoma cells (SH–SY5Y) were treated with SIN-1 (0.5–2 m*M*) for 7 to 24 h. At the end of incubation, cells were collected, lysed, and immunoblotted using a specific antibody against DNPH as described in the text. Each experiment was performed at least three times. A representative gel is shown.

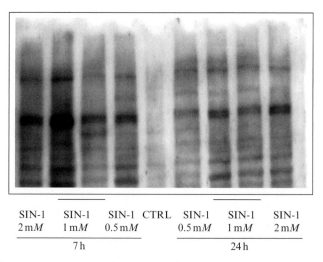

SIN-1   SIN-1   SIN-1   CTRL   SIN-1   SIN-1   SIN-1
2 m*M*   1 m*M*   0.5 m*M*   0.5 m*M*   1 m*M*   2 m*M*

         7 h   24 h

Figure 6.6   SIN-1 increases lipid oxidation in neuronal cells. Human neuroblastoma cells (SH–SY5Y) were treated with SIN-1 (0.5–2 m*M*) for 7 to 24 h. At the end of incubation, cells were collected, lysed, and immunoblotted using a specific antibody against 4-HNE as described in the text. Each experiment was performed at least three times. A representative gel is shown.

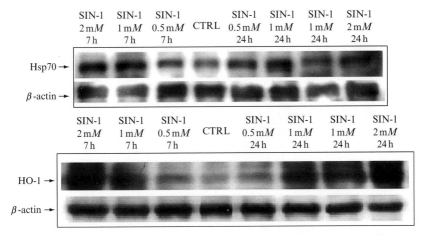

Figure 6.7 SIN-1 upregulates vitagenes. Human neuroblastoma cells (SH-SY5Y) were treated with SIN-1 (0.5–2 mM) for 7 to 24 h. At the end of incubation, cells were collected, lysed, and immunoblotted using specific antibodies against HO-1 and Hsp70 as described in the text. Each experiment was performed at least three times. A representative gel is shown.

may be a mechanism for NO-mediated neurotoxicity arises from different studies that indicate excessive production of NO, a free radical that has several important messenger functions within the CNS, leads to the formation of peroxynitrite anion (ONOO⁻) by reacting with the superoxide anion (Calabrese *et al.*, 2007b). This extremely potent oxidizing agent can interact at the binuclear center of cytochrome oxidase, leading to inhibition of the respiratory rate and ATP stores (Calabrese *et al.*, 2001). NO· can also stimulate the *S*-nitrosylation of protein and nonprotein thiols and also binds to the nonheme iron of ribonucleotide reductase to inhibit DNA synthesis (Gegg *et al.*, 2003).

Efficient functioning of the maintenance and repair process seems to be crucial for both survival and physical quality of life. This is accomplished by a complex network of the so-called longevity assurance processes, which are composed of several genes termed *vitagenes* (Calabrese *et al.*, 2004a, 2006a, 2007a). Among these, chaperones are highly conserved proteins responsible for the preservation and repair of the correct conformation of cellular macromolecules, such as proteins, RNAs. and DNA. Heat shock proteins and molecular chaperones have been known to protect cells against a wide variety of toxic conditions, including extreme temperatures, oxidative stress, virus infection, and exposure to heavy metals or cytotoxic drugs (Calabrese *et al.*, 2004a, 2006a, 2007a). Chaperone-buffered silent mutations may be activated during the aging process and lead to the phenotypic exposure of previously hidden features and contribute to the onset of polygenic diseases, such as age-related disorders, atherosclerosis, and cancer

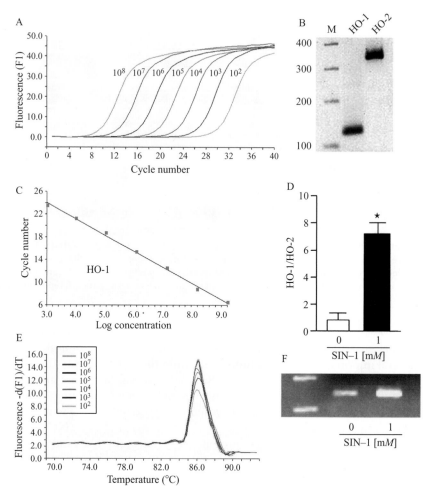

Figure 6.8 Real-time quantification of HO-1 and HO-2 mRNA levels by RT-PCR in SH-SY5Y cells treated with SIN-1. Specific primers for HO-1 and HO-2 were used to amplify human brain RNA (Fig. 6.8 A–E). Total RNA from different samples and known amounts of external standards (purified PCR product, 102 to 108 copies) were amplified in parallel reaction. Fluorimetric intensity of SYBR Green I, indicating the amount of PCR product formed, was measured at the end of each elongation phase. Quantification was performed by comparing the fluorescence of PCR products of unknown concentration with the fluorescence of the external standards (A). Fluorescence values measured in the log-linear phase of amplification were measured by the second derivative maximum method and used to produce standard curves (C) that were used to estimate the concentration of unknown samples. The specificity of the products amplified was evaluated by melting curve analysis (E). Cellular expression of HO-1 transcript relative to the expression of HO-2 (mean ± SEM) after treatment with SIN-1 1mM is shown (D). *$p < 0.05$ versus 0 mM SIN-1. The amplified fragments for HO-1 was run on 2% agarose gel (B). A 100-bp DNA ladder is shown at the left of the gel (M), with bands labelled in bp units. Exposure of SH-SY5Y cells to SIN-1 significantly ($p < 0.05$) up-regulates mRNA levels measured 3 hours after SIN-1 addition (panels D,F).

(Soti and Csermely, 2003). Hence, Hsp induction is not only a signal for the detection of physiological stress, but is utilized by the cells in the repair process following a wide range of injuries to prevent damage resulting from the accumulation of nonnative proteins (Kelly and Yenari, 2002).

Involvement of the heme oxygenase pathway in antidegenerative mechanisms, especially those operating in Alzheimer's disease, has been demonstrated: overexpression of HO-1 has been observed in association with neurofibrillary tangles and senile plaques (Premkumar et al., 1995; Schipper, 2000; Takeda et al., 2000). HO induction, which occurs together with the induction of other Hsp during various physiopathological conditions by generating the vasoactive molecule CO and the potent antioxidant BR, represents a protective system potentially active against brain oxidative injury (Mancuso et al., 2004, 2006a,b, 2007). The HO-1 gene is redox regulated; this is supported by the fact that the HO-1 gene has a heat shock consensus sequence as well as AP1, AP2, and NF-κB binding sites in its promoter region. In addition, heme oxygenase-1 is upregulated rapidly by oxidative and nitrosative stress, as well as by glutathione depletion (Mancuso et al., 2007; Motterlini et al, 2002).

All this evidence emphasizes the well-established concept of the cellular stress response to oxidative insults as a crucial mechanism operating against neurodegenerative damage (Calabrese et al., 2004b, 2006a, 2007a; Mancuso et al., 2007). However, relatively new is the notion that pharmacological or nutritional intervention can lead to the same cytoprotective cellular responses (Butterfield et al., 2002; Calabrese et al., 2003a,b). Acetyl-l-carnitine(LAC) is proposed as a therapeutic agent for several neurodegenerative disorders, as well as an agent protective in numerous disease paradigms; however, the mechanism of protection in brain disorders still remains elusive (Calabrese et al., 2007b). Furthermore, experimental evidence exists that upregulation of HO-1 involves the transcription factor Nrf 2 (see earlier discussion). These data, along with evidence that LAC induces Nrf 2 in rat astrocytes (Calabrese et al., 2005b), suggest the potential impact on dietary antioxidants in the HO-1/Nrf 2 axis. Therefore, this highly inducible system should be seriously considered as a target for novel therapeutic interventions focusing on the capability that compounds such as antioxidant polyphenols or acetylcarnitine have to upregulate the vitagene system as a means to limit the deleterious consequences of oxidative and nitrosative stress associated with aging and age-related disorders.

ACKNOWLEDGMENT

This work was supported by grants of PRIN 2005, FIRB RBRN07BMCT, and ICT-E1 Comune di Catania.

REFERENCES

Abuarqoub, H., Green, C. J., Foresti, R., and Motterlini, R. (2007). Curcumin reduces cold storage-induced damage in human cardiac myoblasts. *Exp. Mol. Med.* **39,** 139–148.

Andreadi, C. K., Howells, L. M., Atherfold, P. A., and Manson, M. M. (2006). Involvement of Nrf 2, p38, B-Raf, and nuclear factor-kappaB, but not phosphatidylinositol 3-kinase, in induction of hemeoxygenase-1 by dietary polyphenols. *Mol. Pharmacol.* **69,** 1033–1040.

Balogun, E., Foresti, R., Green, C. J., and Motterlini, R. (2003). Changes in temperature modulate heme oxygenase-1 induction by curcumin in renal epithelial cells. *Biochem. Biophys. Res. Commun.* **308,** 950–955.

Barañano, D. E., and Snyder, S. H. (2001). Neural roles for heme oxygenase: Contrasts to nitric oxide synthase. *Proc. Natl. Acad. Sci. USA* **98,** 10996–11002.

Belay, H. T., and Brown, I. R. (2006). Cell death and expression of heat-shock protein Hsc70 in the hyperthermic rat brain. *J. Neurochem.* **97,** 116–119.

Bellomo, F., Piccoli, C., Cocco, T., Scacco, S., Papa, F., Gabello, A., Boffoli, D., Signorile, A., D'Aprile, A., Scrima, R., Sardanelli, A. M., Capitanio, N., *et al.* (2006). Regulation by the cAMP cascade of oxygen free radical balance in mammalian cells. *Antioxid. Redox Signal.* **8,** 495–502.

Bergamini, C. M., Gambetti, S., Dondi, A., and Cervellati, C. (2004). Oxygen, reactive oxygen species and tissue damage. *Curr. Pharm. Des.* **10,** 1611–1626.

Butterfield, D., Castegna, A., Pocernich, C., Drake, J., Scapagnini, G., and Calabrese, V. (2002). Nutritional approaches to combat oxidative stress in Alzheimer's disease. *J. Nutr. Biochem.* **13,** 444.

Calabrese, V., Boyd-Kimball, D., Scapagnini, G., and Butterfield, D. A. (2004a). Nitric oxide and cellular stress response in brain aging and neurodegenerative disorders: The role of vitagenes. *In Vivo* **18,** 245–267.

Calabrese, V., Butterfield, D. A., Scapagnini, G., Stella, A. M., and Maines, M.D (2006a). Redox regulation of heat shock protein expression by signaling involving nitric oxide and carbon monoxide: Relevance to brain aging, neurodegenerative disorders, and longevity. *Antioxid. Redox Signal.* **8,** 444–477.

Calabrese, V., Butterfield, D. A., and Stella, A. M. (2003a). Nutritional antioxidants and the heme oxygenase pathway of stress tolerance: Novel targets for neuroprotection in Alzheimer's disease. *Ital. J. Biochem.* **52,** 177–181.

Calabrese, V., Colombrita, C., Guagliano, E., Sapienza, M., Ravagna, A., Cardile, V., Scapagnini, G., Santoro, A. M., Mangiameli, A., Butterfield, D. A., Giuffrida Stella, A. M., and Rizzarelli, E. (2005a). Protective effect of carnosine during nitrosative stress in astroglial cell cultures. *Neurochem. Res.* **30,** 797–807.

Calabrese, V., Copani, A., Testa, D., Ravagna, A., Spadaro, F., Tendi, E., Nicoletti, V. G., and Giuffrida Stella, A. M. (2000). Nitric oxide synthase induction in astroglial cell cultures: Effect on heat shock protein 70 synthesis and oxidant/antioxidant balance. *J. Neurosci. Res.* **60,** 613–622.

Calabrese, V., Guagliano, E., Sapienza, M., Panebianco, M., Calafato, S., Puleo, E., Pennisi, G., Mancuso, C., Butterfield, D. A., and Stella, A. G. (2007a). Redox regulation of cellular stress response in aging and neurodegenerative disorders: Role of vitagenes. *Neurochem. Res.* **32,** 757–773.

Calabrese, V., Mancuso, C., Calvani, M., Rizzarelli, E., Butterfield, D. A., and Giuffrida Stella, A. M. (2007b). Nitric oxide in the central nervous system: Neuroprotection versus neurotoxicity. *Nat. Rev. Neurosci.* **8,** 766–775.

Calabrese, V., Ravagna, A., Colombrita, C., Scapagnini, G., Guagliano, E., Calvani, M., Butterfield, D. A., and Giuffrida Stella, A. M. (2005b). Acetylcarnitine induces heme oxygenase in rat astrocytes and protects against oxidative stress: Involvement of the transcription factor Nrf 2. *J. Neurosci. Res.* **79**, 509–521.

Calabrese, V., Scapagnini, G., Colombrita, C., Ravagna, A., Pennisi, G., Giuffrida Stella, A. M., Galli, F., and Butterfield, D. A. (2003b). Redox regulation of heat shock protein expression in aging and neurodegenerative disorders associated with oxidative stress: A nutritional approach. *Amino Acids* **25**, 437–444.

Calabrese, V., Scapagnini, G., Giuffrida Stella, A. M., Bates, T. E., and Clark, J. B. (2001). Mitochondrial involvement in brain function and dysfunction: Relevance to aging, neurodegenerative disorders and longevity. *Neurochem. Res.* **26**, 739–764.

Calabrese, V., Scapagnini, G., Ravagna, A., Colombrita, C., Spadaro, F., Butterfield, D. A., and Giuffrida Stella, A. M. (2004c). Increased expression of heat shock proteins in rat brain during aging: Relationship with mitochondrial function and glutathione redox state. *Mech. Aging Dev.* **125**, 325–335.

Calabrese, V., Scapagnini, G., Ravagna, A., Fariello, R. G., Giuffrida Stella, A. M., and Abraham, N. G. (2002). Regional distribution of heme oxygenase, HSP70, and glutathione in brain: Relevance for endogenous oxidant/antioxidant balance and stress tolerance. *J. Neurosci. Res.* **68**, 65–75.

Calabrese, V., Stella, A. M., Butterfield, D. A., and Scapagnini, G. (2004b). Redox regulation in neurodegeneration and longevity: Role of the heme oxygenase and HSP70 systems in brain stress tolerance. *Antioxid. Redox Signal.* **6**, 895–913.

Chan, J. Y., Ou, C. C., Wang, L. L., and Chan, S. H. (2004). Heat shock protein 70 confers cardiovascular protection during endotoxemia via inhibition of nuclear factor-kappaB activation and inducible nitric oxide synthase expression in the rostral ventrolateral medulla. *Circulation* **110**, 3560–3566.

Chen, Y. C., Kuo, T. C., Lin-Shiau, S. Y., and Lin, J. K. (1996). Induction of HSP70 gene expression by modulation of Ca(+2) ion and cellular p53 protein by curcumin in colorectal carcinoma cells. *Mol. Carcinog* **17**, 224–234.

Chen, Y. C., Tsai, S. H., Shen, S. C., Lin, J. K., and Lee, W. R. (2001). Alternative activation of extracellular signal-regulated protein kinases in curcumin and arsenite-induced HSP70 gene expression in human colorectal carcinoma cells. *Eur. J. Cell. Biol.* **80**, 213–221.

Churg, A. (2003). Interactions of exogenous or evoked agents and particles: The role of reactive oxygen species. *Free Radic. Biol. Med.* **34**, 1230–1235.

Dinkova-Kostova, A. T., Holtzclaw, W. D., Cole, R. N., Itoh, K., Wakabayashi, N., Katoh, Y., Yamamoto, M., and Talalay, P. (2002). Direct evidence that sulfhydryl groups of Keap1 are the sensors regulating induction of phase 2 enzymes that protect against carcinogens and oxidants. *Proc. Natl. Acad. Sci. USA* **99**, 11908–11913.

Dinkova-Kostova, A. T., Holtzclaw, W. D., and Kensler, T. W. (2005). The role of Keap1 in cellular protective responses. *Chem. Res. Toxicol.* **18**, 1779–1791.

Dinkova-Kostova, A.T, and Talalay, P (1999). Relation of structure of curcumin analogs to their potencies as inducers of phase 2 detoxification enzymes. *Carcinogenesis.* **20**, 911–914.

Durante, W., Kroll, M. H., Christodoulides, N., Peyton, K. J., and Schafer, A. I. (1997). Nitric oxide induces heme oxygenase-1 gene expression and carbon monoxide production in vascular smooth muscle cells. *Circ. Res.* **80**, 557–564.

Ewing, J. F., and Maines, M. D. (1992). *In situ* hybridization and immunohistochemical localization of heme oxygenase-2 mRNA and protein in normal rat brain: Differential distribution of isozyme 1 and 2. *Mol. Cell. Neurosci.* **3**, 559–570.

Feinstein, D. L., Galea, E., Aquino, D. A., Li, G. C., Xu, H., and Reis, D. J. (1996). Heat shock protein 70 suppresses astroglial-inducible nitric-oxide synthase expression by decreasing NFkappaB activation. *J. Biol. Chem.* **271**, 17724–17732.

Fink, S. L., Chang, L. K., Ho, D. Y., and Sapolsky, R. M. (1997). Defective herpes simplex virus vectors expressing the rat brain stress-inducible heat shock protein 72 protect cultured neurons from severe heat shock. *J. Neurochem.* **68,** 961–969.

Foresti, R., Sarathchandra, P., Clark, J. E., Green, C. J., and Motterlini, R. (1999). Peroxynitrite induces haem oxygenase-1 in vascular endothelial cells: A link to apoptosis. *Biochem. J.* **339,** 729–736.

Gegg, M. E., Beltran, B., Salas-Pino, S., Bolanos, J. P., Clark, J. B., Moncada, S., and Heales, S. J. (2003). Differential effect of nitric oxide on glutathione metabolism and mitochondrial function in astrocytes and neurones: Implications for neuroprotection/neurodegeneration? *J. Neurochem.* **86,** 228–237.

Giffard, R. G., and Yenari, M. A. (2004). Many mechanisms for hsp70 protection from cerebral ischemia. *J. Neurosurg. Anesthesiol.* **16,** 53–61.

Hara, E., Takahashi, K., Takeda, K., Nakayama, M., Yoshizawa, M., Fujita, H., Shirato, K., and Shibahara, S. (1999). Induction of heme oxygenase-1 as a response in sensing the signals evoked by distinct nitric oxide donors. *Biochem. Pharmacol.* **58,** 227–236.

Heneka, M. T., Sharp, A., Klockgether, T., Gavrilyuk, V., and Feinstein, D. L. (2000). The heat shock response inhibits NF-kappaB activation, nitric oxide synthase type 2 expression, and macrophage/microglial activation in brain. *J. Cereb. Blood Flow Metab.* **20,** 800–811.

Heneka, M. T., Sharp, A., Murphy, P., Lyons, J. A., Dumitrescu, L., and Feinstein, D. L. (2001). The heat shock response reduces myelin oligodendrocyte glycoprotein-induced experimental autoimmune encephalomyelitis in mice. *J. Neurochem.* **77,** 568–579.

Ishikawa, M., Kajimura, M., Adachi, T., Maruyama, K., Makino, N., Goda, N., Yamaguchi, T., Sekizuka, E., and Suematsu, M. (2005). Carbon monoxide from heme oxygenase-2 is a tonic regulator against NO-dependent vasodilatation in the adult rat cerebral microcirculation. *Circ. Res.* **97,** e104–e114.

Jeong, G. S., Oh, G. S., Pae, H. O., Jeong, S. O., Kim, Y. C., Shin, M. K., Seo, B. Y., Han, S. Y., Lee, H. S., Jeong, J. G., Koh, J. S., and Chung, H. T. (2006). Comparative effects of curcuminoids on endothelial heme oxygenase-1 expression: Ortho-methoxy groups are essential to enhance heme oxygenase activity and protection. *Exp. Mol. Med.* **38,** 393–400.

Kaneto, H., Katakami, N., Kawamori, D., Miyatsuka, T., Sakamoto, K., Matsuoka, T. A., Matsuhisa, M., and Yamasaki, Y. (2007). Involvement of oxidative stress in the pathogenesis of diabetes. *Antioxid. Redox Signal.* **9,** 355–366.

Kanski, J., Aksenova, M., Stoyanova, A., and Butterfield, D. A. (2002). Ferulic acid antioxidant protection against hydroxyl and peroxyl radical oxidation in synaptosomal and neuronal cell culture systems *in vitro*: Structure-activity studies. *J. Nutr. Biochem.* **13,** 273–281.

Kato, K., Ito, H., Kamei, K., and Iwamoto, I. (1998). Stimulation of the stress-induced expression of stress proteins by curcumin in cultured cells and in rat tissues *in vivo*. *Cell Stress Chaperones* **3,** 152–160.

Kelly, S., and Yenari, M. A. (2002). Neuroprotection: Heat shock proteins. *Curr. Med. Res. Opin.* **18**(Suppl. 2), 55–60.

Kelly, S., Zhang, Z. J., Zhao, H., Xu, L., Giffard, R. G., Sapolsky, R. M., Yenari, M. A., and Steinberg, G. K. (2002). Gene transfer of HSP72 protects cornu ammonis 1 region of the hippocampus neurons from global ischemia: Influence of Bcl-2. *Ann. Neurol.* **52,** 160–167.

Kensler, T. W., Wakabayashi, N., and Biswal, S. (2007). Cell survival responses to environmental stresses via the Keap1-Nrf 2-ARE pathway. *Annu. Rev. Pharmacol. Toxicol.* **47,** 89–116.

Kinobe, R., Ji, Y., and Nakatsu, K. (2004). Peroxynitrite-mediated inactivation of heme oxygenases. *BMC Pharmacol.* **4,** 26.

Kitamura, Y., Furukawa, M., Matsuoka, Y., Tooyama, I., Rimura, H., Nomura, Y., and Taniguchi, T. (1998a). *In vitro* and *in vivo* induction of heme oxygenase-1 in rat glial cells: Possible involvement of nitric oxide production from inducible nitric oxide synthase. *Glia.* **22**, 138–148.

Kitamura, Y., Matsuoka, Y., Nomura, Y., and Taniguchi, T. (1998b). Induction of inducible nitric oxide synthase and heme oxygenase-1 in rat glial cells. *Life Sci.* **62**, 1717–1721.

Kobayashi, M., and Yamamoto, M. (2006). Nrf 2-Keap1 regulation of cellular defense mechanisms against electrophiles and reactive oxygen species. *Adv. Enzyme Regul.* **46**, 113–140.

Kroncke, K. D. (2003). Nitrosative stress and transcription. *Biol. Chem.* **384**, 1365–1377.

Lai, Y., Du, L., Dunsmore, K. E., Jenkins, L. W., Wong, H. R., and Clark, R. S. (2005). Selectively increasing inducible heat shock protein 70 via TAT-protein transduction protects neurons from nitrosative stress and excitotoxicity. *J. Neurochem.* **94**, 360–366.

Liu, H., Lightfoot, R., and Stevens, J. L. (1996). Activation of heat shock factor by alkylating agents is triggered by glutathione depletion and oxidation of protein thiols. *J. Biol. Chem.* **271**, 4805–4812.

Loh, K. P., Huang, S. H., De Silva, R., Tan, B. K., and Zhu, Y. Z. (2006). Oxidative stress: Apoptosis in neuronal injury. *Curr. Alzheimer Res.* **3**, 327–337.

Madamanchi, N. R., and Runge, M. S. (2007). Mitochondrial dysfunction in atherosclerosis. *Circ. Res.* **100**, 460–473.

Maines, M. D. (1997). The heme oxygenase system: A regulator of second messenger gases. *Annu. Rev. Pharmacol. Toxicol.* **37**, 517–554.

Maines, M. D. (2005). The heme oxygenase system: Update 2005. *Antioxid. Redox Signal.* **7**, 1761–1766.

Mancuso, C. (2004). Heme oxygenase and its products in the nervous system. *Antioxid. Redox Signal.* **6**, 878–887.

Mancuso, C., Bonsignore, A., Capone, C., Di Stasio, E., and Pani, G. (2006b). Albumin-bound bilirubin interacts with nitric oxide by a redox mechanism. *Antioxid. Redox Signal.* **8**, 487–494.

Mancuso, C., Bonsignore, A., Di Stasio, E., Mordente, A., and Motterlini, R. (2003). Bilirubin and S-nitrosothiols interaction: Evidence for a possible role of bilirubin as a scavenger of nitric oxide. *Biochem. Pharmacol.* **66**, 2355–2363.

Mancuso, C., Pani, G., and Calabrese, V. (2006a). Bilirubin: An endogenous scavenger of nitric oxide and reactive nitrogen species. *Redox Rep.* **11**, 207–213.

Mancuso, C., Scapagnini, G., Currò, D., Giuffrida Stella, A. M., De Marco, C., Butterfield, D. A., and Calabrese, V. (2007). Mitochondrial dysfunction, free radical generation and cellular stress response in neurodegenerative disorders. *Front. Biosci.* **12**, 1107–1123.

Maroni, P., Bendinelli, P., Tiberio, L., Rovetta, F., Piccoletti, R., and Schiaffonati, L. (2003). *In vivo* heat-shock response in the brain: Signalling pathway and transcription factor activation. *Brain Res. Mol. Brain Res.* **119**, 90–99.

Matsumori, Y., Northington, F. J., Hong, S. M., Kayama, T., Sheldon, R. A., Vexler, Z. S., Ferriero, D. M., Weinstein, P. R., and Liu, J. (2006). Reduction of caspase-8 and -9 cleavage is associated with increased c-FLIP and increased binding of Apaf-1 and Hsp70 after neonatal hypoxic/ischemic injury in mice overexpressing Hsp70. *Stroke.* **37**, 507–512.

McCoubrey, W. K., Jr. Huang, T. J., and Maines, M. D. (1997). Isolation and characterization of a cDNA from the rat brain that encodes hemoprotein heme oxygenase-3. *Eur. J. Biochem.* **247**, 725–732.

McNally, S. J., Harrison, E. M., Ross, J. A., Garden, O. J., and Wigmore, S. J. (2006). Curcumin induces heme oxygenase-1 in hepatocytes and is protective in simulated cold preservation and warm reperfusion injury. *Transplantation* **81**, 623–626.

McNally, S. J., Harrison, E. M., Ross, J. A., Garden, O. J., and Wigmore, S. J. (2007). Curcumin induces heme oxygenase 1 through generation of reactive oxygen species, p38 activation and phosphatase inhibition. *Int. J. Mol. Med.* **19,** 165–172.

Motohashi, H., and Yamamoto, M. (2004). Nrf 2-Keap1 defines a physiologically important stress response mechanism. *Trends Mol. Med.* **10,** 549–557.

Motterlini, R., Foresti, R., Bassi, R., Calabrese, V., Clark, J. E., and Green, C. J. (2000). Endothelial heme oxygenase-1 induction by hypoxia: Modulation by inducible nitric-oxide synthase and S-nitrosothiols. *J. Biol. Chem.* **275,** 13613–13620.

Motterlini, R., Foresti, R., Intaglietta, M., and Winslow, R. M. (1996). NO-mediated activation of heme oxygenase: Endogenous cytoprotection against oxidative stress to endothelium. *Am. J. Physiol.* **270,** H107–H114.

Motterlini, R., Foresti, R., Vandegriff, K., Intaglietta, M., and Winslow, R. M. (1995). Oxidative-stress response in vascular endothelial cells exposed to acellular hemoglobin solutions. *Am. J. Physiol.* **269,** H648–H655.

Motterlini, R., Green, C. J., and Foresti, R. (2002). Regulation of heme oxygenase-1 by redox signals involving nitric oxide. *Antioxid. Redox Signal.* **4,** 615–624.

Narasimhan, P., Swanson, R. A., Sagar, S. M., and Sharp, F. R. (1996). Astrocyte survival and HSP70 heat shock protein induction following heat shock and acidosis. *Glia* **17,** 147–159.

Naughton, P., Foresti, R., Bains, S. K., Hoque, M., Green, C. J., and Motterlini, R. (2002). Induction of heme oxygenase 1 by nitrosative stress: A role for nitroxyl anion. *J. Biol. Chem.* **277,** 40666–40674.

Papa, S., Capitanio, G., and Luca Martino, P. (2006). Concerted involvement of cooperative proton-electron linkage and water production in the proton pump of cytochrome c oxidase. *Biochim. Biophys. Acta* **1757,** 1133–1143.

Premkumar, D. R., Smith, M. A., Richey, P. L., Petersen, R. B., Castellani, R., Kutty, R. K., Wiggert, B., Perry, G., and Kalaria, R. N. (1995). Induction of heme oxygenase-1 mRNA and protein in neocortex and cerebral vessels in Alzheimer's disease. *J. Neurochem.* **65,** 1399–1402.

Prestera, T., Talalay, P., Alam, J., Ahn, Y. I., Lee, P. J., and Choi, A. M. (1995). Parallel induction of heme oxygenase-1 and chemoprotective phase 2 enzymes by electrophiles and antioxidants: Regulation by upstream antioxidant-responsive elements (ARE). *Mol. Med.* **1,** 827–837.

Raps, S. P., Lai, J. C., Hertz, L., and Cooper, A. J. (1989). Glutathione is present in high concentrations in cultured astrocytes but not in cultured neurons. *Brain Res.* **493,** 398–401.

Rashmi, R., Santhosh Kumar, T. R., and Karunagaran, D. (2003). Human colon cancer cells differ in their sensitivity to curcumin-induced apoptosis and heat shock protects them by inhibiting the release of apoptosis-inducing factor and caspases. *FEBS Lett.* **538,** 19–24.

Ridnour, L. A., Thomas, D. D., Mancardi, D., Espey, M. G., Miranda, K. M., Paolocci, N., Feelisch, M., Fukuto, J., and Wink, D. A. (2004). The chemistry of nitrosative stress induced by nitric oxide and reactive nitrogen oxide species: Putting perspective on stressful biological situations. *Biol. Chem.* **385,** 1–10.

Rokutan, K., Miyoshi, M., Teshima, S., Kawai, T., Kawahara, T., and Kishi, K. (2000). Phenylarsine oxide inhibits heat shock protein 70 induction in cultured guinea pig gastric mucosal cells. *Am. J. Physiol. Cell Physiol.* **279,** C1506–C1515.

Rushworth, S. A., Ogborne, R. M., Charalambos, C. A., and O'Connell, M. A. (2006). Role of protein kinase C delta in curcumin-induced antioxidant response element-mediated gene expression in human monocytes. *Biochem. Biophys. Res. Commun.* **341,** 1007–1016.

Sammut, I. A., Foresti, R., Clark, J. E., Exon, D. J., Vesely, M. J., Sarathchandra, P., Green, C. J., and Motterlini, R. (1998). Carbon monoxide is a major contributor to the regulation of vascular tone in aortas expressing high levels of haeme oxygenase-1. *Br. J. Pharmacol.* **125,** 1437–1444.

Scacco, S., Petruzzella, V., Bestini, E., Luso, A., Papa, F., Bellomo, F., Signorile, A., Torraco, A., and Papa, S. (2006). Mutations in structural genes of complex I associated with neurological diseases. *Ital. J. Biochem.* **55,** 254–262.

Scapagnini, G., D'Agata, V., Calabrese, V., Pascale, A., Colombrita, C., Alkon, D., and Cavallaro, S. (2002). Gene expression profiles of heme oxygenase isoforms in the rat brain. *Brain Res.* **954,** 51–59.

Scapagnini, G., Colombrita, C., Amadio, M., D'Agata, V., Arcelli, E., Sapienza, M., Quattrone, A., and Calabrese, V. (2006). Curcumin activates defensive genes and protects neurons against oxidative stress, *Antioxid. Redox Signal.* **8,** 395–403.

Schipper, H. M. (2000). Heme oxygenase-1: Role in brain aging and neurodegeneration. *Exp. Gerontol.* **35,** 821–830.

Slivka, A., Mytilineou, C., and Cohen, G. (1987). Histochemical evaluation of glutathione in brain. *Brain Res.* **409,** 275–284.

Snyder, S. H., and Baranano, D. E. (2001). Heme oxygenase: A font of multiple messengers. *Neuropsychopharmacology* **25,** 294–298.

Son, E., Jeong, J., Lee, J., Jung, D. Y., Cho, G. J., Choi, W. S., Lee, M. S., Kim, S. H., Kim, I. K., and Suk, K. (2005). Sequential induction of heme oxygenase-1 and manganese superoxide dismutase protects cultured astrocytes against nitric oxide. *Biochem. Pharmacol.* **70,** 590–597.

Sood, A., Mathew, R., and Trachtman, H. (2001). Cytoprotective effect of curcumin in human proximal tubule epithelial cells exposed to shiga toxin. *Biochem. Biophys. Res. Commun.* **283,** 36–41.

Soti, C., and Csermely, P. (2003). Aging and molecular chaperones. *Exp. Gerontol.* **38,** 1037–1040.

Srisook, K., Kim, C., and Cha, Y. N. (2005). Role of NO in enhancing the expression of HO-1 in LPS-stimulated macrophages. *Methods Enzymol.* **396,** 368–377.

Sultana, R., Ravagna, A., Mohmmad-Abdul, H., Calabrese, V., and Butterfield, D. A. (2005). Ferulic acid ethyl ester protects neurons against amyloid beta-peptide(1-42)-induced oxidative stress and neurotoxicity: Relationship to antioxidant activity. *J. Neurochem.* **92,** 749–758.

Sun, X., Shih, A. Y., Johannssen, H. C., Erb, H., Li, P., and Murphy, T. H. (2006). Two-photon imaging of glutathione levels in intact brain indicates enhanced redox buffering in developing neurons and cells at the cerebrospinal fluid and blood-brain interface. *J. Biol. Chem.* **281,** 17420–17431.

Takeda, A., Perry, G., Abraham, N. G., Dwyer, B. E., Kutty, R. K., Laitinen, J. T., Petersen, R. B., and Smith, M. A. (2000). Overexpression of heme oxygenase in neuronal cells, the possible interaction with Tau. *J. Biol. Chem.* **275,** 5395–5399.

Wakabayashi, N., Dinkova-Kostova, A. T., Holtzclaw, W. D., Kang, M. I., Kobayashi, A., Yamamoto, M., Kensler, T. W., and Talalay, P. (2004). Protection against electrophile and oxidant stress by induction of the phase 2 response: Fate of cysteines of the Keap1 sensor modified by inducers. *Proc. Natl. Acad. Sci. USA* **101,** 2040–2045.

Willis, D., Tomlinson, A., Frederick, R., Paul-Clark, M. J., and Willoughby, D. A. (1995). Modulation of heme oxygenase activity in rat brain and spleen by inhibitors and donors of nitric oxide. *Biochem. Biophys. Res. Commun.* **214,** 1152–1156.

Winterbourn, C. C., Hampton, M. B., Livesey, J. H., and Kettle, A. J. (2006). Modeling the reactions of superoxide and myeloperoxidase in the neutrophil phagosome: Implications for microbial killing. *J. Biol. Chem.* **281,** 39860–39869.

Yenari, M. A., Giffard, R. G., Sapolsky, R. M., and Steinberg, G. K. (1999). The neuroprotective potential of heat shock protein 70 (HSP70). *Mol. Med. Today.* **5,** 525–531.

Zhang, X., Lu, L., Dixon, C., Wilmer, W., Song, H., Chen, X., and Rovin, B. H. (2004).
Stress protein activation by the cyclopentenone prostaglandin 15-deoxy-delta12,
14-prostaglandin J2 in human mesangial cells. *Kidney Int.* **65,** 798–810.

CHAPTER SEVEN

Monitoring Oxidative Stress in Vascular Endothelial Cells in Response to Fluid Shear Stress: From Biochemical Analyses to Micro- and Nanotechnologies

Mahsa Rouhanizadeh, Wakako Takabe, Lisong Ai, Hongyu Yu, *and* Tzung Hsiai

Contents

Department of Biomedical Engineering and Division of Cardiovascular Medicine, School of Engineering & School of Medicine, University of Southern California, Los Angeles, California

Methods in Enzymology, Volume 441
ISSN 0076-6879, DOI: 10.1016/S0076-6879(08)01207-X

Abstract

Hemodynamics, specifically, fluid shear stress, modulates the focal nature of atherosclerosis. Shear stress induces vascular oxidative stress via the activation of membrane-bound NADPH oxidases present in vascular smooth muscle cells, fibroblasts, and phagocytic mononuclear cells. Shear stress acting on the endothelial cells at arterial bifurcations or branching points regulates both NADPH oxidase and nitric oxide (NO) synthase activities. The former is considered a major source of oxygen-centered radicals (i.e., superoxide anion [$O_2^{\cdot-}$]) that give rise to oxidative stress; the latter is a source of nitrogen-centered radicals (i.e., nitric oxide [NO]) that give rise to nitrative/nitrosative stress. In addition to conventional biochemical analyses, the emerging microelectromechanical systems (MEMS) provide spatial and temporal resolutions to investigate the mechanisms whereby the characteristics of shear stress regulate the biological activities of endothelial cells at the complicated arterial geometry. In parallel, the development of MEMS liquid chromatography (LC) provides a new venue to measure circulating oxidized low-density lipoprotein (ox-LDL) particles as a lab-on-a chip platform. Nanowire-based field effect transistors further pave the way for a high throughput approach to analyze the LDL redox state. Integration of MEMS with oxidative biology is synergistic in assessing vascular oxidative stress. The MEMS LC provides an emerging lab-on-a-chip platform for ox-LDL analysis. In this context, this chapter has integrated expertise from the fields of vascular biology and oxidative biology to address the dynamics of inflammatory responses.

1. INTRODUCTION

1.1. Vascular oxidative stress

Several lines of evidence support the role for vascular oxidative stress and inflammation (Griendling, 2003; Griendling and FitzGerald, 2003). Multidisciplinary evidence indicates that enhanced lipid peroxidation is associated with accelerated atherogenesis (Salonen *et al.*, 1992). Epidemiological studies suggest that a low level of antioxidants is associated with an increased risk for cardiovascular disease and that increased intakes appear to be protective (Jialal and Devaraj, 2003).

Reactive oxygen species (ROS) and reactive nitrogen species, including superoxide ($O_2^{\cdot-}$), hydrogen peroxide (H_2O_2), and peroxynitrite, have been implicated in endothelial dysfunction. Within the vascular wall, there are several sources of ROS generation, including NADPH oxidases, cytochrome P-450, xanthine oxidase, uncoupled nitric oxide synthase (NOS), and mitochondria (Griendling *et al.*, 2000). NADPH oxidase, a major source of $O_2^{-\bullet}$ generation in the vascular wall (Hwang *et al.*, 2003d), is a multicomponent, membrane-associated enzyme that catalyzes the one-electron reduction of oxygen to $O_2^{-\bullet}$ using NADPH as the electron donor (Hilenski *et al.*, 2004). NADPH oxidase components include the Nox homologs and p22[phox] in the membrane and p47[phox], p67[phox], and Rac in the cytosol. Among the Nox homologs, human endothelial cells express Nox4 at relatively high levels, gp91[phox] (Nox2) to a lesser extent, and Nox1 much less abundantly (Ago *et al.*, 2004; Sorescu *et al.*, 2002). Therefore, NADPH oxidase expression, assembly, and localization are important regulatory steps in endothelial $O_2^{-\bullet}$ production.

1.2. Fluid shear stress

Atherosclerosis is a systemic disease; however, its manifestations tend to be focal and eccentric (Ku *et al.*, 1985; Malek *et al.*, 1999). Hemodynamics, namely, fluid shear stress, is intimately involved in the focal nature of atherosclerosis. Atherosclerotic lesions preferentially develop in the lateral walls of vessel bifurcations and curvatures (Fung, 1997). Predilection sites for atherosclerosis are modulated by flow patterns to which the endothelium is exposed (Davies *et al.*, 1984). At the lateral wall of arterial bifurcations, a complex flow profile develops; flow separation and migrating stagnation points create oscillating shear stress (i.e., bidirectional with no net forward flow) (Ku, 1997) (Fig. 7.1). Oscillatory shear stress (OSS) increases oxidative stress, which promotes the development of atherosclerotic plaques (De Keulenaer *et al.*, 1998; Landmesser and Harrison, 2001; Sorescu *et al.*, 2002; Witztum, 1994; Ziegler *et al.*, 1998). In contrast, in the medial wall of bifurcations or relatively straight segments, pulsatile shear stress (PSS) downregulates adhesion molecules and reactive species. These studies provide strong evidence that endothelial cells (EC) have the ability to both "sense" hemodynamic forces and to differentiate among different types of biomechanical stimuli.

1.3. Linking shear stress with oxidative stress and nitrosative/nitrative stress

The spatial and temporal components of shear stress largely determine the focal character of atherosclerosis (Hsiai *et al.*, 2007). Shear stress acting on the endothelial cells at arterial bifurcations or branching points regulates both NADPH oxidase (Griendling, 2003; Hwang *et al.*, 2003a) and nitric oxide synthase activities (Topper *et al.*, 1996; Ziegler *et al.*, 1998). The

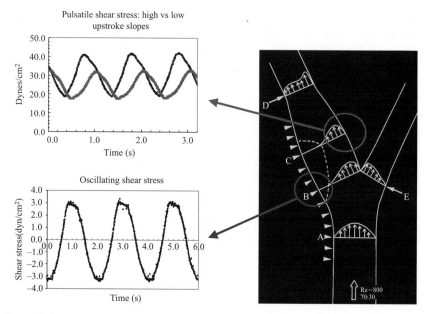

Figure 7.1 Schematic diagram showing velocity profiles in arterial bifurcation. Pulsatile flow, which is circled in red, occurs in the medial wall, whereas oscillating flow occurs at the reattachment points of the lateral wall (circled in blue). The reattachment point is the site at which flow separation occurs. At this point, the mean shear stress is known to be zero while the instantaneous shear stress oscillates in response to cardiac contraction.

former is considered a major source of ROS [i.e., superoxide anion $(O_2^{-\bullet})$] that give rise to oxidative stress; the latter is a source of reactive nitrogen species (RNS) [i.e., nitric oxide (NO)] that give rise to nitrative/nitrosative stress.

Shear stress induces vascular oxidative stress via the activation of membrane-bound NADPH oxidases present in vascular smooth muscle cells (SMC), fibroblasts, and phagocytic mononuclear cells (Griendling *et al.*, 2000) (Fig. 7.2). In coronary arteries, NAD(P)H oxidases of the Nox family are a predominant source of $O_2^{-\bullet}$ from noninflammatory cells such as the endothelium (Görlach *et al.*, 2000; Sorescu *et al.*, 2002). NADPH oxidases are a family of enzymes that generate $O_2^{-\bullet}$ by transferring electrons from NADPH and/or NADH to molecular oxygen via flavins within their structure. NADPH oxidases are made up of three cytosolic protein subunits, p47phox, p67phox, and the G-protein, Rac, as well as a membrane-bound cytochrome reductase domain (Görlach *et al.*, 2000). The increased vascular activities of NADPH oxidases enhance the production of $O_2^{-\bullet}$ (Irani, 2000). The formation of $O_2^{-\bullet}$ by the endothelial NADPH oxidase accounts for reduced ˙NO bioavailability and development of endothelial dysfunction

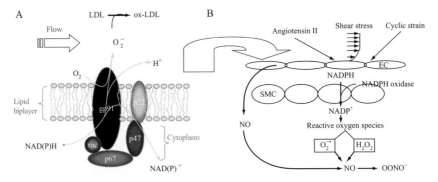

Figure 7.2 (A) NADPH oxidase system consists of membranous subunits, namely, Gp91phox or its homolog Nox4, and p22^{phox}, as well as the cytosolic subunits, Rac, p67, and p 47. O_2^- formation in response to shear stress may oxidatively modify LDL to ox-LDL. (B) Biomechanical forces in the generation of ROS in the vessel wall. In the presence of O_2^-, $\cdot$NO is converted to ONOO$^-$. SMC, smooth muscle cells.

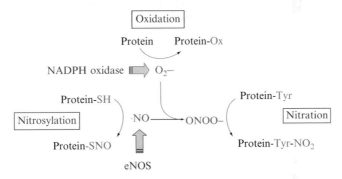

Figure 7.3 Relation of oxidation, nitrosylation, and nitration.

from the aortic ring of gp91^{phox}-knockout mice (gp91^{phox}-/-) (Jung et al., 2004), as evidenced from Nox4 upregulation in gp91^{phox}-/- mice in response to angiotensin II (Byrne et al., 2003). Our group and others have shown increased O_2^- production in response to oscillatory flow (Hwang et al., 2003a,b).

O_2^- generated by NADPH oxidases and $\cdot$NO generated by eNOS react with each other at diffusion–controlled rates ($O_2^- + \cdot$NO → ONOO$^-$) to yield a potent oxidant, peroxynitrite (ONOO$^-$) (Fig. 7.3) (Harrison, 2003). Peroxynitrite reacts rather specifically with tyrosine (Tyr) residues in proteins, leading to the formation of nitrotyrosine, which may be considered a fingerprint of ONOO$^-$ reactivity in a cellular setting. $\cdot$NO/O_2^- imbalance resulting from increased NADPH oxidase activity at the vascular wall may contribute to inflammatory responses. The relative flux of $\cdot$NO and O_2^-—a function of abundance and location of both NOSs and NADPH oxidases in the heart—determines the chemical fate of their interactions: S-nitrosylation, oxidation reactions, and/or protein nitration (Foster *et al.*, 2003; Wink *et al.*, 1997). Thus, controlled production of RNS and ROS preserves a redox environment that is not conducive to protein modifications relevant for atherogenesis. The balance between $^{\cdot}NO$ and $O_2^{-\cdot}$ production also plays a pivotal role in cell/organ function at key sites in the cardiovascular system, including the heart, large- and medium-sized conductance blood vessels, and the microvasculature. Thus, $O_2^{-\cdot}$ and derived species may contribute to cardiac injury (Cesselli *et al.*, 2001; Siwik *et al.*, 1999), both by oxidizing cellular constituents and by diminishing $^{\cdot}NO$ bioactivity (Khan *et al.*, 2004). The extent to which interactions between $^{\cdot}NO$ and $O_2^{-\cdot}$ constitute a pathophysiologic mechanism, however, remains undefined.

Vascular cells utilize ROS and RNS to modify low-density lipoprotein (LDL). RNS and ROS can modify LDL to "atherogenic particles," and $ONOO^-$ is a prime oxidant. EC produce ROS and RNS from the enzymes, NOS, and by specific homologs of NADPH oxidase (Nox1 and Nox4) (Sorescu *et al.*, 2002). Shear force and angiotensin are two factors that are able to influence the amount of $O_2^{-\cdot}$ and $^{\cdot}NO$ generated by EC (Griendling, 2003). Thus, the relative production of $O_2^{-\cdot}$ versus $^{\cdot}NO$ influences the capacity of EC to either prevent or promote the oxidation of LDL.

In this context, developing methodology to investigate the interplay between fluid shear stress and vascular oxidative stress is important to understand mechanotransduction signaling and vascular disease. The first part of this chapter focuses on the *in vitro* approach used to measure ROS and RNS in response to fluid shear stress. The second part introduces the emerging micro- and nanotechnologies to investigate oxidative modification of LDL.

2. *IN VITRO* MONITORING OF REACTIVE SPECIES

2.1. Dynamic flow apparatus

Wall shear stress, the tangential drag force of blood passing along the surface of the endothelium (Fung and Liu 1993), has metabolic as well as mechanical effects on EC function (DePaola *et al.*, 1992; Frangos *et al.*, 1996; Helmlinger *et al.*, 1991). Two commonly used flow apparatus have been used to assess the effects of fluid shear stress on vascular endothelial cells: (1) parallel-plate and (2) cone-and-plate flow channels. Both apparatus have been modified to generate dynamic flow profiles (Dewey *et al.*, 1981; Nerem *et al.*, 1998).

The newly designed modified parallel-plate and pulsatile flow system provide precise and well-defined flow profiles across the width of the chamber to simulate cardiac contractility at various temporal gradients ($\partial \tau / \partial t$), frequency, and amplitude (Hsiai *et al.*, 2002a) (Fig. 7.4). Many hemodynamic flow properties, which have been difficult or impossible to assess in vessels *in vivo*, can potentially be realized with microelectrical mechanical system (MEMS) sensors, which provide both spatial and temporal resolutions for shear stress measurement over a dynamic range of Reynolds numbers from the right coronary artery (at 150) to the internal carotid artery (at 330) as well as extreme variations in response to exercise (at 4000). The pulsatile flow system is capable of generating magnitudes of mean shear stress, which have not commonly been reported in the literature, to simulate flow patterns at regions of arterial branching points as well as the extreme variations that exist in regions of branching, stenotic arteries (Nerem *et al.*, 1998) and the flow divider where shear stress may exceed 100 dynes/cm^2 (Fung, 1997).

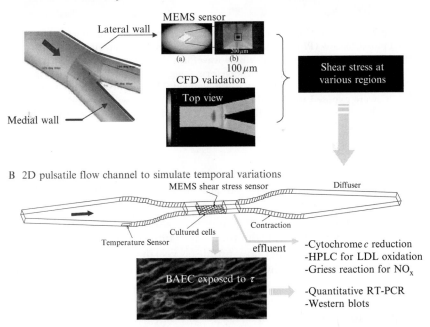

Figure 7.4 (A) An upscaled 3D bifurcation model consists of MEMS sensors to resolve $\partial \tau / \partial x$. A computational fluid dynamics (CFD) solution will validate the direct shear stress measurements. (B) The 2D model will be used to implement the specific shear stress values of interest. The effluents will be collected for biochemical analyses. The recovered BAEC will be used for gene and protein expression. The sensors will be embedded in the upper wall while EC will be seeded on the bottom. Based on symmetry, the shear stress acting on the upper wall is a mirror image on the bottom plate.

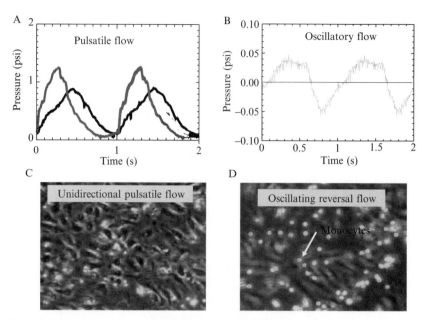

Figure 7.5 Monitoring vascular oxidative stress in response to shear stress **(A)** Two distinct pulsatile flow (PF) profiles that simulated arterial flow profiles were delivered by the programmed step motor. The blue waveform simulates pulsatile flow with a high slew rate (*dP/dt*), whereas the black waveform represents a low slew rate. **(B)** Oscillatory flow (OF) or back-and-forth flow was generated. The net forward flow was zero. **(C)** BAEC were exposed to pulsatile shear stress. **(D)** BAEC were exposed to oscillatory shear stress. Compared to BAEC exposed to PF, BAEC exposed to OF induced an increase in monocyte adhesion.

The dynamic parallel flow system is not only capable of delivering two distinct flow profiles, namely, pulsatile and oscillatory shear stress, but also high and low slew rates, specifically *dP/dt* at which pulsatile flow is generated to simulate flow during physical activities (Hsiai *et al.*, 2002a) (Figs. 7.5A and 7.5B). Furthermore, real-time monitoring of endothelial cells in response to distinct flow profiles is visualized (Figs. 7.5C and 7.5D). Also shown in Figs. 7.5C and 7.5D are human monocytes binding to bovine aortic endothelial cells (BAEC). Oscillatory flow induced NF-κB-mediated adhesion molecules and cytokines that recruit monocytes in response to oscillatory shear stress (Hsiai *et al.*, 2001, 2003).

2.2. Cone-and-plate model

The cone-and-plate model has contributed to the understanding of shear stress and endothelial cell biology (Dewey *et al.*, 1981). The newly developed cone-and-plate flow channel was introduced to deliver two distinct

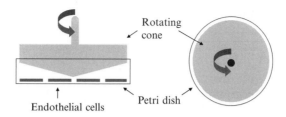

Figure 7.6 The *in vitro* cone-and-plate model. The cone is directly driven by the motor. Endothelial cells were cultured on the petri dish. Stationary control cells were kept in petri dishes filled with incubation medium. In other experiments, endothelial mono-layers were sheared in a cone-and-plate viscometer. The cone angle of this device was varied around 0.5°.

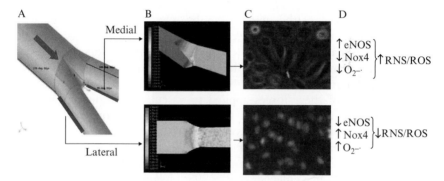

Figure 7.7 (A) Oblique view of the bifurcation region. (B) CFD solutions expressed in skin friction coefficient (C_f), which is proportional to fluid shear stress. (C) Real-time intracellular O_2^- in response to flow measured by dihydroethidium (DHE). Real-time merged images of phase and fluorescence at 4 h demonstrate the localization of red fluorescence in the nuclei as ethidium intercalated into the double-stranded DNA. (D) We anticipate that the characteristics of shear stress at the medial vs lateral wall will regulate the relative production of ROS to RNS, which in turn influences oxidative medication of LDL.

arterial waveforms, "athero-prone" and "athero protective" (Fig. 7.6.) (Blackman *et al.*, 2002). These profiles were representative of the wall shear stress profiles in the corresponding atherosclerosis-susceptible and atherosclerosis-resistant regions of the human carotid artery. The pH of the circulating medium is maintained at 7.4 by bubbling 5% CO_2/95% air in the reservoir. Both the flow chamber and the medium reservoir are maintained at 37 °C.

In summary, development of the 3D bifurcation model integrated with the MEMS sensors provides a means to resolve the spatial variations in shear stress ($\partial\tau/\partial x$) over a range of Reynolds number at steady state (Fig. 7.7A). The newly developed 2D system will permit analysis of biochemical responses of BAEC in response to temporal variations ($\partial\tau/\partial t$). The combination of 2D and 3D models, as well as real-time assessment of shear stress, would enable us to demonstrate pulsatile vs oscillatory flow pattern-mediated $O_2^{-\bullet}$ and $^{\bullet}NO$ production, as well as LDL oxidation. These *in vitro* studies pave the way to uncover a link between local flow characteristics (namely, magnitude, frequency, direction, and spatial/temporal gradients) and oxidation of LDL for *in vivo* investigation (Rouhanizadeh *et al.*, 2005b).

3. Measurement of Endothelial ROS Generation in Response to Shear Stress

Investigation of the mechanisms by which shear stress regulate the oxidative stress-mediated responses under physiologically arterial geometric conditions is made possible by the emerging MEMS sensors that provide a spatial resolution comparable to the individually elongated endothelial cells and temporal resolution in the Kilohertz range. Findings show that the characteristics of shear forces applied to vascular endothelium affect the production of $O_2^{-\bullet}$ as well as the $O_2^{-\bullet}$–dependent oxidation of LDL (Hwang *et al.*, 2003a, 2006; Rouhanizadeh *et al.*, 2005a). The vascular regions of moderate to high shear stress—where flow remains unidirectional and axially aligned—experience relatively little oxidative stress and focal LDL oxidation (De Keulenaer *et al.*, 1998; Hwang *et al.*, 2003b). Excessive production of reactive oxygen and nitrogen species develops largely in regions of relative low shear stress, flow separation, and departure from axially aligned and unidirectional flow profiles (Passerini *et al.*, 2004; Sorescu *et al.*, 2004). Hence, we hereby introduce the methodology to assess that spatial and temporal variations in shear stress differentially regulate the endothelial production of $O_2^{\bullet-}$ and $^{\bullet}NO$, leading to LDL oxidative modifications relevant for the initiation of atherosclerotic lesions.

3.1. Extracellular superoxide ($O_2^{-\bullet}$) formation

Bovine aortic endothelial cells monolayers on the microscope slides are exposed to three conditions: (1) control (no flow, 37 °C incubator), (2) oscillatory flow, and (3) pulsatile flow with the identical physical parameters as described earlier. In each case, the medium contains 100 μM acetylated ferricytochrome c (Sigma-Aldrich). Control samples are maintained in a cell culture dish with media containing cytochrome c (100 μM) and incubated at 37 °C. Aliquots of culture medium (300 μl) are collected at 0, 1, 2, 3, and 4 h for absorbance measurements at 550 nm (Beckman DU 640 spectrophotometer) to detect the formation of ferrocytochrome c. For the oscillatory and pulsatile flow conditions, BAEC monolayers are placed in the flow system. At 0, 1, 2, 3, and 4 h, aliquots of medium bathing the BAEC in the flow apparatus are aspirated into a media solution containing acetylated ferricytochrome c (1 mM) and measured for absorbance at 550 nm. The specificity of reduction by $O_2^{-\bullet}$ is established by comparing reduction rates in the presence and absence of SOD (60 μg/ml). The corrected rates for SOD-inhibited cytochrome c reduction are plotted after computing $O_2^{-\bullet}$ formation using the following extinction coefficient: $E_{550} = 2.1 \times 10^4$ M^{-1} cm^{-1} (Hsiai et al., 2007; Hwang et al., 2003a).

Values of $O_2^{-\bullet}$ production under oscillatory flow conditions are similar to those described in the literature (Landmesser and Harrison, 2001; Landmesser et al., 2002) (Fig. 7.8, top). Thus, temporal variations in shear stress, namely, pulsatile versus oscillatory flow patterns, regulate the rates of extracellular $O_2^{-\bullet}$ production. One of the molecular mechanisms underlying differential rates of $O_2^{-\bullet}$ production is NADPH oxidase subunits, such as Nox 2 and Nox 4 (Fig. 7.8, bottom) (Hwang et al., 2003a).

3.2. Intracellular $O_2^{-\bullet}$ production measurements

We demonstrated that oxidative stress develops at the lateral wall of arterial bifurcation where disturbed flow, including oscillatory shear stress, occurs. We used dihydroethidium (DHE), which specifically reacts with $O_2^{-\bullet}$, to form the red fluorescent compound ethidium (Görlach et al., 2000). Representative fluorescent photomicrographs of DHE-stained BAEC revealed that oscillatory flow conditions promoted the production of reactive oxygen species (Fig. 7.9). DHE-stained fluorescent images show that increased intracellular $O_2^{-\bullet}$ levels are present at significantly higher intensity in the BAEC exposed to oscillatory flow than to pulsatile flow. This is consistent with observations showing significant extracellular $O_2^{-\bullet}$ production measured by the cytochrome c assay in response to oscillatory flow (Hwang et al., 2003a). In parallel, the addition of polyethylene glycol–superoxide dismutase (PEG-SOD) at 60 μg/ml attenuated oscillatory flow–mediated red fluorescent intensity (data not shown), suggesting the important role of $O_2^{-\bullet}$ formation. DHE was used to localize intracellular $O_2^{-\bullet}$ production as described previously (Rouhanizadeh et al., 2005a).

In brief, cells are permeable to DHE, which is oxidized to fluorescent ethidium in the presence of $O_2^{-\bullet}$. The fluorescent ethidium is trapped by intercalation within the double-stranded DNA in the nuclei. Confluent BAEC on the microscope slides are incubated with 30 μg/ml of DHE (Sigma-Aldrich). After 1 h, BAEC are washed twice with phosphate-buffered saline and then subjected to the flow conditions as described earlier. Real-time fluorescence microscopy (Olympus) is performed to evaluate the localization of DHE fluorescence. DHE-stained fluorescent images show that increased intracellular $O_2^{-\bullet}$ formation is present at a

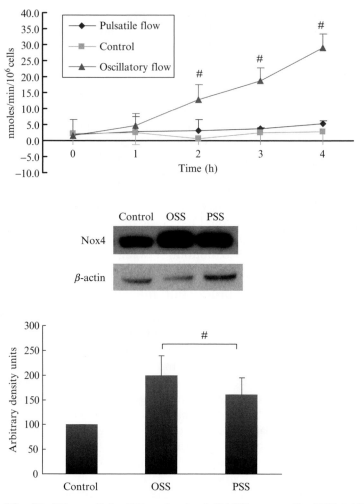

Figure 7.8 (Top) Extracellular $O_2^{-\cdot}$ production in BAEC measured by SOD-inhibited acetylated cytochrome c reduction assay. Rates of superoxide production remained unchanged under the static state (control). However, rates in response to oscillatory vs pulsatile shear stress diverged starting at 2-h exposure ($P < 0.01$, $n = 4$). (Bottom) Nox4 protein expression in response to shear stress. OSS upregulated Nox4 protein by 2 ± 0.8-fold, whereas PSS upregulated Nox4 protein by 1.60 ± 0.7-fold at 4 h ($n = 4$, $P < 0.05$). Nox4 protein was normalized to β-actin. Controls were performed under static conditions.

significantly higher intensity in the BAEC exposed to oscillatory flow than to pulsatile flow, $O_2^{-\cdot}$ production is largely dependent on the availability of reducing equivalents (NADPH) for NADPH oxidase provided by the pentose shunt (Hwang *et al.*, 2003a). The known coupling of pentose

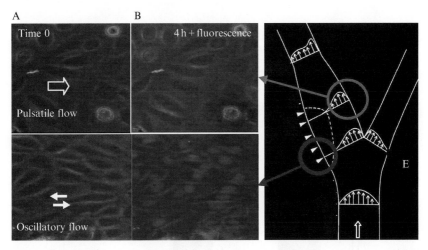

Figure 7.9 Real-time $O_2^{-\cdot}$ production: (A) At time zero, column (A) illustrates BAEC under two conditions: pulsatile flow and oscillatory flow. The static state/control is not shown. (B) Merged images of phase and fluorescence at 4 h demonstrate the localization of red fluorescence in nuclei. In the presence of $O_2^{-\cdot}$, DHE was converted to ethidium, which intercalated into the double-stranded DNA.

shunt-derived NADPH production with the supply of NADPH to NADPH oxidase and eNOS suggests the importance of modulating the activity of the ROS/RNS-producing enzymes by shear force and its impact on vascular oxidant production. These findings were reported in the context of shear force-dependent oxidant production by endothelial cells and its effect on LDL modification (Hwang *et al.*, 2003a).

3.3. Analysis of nitrite (NO_2^-) and nitrate (NO_3^-)

To demonstrate whether shear stress influences the relative formation of $O_2^{-\cdot}$ and $\cdot NO$, we quantitatively measured the production of $O_2^{-\cdot}$ and $\cdot NO$ (total NO_2^- and NO_3^-) from effluents as obtained from BAEC exposed to static and flow conditions at medial versus lateral walls. Preliminary results show that oscillatory shear stress acting on BAEC induced a higher rate of $O_2^{-\cdot}$ production than pulsatile shear stress at 4 h (control = 1.2 ± 0.8 nm/min/10^6 cells; OSS = 25.7 ± 5.1; PSS = 5.4 ± 4.2, $n = 4$, $P < 0.05$) (Fig. 7.10A). PSS induced a higher level of $\cdot NO$ than $O_2^{-\cdot}$ production (NO = 37.5 ± 4.1 vs $O_2^{-\cdot}$ = 5.4 ± 4.2 nmol/min/10^6 cells, $n = 4$, $P < 0.05$) (Fig. 7.10A). OSS also induced a higher level of $\cdot NO$ than $O_2^{-\cdot}$ production ($\cdot NO = 48.6 \pm 5.1$ vs $O_2^{-\cdot}$ = 25.7 ± 4.7 nmol/min/10^6 cells, $n = 4$, $P < 0.05$). Furthermore, OSS induced a higher ratio of $O_2^{-\cdot}/\cdot NO$ than PSS (control: 0.63 ± 0.05; lateral wall: 0.53 ± 13; medial: 0.14 ± 11, $n = 4$, $P < 0.05$) (Fig. 7.10B). These observations correlated with

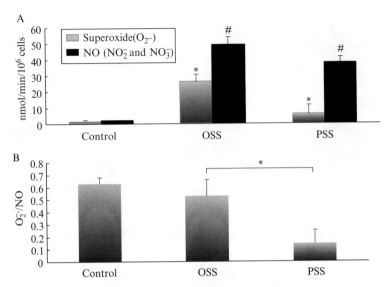

Figure 7.10 Relative rates of $O_2^{-\bullet}$ and NO production. **(A)** OSS promotes the rates of $O_2^{-\bullet}$ and NO (total NO_2^- and NO_3^-) production relative to PSS at 4 h. Rates of $O_2^{-\bullet}$ were 1.2 ± 0.8 for control, 25.7 ± 5.1 at the lateral wall, and 5.4 ± 4.2 nm/min/million cells at the medial wall ($n = 4$, $P < 0.05$). Total NO formation was 37.5 ± 4.1 at the medial wall and 48.6 ± 5.1 nm/min/million cells at the lateral wall ($n = 4$, $P < 0.05$). Rates of superoxide and RNS production remained steady under the static state. While production rates of $O_2^{-\bullet}$, NO, and RNS (reflected by total nitrite and nitrate levels) were higher in response to oscillatory at the lateral wall compared to pulsatile flow at the medial wall, rates of RNS were higher than ROS in response to both flow conditions. Oscillatory flow induced a higher relative ratio of ROS/RNS production than that of pulsatile flow ($n = 4$, $P < 0.05$). **(B)** The ratio of $O_2^{-\bullet}$ to $^\bullet$NO is higher in response to OSS than to PSS.

the differential upregulation of gp91^{phox}/Nox4 and eNOS mRNA expression. The increase in $O_2^{-\bullet}$ and $^\bullet$NO production is consistent with the focal nature of atherosclerosis at the lateral walls of vascular bifurcations in which the presence of ROS and RNS likely subserves oxidative and nitrative modifications of LDL.

Quantitative measurements of NO_2^- ($^\bullet$NO derived) and NO_3^- (ONOO$^-$ derived) are performed as an index of global nitric oxide ($^\bullet$NO) production following methods described previously (Braman and Hendrix, 1989). $^\bullet$NO is metabolized or decomposes via various reactions to the major product NO_2^- and ONOO$^-$ decomposes to NO_3^- (superoxide mediated nitric oxide decay) *in vivo*; together they serve as a useful measure of overall $^\bullet$NO production and metabolism (Pietraforte *et al.*, 2004). Briefly, the analytical procedure is based on the acidic reduction of NO_2^- and NO_3^- to $^\bullet$NO by vanadium(III) and purging of $^\bullet$NO with helium into a stream of ozone and detected by an Antek 7020 chemiluminescence $^\bullet$NO detector (Antek Instruments, Houston, TX). At room temperature, vanadium(III) only reduces NO_2^-, whereas NO_3^- and other redox forms of $^{\bullet}NO$ (such as S- nitrosothiols) are reduced only if the solution is heated to 90–100 °C so that both NO_2^- and total NO_x can be measured, and NO_3^- is determined by the difference between heated and unheated solutions. This allows for determination of the relative levels of NO_2^- and NO_3^-, which are indicative of the differential oxidative metabolism of $^{\bullet}NO$. Quantification is performed by comparison with standard solutions of NO_2^- and NO_3^-.

4. Protein Oxidation and Nitration in Response to Shear Stress

At arterial bifurcations where oscillatory shear stress is prevalent, the increase in $O_2^{-\bullet}$ production relative to $^{\bullet}NO$ production likely limits NO bioavailability through formation of the potent oxidant $ONOO^-$. To determine whether enhanced production of $O_2^{-\bullet}$ and $^{\bullet}NO$ by OSS promotes the formation of RNS (e.g., $ONOO^-$), the nitration of protein tyrosine residues from both culture medium and explants of human coronary arteries is examined by liquid chromatography/electron spray ionization/mass spectroscopy/mass spectroscopy (LC/ESI/MS/MS). Integrating the *in vitro* flow model with proteomics allows for the assessment of whether localized production of $O_2^{-\bullet}$ relative to $^{\bullet}NO$ under the oscillatory sheer stress conditions in the lateral versus medial wall of arterial bifurcations resulted in enhanced localized oxidative stress in these atherosclerosis prone regions of the vasculature (Hsiai *et al.*, 2007).

4.1. Oxidatively modified LDL

Oxidative modification of LDL is evidenced by alternations in both protein and lipid components of the LDL particle. Progressive oxidation of the apoprotein is associated with loss of specific amino acids sensitive to oxidation, such as lysine, tyrosine, and cysteine, and a gradual increase in electronegativity (Sevanian *et al.*, 1997). Electronegative LDL has been isolated from human plasma (LDL^-) by several groups using liquid chromatographic techniques and appears to be oxidized based on increased lipid peroxide levels and cholesterol oxidation products (ChOx). Formation of LDL^- also takes place following Cu^{2+}-induced oxidation (Sevanian *et al.*, 1994) (Fig. 7.11). Cu^{2+}-induced oxidation causes a small fraction of the normal unoxidized LDL (n-LDL) to convert to LDL^- during the oxidative lag phase while minimal increases in conjugated dienes are apparent. After the lag phase, there is a further increase in LDL^-, a rapid accumulation of conjugated dienes, and another more electronegative particle is formed (LDL^{2-}). By the end of the lag phase, approximately 30 and 12% of the

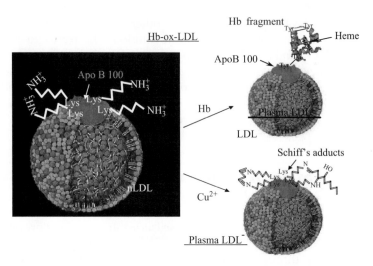

Figure 7.11 The upper right part of the scheme shows the formation of protein conjugates between the Hb fragment and ApoB 100 via dityrosine adducts. Oxidation of LDL protein takes place through tyrosol radical formation on Hb, radical attachment of ApoB 100 tyrosyl residues, and subsequent tyrosyl radical reactions to form dityrosine adducts. The participation of lipid peroxidation is minimal (Sevanian *et al.*, 1999b). Oxidation of LDL by means of Cu^{2+} involves the formation of reactive aldehydes (MDA and HNE) and adduct formation with lysine residues of ApoB 100. The major pathway for oxidation is thought to involve lipid oxidation, which in turn mediates modification of the apoprotein through the formation of reactive aldehyde decomposition products (glycosylation or modification with amino acid-derived aldehydes proceeds by a principally similar mechanism). There is high reactivity of aldehydes toward positively charged amino groups; in particular, lysine increases the relative electronegativity in LDL.

total LDL convert to LDL^- and LDL^{2-}, respectively. Nearly 40% of the total ChOx formed is present by the end of the lag period, accompanied by small increases in conjugated dienes. The major products accumulating during this time are 7-ketocholesterol, cholesterol-β-epoxide, and 7-α-hydroxycholesterol. At the end of propagation phase, there is a sixfold increase in conjugated dienes and total ChOx increases by eightfold. However, the levels of LDL^- decrease markedly and essentially disappear at the end of the propagation phase. This indicates that LDL^- particles are transiently and minimally modified and decompose to more extensively oxidized and aggregated particles. It appears that a subpopulation of LDL rapidly converts to LDL^-, representing a mildly oxidized but oxidant-sensitive LDL population.

The major fraction, native LDL (nLDL), represents ~90–99% of total LDL. LDL^-, which is found in plasma *in vivo*, is a minimally oxidized subspecies of LDL, characterized by its greater electronegativity and oxidative status (Sevanian and Hodis, 1997; Sevanian *et al.*, 1997). LDL^{-2} is more

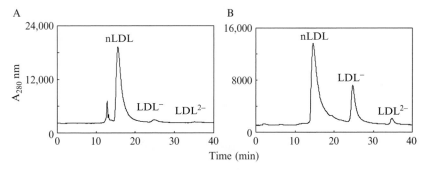

Figure 7.12 An HPLC chromatogram of LDL isolated from human plasma. Three peaks are typically seen, consisting of normal LDL (nLDL), LDL$^-$, and an even more electronegative peak (LDL^{2-}). (A) Chromatograms of LDL isolated from plasma after 4 h from noncirculated blood. (B) Accentuated peaks of blood circulation through the hemodialysis model system (Sevanian *et al.*, 1999a).

electronegative than LDL$^-$. LDL^{-2} constitutes 0.1–1% of total LDL and appears to be a highly oxidized subfraction of LDL (Fabjan *et al.*, 2001) (Fig. 7.12). LDL$^-$ represents 0.2–8% of LDL and is found predominantly in the small, dense LDL fraction, strongly associated with an increased risk of atherosclerosis (Austin, 1994; Sevanian *et al.*, 1996). Three possible sources of LDL$^-$ are (1) oxidation of LDL entrapped in the arterial wall (Henriksen *et al.*, 1981), (2) ingestion of oxidants or generation from postprandial lipoprotein remnants (Wesley *et al.*, 1998), and (3) oxidation in plasma (Stone *et al.*, 1994).

Pulsatile flow significantly reduced the ratios of oxidatively modified LDL species relative to static conditions by 51 $\pm$ 12% for LDL$^-$ and 30 $\pm$ 7% for LDL^{-2} (a more electronegative LDL), whereas oscillatory shear stress increased these modified LDLs by 67 $\pm$ 17 and 30 $\pm$ 7%, respectively ($P < 0.05$, $n = 5$) (Fig. 7.13). The change in LDL modification coincided with the downregulation of GP91^{phox} and Nox4 mRNA expression and the relative rates of O$_2^-$ $\cdot$ production in response to PF and upregulation in response to OF conditions (Hwang *et al.*, 2003a).

4.2. Analysis of protein nitration

We demonstrated that at the lateral walls of vascular bifurcations, the increase in O$_2^-$$\cdot$ relative to $\cdot$NO production likely limits $\cdot$NO bioavailability by supporting an $\cdot$NO decay pathway leading to ONOO$^-$ formation. LDL particles isolated from fasting adult human volunteers are added to BAEC that are exposed to PSS and OSS conditions. Culture medium is collected at 4 h, and modifications of apoB-100 tyrosine residues are examined by LC/ESI/MS/MS. Modifications of LDL provide a distinct modified species that reflects the nature and extent of ROS/RNS production.

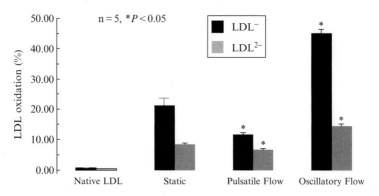

Figure 7.13 Flow regulation of native LDL oxidation. Pulsatile flow significantly reduced the ratios of oxidatively modified forms of LDL relative to static conditions by $51 \pm 12\%$ for LDL^- and $30 \pm 7\%$ for LDL^{2-}, whereas oscillatory flow increased LDL oxidation by $67 \pm 17\%$ and $30 \pm 7\%$, respectively ($P < 0.05$, $n = 5$).

4.2.1. Flow experiments to analyze LDL protein nitration

Venous blood is obtained from fasting adult human volunteers under institutional review board approval from the Atherosclerosis Research Unit at the University of Southern California. Plasma is pooled and immediately separated by centrifugation at $1500g$ for 10 min at $4\,^{\circ}C$. LDL ($\delta = 1.019$ to 1.063 g/ml). The technique used for separating LDL is similar to that described previously (Hwang *et al.*, 2003c; Sevanian *et al.*, 1994).

A dynamic flow system is used to implement temporal variations in shear stress ($\partial\tau/\partial t$); namely, PSS and OSS (Hsiai *et al.*, 2004). Confluent BAEC are subjected to the flow conditions in the absence and presence of LDL at 50 μg/ml: (1) control, using cells grown under static conditions ($\tau_{ave} = 0$ dyn·cm^{-2} at $\partial\tau/\partial t = 0$); (2) PSS at a mean shear stress (τ_{ave}) of 23 dyn·cm^{-2} with a temporal variation ($\partial\tau/\partial t$) at 71 dyn·cm^{-2}·s^{-1}; and (3) OSS at $\tau_{ave} = 0.02$ with $\partial\tau/\partial t$ at ± 3 dyn·cm^{-2}·s^{-1}. After 4 h, BAEC are collected for quantitative RT-PCR and Western blots. In the absence of LDL, the culture medium is collected to identify the differential production of $O_2^{-\cdot}$ and NO_2^-/NO_3^- in response to PSS and OSS, respectively. In the presence of LDL, the culture medium is used to determine apo-B 100 posttranslational modifications.

4.2.2. Analyses of LDL protein nitration by LC/ESI/MS/MS

After shear stress exposure, LDL suspended in medium is collected for LDL analysis of protein nitration. The extent of total protein-bound nitrotyrosine, *o*-hydroxyphenylalanine, and dityrosine formation in recovered media is determined by stable isotope dilution liquid chromatography–tandem mass spectrometry on a triple quadruple mass spectrometer (API 4000,

Applied Biosystems, Foster City, CA) interfaced with a Cohesive Technologies Aria LX series HPLC multiplexing system (Franklin, MA) (Brennan *et al.*, 2002). Synthetic $^{13}C_6$-labeled nitrotyrosine and $^{13}C_{12}$-labeled dityrosine internal standard are added to samples for the quantification of natural abundance analytes. Simultaneously, a universal labeled precursor amino acid, $(^{13}C_9, {}^{15}N_1)$ tyrosine (for nitrotyrosine and dityrosine) or $(^{13}C_9, 15N_1)$ phenylalanine (*o*-Phe), is added to both quantify the precursors and assess potential intrapreparative artifactual oxidation during sample handling and analysis. Proteins are hydrolyzed under an argon atmosphere in methane sulfonic acid, and then samples are passed over mini solid-phase C18 extraction columns (Supelclean LC-C18-SPE minicolumn; Supelco, Inc., Bellefone, PA) prior to mass spectrometry analysis. Results are normalized to the content of the appropriate precursor amino acid, which is monitored within the same injection. Intrapreparative formation of $(^{13}C_9, {}^{15}N)$-labeled nitrotyrosine, *o*-hydroxyphenylalanine, and dityrosine is routinely monitored and is negligible (i.e., <5% of the level of the natural abundance product observed) under the conditions employed.

4.3. Pulsatile and oscillatory shear stress differentially influenced the formation of peroxynitrite

$ONOO^-$ reacts rather specifically with tyrosine residues in proteins to yield 3-nitrotyrosine. To measure the production of $ONOO^-$ as protein-bound nitrotyrosine, confluent BAEC are exposed to PSS and OSS in the presence of LDL at 50 $\mu g/ml$. After shear stress exposure, LDL particles suspended in medium are collected for apoB-100 protein modifications. The levels of nitration in nitrotyrosine residues of apoB-100 are higher in response to OSS than to PSS (Table 7.1). LC/ESI/MS/MS analyses elicit the differential concentration of protein modifications in response to PSS versus OSS: 3-nitrotyrosine << dityrosine < *o*-Phe. *o*-Phe appears to be the predominant

Table 7.1   PSS and OSS influenced LDL protein nitration[a]

	Control	OSS	PSS
3-Nitrotyrosine [mmol/mol]	0.15 ± 0.08	$0.17 \pm 0.09^{\star}$	$0.09 \pm 0.04^{\star}$
Dityrosine [mmol/mol]	16.0 ± 4.4	$21.1 \pm 3.4^{\#}$	$13.0 \pm 2.7^{\#}$
o-Hydroxyphenylalanine [mmol/mol]	54.5 ± 4.1	$82.0 \pm 4.4^{+}$	$52.0 \pm 3.1^{+}$

[a] LC/ESI/MS/MS analyses of LDL apoB-100 revealed that OSS induced higher levels of LDL nitrotyrosine, dityrosine, and *o*-hydroxyphenylalanine (*o*-Phe) than PSS. Dityrosine and *o*-Phe appeared to be the predominant forms of protein nitration. Differences between control and PSS were statistically insignificant ($\star$, #, +, $P < 0.05$, $n = 3$).

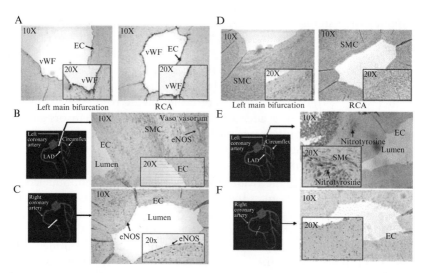

Figure 7.14 Immunostaining of representative sections of coronary arteries. (A) Endothelial cells (EC) were stained with von Willebrand factor in both the lateral wall of left main bifurcation (OSS-exposed region) and straight regions (PSS-exposed regions) of the right coronary artery (RCA). Inserts (20X) detail EC lining the inner lumens. *En face* staining of the human left main bifurcation was beyond the field of view at the lower magnification (10X). (B) A section of left main bifurcation revealed that eNOS staining was absent in the luminal EC, but was present in both smooth muscle cells and the vaso vasorum. (C) A representative PSS-exposed section of RCA revealed that eNOS staining was prevalent throughout the entire luminal EC. (D) Smooth muscle cells in media of left main bifurcation were counterstained with β-actin. β-Actin was observed in the intima and media of the OSS-exposed section, suggesting SMC migration (data not shown). (E) OSS-exposed section of the left main bifurcation revealed nitrotyrosine staining in media. The insert (20X) further showed nitrotyrosine staining in the SMC. (F) A representative PSS-exposed section in RCA revealed a lack of nitrotyrosine staining.

form. Together with immunohistochemistry analyses (Fig. 7.14) and the imbalanced production of $O_2^{-\bullet}$ and $^{\bullet}NO$ (Fig. 7.10), the higher levels of apoB-100 protein nitration in response to OSS suggest that vascular nitrative stress is likely to develop within the lateral walls of arterial bifurcations and curvatures. The pathophysiologic implication of dityrosine and *o*-hydroxyphenylalanine remains to be determined.

4.4. Proteomics analysis of LDL apo B-100 nitration

Samples of LDL (0.2 mg/ml) treated with $ONOO^-$ (100 μM) are processed for measurements of tyrosine nitration by LC/MS/MS. The tertiary structure of LDL is suspended in 10 μl 60% formic acid. Chromatographic separation is achieved using a ThermoFinnigan Surveyor MS-pump with a BioBasic-18 100-mm C 0.18-mm reversed-phase capillary column. Mass analysis is performed with a ThermoFinnigan LQ Deca XP Plus ion trap mass spectrometer equipped with a nanospray ion source employing a 4.5-cm-long metal needle in the data-dependent acquisition mode. Electrical contact and voltage application to the probe tip take place via the nanoprobe assembly. Spray voltage is set to 2.9 kV and heated capillary temperature at 190 °C. The column is equilibrated for 5 min with 95% solution A, 5% solution B (A, 0.1% formic acid in H_2O; B, 0.1% formic acid in acetonitrile) prior to sample injection. A linear gradient is initiated 5 min after sample injection, ramping to 35% A, 65% B after 50 min and 20% A, 80% B after 60 min. Mass spectra are acquired in the m/z 400–800 range.

Protein identification is carried out with the MS/MS search software Mascot 1.9 (Matrix Science) with confirmatory or complementary analyses with TurboSequest as implemented in the Bioworks Browsers 3.2, build 41 (ThermoFinnigan). NCBI Sus scrofa protein sequences are used as the primary search database; searches are complemented with the NCBI nonredundant protein database.

LC/MS/MS analyses after treating LDL (0.2 mg/ml) with 100 μM of $ONOO^-$ reveal the molecular mechanisms by which protein undergoes nitration in the presence of $ONOO^-$ (Table 7.2).

The tyrosine residues of LDL apoB-100 were susceptible to nitration in the α-1, α-2, α-3, and β-2 helices; specifically, α-1 (Tyr^{144}), α-2 (Tyr^{2524}), α-3 (Tyr^{4116}), β-2 (Tyr^{3295}), and β-2 (Tyr^{4211}). Relative Mascot and Sequest scores were obtained by software that demonstrated LDL modifications. Mascot is the score obtained using matrix science software to analyze the actual observed masses, followed by searching the database of proteins to determine the sequence of the digested protein (Hsiai et al., 2007; Perkins et al., 1999), whereas Sequest scores were obtained using Thermo Finnigan (Excaliber) software to determine the most likely peptide from actual observed peptide masses by searching a protein database. The representative MS/MS spectrum illustrates a tryptic peptide, NLQNNAEWVYQGAIR, which was modified with tyrosine residue 14 (Fig. 7.15). The Y and B ion series reflected the direction in which the mass of the ion was observed. The Y6 ion in the MS/MS spectrum showed the presence of tyrosine nitration as evidenced by an additional mass of 45 Da (NO_2 = 46 Da, but replacement of a hydrogen atom resulted in a mass of 45 Da). Further evidence for an additional 45 Da was identified in ions Y7, Y9, Y10, Y11, Y13, B13, and B14. The mass of modified and nonmodified peptides bears a mass one-half of the theoretical for doubly charged ions and one-third of the theoretical for triply charged ions. A mass difference of 22.5 Da was observed for the nitrotyrosine-containing peptide as compared to the unmodified peptide. The ion mass to charge ratio (m/z) of the entire peptide demonstrated that the observed mass of the modified peptide NLQNNAEWVYQGAIR for the doubly charged ion was 911.23 Da, whereas the unmodified peptide mass was 888.57 Da, consistent with the presence of nitrotyrosine.

Table 7.2 Tyrosine nitration of LDL apoB-100: Six tryptic peptides were identified to have undergone tyrosine nitration[a]

AA	Peptide	Charge	Sequence	M	Xcor	Δcn
Y144	140–157	3	QVFLY*PEKDEPTYILNIK	42	3.2	0.328
Y2524	2523–2534	2	M*Y*QM*DIQQELQR	71	4.4	0.264
Y2524	2523–2534	2	M*Y*QM*DIQQELQR	79	4.5	0.286
Y3295	3292–3311	3	VPSY*TLILPSLELPVLHVPR	78	5.9	0.558
Y4116	4107–4121	2	NLQNNAEWVY*QGAIR	97	4.7	0.478
Y4211	4202–4213	3	FQFPGKPGIY*TR	34	2.5	0.271

[a] Nitrotyrosines are denoted by asterisks. "AA" represents modified tyrosine residue, "Peptide" the numerical sequence of amino acid in the tryptic fragment, "Charge" the ion precursor charge of the tryptic fragment, "Sequence" the amino acid sequences of the tryptic fragment, "M" the Mascot score, "Xcor" the Sequest score, and "Δcn" the difference between two closely matched peptides.

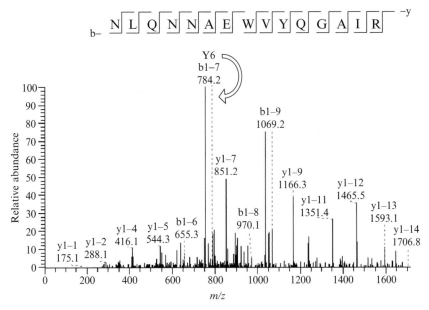

Figure 7.15 A representative MS/MS spectrum of nitro-modified peptide: The tryptic peptide, NLQNNAEWVYQGAIR, revealed that Y6 ion is the largest blue peak. The molecular mass of Y6 is greater in the modified peptide (752.4 Da) than in the nonmodified (707.3 Da) by 45 Da. Only representative peaks were labeled in the spectrum.

In summary, an imbalance in $O_2^{-\bullet}$ and $\bullet NO$ production in the PSS- and OSS-exposed regions of vasculatures influenced the formation of peroxynitrite, implicating nitrative modifications of the apoB-100 protein in LDL. By immunohistochemistry staining of human coronary arteries and applications of a dynamic flow system (Hsiai *et al.*, 2002b), OSS-exposed regions were prevalent areas of nitrative stress. Using a combination of LC/ESI/MS/MS, we showed that OSS favored the formation of LDL apoB-100 protein nitrotyrosine and that $ONOO^-$ specifically modified LDL protein tyrosine residues mostly in α-helices and in a minimal portion of β-2.

5. Emerging Technology: Micro- and Nanotechnology

Microelectrical mechanical systems technology explores the science of microrealm, in which surface tension and viscous force, rather than the force of gravity, influence the design and operation of sensors. Using tools originally developed for the integrated circuit (IC) industry, one is able to fabricate miniaturized transducers at cellular levels. MEMS devices enable real-time control of time-varying events. By virtue of their small dimension, direct interactions between MEMS components and biological entities at single cellular level are possible.

Likewise, nanoscale materials and systems rank among the most exciting developments in modern science and engineering. The study of nanomaterials and nanoscale science paves a pathway for an ambitious but realistic goal: the use of synthesis, assembly, and miniaturization down to the nanometer scale to create novel nano-structured materials and devices with unique properties and superior performance.

The first section discusses MEMS liquid chromatography to analyze Ox-LDL, which is considered as an emergent marker for unstable angina (Ehara *et al.*, 2001b). The second section focuses on nanowire-based field effect transistors (FET) to detect the redox state of LDL particles.

5.1. MEMS liquid chromatography

5.1.1. Introduction

Chromatography has made separation of molecules possible in a fast and efficient manner. Protein molecules are separated according to their physical properties, such as their size, shape, charge, hydrophobicity, and affinity for other molecules. High-performance liquid chromatography (HPLC) enables the separation of molecules under high pressure in a stainless steel column filled with beads. Although the LC column is normally made of capillary tubes due to fluidics limitations, miniaturization of the column can enhance separation performance. The first MEMS LC was reported by Y. C. Tai's group at the California Institute of Technology (Meyer, 1999) (Fig. 7.16). When identical separation chemistry applies, separated peak width is independent of column internal diameter. The "band width" of peaks is more distinct for smaller columns with smaller beads (Meyer, 1999). The MEMS-based LC provides high- pressure compatibility in the microfluidic devices (Fig. 7.16) (Qing *et al.*, 2004). The column is packed with conventional LC stationary phase support materials such as microbeads with surface functional groups. A self-aligned, channel-anchoring technique has been developed to increase the pressure rating of the Parylene microfluidic devices from 30 to at least 800 psi. MEMS-based LC also minimizes the input sample volume to less than 1 μl from the patients. Because of the MEMS LC, it is possible to measure the oxidative modification of LDL particles with sensitivity and selectivity.

5.1.2. MEMS LC for separation of LDL particles

Ox-LDL is considered an emergent marker for unstable angina (Ehara *et al.*, 2001a). Currently, conventional desktop HPLC has been used as a gold standard to obtain LDL subspecies through chromatograms (Sevanian *et al.*, 1994).

The goal is to develop a lab-on-a-chip HPLC system with which measurement of circulating ox-LDL can be achieved with a shorter sample

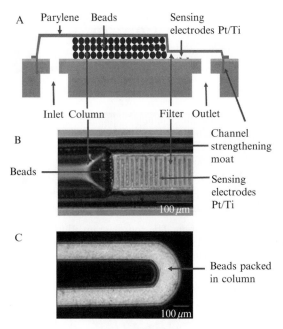

Figure 7.16 (A) Cross section of the device after packing. (B) Picture of column, packed beads, filter, detector, and channel-strengthening moat. (C) Fluorescent picture of a densely packed column.

time, higher throughput, less sample volume consumption, improved sensitivity, and, finally, lower cost. The MEMS HPLC system is comprised of a high-pressure parylene LC column, interdigital conductivity sensor, and integrated heater for on- chip temperature control (Fig. 7.17) (Qing *et al.*, 2004). The system for on-chip chromatography tasks is packaged with a PCB board for electric contacts, a MEMS LC chip with backside holes, a mini O ring for sealing, and the chemically inert PEEK jig to integrate the fitting with the MEMS LC chip (Fig. 7.18) (Qing *et al.*, 2004). Channel-anchoring techniques have been developed to increase the pressure compatibility of the Parylene microfluidic system to at least 800 psi (Qing *et al.*, 2004). When the MEMS LC column is packed with ion-exchange chromatography beads, LDL subspecies are separated by anion-exchange chromatography into native LDL (nLDL- reduced state) and ox-LDL in terms of LDL^- and LDL^{2-}, with LDL^{2-} being more electronegative (Sevanian *et al.*, 1997). The LDL chromatogram is obtained using isocratic elution and on-chip conductivity sensing (Fig. 7.19). The conductivity of the mobile phase solution provides a baseline signal. When separated LDL subspecies pass by the detector, changes of solution conductivity are detected.

The LDL sample contains 680 ng of nLDL and 120 ng of LDL^- and LDL^{2-} as validated by HPLC. The result is compared with the

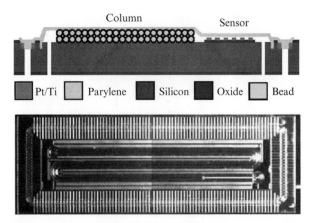

Figure 7.17 MEMS LC system. The top diagram illustrates the key components, including the column packed with beads and on-chip conductivity sensor. The photograph shows column. The column inner diameter is 50 μm and the length is 1 cm.

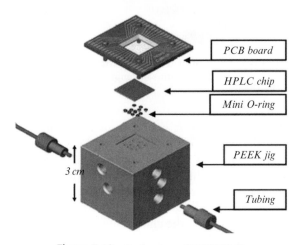

Figure 7.18 Packaging of MEMS LC.

chromatogram obtained from HPLC with salt gradient elution and UV detection (Fig. 7.20). Results demonstrate the feasibility to separate native LDL particles from the oxidized LDL at 680 and 120 ng, respectively.

5.2. Development of the MEMS LC

The LC-on-a-chip device is fabricated from parylene MEMS technology. The chip consists of four access ports (Fig. 7.21A). The LC mobile phase starts from port #1 to #2. Injection at a small sample volume was achieved by a cross-channel injection method from #3 to #4. Beads are packed externally

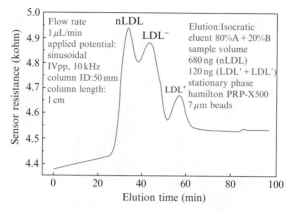

Figure 7.19 Chromatogram obtained from MEMS LC. The flow rate was controlled by applying sinusoidal potential at 1 Vpp and 10 kHz. The elution was isocratic. The eluent contains 80% of buffer A (20 mM Tris, pH 7.2) and buffer B (20 mM Tris, 1 M NaCl, pH 7.2). The separation column was packed with 7-μm beads.

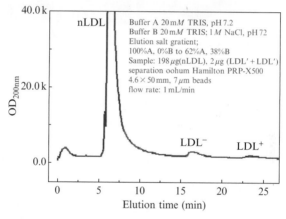

Figure 7.20 Chromatogram obtained from HPLC with the identical LDL sample and buffer solutions A and B. The elution contains a salt gradient from 100% A and 0% B to 62% A and 38% B. The flow rate was 1 ml/min. Identical beads used for the MEMS LC were packed in the Hamilton PRP-X500 (4.6 × 50-mm) separation column.

into the on-chip column from port #1 (O'Neill et al., 2002). Filters (channel with a height smaller than the bead size) trap beads in the main channel and prevent beads from entering the side injection channels. The bead-packed column inlet is near sample output port #4, which is the plug front of the injected sample. The outlet is designed at the filter/sensor edge (Fig. 7.21B). This integrated design minimizes dead volume and extra-column peak broadening. Interdigital electrodes are patterned to monitor liquid conductivity (He et al., 2004). When separated ionic LDL subspecies pass by the detector,

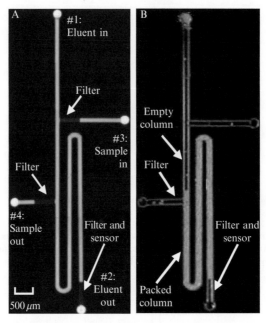

Figure 7.21   (A) Fluorescent overview of the integrated separation system. (B) Optical picture after packing beads.

changes of conductivity are detected. A chromatogram is obtained by recording the conductivity of the solution past the detection cell over time.

5.3. Performance of MEMS LC

5.3.1. Selectivity of conductive electrodes to detect LDL subspecies

Data (Figs. 7.19 and 7.20) demonstrated the feasibility of MEMS LC for separation of native LDL from oxidized LDL particles at 680 and 120 ng, respectively. The application of electrode conductivity is ideal for detecting charged particles, as native LDL is in a reduced state, whereas LDL^- and LDL^{2-} (ox-LDL) are electronegative in a oxidized state. The baseline conductance signal is generated by the eluent. When the sample peak enters the detector, a change in conductance can be measured. For anion-exchange ion chromatography, the magnitude of this change is expressed in the following equation, assuming that the samples and eluent are ionized completely (Fritz and Gjerde, 2000):

$$\Delta G = \frac{1}{10^{-3}K_{cell}}(\lambda_{S^-} - \lambda_{E^-})C_S, \qquad (7.1)$$

where λ_{S^-} and λ_{E^-} are the equivalent conductance for the sample and eluent anions, C_S is the concentration of the sample anion, and K_{cell} is the conductivity sensor cell constant. The conductance change ratio on sample elution is usually very small, only a few percent change in relation to the background conductivity:

$$\frac{\Delta G}{G_{Background}} = \frac{(\lambda_{S^-} - \lambda_{E^-})C_S}{(\lambda_{E^+} + \lambda_{E^-})C_E} \approx \frac{C_S}{C_E} \approx 1\%, \tag{7.2}$$

where C_E is the concentration of eluent anions. The sensitivity of conductivity detection is affected by the conductance noise level, sample injection volume, system dead volume, and inherent detector sensitivity. The conductance noise is mainly contributed from the temperature fluctuation-induced background conductance change. In general, a change in 1 °C temperature will result in a conductance change of 2%. Several alternatives are available to increase the sensitivity for conductivity detection: (1) suppressing the background conductance using a suppressor column inserted after the separation column (Fritz and Gjerde, 2000); (2) increasing the sample plug concentration using a sample preconcentration technique (Dionex Corp., Sunnyvale, CA); (3) reducing the sample injection volume and system dead volume; and (4) reducing the sensor cell constant, K_{cell}, which is especially achievable using MEMS technology (Qing et al., 2004). The state-of-the-art conductivity detection provides a detection limit of 100 ppb ($\sim$10 nM) for unsuppressed detection and 10 ppb ($\sim$1 nM) for suppressed detection using a 50-μl sample injection for separation (Fritz and Gjerde 2000).

5.3.2. Selectivity for blood plasma

Conductivity detection detects the overall sample conductance and cannot resolve the conductance contribution from each analyte if more than one analyte is present in the cell at the same time. The selectivity of LDL particles from other protein molecules in the blood plasma is established from the following separation steps: (1) the centrifugation step to separate LDL from the majority of plasma components and (2) the chromatography step to further purify LDL followed by LDL subspecies separation. Modern separation science has entered a regime where it is common to have more than 1 million theoretical plates in a separation column (Neue). This indicates the possibility that any mixture of analytes can be separated as long as the retention factor of each analyte is slightly different from each other. Alternatively, the selectivity between analytes can be adjusted by adding functional groups to the analytes through chemical derivatization (Boyd et al., 2000) or using multidimensional separation techniques (Houdiere et al., 1997), which are commonly used in the field of liquid chromatography.

5.3.3. Elution time

Using MEMS technology, the lab-on-a-chip LC system offers a great degree of flexibility in reducing (a) the separation column inner diameter ($<10 \mu m$), (b) the system dead volume ($<10 nl$), and (c) the sample injection volume ($<10 nl$) to levels theretofore impossible by conventional HPLC technologies (Ziouzenkova et al., 1998). This flexibility implies a higher magnitude of achievable separation efficiency. With higher separation efficiency, the column length can be shortened while the number of theoretical plates remains unchanged. Accordingly, a shorter column results in a reduced elution time. Although we demonstrated the isocratic elution of the LDL subfraction with the MEMS LC system in preliminary data (see earlier discussion), alternatively and preferably, we can carry out on-chip gradient elution to further reduce the elution time and the broadening effect.

5.3.4. Sample volume and minimal concentration for sensitivity

From preliminary data of isocratic LDL subspecies elution (Fig. 7.19), we demonstrated separation and detection of LDL particles at 800 ng LDL (nLDL at 680 ng and ox-LDL at 120 ng). The limit of detection (LOD), calculated based on a signal/noise ratio around 50, was 80 ng, assuming that a signal/noise ratio of 50 is acceptable for detection. The LOD of 80 ng corresponds to an original sample concentration LOD of 20 $\mu g/ml$ if the sample injection volume remains at 20 μl. By further reduction in system dead volume and sample injection volume from 10 μl to submicroliters in the MEMS LC, it will be feasible to significantly decrease the band-broadening effect and increase the signal/noise ratio, thereby reducing LOD down to 10 ng or 10 fmol of LDL particles.

5.4. Nanotechnology to monitor oxidative stress

Nanowire-based field effect transistors were developed to assess the flow regulation of LDL oxidation mediated by vascular endothelial cells. Indium oxide nanowires were synthesized with controlled diameters around 10 nm and lengths exceeding 5 μm. FET were constructed on the basis of a multitude of conducting channels between sources and drains using photo- and e-beam lithography. Insulating nanocircuitry in the presence of aqueous material was addressed by deposition of the parylene C polymer. HPLC was performed to validate the effluent fort the extent of LDL oxidation in parallel with detection by the nanowire-based FETs (Rouhanizadeh et al., 2004).

5.4.1. Nanowire sensors

Nanoscale materials and systems rank among the most exciting developments in modern science and engineering. The study of nanomaterials and nanoscale science paves a pathway for an ambitious but realistic goal: the use of

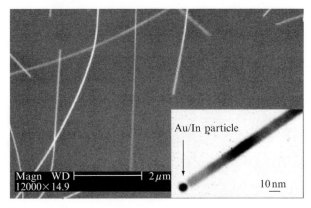

Figure 7.22 A representative SEM image of In_2O_3 nanowires synthesized by laser ablation on a Si/SiO_2 substrate using mono-dispersed 10-nm Au clusters as the catalyst. The inset is a TEM image of an In_2O_3 nanowire with a catalyst particle at the tip. The scale bar is 10 nm.

synthesis, assembly, and miniaturization down to the nanometer scale to create novel nano- structured materials and devices with unique properties and superior performance (Fig. 7.22. Despite its utmost role in the new generation of electronics, applications of nanomaterials and nanodevices in the biological realm remain an emerging field with numerous challenges but great potential. Therefore, biosensors based on individual semiconductive nanowires have the advantage of the enormous surface-to-volume ratios of such nanostructures and hold great promise to offer unprecedented sensitivity and response time, as well as the capability of directly converting biological signals to electrical signals. Because of their extremely small sizes, nanosensors will also have important applications in emerging research areas, such as *in vivo* monitoring and clinical diagnostics. We have demonstrated the feasibility of detecting LDL particles by an individual In_2O_3 (indium oxide) nanowire-based FET *in vitro* (Li *et al.*, 2003). The possibility of applying nanowire sensors to detect oxidative stress is of particular interest.

5.4.2. Selective detection of ox-LDL

At the heart of any sensing technology is selective detection of the targeted species. A major effort has been focused on the selectivity of nanowire/nanotube sensors by functionalization of the nanowire/nanotube surfaces with antibodies specific for LDL species. The first step is to decorate the nanowire/nanotube transistor surface with a layer of polyethylene imine (PEI) and polyethylene glycol (PEG), followed by conjugation with the ox-LDL antibody to the PEI/PEG surface. Hydrogen bonding between the carboxyl group at the Y end of the antibodies and the primary amines available in PEI is illustrated in Fig. 7.23. These devices are then loaded

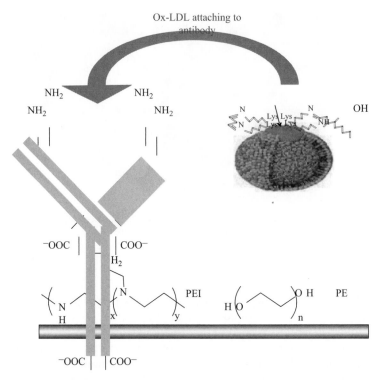

Figure 7.23   The nanosensor was coated with a layer of PEI/PEG polymer, followed by the attachment of antibodies to the primary amine of PEI. This sensor was then exposed to various biospecies, including ox–LDL and oxidized cytochrome *c*. Selectivity is achieved via the selective binding of ox–LDL to the antibodies.

into the fluid channel shown in Fig. 7.17 or sometimes in a simple vial. The sensing events are carried out by first exposing the devices to a buffer solution, followed by adding a small amount of ox–LDL diluted in buffer. The electrical response before and after the ox–LDL addition is thus linked directly to the attachment and chemical gating of the LDL species to the semiconducting nanosensor.

Networked nanotube devices are tested for exposure of ox–LDL species (Figs. 7.24 and 7.25). A reduction in current occurs upon sensor exposure to ox–LDL. When oxidized cytochrome is added, the change in the current is small.

5.5. Selectivity by individual nanowire-based FETs

In the presence of Fe^{3+} cytochrome *c*, the In_2O_3-based FET exhibits a distinct conductivity from that of ox–LDL (Fig. 7.26). DI water at 10 μl min is used to establish the based–line I-V$_{DS}$ curve. Figure 7.25 demonstrates the

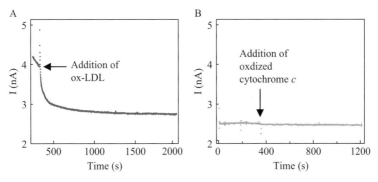

Figure 7.24 Current response of a carbon nanotube network sample functionalized with antibodies upon exposure to ox-LDL (A) and oxidized cytochrome (B). The selective response to ox-LDL is clearly observed.

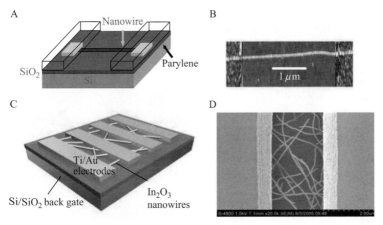

Figure 7.25 Nanowire-based FET. (A) In$_2$O$_3$ nanowire transistor with metal electrode passivated by parylene. (B) Scanning electron microscope (SEM) of nanowire-based FET. (C) Concept of In$_2$O$_3$ nanowire networked transistor. (D) SEM image of the active area of one device showing multiple In$_2$O$_3$ nanowires bridging a pair of electrodes.

feasibility of developing In$_2$O$_3$ nanowires to detect redox state of protein molecules. Using both I$_D$–V$_{DS}$ (current versus drain-source voltage) and I$_D$–V$_{GS}$ (current versus gate-source voltage) (Fig. 7.26), we found that the sample containing more oxidized LDL particles, and consequently more free electrons, increases the conductivity of the nanowire-based FET. This increase in conductivity is distinct from the presence of ferrocytochrome c (Rouhanizadeh et al., 2004).

The feasibility of developing In$_2$O$_3$ nanowires to detect the redox state of protein molecules is demonstrated in Fig. 7.27. Two factors may account

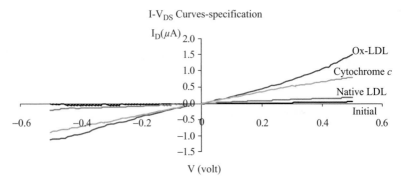

Figure 7.26 Individual I-V_d curves demonstrate distinct differences in conductivities among ox-LDL, nLDL, Fe^{3+}-cytochrome *c*, and initial state.

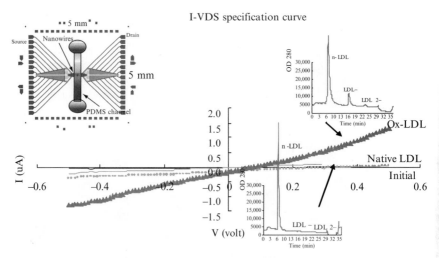

Figure 7.27 I-V_{DS} selectivity curve under dynamic condition. The presence of ox-LDL (15.1%) gave rise to the highest conductivity compared with nLDL (4.4%). HPLC chromatograms of LDL were used to validate the results. Three peaks are seen, starting with the most abundant native LDL (nLDL) and oxidized LDL. The top left insert shows the electrodes of source and drain in relation to the nanowires.

for the increased electron concentration in the nanowires. The first was that the amino groups carried by the apoB-100 protein in LDL particles may function as reductive species and hence donate electrons to the nanowires. The second concomitant factor was due to positive charges carried by the amino groups, which can function as a positive gate bias to our nanowires, thus leading to the enhanced carrier concentration (Li *et al.*, 2003). Rouhanizadeh *et al.* (2004) reported that In_2O_3 nanowire-based FET as an emergent sensor to detect ox-LDL. Using both the I_D-V_{DS} (current versus drain-source voltage) and I_D-V_{GS} (current versus gate-source voltage), the investigators found that the sample containing more oxidized LDL particles, and consequently more free electrons, increased the conductivity of the nanowire-based FET. This increase in conductivity was distinct from the presence of ferrocytochrome c.

6. SUMMARY

Atherosclerosis, which involves complex plaque formation in the arterial vessels, is considered to be an inflammatory disease. Ox-LDL has been considered important in the pathogenesis of these inflammatory processes with the seminal observation that LDL must be modified oxidatively for it to be taken up by macrophages (Ross, 1999; Witztum, 1994). At arterial bifurcations or branching points, endothelial cells become hyperpermeable in the presence of hyperlipidemia. This may favor intimal uptake and retention of LDL, resulting in local oxidative degradation of trapped LDL. Furthermore, the characteristics of shear stress acting on endothelial cells at these regions is known to regulate NADPH oxidase activities, which is considered a major source of endothelial superoxide $\left(O_2^{-}\right)$ production. Therefore, hemodynamics, particularly shear stress, plays an important role in regulating the biological activities of endothelial cells.

Emerging MEMS sensors provide a spatial and temporal resolution to investigate the mechanisms, whereby the characteristics of shear stress regulate the biological activities of endothelial cells at the complicated arterial geometry. In parallel, the development of MEMS liquid chromatography provides a new venue to measure circulating ox-LDL particles as a lab-on-a chip platform. Nanowire-based FETs further pave the way for a high throughput approach to analyze the LDL redox state.

Integration of MEMS with oxidative biology is synergistic to assess vascular oxidative stress. The MEMS LC provides an emerging lab-on-a-chip platform for ox-LDL analysis. In this context, this chapter has integrated expertise from the fields of vascular biology and oxidative biology to assess the dynamics of inflammatory responses.

REFERENCES

Ago, T., Kitazono, T., Ooboshi, H., Iyama, T., Han, Y. H., Takada, J., Wakisaka, M., Ibayashi, S., Utsumi, H., and Lida, M. (2004). Nox4 as the major catalytic component of an endothelial NAD(P)H oxidase. *Circulation* **109**, 227–233.
Austin, M. A. (1994). Small, dense low-density lipoprotein as a risk factor for coronary heart disease. *Int. J. Clin. Lab. Res.* **24**, 187–192.

Blackman, B. R., Garcia-Cardena, G., and Gimbrone, M. A., Jr. (2002). A new *in vitro* model to evaluate differential responses of endothelial cells to simulated arterial shear stress waveforms. *J. Biomech. Eng.* **124,** 397–407.

Boyd, B., Witowski, S., and Kennedy, P. (2000). Trace-level amino acid analysis by capillary liquid chromatography and application to *in vivo* microdialysis sampling with 10-s temporal resolution. *Anal. Chem.* **72,** 865–871.

Braman, R. S., and Hendrix, S. A. (1989). Nanogram nitrite and nitrate determination in environmental and biological materials by vanadium (III) reduction with chemilumines-cence detection. *Anal. Chem.* **61,** 2715–2718.

Brennan, M. L., Wu, W., Fu, X., Shen, Z., Song, W., Frost, H., Vadseth, C., Narine, L., Lenkiewicz, E., Borchers, M. T., Lusis, A. J., Lee, J. J., *et al.* (2002). A tale of two controversies: Defining both the role of peroxidases in nitrotyrosine formation *in vivo* using eosinophil peroxidase and myeloperoxidase-deficient mice, and the nature of peroxidase-generated reactive nitrogen species. *J. Biol. Chem.* **277,** 17415–17427.

Byrne, J. A., Grieve, D. J., Bendall, J. K., Li, J. M., Gove, C., Lambeth, J. D., Cave, A. C., and Shah, A. M. (2003). Contrasting roles of NADPH oxidase isoforms in pressure-overload versus angiotensin II-induced cardiac hypertrophy. *Circ. Res.* **93,** 802–805.

Cesselli, D., Jakoniuk, I., Barlucchi, L., Beltrami, A. P., Hintze, T. H., Nadal-Ginard, B., Kajstura, J., Leri, A., and Anversa, P. (2001). Oxidative stress-mediated cardiac cell death is a major determinant of ventricular dysfunction and failure in dog dilated cardiomyop-athy. *Circ. Res.* **89,** 279–286.

Davies, P. F., Dewey, C. F., Jr., Bussolari, S. R., Gordon, E. J., and Gimbrone, M. A., Jr. (1984). Influence of hemodynamic forces on vascular endothelial function: *In vitro* studies of shear stress and pinocytosis in bovine aortic cells. *J. Clin. Invest.* **73,** 1121–1129.

De Keulenaer, G. W., Chappell, D. C., Ishizaka, N., Nerem, R. M., Alexander, R. W., and Griendling, K. K. (1998). Oscillatory and steady laminar shear stress differentially affect human endothelial redox state: Role of a superoxide-producing NADH oxidase. *Circ. Res.* **82,** 1094–1101.

DePaola, N., Gimbrone, M. A., Jr., Davies, P. F., and Dewey, C. F., Jr. (1992). Vascular endothelium responds to fluid shear stress gradients. *Arterioscler. Thromb.* **12,** 1254–1257.

Dewey, C. F., Bussolari, S. R., Gimbrone, M. A., Jr., and Davies, P. F. (1981). The dynamic response of vascular endothelial cells to fluid shear stress. *J. Biomech. Eng.* **103,** 177–185.

Ehara, S., Ueda, M., Naruko, T., Haze, K., Itoh, A., Otsuka, M., Komatsu, R., Matsuo, T., Itabe, H., Takano, T., Tsukamoto, Y., Yoshiyama, M., *et al.* (2001). Elevated levels of oxidized low density lipoprotein show a positive relationship with the severity of acute coronary syndromes. *Circulation* **103,** 1955–1960.

Fabjan, J. S., Abuja, P. M., Schaur, R. J., and Sevanian, A. (2001). Hypochlorite induces the formation of LDL(-), a potentially atherogenic low density lipoprotein subspecies. *FEBS Lett.* **499,** 69–72.

Foster, M. W., McMahon, T. J., and Stamler, J. S. (2003). S-nitrosylation in health and disease. *Trends Mol. Med.* **9,** 160–168.

Frangos, J. A., Huang, T. Y., and Clark, C. B. (1996). Steady shear and step changes in shear stimulate endothelium via independent mechanisms-superposition of transient and sustained nitric oxide production. *Biochem. Biophys. Res. Commun.* **224,** 660–665.

Fritz, J., and Gjerde, D. (2000). "Ion Chromatography," 3rd Ed. Wiley-VCH, New York.

Fung, Y. C. (1997). "Biomechanics: Circulation," 2nd Ed. Springer, New York.

Fung, Y. C., and Liu, S. Q. (1993). Elementary mechanics of the endothelium of blood vessels. *J. Biomech. Eng.* **115,** 1–12.

Görlach, A., Brandes, R. P., Nguyen, K., Amidi, M., Dehghani, F., and Busse, R. (2000). A gp91phox containing NADPH oxidase selectively expressed in endothelial cells is a major source of oxygen radical generation in the arterial wall. *Circ. Res.* **87,** 26–32.

Griendling, K. K., and FitzGerald, G. A. (2003). Oxidative stress and cardiovascular injury. II. Animal and human studies. *Circulation* **108**, 2034–2040.

Griendling, K. K., and FitzGerald, G. A. (2003). Oxidative stress and cardiovascular injury. I. Basic mechanisms and *in vivo* monitoring of ROS. *Circulation* **108**, 1912–1916.

Griendling, K. K., Sorescu, D., and Ushio-Fukai, M. (2000). NAD(P)H oxidase: Role in cardiovascular biology and disease. *Circ. Res.* **86**, 494–501.

Harrison, D., Greindling, K. K., Landmesser, U., Hornig, B., and Drexler, H. (2003). Role of oxidative stress in atherosclerosis. *Am. J. Cardiol.* **91**, 7A–11A.

He, Q., Liu, J., Sun, X., and Zhang, Z. R. (2004). Preparation and characteristics of DNA-nanoparticles targeting to hepatocarcinoma cells. *World J. Gastroenterol.* **10**(5), 660–663.

Helmlinger, G., Geiger, R. V., Schreck, S., and Nerem, R. M. (1991). Effects of pulsatile flow on cultured vascular endothelial cell morphology. *J. Biomech. Eng.* **113**, 123–131.

Henriksen, T., Mahoney, E. M., and Steinberg, D. (1981). Enhanced macrophage degradation of low density lipoprotein previously incubated with cultured endothelial cells: Recognition by receptors for acetylated low density lipoproteins. *Proc. Natl. Acad. Sci. USA* **78**, 6499–6503.

Hilenski, L. L., Clempus, R. E., Quinn, M. T., Lambeth, J. D., and Griendling, K. K. (2004). Distinct subcellular localizations of Nox1 and Nox4 in vascular smooth muscle cells. *Arterioscler. Thromb. Vasc. Biol.* **24**, 677–683.

Houdiere, F., Fowler, P., and Djordjevic, N. (1997). Combination of column temperature gradient and mobile phase flow gradient in microcolumn and capillary column high-performance liquid chromatography. *Anal. Chem.* **69**, 2589–2593.

Hsiai, T. K., Cho, S. K., Honda, H. M., Hama, S., Navab, M., Demer, L. L., and Ho, C. M. (2002). Endothelial cell dynamics under pulsating flows: Significance of high versus low shear stress slew rates (delta tau/delta t). *Annals of Biomedical Engineering* **30**, 646–656.

Hsiai, T. K., Cho, S. K., Reddy, S., Hama, S., Navab, M., Demer, L. L., Honda, H. M., and Ho, C. M. (2001). Pulsatile flow regulates monocyte adhesion to oxidized lipid-induced endothelial cells. *Arteriosclerosis Thrombosis and Vascular Biology* **21**, 1770–1776.

Hsiai, T. K., Cho, S. K., Ing, M., Navab, M., Reddy, S., and Ho, C. M. (2002b). "Microsensors to Characterize Shear Stress Regulating Inflammatory Responses in the Arterial Bifurcations." Proceedings of the Second Joint Meeting of the IEEE Engineering in Medicine and Biology and the Biomedical Engineering Society, Vol. 7, pp. 2–5. Houston, TX.

Hsiai, T. K., Cho, S. K., Wang, P. K., Ing, M. H., Salazar, A., Hama, S., Navab, M., Demer, L. L., and Ho, C. H. (2004). Micro sensors: Linking vascular inflammatory responses with real-time oscillatory shear stress. *Ann. Biomed. Eng.* **32**, 189–201.

Hsiai, T. K., Cho, S. K., Wong, P. K., Ing, M., Salazar, A., Sevanian, A., Navab, M., Demer, L. L., and Ho, C. M. (2003). Monocyte recruitment to endothelial cells in response to oscillatory shear stress. *FASEB J.* **17**, 1648–1657.

Hsiai, T. K., Hwang, J., Barr, M. L., Correa, A., Hamilton, R., Alavi, M., Rouhanizadeh, M., Cadenas, E., and Hazen, S. L. (2007). Hemodynamics influences vascular peroxynitrite formation: Implication for low-density lipoprotein apo-B-100 nitration. *Free Radic. Biol. Med.* **42**, 519–529.

Hwang, J., Ing, M. H., Salazar, A., Lassegue, B., Griendling, K., Navab, M., Sevanian, A., and Hsiai, T. K. (2003a). Pulsatile versus oscillatory shear stress regulates NADPH oxidase subunit expression: Implication for native LDL oxidation. *Circ. Res.* **93**, 1225–1232.

Hwang, J., Rouhanizadeh, M., Hamilton, R. T., Lin, T. C., Eiserich, J. P., Hodis, H. N., and Hsiai, T. K. (2006). 17beta-Estradiol reverses shear-stress-mediated low density lipoprotein modifications. *Free Radic. Biol. Med.* **41**, 568–578.

Hwang, J., Saha, A., Boo, Y. C., Sorescu, G. P., McNally, J. S., Holland, S. M., Dikalov, S., Giddens, D. P., Griendling, K. K., Harrison, D. G., and Jo, H. (2003b). Oscillatory shear stress stimulates endothelial production of O2- from p47phox-dependent NAD(P)H oxidases, leading to monocyte adhesion. *J. Biol. Chem.* **278,** 47291–47298.

Hwang, J., Wang, J., Morazzoni, P., Hodis, H. N., and Sevanian, A. (2003c). The phytoestrogen equol increases nitric oxide availability by inhibiting superoxide production: An antioxidant mechanism for cell-mediated LDL modification. *Free Radic. Biol. Med.* **34,** 1271–1282.

Irani, K. (2000). Oxidant signaling in vascular cell growth, death, and survival: A review of the roles of reactive oxygen species in smooth muscle and endothelial cell mitogenic and apoptotic signaling. *Circ. Res.* **87,** 179–183.

Jialal, I., and Devaraj, S. (2003). Antioxidants and atherosclerosis: Don't throw out the baby with the bath water. *Circulation* **107,** 926–928.

Jung, O., Schreiber, J. G., Geiger, H., Pedrazzini, T., Busse, R., and Brandes, R. P. (2004). gp91phox-containing NADPH oxidase mediates endothelial dysfunction in renovascular hypertension. *Circulation* **109,** 1798–1801.

Khan, S. A., Lee, K., Minhas, K. M., Gonzalez, D. R., Raju, S. V., Tejani, A. D., Li, D., Berkowitz, D. E., and Hare, J. M. (2004). Neuronal nitric oxide synthase negatively regulates xanthine oxidoreductase inhibition of cardiac excitation-contraction coupling. *Proc. Natl. Acad. Sci. USA* **101,** 15944–15948.

Ku, D. N. (1997). Blood flow in arteries. *Annu. Rev. Fluid Mech.* **29,** 399–434.

Ku, D. N., Giddens, D. P., Zarins, C. K., and Glagov, S. (1985). Pulsatile flow and atherosclerosis in the human carotid bifurcation: Positive correlation between plaque location and low oscillating shear stress. *Arteriosclerosis* **5,** 293–302.

Landmesser, U., and Harrison, D. G. (2001). Oxidant stress as a marker for cardiovascular events: Ox marks the spot. *Circulation* **104,** 2638–2640.

Landmesser, U., Spiekermann, S., Dikalov, S., Tatge, H., Wilke, R., Kohler, C., Harrison, D. G., Hornig, B., and Drexler, H. (2002). Vascular oxidative stress and endothelial dysfunction in patients with chronic heart failure: Role of xanthine-oxidase and extracellular superoxide dismutase. *Circulation* **106,** 3073–3078.

Li, C., Zhang, X., Liu, X., Han, S., Tang, T., Han, J., and Zhou, C. (2003). In$_2$O$_3$ nanowires as chemical sensors. *Appl. Phys. Lett.* **82**(10), 1613–1615.

Malek, A. M., Alper, S. L., and Izumo, S. (1999). Hemodynamic shear stress and its role in atherosclerosis. *J. Am. Med. Assoc.* **282,** 2035–2042.

Meyer, V. (1999). Practical high-performance liquid chromatography. 3rd ed. New York: John Wiley and Sons.

Nerem, R. M., Alexander, R. W., Chappell, D. C., Medford, R. M., Varner, S. E., and Taylor, W. R. (1998). The study of the influence of flow on vascular endothelial biology. *Am. J. Med. Sci.* **316,** 169–175.

O'Neill, A., O'Brien, P., and Verpoorte, E. (2002). An integrated fritless column for on-chip capillary electrochromatography with conventional stationary phases. *Anal. Chem.* **74,** 639–647.

Passerini, A. G., Polacek, D. C., Shi, C. Z., Francesco, N. M., Manduchi, E., Grant, G. R., Pritchard, W. F., Powell, S., Chang, G. Y., Stoeckert, C. J., and Davies, P. F. (2004). Coexisting proinflammatory and antioxidative endothelial transcription profiles in a disturbed flow region of the adult porcine aorta. *Proc. Natl. Acad. Sci. USA* **101,** 2482–2487.

Perkins, D. N., Pappin, D. J. C., Creasy, D. M., and Cottrell, J. S. (1999). Probability-based protein identification by searching sequence databases using mass spectrometry data. *Electrophoresis* **20,** 3551–3567.

Pietraforte, D., Salzano, A. M., Scorza, G., and Minetti, M. (2004). Scavenging of reactive nitrogen species by oxygenated hemoglobin: Globin radicals and nitrotyrosines distinguish nitrite from nitric oxide reaction. *Free Radic Biol. Med.* **37.**

Qing, H. C., Tai, Y.-C., and Lee, T. (2004). "Ion Liquid Chromatography On-a-Chip with Beads-Packed Parylene Column. The 17th IEEE International Conference on Micro Electro Mechanical Systems (MEMS 2004). Maastricht, The Netherlands, pp. 212–215.

Ross, R. (1999). Atherosclerosis is an inflammatory disease. *Am. Heart J.* **138**, S419–S420.

Rouhanizadeh, M., Hwang, J., Clempus, R. E., Marcu, L., Lassegue, B., Sevanian, A., and Hsiai, T. K. (2005a). Oxidized-1-palmitoyl-2- arachidonoyl-sn-glycero-3-phosphorylcholine induces vascular endothelial superoxide production: Implication of NADPH oxidase.. *Free Radic. Biol. Med.* **39**, 1512–1522.

Rouhanizadeh, M., Lin, T. C., Arcas, D., Hwang, J., and Hsiai, T. K. (2005b). Spatial variations in shear stress in a 3-d bifurcation model at low reynolds numbers. *Ann. Biomed. Eng.* **33**, 1360–1374.

Rouhanizadeh, M., Tang, T., Li, C., Soundararajan, G., Zhou, C., and Hsiai, T. K. (2004). "Applying Indium Oxide Nanowires as Sensitive and Selective Redox Protein Sensors." 17th IEEE International Conference on Micro Electro Mechanical Systems (MEMS 2004), Vol 1, Maastricht, The Netherlands, pp. 431–434.

Sevanian, A., Asatryan, L., and Ziouzenkova, O. (1999). Low density lipoprotein (LDL) modification: Basic concepts and relationship to atherosclerosis. *Blood Purif.* **17**, 66–78.

Sevanian, A., and Hodis, H. (1997). Antioxidants and atherosclerosis: An overview. *Biofactors* **6**, 385–390.

Sevanian, A., BittoloBon, G., Cazzolato, G., Hodis, H., Hwang, J., Zamburlini, A., Maiorino, M., Ursini, F. (1997). LDL- is a lipid hydroperoxide-enriched circulating lipoprotein. *J. Lipid Res.* **38**, 419–428.

Sevanian, A., Hwang, J., Hodis, H., Cazzolato, G., Avogaro, P., and Bittolo-Bon, G. (1996). Contribution of an *in vivo* oxidized LDL to LDL oxidation and its association with dense LDL subpopulations. *Arterioscler. Thromb. Vasc. Biol.* **16**, 784–793.

Sevanian, A., Seraglia, R., Traldi, P., Rossato, P., Ursini, F., and Hodis, H. N. (1994). Analytical approaches to the measurement of plasma cholesterol oxidation products using gas- and high-performance liquid chromatography/mass spectrometry. *Free Radic, Biol. Med.* **17**, 397–410.

Salonen, J. T., Yla-Herttuala, S., Yamamoto, R., Butler, S., Korpela, H., Salonen, R., Nyyssonen, K., Palinski, W., and Witztum, J. L. (1992). Autoantibody against oxidised LDL and progression of carotid atherosclerosis. *Lancet* **339**, 883–887.

Siwik, D. A., Tzortzis, J. D., Pimental, D. R., Chang, D. L., Pagano, P. J., Singh, K., Sawyer, D. B., and Colucci, W. S. (1999). Inhibition of copper-zinc superoxide dismutase inhibits cell growth, hypertrophic phenotype, and apoptosis in neonatal rat cardiac myocytes *in vitro*. *Circ. Res.* **85**, 147–153.

Sorescu, D., Weiss, D., Lassegue, B., Clempus, R. E., Szocs, K., Sorescu, G. P., Valppu, L., Quinn, M. T., Lambeth, J. D., Vega, J. D., Taylor, W. R., and Griendling, K. K. (2002). Superoxide production and expression of nox family proteins in human atherosclerosis. *Circulation* **105**, 1429–1435.

Sorescu, G. P., Song, H., Tressel, S. L., Hwang, J., Dikalov, S., Smith, D. A., Boyd, N. L., Platt, M. O., Lassegue, B., Grindling, K. K., and Jo, H. (2004). Bone morphogenic protein 4 produced in endothelial cells by oscillatory shear stress induces monocyte adhesion by stimulating reactive oxygen species production from a nox1-based NADPH oxidase. *Circ Res.* **95**, 773–779.

Stone, W. L., Heimberg, M., Scott, R. L., LeClair, I., and Wilcox, H. G. (1994). Altered hepatic catabolism of low-density lipoprotein subjected to lipid peroxidation *in vitro*. *Biochem. J.* **297**(Pt 3), 573–579.

Topper, J. N., Cai, J. X., Falb, D., and Gimbrone, M. A. (1996). Identification of vascular endothelial genes differentially responsive to fluid mechanical stimuli: Cyclooxygenase-2, manganese superoxide dismutase, and endothelial cell nitric oxide synthase are selectively up-regulated by steady laminar shear stress. *Proc. Natl. Acad. Sci. USA* **93,** 10417–10422.

Wesley, R. B., 2nd, Meng, X., Godin, D., and Galis, Z. S. (1998). Extracellular matrix modulates macrophage functions characteristic to atheroma: Collagen type I enhances acquisition of resident macrophage traits by human peripheral blood monocytes *in vitro*. *Arterioscler. Thromb. Vasc. Biol.* **18,** 432–440.

Wink, D. A., Cook, J. A., Kim, S. Y., Vodovotz, Y., Pacelli, R., Krishna, M. C., Russo, A., Mitchell, J. B., Jourd'heuil, D., Miles, A. M., and Grisham, M. B. (1997). Superoxide modulates the oxidation and nitrosation of thiols by nitric oxide-derived reactive intermediates: Chemical aspects involved in the balance between oxidative and nitrosative stress. *J. Biol. Chem.* **272,** 11147–11151.

Witztum, J. R. (1994). The oxidation hypothesis of atherosclerosis. *Lancet* **344,** 793–795.

Ziegler, T., Bouzourene, K., Harrison, V. J., Brunner, H. R., and Hayoz, D. (1998). Influence of oscillatory and unidirectional flow environments on the expression of endothelin and nitric oxide synthase in cultured endothelial cells. *Arterioscler. Thromb. Vasc. Biol.* **18,** 686–692.

Ziouzenkova, O., Sevanian, A., Abuja, P. M., Ramos, P., and Esterbauer, H. (1998). Copper can promote oxidation of LDL by markedly different mechanisms. *Free Radic. Biol. Med.* **24,** 607–623.

CHAPTER EIGHT

DETERMINATION OF *S*-NITROSOTHIOLS IN BIOLOGICAL AND CLINICAL SAMPLES USING ELECTRON PARAMAGNETIC RESONANCE SPECTROMETRY WITH SPIN TRAPPING

Paul G. Winyard,* Iona A. Knight,* Frances L. Shaw,* Sophie A. Rocks,[†] Claire A. Davies,[‡] Paul Eggleton,* Richard Haigh,*,[§] Matthew Whiteman,* *and* Nigel Benjamin*

Contents

Abstract

S-Nitroso moieties, such as the *S*-nitroso group within *S*-nitrosated albumin, constitute a potential endogenous reservoir of nitric oxide (NO·) in human tissues and other biological systems. Moreover, *S*-nitroso compounds are under investigation as therapeutic agents in humans. Therefore, it is important

* Peninsula Medical School, Universities of Exeter and Plymouth, St. Luke's Campus, Exeter, United Kingdom
[†] Microsystems and Nanotechnology Centre, Department of Materials, School of Applied Sciences, Cranfield University, Cranfield, United Kingdom
[‡] Genzyme Corporation, Framingham, Massachusetts
[§] Department of Rheumatology, Princess Elizabeth Orthopaedic Centre, Royal Devon and Exeter NHS Foundation Trust (Wonford), Exeter, United Kingdom

Methods in Enzymology, Volume 441
ISSN 0076-6879, DOI: 10.1016/S0076-6879(08)01208-1

to be able to detect *S*-nitrosothiols (RSNOs) in human extracellular fluids, such as plasma and synovial fluid, as well as other biological samples. This chapter describes a method for the determination of *S*-nitrosothiols in biofluids. The method is based on electron paramagnetic resonance (EPR) spectrometry, in combination with spin trapping using a ferrous ion complex of the iron chelator *N*-methyl-D-glucamine dithiocarbamate under alkaline conditions. This iron complex mediates the decomposition of RSNO to NO·, as well as spin trapping the generated NO·. The resulting spin adduct has a unique EPR signal that can be quantified.

1. INTRODUCTION

A number of compounds have been suggested as reservoirs of nitric oxide (NO·) in biological systems, including nitrite (Cosby *et al.*, 2003; Webb *et al.*, 2004), nitrated lipids (Lim *et al.*, 2002), and *S*-nitrosothiols (Foster *et al.*, 2003). As NO· is short-lived in biological environments—and therefore difficult to measure directly in complex biological matrices—these reservoirs can be used as indicators of NO· availability. Protein *S*-nitrosothiol (RSNO) formation (*S*-nitrosation) is also thought to represent a key posttranslational modification with a role in cell signaling in a manner analogous to protein *O*-phosphorylation (Lane *et al.*, 2001). Dysregulation of NO· metabolism that leads to changes in blood *S*-nitrosothiol concentrations has been implicated in a wide variety of pathophysiological situations (Carver *et al.*, 2005; Foster *et al.*, 2003; Taylor and Winyard, 2007). Moreover, *S*-nitrosothiol-containing compounds, such as *S*-nitrosated albumin and *S*-nitrosated nonsteroidal anti-inflammatory drugs, have been developed with a view to therapeutic applications (Hallstrom *et al.*, 2006; Richardson and Benjamin, 2002).

There are a number of methods for measuring *S*-nitrosothiols, each with strengths and weaknesses. Many of these methods are described in other parts of this volume. They include the Saville assay (Saville, 1958), the "biotin-switch" method (Jaffrey *et al.*, 2001), chemiluminescence (MacArthur *et al.*, 2007), and gas chromatography–mass spectrometry (Tsikas *et al.*, 2002). This field of bioanalysis has also been controversial (Giustarini *et al.*, 2007). Reported concentrations of RSNOs in human plasma (and other biofluids) have varied greatly among different laboratories and methods of RSNO detection, with reported values ranging from less than 50 nM (e.g., Marley *et al.*, 2000; Rassaf *et al.*, 2002) to several micromolar (e.g, Stamler *et al.*, 1992; Hilliquin *et al.*, 1997). In general, reports using apparently improved methodologies have provided the lower values (Gladwin *et al.*, 2006). Nevertheless, many of the commonly used methods involve multiple steps and sample clean-up, making some of them unsuitable for semiroutine clinical analysis.

In order to provide a rapid method for the detection of *S*-nitrosothiols in clinical and biological samples, we have developed a method (Rocks *et al.*, 2005) that relies on the use of a ferrous ion complex of the iron chelator *N*-methyl-D-glucamine dithiocarbamate (MGD) at high pH. This complex—Fe^{2+}–$(MGD)_2$—both mediates the decomposition of RSNO to NO$^\cdot$ and spin traps the liberated NO$^\cdot$ (Arnelle *et al.*, 1997; Tsuchiya *et al.*, 2002). The resulting spin adduct has a unique electron paramagnetic resonance (EPR) signal that can be quantified. Deproteinization of extracellular fluid samples is not required. The assay has been applied to the detection of *S*-nitrosothiols in human plasma and synovial fluid (Rocks *et al.*, 2005), neonatal calf plasma (Christen *et al.*, 2007), and rabbit plasma in a model of liver ischemia/reperfusion injury (Glantzounis *et al.*, 2007). The coefficient of variation of the assay was 4.7%, the recovery of *S*-nitrosoglutathione (GSNO) spiked into synovial fluid was 71%, and the limit of detection of the assay was about 50 n*M* (Rocks *et al.*, 2005), which is consistent with the reported limit of detection of *S*-nitrosoglutathione by Fe^{2+}–$(MGD)_2$ (Tsuchiya *et al.*, 2002). Nitrite, nitrate, and 3-nitrotyrosine did not interfere with the assay under the conditions used (Rocks *et al.*, 2005). The median concentrations of RSNO in the human fluids tested were healthy human plasma, 0 n*M*; synovial fluid from knee joints of rheumatoid arthritis patients (i.e., a site of inflammation), 309 n*M*; and plasma from rheumatoid arthritis patients, 109 n*M* (Rocks *et al.*, 2005).

The main contributor (about 80%) of total free thiols in human plasma is albumin, since this protein is present at a concentration of about 0.6 m*M* and each molecule of albumin has a single free thiol associated with the cysteine-34 residue. This free cysteine thiol group is thought to be the main site of *S*-nitrosation in plasma. It is therefore generally assumed that RSNOs detected in human plasma mainly reflect the presence of *S*-nitrosoalbumin (Stamler *et al.*, 1992). In a study of newborn calf plasma (Christen *et al.*, 2007), changes in RSNO measured by the EPR assay were in good agreement with the semiquantitative determination of protein RSNO by Western blotting using an antibody to *S*-nitrosocysteine. Ultrafiltration of the plasma (Christen *et al.*, 2007) showed that the majority of the RSNO-derived signal, detected by the EPR assay, was associated with a high molecular weight protein fraction.

2. MATERIALS AND METHODS

2.1. Materials

All chemicals and glass vials are from Sigma–Aldrich (Dorset, UK), except GSNO, which is from Axxora Ltd. (Nottingham, UK), and MGD, which is from Dr. L. Hamilton (Randox Laboratories Ltd., County Antrim, UK).

MGD is available commercially (e.g., Acros Organics and Axxora). The vacutainers (sterile, no additives, 10 ml volume) are from Becton-Dickinson, Oxford, UK.

2.2. Assay protocol

The present assay for RSNOs employs EPR spectrometry in conjunction with spin trapping: the RSNOs in biofluids such as plasma are degraded using an alkaline pH (pH 10.6) in the presence of the spin trap complex $Fe^{2+}-(MGD)_2$. When collecting clinical samples for the assay, blood samples (1 ml) are added to EDTA-coated tubes containing 10 μl of 250 mM N-ethylmaleimide (NEM). The samples are centrifuged immediately at 3000 rpm for 10 min. The supernatant is aliquoted into small centrifuge tubes, snap frozen in liquid nitrogen, and kept at $-80\,°C$. NEM is added, as it is known to alkylate free thiol groups, thereby preventing the reactive nitrogen species released from RSNOs from reforming RSNOs.

3-(Cyclohexylamino)-1-propanesulfonic acid (CAPS) buffer (10 ml; 0.2 M; pH 10.6) is sparged for 30 min with N_2 (g) and then 5 ml is added to ammonium ferrous sulfate (78 mg) in an evacuated vacutainer. A portion (500 μl) of this green solution is added to MGD (30 mg) in an airtight vial, producing a stock solution of the spin trap $Fe^{2+}-(MGD)_2$ (10 mM Fe:50 mM MGD). The spin trap is stored on ice until needed. Gas-tight syringes and glass vials with a rubber septum in the cap are used to transfer solutions in order to prevent air entering the system. Vials are evacuated using a 50-ml syringe and needle.

As required, $Fe^{2+}-(MGD)_2$ is added to an equal volume (100 μl) of sample in an evacuated glass vial and incubated for 5 min at room temperature. The samples are then analyzed at room temperature in a WG-LC-11 quartz flat cell (Wilmad Glass, Buena, NJ, USA) using a JEOL FR30 EPR spectrometer or a JEOL JES RE1X spectrometer equipped with an ES-UCX2 cylindrical mode X-band cavity (JEOL (UK) Ltd., Welwyn Garden City, Hertfordshire, UK). Spectra are recorded, stored, and analyzed using a personal computer with SpecESR software [version 1.0, 1998; JEOL (UK) Ltd.]. Water is taken through the same procedure to act as a control. Optimal spectral acquisition parameters for the JEOL JES RE1X spectrometer are a microwave frequency of 9.45 GHz, a microwave power of 20 mW, a center field of 330.0 mT, a sweep width of 4 mT, a time constant of 1 s, a sweep time of 80 s, a modulation frequency of 100 kHz, and a modulation width of 1 mT. We have also used a JEOL FR30 instrument to perform the assay in our laboratory, and the spectral acquisition parameters used with this instrument are given in the legend to Fig. 8.1. Three sweeps are averaged to give the final signal. Samples containing detectable RSNO give a characteristic triplet signal [g = 2.035, a_N = 1.28, corresponding to the $(MGD)_2-Fe^{2+}-NO^{\cdot}$ spin adduct] from which

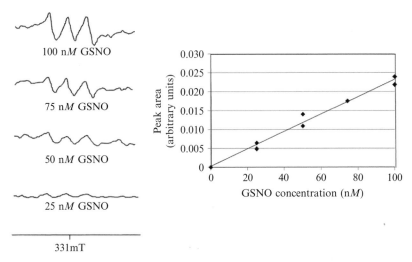

Figure 8.1 Construction of a standard curve using S-nitrosoglutathione (GSNO): (left) EPR spectra obtained from different concentrations of GSNO standards when analyzed by the method described in the text using Fe^{2+} and MGD solutions, which were combined immediately prior to each sample analysis. (Right) Typical standard curve ($R^2 = 0.97$) obtained by plotting the area of the middle peak of the $(MGD)_2$–Fe^{2+}–NO· triplet signal against GSNO concentration. Duplicate determinations were carried out at each concentration. Spectra were obtained using a JEOL FR30 spectrometer with the following instrument parameters: microwave frequency, 9.43 GHz; microwave power, 4 mW; center field, 331.3 mT; sweep width, 4 mT; time constant, 1 s; sweep time, 80 s; modulation frequency, 100 KHz; and modulation width, 0.1 mT.

the area of the middle peak is calculated. RSNO concentrations are calculated from a standard curve constructed using S-nitrosoglutathione (GSNO; see Fig. 8.1). The concentration of the stock GSNO solution is determined by UV/VIS spectrophotometry: ε at 334 nm $= 767$ M^{-1} cm^{-1} (Ji *et al.*, 1996). The GSNO stock solution is diluted in water to the appropriate dilutions (e.g., 0–1000 nM standards, the concentration range being dependent on the expected RSNO concentrations in the samples to be analyzed).

A high pH of reaction mixture is maintained in order to ensure that there is no reduction of nitrite, present in biological fluids, to NO·. In the presence of oxygen and at high pH, ferrous ions will readily autooxidize to ferric ions, changing the color of the solution from blue-green to brown. Careful attention must be paid to ensuring that the reaction mixture is kept completely oxygen free. Batches of CAPS-buffered Fe^{2+}–$(MGD)_2$ that turn brown should be discarded, as in our experience these will be unsuitable for the sensitive detection of RSNOs. As usual with such an analytical procedure, it is crucial to ensure that all reusable glassware, and other surfaces that come into contact with the sample, are washed thoroughly between sample analyses. We wash the syringes and needles five times with water and three times with 1% Decon before rinsing 10 times with water. Also, the quartz cell is washed rigorously with water and dried with nitrogen gas between samples.

With a view to improving the protocol by minimizing autoxidation of the spin trap, we have found that the protocol may be modified successfully by preparing separate solutions of ferrous ion and MGD, which are combined immediately prior to use for the assay of each RSNO-containing sample. Smaller glass vials with PTFE-silicon septa (Supelco/Sigma-Aldrich, Poole, Dorset, UK) are used instead of clinical-type vacutainers, as it is easier to ensure a complete seal after the vial septum is punctured by a needle. Solutions of ammonium ferrous sulfate (20 mM) and MGD (100 mM) are each prepared separately in nitrogen-sparged CAPS buffer (0.2 M, pH 10.6). Portions of each solution (25 μl of ferrous ion solution and 25 μl of MGD) are then combined in an evacuated glass vial immediately before the RSNO-containing sample (50 μl) is added. The mixture is then incubated for 5 min and analyzed as described earlier.

3. RELATIVE YIELD OF SPIN-TRAPPED NO$^{\cdot}$ FROM S-NITROSOALBUMIN IN COMPARISON WITH S-NITROSOGLUTATHIONE

We have carried out further experiments to determine the relative yield of (MGD)$_2$–Fe^{2+}–NO$^{\cdot}$ from S-nitrosoalbumin (ASNO), in comparison with GSNO. We prepared two batches of ASNO, containing different concentrations of S-nitrosothiol, by the S-nitrosation of human albumin using acidified nitrite. Each sample was analyzed in triplicate for RSNO content by the Saville reaction (Saville, 1958; Stamler and Feelisch, 1996) using a standard curve constructed from the low molecular weight S-nitrosothiol, GSNO (Gladwin *et al.*, 2002). Each sample was also analyzed in duplicate for RSNO content using the EPR assay calibrated using a GSNO standard curve as described previously. The protein concentration was also measured using the Bradford assay. For the first ASNO preparation the S-nitroso content was low relative to the protein concentration: for 0.440 mM protein, the Saville assay gave 7.3 μM RSNO and EPR gave 7.2 μM RSNO. In a second ASNO preparation (0.366 mM protein), the Saville assay gave 153 μM RSNO and EPR gave 176 μM RSNO. Thus, results from the Saville reaction are in reasonable agreement with results from the EPR method, confirming that the yield of NO$^{\cdot}$ from the high molecular weight S-nitrosothiol, ASNO, is similar to the yield from the low molecular weight S-nitrosothiol, GSNO.

4. S-Nitrosoalbumin Recovery

To examine the recovery of ASNO, a known amount of ASNO was spiked into a sample of whole human blood (immediately after collection onto EDTA and addition of NEM) to give a final concentration of 1 μM of added ASNO. The sample was centrifuged as described previously, and the resulting spiked plasma sample was analyzed by EPR spectrometry, providing a recovery of the added S-nitrosothiol of 71%. This result is in agreement with a similar experiment, reported previously, in which a synovial fluid sample was spiked with 1 μM (final concentration) of the low molecular weight S-nitrosothiol, GSNO, from which 71% was recovered (Rocks et al., 2005).

5. Potential Signal Contribution from N-Nitrosamines and Other Possible Sources of NO·

Just as interference in the chemiluminescence-based detection of S-nitrosothiols can occur (Rassaf et al., 2002), it is possible that compounds other than S-nitrosothiols could contribute to the detected signal when using the present EPR-based detection method. It has been shown that N-nitrosamines (RNNO), at pH values close to neutral, decay in the presence of $Fe^{2+}-(MGD)_2$ to produce the paramagnetic $(MGD)_2-Fe^{2+}-NO·$ adduct (Hiramoto et al., 2002; Peyrot et al., 2005). To test whether RNNOs could contribute to the $(MGD)_2-Fe^{2+}-NO·$ signal under the alkaline conditions of the present assay, solutions of nitrosodimethylamine (NDMA) and GSNO (both 100 μM) were each made up in water immediately prior to addition to the $Fe^{2+}-(MGD)_2$ spin trap. Both NDMA and GSNO were incubated for 5 min with the spin trap, as described in the aforementioned protocol. The experiment was carried out four times in duplicate. Typical spectra obtained from NDMA and GSNO are shown in Fig. 8.2.

The percentage signal strength of the NDMA relative to the GSNO signal strength (mean $\pm$ 1 standard deviation) was 8.9 $\pm$ 1.8% ($n = 4$). Clearly, under our assay conditions, NDMA provided about a 10-fold lower yield of the $(MGD)_2-Fe^{2+}-NO·$ adduct compared with GSNO. However, it has been reported (Rassaf et al., 2002) that mean concentrations of RNNOs in normal human plasma are about 32 nM, which is about five times the average concentration of plasma RSNOs that these workers determined (7 nM). Therefore, in human plasma measurements, the possibility of a contribution to the $(MGD)_2-Fe^{2+}-NO·$ signal from N-nitroso compounds in biological samples certainly cannot be ruled out. It is possible to determine the mercury-stable portion of the signal—as has been done

A   B

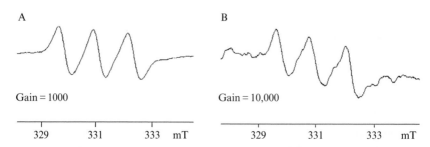

Gain = 1000   Gain = 10,000

    329   331   333   mT   329   331   333   mT

Figure 8.2 Relative signal intensities of EPR spectra obtained from the *S*-nitrosothiol, GSNO, in comparison with the *N*-nitrosamine, nitrosodimethylamine (NDMA). Analysis was carried out according to the method in the text using Fe^{2+} and MGD solutions, which were combined immediately prior to each sample analysis. Spectral acquisition parameters were as given in the legend to Fig. 8.1: (A) an example of the EPR spectrum from 100 μM GSNO (gain, 1000; peak area, 0.26 arbitrary units) and (B) an example of the EPR spectrum from 100 μM NDMA (gain, 10,000; peak area, 0.026 arbitrary units). Note the different signal gains for the two spectra.

for the chemiluminescence assay by some workers (Rassaf *et al.*, 2002)—as a measure of the contribution from RNNOs or other potential $NO^{\cdot}$ reservoirs, such as plasma nitrated lipids (Baker *et al.*, 2004). However, in the EPR-based assay, care must be taken to ensure that the $(MGD)_2$–Fe^{2+}–$NO^{\cdot}$ signal is not diminished as a result of Hg^{2+} displacing Fe^{2+} from the Fe^{2+}–$(MGD)_2$ complex. We have successfully applied the approach of preincubation with mercuric chloride to the determination of the mercury-resistant and mercury-labile signal in preparations of *S*-nitrosated human albumin (I. Knight and P. G. Winyard, manuscript in preparation).

6. Use of the EPR Assay for *S*-Nitrosothiols in Combination with Isotopic Labeling

We have also employed the assay in combination with stable isotope labeling to investigate an aspect of $NO^{\cdot}$ metabolism (Rocks *et al.*, 2005). This isotopic labeling approach, which cannot be applied when using chemiluminescence detection, allowed us to investigate the conversion of ingested ^{15}N-labeled nitrate to *S*-nitrosothiols in human subjects. Human subjects took an oral dose of ^{15}N-labeled sodium nitrate, and a nasogastric tube was used to sample gastric juice at several time points thereafter. The gastric juice was then assayed for *S*-nitrosothiols using the EPR-based assay as described. The resulting $(MGD)_2$–Fe^{2+}–[14]$NO^{\cdot}$ complex gives a characteristic triplet signal when analyzed by EPR spectrometry with g = 2.04 and a_N = 1.28 mT. The $(MGD)_2$–Fe^{2+}–[15]$NO^{\cdot}$ complex has a doublet signal also centred at g = 2.04 with a_{N15} = 1.90 mT. *S*-Nitrosothiols arising from the administered ^{15}N-labeled nitrate can thereby be distinguished from "background" S-nitrosothiols.

ACKNOWLEDGMENTS

We thank Bio-Products Laboratory (Elstree, UK), the Diving Diseases Research Centre (Plymouth, UK), and the Peninsula Medical School for financial support. We are grateful to Dr Lynne Hamilton (Randox Laboratories Ltd, Crumlin, Co. Antrim, UK) for providing MGD.

REFERENCES

Arnelle, D. R., Day, B. J., and Stamler, J. S. (1997). Diethyl dithiocarbamate-induced decomposition of S-nitrosothiols. *Nitric Oxide Biol. Chem.* **1,** 56–64.

Baker, P. R. S., Schopfer, F. J., Sweeney, S., and Freeman, B. A. (2004). Red cell membrane and plasma linoleic acid nitration products: Synthesis, clinical identification, and quantitation. *Proc. Natl. Acad. Sci. USA* **101,** 11577–11582.

Carver, J., Doctor, A., Zaman, K., and Gaston, B. (2005). S-nitrosothiol formation. "Nitric Oxide, Pt E," Vol. 396, pp. 95–105.

Christen, S., Cattin, I., Knight, I., Winyard, P. G., Blum, J. W., and Elsasser, T. H. (2007). Plasma S-nitrosothiol status in neonatal calves: Ontogenetic associations with tissue-specific S-nitrosylation and nitric oxide synthase. *Exp. Biol. Med.* **232,** 309–322.

Cosby, K., Partovi, K. S., Crawford, J. H., Patel, R. P., Reiter, C. D., Martyr, S., Yang, B. K., Waclawiw, M. A., Zalos, G., Xu, X. L., Huang, K. T., Shields, H., *et al.* (2003). Nitrite reduction to nitric oxide by deoxyhemoglobin vasodilates the human circulation. *Nat. Med.* **9,** 1498–1505.

Foster, M. W., McMahon, T. J., and Stamler, J. S. (2003). S-nitrosylation in health and disease. *Trends Mol. Med.* **9,** 160–168.

Giustarini, D., Milzani, A., Dalle-Donne, I., and Rossi, R. (2007). Detection of S-nitrosothiols in biological fluids: A comparison among the most widely applied methodologies. *J. Chromatogr. B Anal. Technol. Biomed. Life Sci.* **851,** 124–139.

Gladwin, M. T., Wang, X. D., and Hogg, N. (2006). Methodological vexation about thiol oxidation versus S-nitrosation: A commentary on "An ascorbate-dependent artifact that interferes with the interpretation of the biotin-switch assay." *Free Radic. Biol. Med.* **41,** 557–561.

Gladwin, M. T., Wang, X. D., Reiter, C. D., Yang, B. K., Vivas, E. X., Bonaventura, C., and Schechter, A. N. (2002). S-nitrosohemoglobin is unstable in the reductive erythrocyte environment and lacks O-2/NO-linked allosteric function. *J. Biol. Chem.* **277,** 27818–27828.

Glantzounis, G. K., Rocks, S. A., Sheth, H., Knight, I., Salacinski, H. J., Davidson, B. R., Winyard, P. G., and Seifalian, A. M. (2007). Formation and role of plasma S-nitrosothiols in liver ischemia-reperfusion injury. *Free Radic. Biol. Med.* **42,** 882–892.

Hallstrom, S., Franz, M., Gasser, H., Kalinowski, L., Vodrazka, M., Semsroth, S., Podesser, B. K., and Malinski, T. (2006). S-nitroso human serum albumin reduces ischemia/reperfusion injury in the pig heart after unprotected warm ischemia. *J. Vasc. Res.* **43,** 554–555.

Hilliquin, P., Borderie, D., Hernvann, A., Menkes, C. J., and Ekindjian, O. G. (1997). Nitric oxide as S-nitrosoproteins in rheumatoid arthritis. *Arthritis Rheumatism* **40,** 1512–1517.

Hiramoto, K., Ryuno, Y., and Kikugawa, K. (2002). Decomposition of N-nitrosamines, and concomitant release of nitric oxide by Fenton reagent under physiological conditions. *Mutat. Res. Genet. Toxicol. Environm. Mutagen.* **520,** 103–111.

Jaffrey, S. R., Erdjument-Bromage, H., Ferris, C. D., Tempst, P., and Snyder, S. H. (2001). Protein S-nitrosylation: A physiological signal for neuronal nitric oxide. *Nat. Cell Biol.* **3,** 193–197.

Ji, Y., Akerboom, T. P., and Sies, H. (1996). Microsomal formation of S-nitrosoglutathione from organic nitrites: Possible role of membrane-bound glutathione transferase. *Biochem. J.* **313,** 377–380.

Lane, P., Hao, G., and Gross, S. S. (2001). S-Nitrosylation is emerging as a specific and fundamental posttranslational protein modification: Head-to-head comparison with O-phosphorylation. *Sci. STKE* **2001,** re1–re1.

Lim, D. G., Sweeney, S., Bloodsworth, A., White, C. R., Chumley, P. H., Krishna, N. R., Schopfer, F., O'Donnell, V. B., Eiserich, J. P., and Freeman, B. A. (2002). Nitrolinoleate, a nitric oxide-derived mediator of cell function: Synthesis, characterization, and vasomotor activity. *Proc. Natl. Acad. Sci. USA* **99,** 15941–15946.

MacArthur, P. H., Shiva, S., and Gladwin, M. T. (2007). Measurement of circulating nitrite and S-nitrosothiols by reductive chemiluminescence. *J. Chromatogr. B Anal. Technol. Biomed. Life Sci.* **851,** 93–105.

Marley, R., Feelisch, M., Holt, S., and Moore, K. (2000). A chemiluminescense-based assay for S-nitrosoalbumin and other plasma S-nitrosothiols. *Free Radic. Res.* **32,** 1–9.

Peyrot, F., Grillon, C., Vergely, C., Rochette, L., and Ducrocq, C. (2005). Pharmacokinetics of 1-nitrosomelatonin and detection by EPR using iron dithiocarbamate complex in mice. *Biochem. J.* **387,** 473–478.

Rassaf, T., Bryan, N. S., Kelm, M., and Feelisch, M. (2002). Concomitant presence of N-nitroso and S-nitroso proteins in human plasma. *Free Radic. Biol. Med.* **33,** 1590–1596.

Richardson, G., and Benjamin, N. (2002). Potential therapeutic uses for S-nitrosothiols. *Clin. Sci.* **102,** 99–105.

Rocks, S. A., Davies, C. A., Hicks, S. L., Webb, A. J., Klocke, R., Timmins, G. S., Johnston, A., Jawad, A. S. M., Blake, D. R., Benjamin, N., and Winyard, P. G. (2005). Measurement of S-nitrosothiols in extracellular fluids from healthy human volunteers and rheumatoid arthritis patients, using electron paramagnetic resonance spectrometry. *Free Radic. Biol. Med.* **39,** 937–948.

Saville, B. (1958). A scheme for the colorimetric determination of microgram amounts of thiols. *Analyst* **83,** 670–672.

Stamler, J. S., and Feelisch, M. (1996). Preparation and detection of S-nitrosothiols. *In* "Methods in Nitric Oxide Research" (M. Feelisch and J. S. Stamler, eds.), pp. 521–539. Wiley, Chichester.

Stamler, J. S., Jaraki, O., Osborne, J., Simon, D. I., Keaney, J., Vita, J., Singel, D., Valeri, C. R., and Loscalzo, J. (1992). Nitric oxide circulates in mammalian plasma primarily as an S-nitroso adduct of serum albumin. *Proc. Natl. Acad. Sci. USA* **89,** 7674–7677.

Taylor, E., and Winyard, P. G. (2007). S-Nitrosothiols and disease mechanisms. *In* "Oxidative Stress: Clinical and Biomedical Implications" (B. Matata, ed.). Nova Science, Hauppauge, NY.

Tsikas, D., Sandmann, J., and Frolich, J. C. (2002). Measurement of S-nitrosoalbumin by gas chromatography-mass spectrometry. III. Quantitative determination in human plasma after specific conversion of the S-nitroso group to nitrite by cysteine and CU2+ via intermediate formation of S-nitrosocysteine and nitric oxide. *J. Chromatogr. B Anal. Technol. Biomed. Life Sci.* **772,** 335–346.

Tsuchiya, K., Kirima, K., Yoshizumi, M., Houchi, H., Tamaki, T., and Mason, R. P. (2002). The role of thiol and nitrosothiol compounds in the nitric oxide-forming reactions of the iron-N-methyl-D-glucamine dithiocarbamate complex. *Biochem. J.* **367,** 771–779.

Webb, A., Bond, R., McLean, P., Uppal, R., Benjamin, N., and Ahluwalia, A. (2004). Reduction of nitrite to nitric oxide during ischemia protects against myocardial ischemia-reperfusion damage. *Proc. Natl. Acad. Sci. USA* **101,** 13683–13688.

NOVEL METHOD FOR MEASURING S-NITROSOTHIOLS USING HYDROGEN SULFIDE

Xinjun Teng,* T. Scott Isbell,*,‖ Jack H. Crawford,*
Charles A. Bosworth,§ Gregory I. Giles,** Jeffrey R. Koenitzer,†
Jack R. Lancaster,‡,§,‖ Jeannette E. Doeller,¶,‖ David W. Kraus,†,‖
and Rakesh P. Patel*,‖

Contents

Abstract

Recent advances in techniques that allow sensitive and specific measurement of
S-nitrosothiols (RSNOs) have provided evidence for a role for these compounds
in various aspects of nitric oxide (NO) biology. The most widely used approach is
to couple reaction chemistry that selectively reduces RSNOs by one electron to

*   Department of Pathology, University of Alabama at Birmingham, Alabama
†   Department of Biology, University of Alabama at Birmingham, Alabama
‡   Department of Anesthesiology, University of Alabama at Birmingham, Alabama
§   Department of Physiology and Biophysics, University of Alabama at Birmingham, Alabama
¶   Department of Environmental Health Sciences, University of Alabama at Birmingham, Alabama
‖   Center for Free Radical Biology, University of Alabama at Birmingham, Alabama
**   Department of Pharmacology and Toxicology, Otago School of Medical Sciences, University of Otago,
   Dunedin, New Zealand

Methods in Enzymology, Volume 441
ISSN 0076-6879, DOI: 10.1016/S0076-6879(08)01209-3

produce NO, with the sensitive detection of the latter under anaerobic conditions using ozone based chemiluminescence in NO analyzers. Herein, we report a novel reaction that is readily adaptable for commercial NO analyzers that utilizes hydrogen sulfide (H_2S), a gas that can reduce RSNO to NO and, analogous to NO, is produced by endogenous metabolism and has effects on diverse biological functions. We discuss factors that affect H_2S based methods for RSNO measurement and discuss the potential of H_2S as an experimental tool to measure RSNO.

1. INTRODUCTION

Many studies have highlighted biological roles for *S*-nitrosothiols (RSNO) ranging from nitric oxide (NO) storage molecules to regulating protein function integral in cell signaling (Hess *et al.*, 2005; Hogg, 2002). Central to our understanding of RSNO biology are methods for the detection of these species specifically and with sufficient sensitivity in biological matrices. Recently developed assays have employed proteomic approaches to identify *S*-nitrosated proteins (Derakhshan *et al.*, 2007; Greco *et al.*, 2006; Jaffrey *et al.*, 2001; Kettenhofen *et al.*, 2007; Yang and Loscalzo, 2005) and chemiluminescence based approaches to measure RSNO concentrations. The latter chemiluminescence based approaches provide the most sensitive method, with limits of detection in the picomole range. The general strategy is to inject the RSNO containing sample into the anaerobic reaction chamber containing a solution that converts RSNOs to NO, which is then carried into the analyzer by an inert gas and then detected by the reaction with ozone. Using this general approach, however, RSNO concentrations have been reported to range over three orders of magnitude; for example, plasma RSNO (attributed to *S*-nitrosoalbumin) concentrations from $<20\,nM$ (below detection limit) to $\approx 7\,\mu M$ (Giustarini *et al.*, 2007; Gow *et al.*, 2007; MacArthur *et al.*, 2007; Marley *et al.*, 2000; Rassaf *et al.*, 2004; Stamler *et al.*, 1992; Tsikas and Frolich, 2004; Tyurin *et al.*, 2001). These discrepant data are likely attributed to both sample processing leading to artifactual formation/decomposition of RSNOs and the different methods used to release NO from the parent RSNO in the reaction chamber. For example, higher (nanomolar to micromolar) RSNO concentrations are reported using the photolysis chemiluminescence methods, whereas concentrations measured by methods that utilize the chemical reduction of RSNO to NO (by triiodide method or copper/cysteine) are lower (nanomolar range) (Marley *et al.*, 2000; Rassaf *et al.*, 2004; Stamler *et al.*, 1992). In addition, some methods detect nitrite, NO–heme adducts, and RSNOs, necessitating sample pretreatment to exclude or include specifically RSNOs. Sample pretreatment, selective reduction of RSNO to NO, sensitivity of this process, and reproducibility from laboratory to laboratory are key criteria in assessing the utility of a given method for measuring RSNO. These issues are critical, as the actual concentration of a given RSNO is an important consideration in assessing biological function, a fact underscored by a number of review articles that discuss how the different methods employed to date may impact on the variance among reported RSNO concentrations (Giustarini *et al.*, 2007; Gow *et al.*, 2007; MacArthur *et al.*, 2007; Marley *et al.*, 2000; Rassaf *et al.*, 2004; Stamler, 2004; Tsikas and Frolich, 2004).

Hydrogen sulfide (H_2S) is produced endogenously by cystathionine β synthase (CBS) and cystathionine γ lyase (CGL), enzymes integral in sulfur-containing amino acid metabolism. Although it has been appreciated that organisms thriving in sulfide-rich habitats consume H_2S as a source of cellular energy (Doeller *et al.*, 1999; Kraus and Doeller, 2004), it has not been recognized until recently that H_2S can function in mammalian signaling pathways, including those that control cardiovascular activity, and that H_2S can induce a state of suspended animation and protect against ischemia–reperfusion injury (Blackstone *et al.*, 2005; Doeller *et al.*, 2005; Elrod *et al.*, 2007; Koenitzer *et al.*, 2007; Pearson *et al.*, 2006). Little is known concerning whether, and if so how, H_2S and NO pathways interact, although it is interesting that parallels in functional end points (e.g., stimulation of vasodilation, inhibition of platelet aggregation) exist. Although a reaction between NO and NaHS has been proposed to produce RSNO (Whiteman *et al.*, 2006), there is no clear mechanism. In contrast, we have found that H_2S readily reduces GSNO, as a model RSNO, to release NO (Bosworth, Kraus and Lancaster *et al.*, 2008), in preparation which is consistent with H_2S acting as a reductant, $H_2S/S°$ $E_h = -270$ mV, and a competent nucleophile (Kraus and Wittenberg, 1990). Based on this observation, we reasoned that H_2S-dependent RSNO reduction in NO chemiluminescence chambers could be used as a novel selective and sensitive method to detect RSNOs without interference from other NO-containing molecules. Herein, we describe protocols used to demonstrate this concept and discuss experimental factors that affect H_2S-dependent RSNO reduction.

2. EXPERIMENTAL MANIPULATION OF H_2S

Experimentally, H_2S can be administered as a gas (H_2S gas tanks are available commercially). Alternatively, and more typically, H_2S is introduced into reaction chambers via the addition of sodium sulfide (Na_2S), a salt that dissociates readily (analogous to a salt, e.g., NaCl) to produce the sulfide anion (S^{2-}), which in turn is in equilibrium with the hydrosulfide anion (HS^-) and H_2S. Reaction 1 shows these equilibria together with pK_a values, which indicate that at pH 7.4 approximately 30% H_2S and 70% HS^- will exist in solution. For the sake of clarity we will use the denotion H_2S in this chapter but note that the composition of the anion versus unionized H_2S will depend on the pH of reaction buffers and that the reactivity of each form is also different. It is also important to note that anaerobic solutions should be used in the preparation of Na_2S stock solutions, described previously, to avoid oxygen-dependent consumption of H_2S and subsequent formation of sulfites that could interfere with NO/RSNO reactions (Doeller *et al.*, 2005; Koenitzer *et al.*, 2007).

$$\text{Reaction 1: } Na_2S \longrightarrow [2Na^+S^{2-}] \longrightarrow S^{2-} \underset{pK_a=17}{\overset{H^+}{\longleftrightarrow}} HS^- \underset{pK_a=6.9}{\overset{H^+}{\longleftrightarrow}} H_2S$$

3. SPECIFICITY FOR H_2S-DEPENDENT DETECTION OF RSNO

To initially test if H_2S could be used to measure RSNO, we used a solution of 10 mM H_2S in 100 mM sodium carbonate buffer, pH 11, containing 100 μM DTPA at 25 °C in a typical NO chemiluminescence analyzer reaction chamber. Addition of a metal chelator (DTPA) is necessary to avoid contaminating transition metal ion-dependent reactions with H_2S and RSNOs. The choice of alkaline pH is discussed further later. Figure 9.1A shows reaction traces indicating proportional NO formation upon addition of increasing concentrations of *S*-nitrosoglutathione (GSNO). Specificity for RSNO and NO formation is indicated by the loss of a signal when GSNO is preexposed to light or pretreated with $HgCl_2$, treatments that decompose RSNOs and inhibition of NO signal when the NO scavenger C-PTIO is added to the reaction chamber prior to GSNO addition. Furthermore, no NO signal is observed when potential products of GSNO metabolism, potential contaminants of RSNO preparations, and other biologically relevant NO metabolites, including reduced or oxidized glutathione (GSH and GSSG, respectively), nitrite (NO_2^-) or nitrate (NO_3^-), are added. These approaches also indicate important control experiments required to show specificity for RSNO measurement. Figure 9.1B (inset) shows a standard curve for GSNO demonstrating sensitivity in the 2- to 5-pmol range. Similar results are obtained using another low molecular weight *S*-nitrosothiol, *S*-nitroso-*N*-acetylpenicillamine (SNAP) (not shown) demonstrating sensitivity for RSNO measurement that is similar to previous protocols including tri-iodide and copper-cysteine based methods. Note that data in Fig. 9.1 were obtained using a 20-ml reaction volume and the addition of between 10- and 100-μl aliquots of RSNO. Figure 9.1C shows the slope (steeper slopes indicating higher sensitivity for GSNO detection) for GSNO standard curves measured using different H_2S concentrations. Interestingly, a bell-shaped relationship

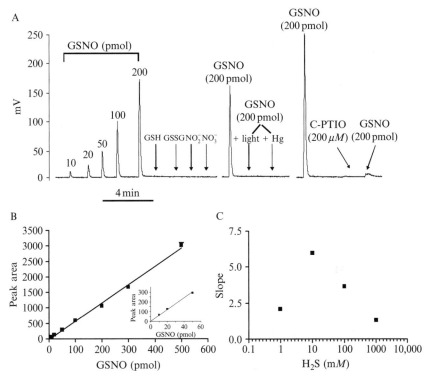

Figure 9.1 (A) Representative traces showing NO formation using a nitric oxide ana-lyzer (NOA, GE Instruments Analytical, Boulder, CO). Sodium carbonate buffer (20 ml, 100 mM, pH 11, 20 °C) containing Na_2S (10 mM) + DTPA (100 μM) was added to the reaction chamber, perfused with helium, and exposed to a vacuum per standard NOA equipment until a consistent baseline signal ($\approx$5 min) was achieved and then either GSNO or indicated compounds added. Using this approach, multiple injections of GSNO (resulting in a cumulative amount of 2000 pmol could be added without loss of signal). (B) GSNO standard curve generated using buffer composition shown in A. Data are mean ± SEM ($n = 3-4$). Line shown best fit calculated by linear regression ($Y = 5.9X - 20.5$, $r^2 = 0.996$). (Inset) Standard curve for lower GSNO concentrations and sensi-tivity for GSNO detection in the low nanomolar range. (C) Effects of Na_2S concentra-tion on sensitivity of NO formation from GSNO. Gradients were calculated by determining standard curves as shown in B at different Na_2S concentrations as indicated in 100 mM sodium carbonate, pH 11, 20 °C + 100 μM DTPA.

is observed, which reflects a complex chemistry underlying NO formation from H_2S–RSNO reactions and which remains under investigation. We present these data to highlight the need to determine optimal condi-tions for RSNO measurements using H_2S, including H_2S concentration, the nature of the RSNO, and the biological matrix in which RSNO will be measured, to name but a few.

4. Comparison of H₂S vs Tri-iodide and Copper/ Cysteine Based Methods for RSNO Detection

To date, cuprous or cupric ion/cysteine (Cu^+ or Cu^{2+}/Cys) or potassium tri-iodide/iodine (KI/I_3^-) based methods have been widely used to measure RSNOs. In these respective cases, the reaction chamber contains a solution of either cupric ion with L-cysteine in a neutral pH buffer or KI/I_3^- in acetic acid. The Cu/Cys system reacts directly with RSNO and is advantageous in that interference from nitrite is not observed, although nitrosylheme is also detected by this method. The I_3 method, however, does detect nitrite in addition to RSNO and therefore the former has to be scavenged by the addition of sulfanilamide under acidic conditions. Both methods are sensitive, have advantages and disadvantages, and are discussed in detail in other chapters in this volume and other articles (Gow *et al.*, 2007; Wang *et al.*, 2006). Indeed, there is an emerging consensus that multiple approaches should be used when possible to validate RSNO measurements. Here we compare the sensitivity for GSNO detection by Cu^{2+}/Cys, I_3, and H_2S based methods (Fig. 9.2). (Note that photolysis based NO analyzers are limited to select laboratories precluding direct comparison with H_2S here.) Under the stated reaction conditions, Cu^{2+}/Cys and KI/I_3 are equivalent in sensitivity, whereas H_2S has an approximately 30% lower sensitivity. Whereas this may preclude using H_2S as a frontline approach to measure RSNO, we suggest that using

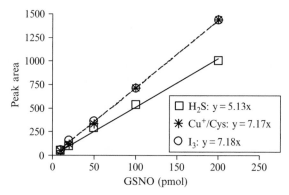

Figure 9.2 Comparison of GSNO detection by Cu/Cys, I_3, and H_2S based methods. GSNO detected using either Na_2S as described in Fig. 9.1 legend, tri-iodide method [using a reaction solution comprising iodine (8.5 m*M*), potassium iodide (66.8 mM) in acetic acid; GSNO was treated with acid sulfanilamide before injected to remove nitrite] or Cu/Cys method [using a reaction solution comprising copper(II) sulfate (100 μ*M*), L-cysteine (1 m*M*) in potassium phosphate buffer (100 m*M*, pH 7.2)]. In all cases, the reaction temperature was 20 °C.

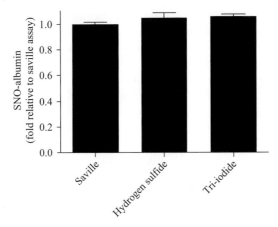

Figure 9.3 Detection of SNO-albumin by H_2S. SNO-albumin was synthesized by transnitrosation using *S*-nitrosocyteine (10:1 molar ratio *S*-nitrosocysteine:albumin) in PBS + 100 μM DTPA at pH 6.0 as described previously (Patel *et al.*, 1999). SNO-albumin was measured using the Saville assay as described previously (Patel *et al.*, 1999); H_2S or tri-iodide based methods as described in Figs. 9.1 and 9.2 legends. Data are shown as fold change relative to Saville assay. For H_2S and tri-iodide methods, GSNO was used to generate the standard curve and calculate SNO-albumin concentration.

H_2S provides a method to detect RSNO under alkaline reaction conditions that can complement existing methods performed under acidic (KI/I_3) or neutral (Cu/Cys) pH values and with a distinct chemistry. As an example, Fig. 9.3 shows a comparison of SNO–albumin detection by the classical Saville assay and chemiluminescence based tri-iodide and H_2S assays. Both chemiluminescence based assays gave comparable results relative to the Saville assay. The H_2S method therefore offers a relatively simple approach to validate RSNO measurements made by other existing methodologies. Moreover, we anticipate that upon a more complete understanding of how H_2S reduces RSNO to NO, more efficient protocols for RSNO detection will be formulated. With this in mind the next section describes factors that influence RSNO detection using H_2S.

5. EFFECTS OF pH ON H_2S-DEPENDENT RSNO DETECTION

The pH dependence of the H_2S-dependent reduction of RSNO to NO has not been defined but will likely influence the rate at which NO is generated by affecting the equilibrium between H_2S and HS^- concentrations. An additional and important factor is the fact that H_2S is a volatile gas whereas HS^- is not. Because reaction chambers are perfused continually with an inert gas to carry NO into the NO analyzer, H_2S will also be carried out of solution, resulting in a continual decrease of solution H_2S concentrations. Therefore, a potential variable is continuous changes in H_2S concentrations, which in turn affects the sensitivity for RSNO detection and limits the addition of samples that will acidify the media. Note that some RSNO preparations are formulated in acidic media and thus may not be suitable without prior pH neutralization. To ensure that the loss of H_2S is not a constantly changing parameter, we suggest using alkaline pH (as shown in Fig. 9.1) in which the presence of H_2S, and hence its loss during solution perfusion, is negligible.

It has been argued that methods evaluating RSNO concentrations should be performed at physiological pH to ensure that the protein bearing RSNO is in its native conformation, as protein denaturation may alter RSNO reactivity and stability, hence concentration detected (Stamler, 2004). It should be noted that the *S*-nitroso group per se is chemically stable at all pH values with comparable sensitivities and concentrations being reported for RSNO measurements performed at acidic or neutral pH values (Fig. 9.2). Nevertheless, if RSNO measurements at neutral pH are required, the H_2S protocol described earlier can be adapted to more neutral pH values by coperfusing H_2S gas (together with helium; pure H_2S gas cylinders or H_2S/He mixtures are available commercially) into the purge vessel containing a phosphate buffer. The strategy is to maintain a constant H_2S concentration in the reaction medium proportional to the partial pressure of H_2S and the pH of the medium, allowing some loss of H_2S gas due to volatization. Two limiting factors to H_2S gas perfusion need to be considered, however. First, the flow rate of H_2S gas into the purge vessel needs to be tightly controlled. We have found that successive experiments using unregulated flow and pressure settings on H_2S tanks can result in up to 30-fold variations in steady-state H_2S concentrations in a phosphate buffer at pH 8. If separate gas cylinders are used to supply He and H_2S, the use of quality mass flow controllers will ensure precise control of flow rates for H_2S gas into the purge vessel. Alternatively, once an optimal mixture of H_2S and He has been determined, a precise gas mix in a single cylinder should suffice. The second limitation is the potential for H_2S itself to generate a signal in the NO analyzers. Indeed, environmental H_2S measurements can be performed using similar chemiluminescence based approaches that couple the reaction between H_2S and ozone to light generation. Because the emission wavelength for H_2S/ozone (300 to 400 nm) is shorter than that of NO/ozone (>640 nm), possible interference can be limited by using specific band-pass filters to block H_2S chemiluminescence. However, not all NOA offer this type of optical fine-tuning. As a consequence, NOA machines should be checked to see if interference from H_2S occurs and, if so, ensure that appropriate control injections are made, especially if changes in local or bulk pH in the reaction vessel are expected. This necessitates measuring solution H_2S concentrations in the purge vessel, which can be achieved using polarographic H_2S sensors coupled with gas-flow controllers that ensure precise control of flow rates for H_2S gas into the purge vessel.

6. EFFECT OF ANTIFOAM

The relatively low concentration of RSNO in biological matrices necessitates the injection of high volumes for detection. An unavoidable consequence of injection of biological material that is high in protein and lipids into a solution exposed to a vacuum and perfused continuously with a gas is excessive foaming in the reaction vessel. This often leads to peak broadening and a decrease in sensitivity. To limit foaming, antifoam reagents are typically used. If this is required, it is important to note that a variety of antifoam reagents exist, but they must be tested systematically for their efficacy in decreasing foaming as well as in not interfering with NO generation and detection. As an example, Fig. 9.4 shows the effects of two antifoam reagents on GSNO and SNAP measurement using H_2S. Silicon antifoam decreased sensitivity for both GSNO and SNAP detection, whereas GE antifoam only reduced H_2S-dependent detection of GSNO. The mechanistic reasons for these effects are not discussed here but these data are presented to highlight the importance of establishing standard curves for RSNO detection in appropriate biological matrices and experimental conditions.

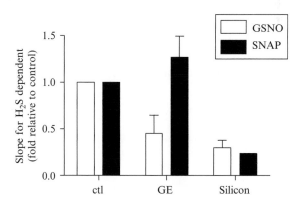

Figure 9.4 Effect of antifoam reagents on H_2S-dependent detection of S-nitrosothiols. Standard curves for GSNO or SNAP detection by H_2S (using conditions shown in Fig. 9.1 legend) was determined in the absence or presence of antifoam reagent (50 μl) of a 1:30 diluted solution supplied by GE with Sievers nitric oxide analyzers or 5 μl silicon antifoam emulsion (obtained from Sigma Chemical Company). Shown are gradients, with higher gradients indicating higher sensitivity for S-nitrosothiol detection by H_2S.

7. SUMMARY AND CONCLUSIONS

Roles for *S*-nitrosothiols from NO storage and donor molecules to mediators of cell signaling have been proposed. The coupling of methods that metabolize RSNO to NO with the sensitive detection of the latter using chemiluminescence has been critical in developing these concepts. Several protocols that reduce RSNO to NO under anaerobic conditions with different chemical mechanisms have been described. Each method has its advantages and disadvantages, but it is important to note that methodological differences are thought to underlie the variation in reported RSNO concentrations (e.g., up to three orders of magnitude for circulating RSNOs). These discrepancies have led to diverse opinions regarding the specific role and biological importance of RSNO and have led to the proposal that more than one method with distinct chemistries that reduce RSNO to NO should be used when possible to validate measured concentrations. With this goal in mind, this chapter described a novel protocol that utilizes H_2S to reduce RSNO to release NO. Although this method is a little less sensitive than existing methods (Cu/Cys, KI/I_3), the H_2S method was specific for RSNO, with no cross-reactivity with other nitrogen oxide containing species that are present in biological solutions nor with oxidized or reduced thiol precursors. A key distinction between H_2S and existing methods is that H_2S can function at alkaline pH values, which we propose can complement acid and neutral pH based methods and collectively therefore provide an array of approaches to measure RSNOs and potentially address criticisms directed at acid based approaches. Finally, we acknowledge that the precise mechanism by which H_2S reduces RSNO to NO remains to be elucidated. This chapter provided preliminary protocols therefore that allow for RSNO detection by H_2S, and we anticipate that as our understanding of the mechanisms improve, sensitivity using H_2S for RSNO measurements will also improve.

ACKNOWLEDGMENTS

This study was supported by grants from the NIH (HL71189 and HL074391 to JRL and RGM073049A to DWK) and from the American Heart Association to RPP (0655312B) and DWK (0455296B). TSI was supported by a NIH Cardiovascular Pathophysiology Training Fellowship.

REFERENCES

Blackstone, E., Morrison, M., and Roth, M. B. (2005). H_2S induces a suspended animation-like state in mice. *Science* **308,** 518.

Derakhshan, B., Wille, P. C., and Gross, S. S. (2007). Unbiased identification of cysteine *S*-nitrosylation sites on proteins. *Nat. Protoc.* **2,** 1685–1691.

Doeller, J. E., Gaschen, B. K., Parrino, V. V., and Kraus, D. W. (1999). Chemolithoheterotrophy in a metazoan tissue: Sulfide supports cellular work in ciliated mussel gills. *J. Exp. Biol.* **202**(Pt 14), 1953–1961.

Doeller, J. E., Isbell, T. S., Benavides, G., Koenitzer, J., Patel, H., Patel, R. P., Lancaster, J. R., Jr., Darley-Usmar, V. M., and Kraus, D. W. (2005). Polarographic measurement of hydrogen sulfide production and consumption by mammalian tissues. *Anal. Biochem.* **341**, 40–51.

Elrod, J. W., Calvert, J. W., Morrison, J., Doeller, J. E., Kraus, D. W., Tao, L., Jiao, X., Scalia, R., Kiss, L., Szabo, C., Kimura, H., Chow, C. W., and Lefer, D. J. (2007). Hydrogen sulfide attenuates myocardial ischemia-reperfusion injury by preservation of mitochondrial function. *Proc. Natl. Acad. Sci. USA* **104**, 15560–15565.

Giustarini, D., Milzani, A., Dalle-Donne, I., and Rossi, R. (2007). Detection of S-nitrosothiols in biological fluids: A comparison among the most widely applied methodologies. *J. Chromatogr. B Analyt. Technol. Biomed. Life Sci.* **851**, 124–139.

Gow, A., Doctor, A., Mannick, J., and Gaston, B. (2007). S-Nitrosothiol measurements in biological systems. *J. Chromatogr. B Analyt. Technol. Biomed. Life Sci.* **851**, 140–151.

Greco, T. M., Hodara, R., Parastatidis, I., Heijnen, H. F., Dennehy, M. K., Liebler, D. C., and Ischiropoulos, H. (2006). Identification of S-nitrosylation motifs by site-specific mapping of the S-nitrosocysteine proteome in human vascular smooth muscle cells. *Proc. Natl. Acad. Sci. USA* **103**, 7420–7425.

Hess, D. T., Matsumoto, A., Kim, S. O., Marshall, H. E., and Stamler, J. S. (2005). Protein S-nitrosylation: Purview and parameters. *Nat. Rev. Mol. Cell Biol.* **6**, 150–166.

Hogg, N. (2002). The biochemistry and physiology of S-nitrosothiols. *Annu. Rev. Pharmacol. Toxicol.* **42**, 585–600.

Jaffrey, S. R., Erdjument-Bromage, H., Ferris, C. D., Tempst, P., and Snyder, S. H. (2001). Protein S-nitrosylation: A physiological signal for neuronal nitric oxide. *Nat. Cell Biol.* **3**, 193–197.

Kettenhofen, N. J., Broniowska, K. A., Keszler, A., Zhang, Y., and Hogg, N. (2007). Proteomic methods for analysis of S-nitrosation. *J. Chromatogr. B Analyt. Technol. Biomed. Life Sci.* **851**, 152–159.

Koenitzer, J. R., Isbell, T. S., Patel, H. D., Benavides, G. A., Dickinson, D. A., Patel, R. P., Darley-Usmar, V. M., Lancaster, J. R., Jr., Doeller, J. E., and Kraus, D. W. (2007). Hydrogen sulfide mediates vasoactivity in an O_2-dependent manner. *Am. J. Physiol. Heart Circ. Physiol.* **292**, H1953–H1960.

Kraus, D. W., and Doeller, J. E. (2004). Sulfide consumption by mussel gill mitochondria is not strictly tied to oxygen reduction: Measurements using a novel polarographic sulfide sensor. *J. Exp. Biol.* **207**, 3667–3679.

Kraus, D. W., and Wittenberg, J. B. (1990). Hemoglobins of the Lucina pectinata/bacteria symbiosis. I. Molecular properties, kinetics and equilibria of reactions with ligands. *J. Biol. Chem.* **265**, 16043–16053.

MacArthur, P. H., Shiva, S., and Gladwin, M. T. (2007). Measurement of circulating nitrite and S-nitrosothiols by reductive chemiluminescence. *J. Chromatogr. B Analyt. Technol. Biomed. Life Sci.* **851**, 93–105.

Marley, R., Feelisch, M., Holt, S., and Moore, K. (2000). A chemiluminescence-based assay for S-nitrosoalbumin and other plasma S-nitrosothiols. *Free Radic. Res.* **32**, 1–9.

Patel, R. P., Hogg, N., Spencer, N. Y., Kalyanaraman, B., Matalon, S., and Darley-Usmar, V. M. (1999). Biochemical characterization of human S-nitrosohemoglobin: Effects on oxygen binding and transnitrosation. *J. Biol. Chem.* **274**, 15487–15492.

Pearson, R. J., Wilson, T., and Wang, R. (2006). Endogenous hydrogen sulfide and the cardiovascular system: What's the smell all about? *Clin. Invest. Med.* **29**, 146–150.

Rassaf, T., Feelisch, M., and Kelm, M. (2004). Circulating NO pool: Assessment of nitrite and nitroso species in blood and tissues. *Free Radic. Biol. Med.* **36**, 413–422.

Stamler, J. S. (2004). *S*-nitrosothiols in the blood: Roles, amounts, and methods of analysis. *Circ. Res.* **94,** 414–417.

Stamler, J. S., Jaraki, O., Osborne, J., Simon, D. I., Keaney, J., Vita, J., Singel, D., Valeri, C. R., and Loscalzo, J. (1992). Nitric oxide circulates in mammalian plasma primarily as an *S*-nitroso adduct of serum albumin. *Proc. Natl. Acad. Sci. USA* **89,** 7674–7677.

Tsikas, D., and Frolich, J. C. (2004). Trouble with the analysis of nitrite, nitrate, *S*-nitrosothiols and 3-nitrotyrosine: Freezing-induced artifacts? *Nitric Oxide* **11,** 209–213. author reply 214–215.

Tyurin, V. A., Liu, S. X., Tyurina, Y. Y., Sussman, N. B., Hubel, C. A., Roberts, J. M., Taylor, R. N., and Kagan, V. E. (2001). Elevated levels of *S*-nitrosoalbumin in pre-eclampsia plasma. *Circ. Res.* **88,** 1210–1215.

Wang, X., Bryan, N. S., MacArthur, P. H., Rodriguez, J., Gladwin, M. T., and Feelisch, M. (2006). Measurement of nitric oxide levels in the red cell: Validation of tri-iodide-based chemiluminescence with acid-sulfanilamide pretreatment. *J. Biol. Chem.* **281,** 26994–27002.

Whiteman, M., Li, L., Kostetski, I., Chu, S. H., Siau, J. L., Bhatia, M., and Moore, P. K. (2006). Evidence for the formation of a novel nitrosothiol from the gaseous mediators nitric oxide and hydrogen sulphide. *Biochem. Biophys. Res. Commun.* **343,** 303–310.

Yang, Y., and Loscalzo, J. (2005). *S*-nitrosoprotein formation and localization in endothelial cells. *Proc. Natl. Acad. Sci. USA* **102,** 117–122.

CHAPTER TEN

KINETIC STUDIES ON PEROXYNITRITE REDUCTION BY PEROXIREDOXINS

Madia Trujillo,[*,†] Gerardo Ferrer-Sueta,[†,‡] *and* Rafael Radi[*,†]

Contents

Abstract

Peroxiredoxins catalytically reduce peroxynitrite to nitrite. The peroxidatic cysteine of peroxiredoxins reacts rapidly with peroxynitrite. The rate constant of that reaction can be measured using a stopped flow spectrophotometer either directly by following peroxynitrite disappearance in the region of 300 to 310 nm using an initial rate approach or steady-state measurements or by competition with a reaction of known rate constant. The reactions used to compete with peroxiredoxins include the oxidation of Mn[III]porphyrins and horseradish peroxidase by peroxynitrite. Additionally, a method is described in which a hydroperoxide competes with peroxynitrite for the oxidation of peroxiredoxin. Moreover, a fluorescent technique for determining the kinetics of thioredoxin-mediated peroxiredoxin reduction, closing the catalytic cycle, is also described. All methods reviewed provide reliable values of rate constants and a combination of them can be used to provide further reassurance; applicability and advantages of the different methodologies are discussed.

[*] Department of Biochemistry, Facultad de Medicina, Universidad de la República, Montevideo, Uruguay
[†] Center for Free Radical and Biomedical Research, Facultad de Medicina, Universidad de la República, Montevideo, Uruguay
[‡] Instituto de Química Biológica, Facultad de Ciencias, Universidad de la República, Montevideo, Uruguay

Methods in Enzymology, Volume 441
ISSN 0076-6879, DOI: 10.1016/S0076-6879(08)01210-X

1. INTRODUCTION

Peroxynitrite, the oxidant formed *in vivo* from the diffusion-controlled reaction between superoxide anion ($O_2^{\cdot-}$) and nitric oxide ($\cdot NO$) radicals (Goldstein and Czapski, 1995; Kissner *et al.*, 1997), can react directly with targets, including thiols, carbon dioxide, heme proteins, iron–sulfur centers, and zinc thiolates (Radi *et al.*, 2000). Thiols represent preferential targets for peroxynitrite reactivity *in vivo* (Radi, 1998; Radi *et al.*, 1991; Trujillo and Radi, 2002). Low molecular weight thiols react with peroxynitrite as shown in the following equations:

$$RS^- + ONOOH \rightarrow RSOH + NO_2^- \qquad (10.1)$$

$$RSOH + RS^- \rightarrow RSSR + OH^- \qquad (10.2)$$

where RS^- represents the thiolate anion and $RSOH$ is the corresponding sulfenic acid, which in the presence of another accessible thiol group leads to formation of a disulfide ($RSSR$), being the overall stoichiometry of the reaction two thiols oxidized by each peroxynitrite (Radi *et al.*, 1991). The apparent rate constant (k'_2) of reaction (1) at a given pH can be calculated from the pH-independent rate constant (k_2) and the pH distribution of thiolate and peroxynitrous acid ($pK_a = 6.8$) as follows (Trujillo and Radi, 2002):

$$k'_2 = k_2 \times \frac{K_{SH}}{K_{SH} + [H^+]} \times \frac{[H^+]}{K_a + [H^+]} \qquad (10.3)$$

We have reported previously that the k'_2 for peroxynitrite-mediated low molecular weight thiol oxidation was higher for those thiols having the lower pK_{SH}, consistent with a greater proportion of the thiols as thiolate at the indicated pH (Trujillo and Radi, 2002). The values for pH-independent k_2, however, increase with thiol pK_{SH}, as shown by a positive Brønsted relationship of the form:

$$\log(k_2) = C + \beta_{nuc} \times pK_{SH}, \qquad (10.4)$$

where C is a constant applicable to the reactions of a particular oxidant with a series of thiolates under specified conditions and β_{nuc} is the Brønsted coefficient (Trujillo *et al.*, 2007a). In the case of peroxynitrite-mediated low molecular weight thiol oxidation, the Brønsted coefficient is 0.4, implying that the factors that favor the Brønsted basicity of the thiolate also favor its nucleophilic displacement on the oxygen in peroxynitrous acid. Thiol oxidation by peroxynitrite can also be indirect, and mediated by hydroxyl, $\cdot NO_2$ and carbonate radicals formed from ONOOH homolysis, or after the reactions between peroxynitrite anion and CO_2 (Denicola *et al.*, 1996;

Quijano *et al.*, 1997; Radi *et al.*, 2000). In this case, thiols are oxidized to thiyl radicals by a one-electron mechanism and the yield of oxidation is lower (Quijano *et al.*, 1997).

Strong evidence of the role of peroxiredoxins Prxs in peroxynitrite detoxification has emerged. Peroxiredoxins are a family of ubiquitously expressed thiol-containing peroxidases that catalyze hydrogen peroxide and organic hydroperoxide reduction using small thiol proteins (such as thioredoxin or tryparedoxin) as the reducing substrate, which in turn are maintained at the reduced state by a reductase at NADPH expense (Hofmann *et al.*, 2002; Wood *et al.*, 2003). According to the number of cysteine residues required for catalysis, Prxs can be classified into one- or two-cysteine Prxs (Wood *et al.*, 2003) (Fig. 10.1).

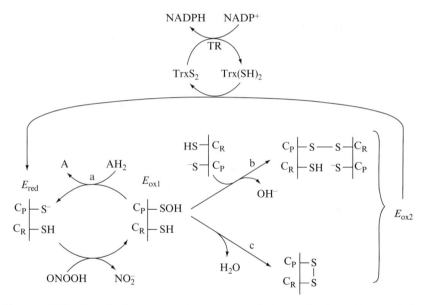

Figure 10.1 Peroxiredoxins catalyze peroxynitrite reduction. Peroxynitrite (ONOOH) rapidly oxidizes the peroxidatic cysteine residue (C_P) of reduced Prxs (E_{red}), usually deprotonated at physiological pH, by a two-electron oxidation mechanism yielding its sulfenic acid derivative (R-SOH). In the case of 1-Cys Prxs (A), this oxidized form of the enzyme (E_{ox1}) is directly reduced by the reducing substrate (AH_2), which has not been unambiguously identified for all 1-Cys Prxs, but thiol-containing compounds as well as ascorbate (Monteiro *et al.*, 2007) have been postulated as candidates. In 2-Cys Prxs, E_{ox1} is unstable, and the sulfenic acid derivative reacts with a second cysteine residue (the resolving cysteine residue, C_R) either in a different and inversely oriented subunit to form an intermolecular disulfide bridge (E_{ox2}) in typical 2-Cys Prxs (B) or in the same subunit to form an intramolecular disulfide (E'_{ox2}) in atypical 2-Cys Prxs (C). In most cases the natural reductant for 2-Cys Prxs is reduced thioredoxin, Trx-$(SH)_2$, which in its oxidized form presents an intramolecular disulfide bridge (Trx-SS) that is reduced by thioredoxin reductase (TR) at NADPH expense.

The role of Prxs in peroxynitrite detoxification was first established by Bryk and colleagues in 2000, where they reported that bacterial alkyl hydroperoxide reductase C (AhpC), which is considered a typical 2-Cys Prx, could catalyze peroxynitrite reduction. Afterward, our group, as well as others, confirmed this peroxynitrite-decomposing activity of different Prxs, either of the 1-Cys or 2-Cys types, such as mammalian Prx 6 (Peshenko *et al.*, 2001); *Mycobacterium tuberculosis* thioredoxin peroxidase (*Mt*TPx) (Jaeger *et al.*, 2004), human Prx 5 (Dubuisson *et al.*, 2004), *Trypanosoma cruzi* and *Trypanosoma brucei* cytosolic tryparedoxin peroxidase (Trujillo *et al.*, 2004), *Plasmodium falciparum* thioredoxin peroxidase 1 (Nickel *et al.*, 2005), *Saccharomyces cerevisiae* thioredoxin peroxidases 1 and 2 (Ogusucu *et al.*, 2007). As Prx-catalyzed reactions follow an enzymatic substitution (ping-pong) mechanism (Hofmann *et al.*, 2002), the oxidative part, as well as the reductive part, of the catalytic cycle can be studied independently by presteady-state techniques, which usually require the application of stopped-flow methodologies (Trujillo *et al.*, 2007a). Moreover, steady-state methodologies that allow studying the kinetics of peroxynitrite reduction by Prxs under catalytic conditions have also been developed.

2. General Considerations for Reagents Used

As has been often pointed out (Koppenol *et al.*, 1996; Radi, 1996), working with peroxynitrite needs special care in the preparation of the working solutions. Because peroxynitrite decays spontaneously in an acid-catalyzed reaction, it must be kept in alkaline solution up to the very moment of use. Peroxynitrite reacts rapidly with carbon dioxide, so every effort has to be made to avoid the contamination of working or stock solutions with atmospheric CO_2. It is advisable to always use freshly made solutions. Peroxynitrite is quantitated spectrophotometrically at 302 nm ($\varepsilon = 1670\ M^{-1}\ cm^{-1}$) by making dilutions in 10 mM NaOH prepared immediately before use.

Peroxiredoxin (or thioredoxin) reduction is performed by incubating 1 ml of the enzyme (4 mg/ml, which assuming a mean monomeric molecular weight of 25,000 would mean a 160 μM enzymatic concentration) with 5 mM dithiothreitol (DTT) for an hour. Excess DTT is removed using a size-exclusion Hitrap 1-ml column (Amersham), using extensively degassed buffer potassium phosphate 100 mM pH 7.0 + 0.2 mM diethylenetriamine pentaacetic acid (DTPA) for elution. With a flow rate of 1 ml/min, the retention time of the protein is approximately 2 min, which can be followed easily at 280 nm, collected in rubber-capped anaerobic flasks, and stored on ice under an argon atmosphere until use (Trujillo *et al.*, 2004). The concentrations of protein and of reduced thiols are measured immediately before the kinetic determinations.

Unless otherwise indicated, all reactions described herein are performed in 50 mM phosphate buffer containing 0.1 mM DTPA; in stopped-flow experiments this buffer is prepared *in situ* by mixing equal volumes of 100 mM phosphate, ≈pH 7.0, in one syringe with NaOH, ≈10 mM, in the other, and final pH is measured at the outlet. In this buffer, peroxynitrite decays spontaneously within seconds at room temperature.

3. Kinetics Studies of the Oxidative Part of the Catalytic Cycle

The kinetics of the oxidative part of the catalytic cycle can be studied by two kinds of approaches: (a) direct and (b) indirect or competitive presteady-state approaches.

3.1. Direct measurements

3.1.1. Background

When peroxynitrite is used as the oxidizing substrate, the oxidative part of the reaction catalyzed by Prxs is described by Eq. (10.1) (see Fig. 10.1), where the peroxidatic cysteine residue in the Prx, usually deprotonated at physiological pH, reacts with peroxynitrous acid to form a sulfenic acid derivative of the enzyme and nitrite. Most direct measurements of the kinetics of peroxynitrite-mediated Prxs oxidation take advantage of the spectral characteristics of peroxynitrite, since under its anionic form it absorbs at 302 nm (Radi *et al.*, 1991), and thus the rate of peroxynitrite decomposition, in the presence or in the absence of the enzyme, can be conveniently followed in a stopped-flow spectrophotometer (Radi, 1996). In order to avoid interferences arising from protein absorption at the mentioned wavelength, the peroxynitrite decomposition rate has been determined at 310 nm ($\varepsilon = 1600\ M^{-1}\ cm^{-1}$)(Trujillo *et al.*, 2004). The lower concentration of peroxynitrite that can be safely used in a conventional stopped- flow apparatus is approximately 10 μM (which would mean an absorbance of 0.0128 using a 1-cm cell path, considering that 80% of the total peroxynitrite is a peroxynitrite anion at pH 7.4). This limitation precludes first-order methodologies from being used in presteady-state kinetic determinations, as enzymatic concentrations of 10-fold excess would be difficult to achieve; even if achieved, assuming a second-order rate constant of peroxynitrite-mediated Prx oxidation of $1 \times 10^6\ M^{-1}\ s^{-1}$, a 100 μM enzyme concentration would mean a k_{obs} value ($k_2 \times$ [Prx]) of 100 s^{-1} and a $t_{1/2} = 6.9$ ms, that is, an important fraction of peroxynitrite decay would occur during the mixing time (generally <2 ms) of the apparatus. Thus, initial rate kinetic approaches are used more conveniently.

3.1.2. Initial rate approach
Materials

Phosphate buffer (100 mM, pH 7, containing 0.2 mM DTPA)
NaOH (approximately 10 mM)
Alkaline stock solution of peroxynitrite
Stock solution of reduced Prx (100 μM)

The kinetics of peroxynitrite-dependent Prx oxidation is performed following the initial rate of peroxynitrite decomposition at 310 nm in 50 mM phosphate buffer, pH 7.4, with 0.1 mM DTPA using a stopped flow spectrometer. Initial rates of reaction are measured more conveniently using an split mode that allows one to obtain half of the data points during the first part of the reaction (typically the first 20 ms) while the other half are obtained up to 20 s, which allows one to measure the amount of remaining peroxynitrite once rapid reaction with the Prx has finished, thus providing an easy way of measuring the amount of reactive thiol present in the enzyme. The initial rate of peroxynitrite decomposition in the absence of enzyme (measured during the first 2 to 10 or 20 ms of reaction) is almost zero, while it increases in the presence of increasing concentrations of the enzyme (Fig. 10.2).

Rates can be calculated by dividing the slope of the absorbance time course by the peroxynitrite molar extinction coefficient at 310 nm (1600) and multiplying by 1.25 to account for the 20% fraction of peroxynitrite that is not deprotonated at pH 7.4. Initial rates of peroxynitrite reduction follow Eq. (10.5):

$$v_0 = k_2' \times [\text{peroxynitrite}]_0 \times [\text{Prx}]_0 \qquad (10.5)$$

Because initial peroxynitrite and reduced Prx concentrations are known, the pH-dependent second-order rate constant for Prx-mediated peroxynitrite reduction (k_2') can be calculated. Caution should be taken in order to make sure that real initial rate velocities are being measured, that is, less than 10% of the reactants are consumed during the time chosen for initial rate measurements. Experiments should be performed at different initial concentrations of enzyme, which should correlate directly with rates. Negative control experiments using the enzyme oxidized previously by equimolar hydrogen peroxide or peroxynitrite and *N*-ethylmaleimide (NEM)-blocked enzyme (treated with a 10-fold excess NEM, subsequently removed by gel filtration) should not lead to any important increase in the peroxynitrite decomposition rate. In order to determine more precisely that peroxynitrite is reacting with the peroxidatic thiol, mutated forms of the enzyme at either the peroxidatic or the resolving cysteine of the enzyme have been used (Bryk *et al.*, 2000; Dubuisson *et al.*, 2004; Trujillo *et al.*, 2004).

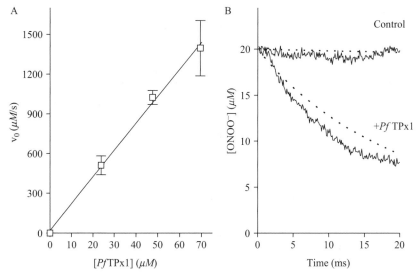

Figure 10.2 Peroxynitrite reduction by *Plasmodium falciparum* thioredoxin peroxidase 1 (*Pf*TPx1) measured by an initial rate approach. (A) Initial rates of peroxynitrite (20 μM) decay in 50 mM potassium phosphate buffer, 0.1 mM DTPA at pH 7.4 and 37 °C in the presence of increasing concentrations of reduced *Pf*TPx1. (B) Experimental traces (continuous lines) and traces obtained by computer-assisted simulations (dotted lines) of peroxynitrite decay in the absence (control) or presence of *Pf*TPx1 (48 μM). Computer-assisted simulations were performed using the following reactions: ONOOH $\rightarrow$ NO$_3^-$ + H$^+$ ($k = 0.6$ s^{-1}); ONOOH $\rightarrow$ ·NO$_2$ + ·OH ($k = 0.3$ s^{-1}); ONOOH + Prx-S$^-$ $\rightarrow$ NO$_2^-$ + oxidized Prx ($k = 1 \times 10^6$ M^{-1} s^{-1}); ·NO$_2$ + Prx-S$^-$ $\rightarrow$ NO$_2^-$ + oxidized Prx ($k = 3 \times 10^8$ M^{-1} s^{-1}). Because of the lack of selectivity of ·OH, and the multiplicity of targets in the Prx sequence, the reaction between this radical and the peroxidatic cysteine residue was not taken into account. Oxidized forms of Prx formed from the reaction with ONOOH and with ·NO$_2$ are different, but for simplicity, subsequent reactions involving those products were not considered in the simulation.

In the case of the atypical 2-Cys human Prx5, its oxidation is accompanied by changes in Trp84 fluorescence in the protein, which consisted in a red shift of emission spectra, as well as an increase in fluorescence intensity when excited at 280 nm, and the kinetics of Prx5 oxidation by different oxidants, including peroxynitrite, has been determined taking advantage of this fluorescence increase in a mutated form of the enzyme lacking the resolving cysteine residue (Trujillo *et al.*, 2007a). However, similar fluorescence changes in other Prxs have not been reported, indicating that, probably, this method could only be useful for selected Prxs. For those proteins, it has the advantage of being a very sensitive method, which allows the measurement of rate constants in the 10^7 M^{-1} s^{-1} range. Additionally, using the fluorescence change has permitted determining the rate constant for the intramolecular disulfide formation (14 s^{-1}), which is the rate-limiting step in the conformational change leading to the increase in fluorescence change in wild-type Prx5 under selected conditions (Trujillo *et al.*, 2007a).

3.1.3. Steady-state approach (enzyme in turnover)
Materials

Phosphate buffer (100 mM, pH 7, containing 0.2 mM DTPA)
NaOH approximately 10 mM
Alkaline stock solution of peroxynitrite
Stock solution of Prx (100 μM, not necessarily reduced)
Reduced thioredoxin 200 μM

Peroxynitrite (10–20 μM) decay in the presence of low concentrations of Prx (0.5–3 μM) that has not been reduced previously is indistinguishable from that in the absence of enzyme, as the only amino acid that reacts with peroxynitrite sufficiently fast to affect its decomposition rate at this low enzyme concentration is the reduced peroxidatic cysteine residue. If the reducing substrate does not react very rapidly with peroxynitrite ($k_2 < 10^4$ M^{-1} s^{-1}), as in the case of *M. tuberculosis* thioredoxin peroxidase (Trujillo *et al.*, 2006) and *Trypanosoma cruzi* tryparedoxin (Trujillo *et al.*, 2004), then up to 50 μM thioredoxin would not lead to an important increase in the peroxynitrite decomposition rate (0.9 s^{-1} vs <1.4 s^{-1}). Then, if Prx reduction occurs more rapidly than its oxidation by peroxynitrite, maintaining a reduced Prx concentration almost constant during the time course of the experiment, it is possible to observe an exponential decay of peroxynitrite, with k_{obs} values that should increase linearly with the concentration of Prx used (Trujillo *et al.*, 2004) (Fig. 10.3).

First, the kinetics of peroxynitrite-mediated reducing substrate oxidation is determined. For that purpose, reduced thioredoxin or alternative reducing substrate in excess (100–200 μM) is mixed rapidly with peroxynitrite (10 μM) in 50 mM potassium phosphate buffer, pH 7.4, at 25 °C, and peroxynitrite decay is followed at 310 nm. Experimental traces are fitted to exponential decays and k_{obs} values are plotted vs the thioredoxin concentration in order to determine k_2 values.

Peroxynitrite (10 μM) decomposition with no other addition (a), in the presence of Prx (0.5–3 μM) (b), in the presence of thioredoxin (or other reducing substrate) (50 μM) (c), or in the presence of both Prx (0.5–3 μM) plus 50 μM thioredoxin (d) is followed at 310 nm in potassium phosphate buffer, pH 7.4, plus 0.1 mM DTPA. If experimental traces in (d) can be fitted to exponential decays, then the rate of reduction is sufficiently high to maintain the reduced Prx constant during the assay and from observed rate constants of peroxynitrite decomposition plotted vs Prx concentration, k_2 values for peroxynitrite-mediated Prx oxidation can be obtained (Fig. 10.3). Moreover, even if reduction is not so fast as to maintain the

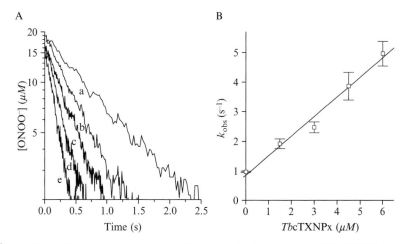

Figure 10.3 *Trypanosoma brucei* cytosolic tryparedoxin peroxidase (*Tbc*TXNPx) cata-lyzes peroxynitrite reduction. (A) Peroxynitrite (18 μM) decomposition in the presence of 70 μM *T. brucei* tryparedoxin (*Tbt*TXN) and increasing *Tbc*TXNPX concentrations (a = 0; b = 1.5 μM; c = 3 μM; d = 4.5 μM; e = 6 μM) in 50 mM potassium phosphate buffer, pH 7.4, and 37 °C was followed at 310 nm. (B) Plot of k_{obs} of peroxynitrite decay versus enzyme concentration. Modified from Trujillo *et al.*, 2004.

reduced Prx concentration constant, experimental traces can be compared with computer-assisted simulations using k_2 values for the reductive and oxidizing part of the catalytic cycle obtained by other approaches, such as initial rate, thus allowing the validation of data obtained by presteady-state conditions (Trujillo *et al.*, 2006).

Direct kinetic measurements have the obvious advantage of being simple, as no other reactions are involved. Moreover, as indicated, they allow an exact measurement of the amount of reactive thiol present in the sample. However, stopped–flow spectrophotometric detection following per-oxynitrite decay has low sensitivity and allows rate constant determinations if $k_2' < 10^7 \, M^{-1}s^{-1}$. For faster reactions, pulse radiolytic techniques have been used (Dubuisson *et al.*, 2004).

3.2. Competition approaches

3.2.1. Background

The kinetic competition between a characterized and an unknown reaction can be used to determine the unknown rate constant. This approach is particularly useful if the known reaction can be monitored by the change in a physical property (e.g., absorbance or fluorescence emission) that reports the advance of the reaction with high sensitivity.

The general scheme of reaction for simple competition is as follows:

$$A_1 + B \rightarrow P_1 \quad k_1 \tag{10.6}$$

$$A_2 + B \rightarrow P_2 \quad k_2 \tag{10.7}$$

Two reactants (A_1 and A_2) compete for reacting with B and, provided one of the rate constants is known along with the initial concentration of reactants and the final concentration of at least one product, the value of the second rate constant can be obtained by the concentration of either remaining reactants ($[A_1]_\infty$, $[A_2]_\infty$) or of formed products ($[P_1]_\infty$, $[P_2]_\infty$) upon completion of the reaction ($[B]_\infty = 0$) or even by determining the apparent rate constant of one of the reactions in the presence of a constant initial concentration of the second reactant. This latter method is not examined here.

In the simplest case, A_1 and A_2 can be used in large excess (10-fold or more) relative to B, thus providing pseudo-first-order conditions for reactions (10.6) and (10.7). It can be shown easily that in this case,

$$k_1 \frac{[A_1]_0 [P_2]_\infty}{[A_2]_0 [P_1]_\infty} = k_2, \tag{10.8}$$

if only one of the product concentrations can be determined, the stoichiometric relationship ($[B]_0 = [P_1]_\infty + [P_2]_\infty$) is used to quantify the other. This consideration leads to the often used equation for competition kinetics:

$$k_1 \frac{[A_1]_0 F}{(1-F)} = k_2 [A_2]_0, \tag{10.9}$$

where F is the fraction of P_1 formation inhibited in the presence of a given $[A_2]_0$. This equation has the practicality of allowing the use of the measurement of any physical property (e.g., absorbance or fluorescence emission) linearly dependent of $[P_1]$. The left-hand side of Eq. (10.9) can be plotted vs $[A_2]_0$ to obtain k_2 as the slope of the resulting line.

In many cases, a large excess of reactants cannot be used because of limitations in the method of quantitation or by availability issues. In such cases, Eq. (10.8) no longer applies and the following relationship should be used instead (Espenson, 1995):

$$\frac{k_1}{k_2} = \frac{\ln\left\{\frac{[A_1]_0}{[A_1]_0 - [P_1]_\infty}\right\}}{\ln\left\{\frac{[A_2]_0}{[A_2]_0 - [P_2]_\infty}\right\}}. \tag{10.10}$$

Once more the stoichiometric relationship ($[B]_0 = [P_1]_\infty + [P_2]_\infty$) can be applied but no simple plot can be constructed and the value of the unknown constant should be calculated for each set of initial concentrations.

3.3. Case 1: One peroxide two reductants

In this case peroxynitrite can react either with Prx, with a rate constant to be determined, or with another compound, whose fast reactivity with peroxynitrite had been determined previously, and whose oxidation should be followed easily spectrophotometrically.

3.3.1. Competition with Mn^{III}porphyrins
Materials

Phosphate buffer (100 mM, pH 7, containing 0.2 mM DTPA)
NaOH approximately 10 mM
Alkaline stock solution of peroxynitrite
Stock solution of reduced Prx (100 μM)
Mn^{III}porphyrins (available commercially from major providers of chemicals and also from specialized providers such as Mid-Century Chemicals). Stock solution concentrations should be measured according to reported ε values (Table 10.1).

The first case to explore uses competition between the peroxidatic thiol of the peroxiredoxin and a synthetic Mn^{III}porphyrin reacting with peroxynitrite. This approach has been used several times in the literature (Jaeger *et al.*, 2004; Trujillo *et al.*, 2004) and constitutes the first quantitative evaluation of rate constants for peroxiredoxin reactions with peroxynitrite by competition kinetics. In the general scheme of Eqs. (10.6) and (10.7), A_1 would be Mn^{III}porphyrin; $P_1 = O=Mn^{IV}$porphyrin; $B_1 =$ reduced Prx; and $P_1 =$ oxidized Prx.

Table 10.1 Rate constants and absorption maxima in the visible range for selected Mn^{III}porphyrins

Mn^{II}porphyrin	k_1, pH indep. (37 °C, $M^{-1}s^{-1}$)	λ_{max} ($\varepsilon M^{-1}cm^{-1}$)
Mn^{III}TE-2-PyP	3.4×10^7	454 (140,000)
Mn^{III}TM-2-PyP	1.9×10^7	454 (129,000)
Mn^{III}TB-2-PyP	1.3×10^7	454 (170,000)
Mn^{III}TM-4-PyP	4.3×10^6	462 (130,000)
Mn^{III}TSPP	3.4×10^5	467 (95,000)

[a] Data from Ferrer-Sueta *et al.* (2003).

Peroxynitrite reactions with Mn^{III}porphyrins

$$Mn^{III}porphyrin + ONOO^- \rightarrow O = Mn^{IV}porphyrin + \cdot NO_2$$
$$(10.11)$$

have been studied thoroughly (Ferrer-Sueta *et al.*, 2003) and can be followed by the absorption in the 450 to 470 nm region. Selected values of rate constants and spectral features are listed for several Mn^{III}porphyrins in Table 10.1. It can be seen from these values that the range of rate constants spans two orders of magnitude with values similar to those reported for peroxiredoxins reacting with peroxynitrite.

Rate constants in Table 10.1 are pH independent and as Mn^{III}porphyrins react with $ONOO^-$ they are similar to the apparent rate constants at pH $>$ 7.4 (Ferrer-Sueta *et al.*, 1999) and higher than those observed at lower pH values. Additionally, k_1 values in Table 10.1 were obtained at 37 °C, and considering an activation enthalpy of about 11 kcal/mol (G. Ferrer-Sueta, unpublished results) the values at 25 °C are approximately 2.2 times lower. In any case, given the sensitivity of the reaction to pH, temperature, and ionic strength, it is highly advisable to determine k_1 under the exact conditions at which the rate constant of the peroxiredoxin is desired; the technique for this experiment has been described previously (Ferrer-Sueta *et al.*, 2002).

When designing the competition experiment it is advisable to have an estimate of the magnitude of k_2 and thus choose the competing Mn^{III}porphyrin accordingly. Usually the experiment is done with a fixed initial concentration of Mn^{III}porphyrins and variable peroxiredoxin. Initial concentrations are chosen so that (1) the proton-catalyzed decomposition of peroxynitrite is almost null; (2) the condition of expected $k_1[Mn^{III}porphyrin]_0 = k_2[Prx]_0$ is included in the range of concentrations; and (3) the expected initial absorbance and Δabsorbance are measurable.

To illustrate the selection of experimental conditions we will take an example from the literature (Trujillo *et al.*, 2007a). *Mt*TPx was characterized previously to react with peroxynitrite with a rate constant of $1.2 \times 10^7 \, M^{-1} s^{-1}$ at 25 °C, pH 7.4 (Jaeger *et al.*, 2004), thus Mn^{III}TB-2-PyP was chosen as the competing reductant; its rate constant was determined under the assay conditions as $4.8 \times 10^6 \, M^{-1} s^{-1}$. A concentration of 7 μM was chosen to yield an initial absorbance of approximately 1.2 and an apparent rate constant of peroxynitrite consumption ($k_1[Mn^{III}$porphyrins]) of 38 s^{-1}, which is more than 100 times larger than the proton-catalyzed decomposition of peroxynitrite at this pH, thus precluding the formation of significant amounts of homolysis products. The initial concentration of peroxynitrite was chosen as 2 μM to have a good ΔAbs, to always have excess Mn^{III}Pophyrin, and not to increase the $[Mt$TPx$]_0$ needed.

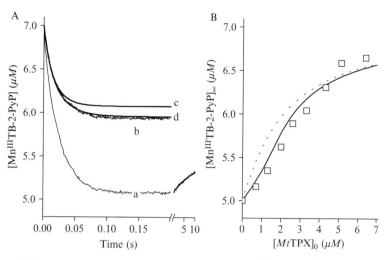

Figure 10.4 Competition reactions of MtTPX and Mn^{III}TB-2-PyP with peroxynitrite. Conditions $[Mn^{III}$TB-2-PyP$]_0 = 7 \mu M$; $[ONOO^-]_0 = 2 \mu M$; $[Mt$TPX$]_0 = 0$–$6.4 \mu M$; 25 °C 100 mM phosphate, pH 7.4. (A) Time courses at 454 nm with $[Mt$TPX$]_0 = 0$ (line a) and 2.6 μM (line b). Lines c and d are simulated results assuming that the critical thiol of MtTPX reacts with ·NO_2 with a rate constant of $3 \times 10^8 \, M^{-1} s^{-1}$ and 0, respectively. (B) Experimental concentration of Mn^{III}TB-2-PyP remaining after 0.2 s of reaction (squares), simulated results avoiding (dotted line) or considering (continuous line) a cross-reaction with ·NO_2 with a rate constant of $3 \times 10^8 \, M^{-1} s^{-1}$.

Some of the results obtained are shown in Fig. 10.4A as time courses at 454 nm. Figure 10.4B summarizes the competition experiment through the remaining concentration of Mn^{III}porphyrin at different initial concentrations of MtTPX. Analysis of the experimental results using Eq. (10.10) yields a k_2 of $2 \pm 0.9 \times 10^7 \, M^{-1} \, s^{-1}$, in good agreement with previously reported values (Jaeger *et al.*, 2004; Trujillo *et al.*, 2007a) and used to simulate the time courses using Gepasi 3.30 (Mendes, 1993, 1997). Some useful pieces of information can be extracted from Fig. 10.5. First, even in the absence of MtTPX (line a), $O = Mn^{IV}$TB-2-PyP is reduced back to Mn^{III}TB-2-PyP after a few seconds, underscoring the need of performing this experiment with a spectrophotometer fitted with a rapid mixing device (e.g., a stopped flow). Second, reaction (10.11) yields ·NO_2 as a by-product and this radical is known to react with thiols very rapidly; second-order rate constants are reported as $3 \times 10^7 \, M^{-1} s^{-1}$ with thiols and $3 \times 10^8 \, M^{-1} \, s^{-1}$ with thiolates (Ford *et al.*, 2002). This cross-reaction between a product and one of the competing reagents may alter the results and lead to underestimation of the value of k_2; this is seen clearly in Fig. 10.4A, as simulation ignoring the reaction with ·NO_2 predicts a more efficient competition of MtTPX with the calculated k_2 than is

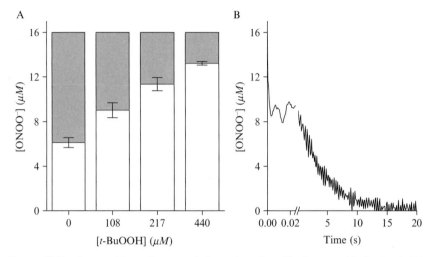

Figure 10.5 Competition of peroxynitrite and *tert*-butyl hydroperoxide for Prx5. (A) Reduced Prx (10 μM) is mixed rapidly with peroxynitrite (16 μM) and variable concentrations of *t*-BuOOH in a stopped-flow spectrophotometer. Peroxynitrite decay has two phases: the reaction with Prx5 is fast ($t < 10$ ms) and is represented by the gray part of the bars and the proton-catalyzed decomposition ($t < 20$ s) is shown as the white part of the bar. (B) Time course of the aforementioned reaction obtained using 108 μM *t*-BuOOH. Reactions were carried out at 25 °C, 50 mM phosphate, 0.1 mM DTPA, pH 7.4.

observed experimentally. As *Mt*TPX is a protein with two other cysteines in addition to peroxidatic C47 (C95, C151), we included in the simulation the reaction of $\cdot NO_2$ with two additional Cys residues; the inclusion of these reactions did not alter the previous results significantly (not shown).

In summary, the determination of rate constants using Mn^{III}porphyrins as competing reagents for peroxynitrite has the advantages of high sensitivity, with $\Delta\varepsilon$ in the range of 10^4 to 10^5 M^{-1} cm^{-1} and a wide range of rate constants (10^5–10^7 M^{-1} s^{-1}) already determined for the competing reaction. As potential drawbacks, the product that needs to be measured is not stable, requiring fast mixing and absorbance measurement; additionally, the formation of $\cdot NO_2$ and its cross-reaction with the thiolate of Prxs can lead to an underestimation of the rate constant of interest. Other than that, the method has yielded results in excellent agreement with rate constants determined by other techniques, requiring less enzyme and having higher sensitivity.

3.3.2. Competition with heme peroxidases
Materials

Phosphate buffer (100 mM, pH 7, containing 0.2 mM DTPA)
NaOH approximately 10 mM
Alkaline stock solution of peroxynitrite
Stock solution of reduced Prx (100 μM)

Heme peroxidase ($250\ \mu M$), stock solution concentrations should be measured according to reported ε value; in the case of HRP, $\varepsilon_{403} = 102{,}000\ M^{-1}\,s^{-1}$ (Schonbaum and Lo, 1972).

The second case of kinetic study using a competing reductant involves the reaction between peroxynitrite and a heme peroxidase. The only report in the literature (Ogusucu et al., 2007) measures the rate constant of two thioredoxin peroxidases from S. cerevisiae (Tsa1 and Tsa2) using the reaction with horseradish peroxidase (HRP):

$$HRP + ONOOH \rightarrow compound\ I + NO_2^-. \qquad (10.12)$$

This reaction has a pH–independent rate constant of $3.2 \times 10^6\ M^{-1}\,s^{-1}$ at $25\,^{\circ}C$ (Floris et al., 1993; Ogusucu et al., 2007) and its spectral change at 398 nm involves a $\Delta\varepsilon$ of $4.2 \times 10^4\ M^{-1}\,cm^{-1}$ (Hayashi and Yamazaki, 1979).

The reported conditions for the kinetic determination were as follows: $[HRP]_0 = 8.2\ \mu M$; $[ONOO^-]_0 = 8\ \mu M$. and $[Prx]_0 = 0$ to $6.8\ \mu M$; the authors obtained rate constants of 7.4 and 5.1×10^5 for Tsa1 and Tsa2, respectively, at $25\,^{\circ}C$ and pH 7.4.

General kinetic considerations and the way of obtaining second-order constants are substantially the same as in the previously discussed case of Mn^{III}porphyrins. Some differences arise from the pH dependence of reaction (12) being different to reaction (11). Additionally, the colored product, compound I, is also unstable and decays first to compound II and then to HRP if reductants are available; this prompts using a system of rapid mixing and absorbance measurement such as a stopped–flow spectrophotometer. Measuring the absorbance at the isosbestic between compound I and compound II (398 nm) provides further confidence in the measurement.

Horseradish peroxidase is clearly the first choice among heme proteins because it is easily available, its relatively low price, and its well-characterized spectra and kinetics, but other heme peroxidases are known to react with peroxynitrite with high rate constants and large $\Delta\varepsilon$; some examples appear in Table 10.2. It is worth noticing that HRP and prostaglandin endoperoxide H synthase-1 produce compound I on reaction with peroxynitrite, while other peroxidases produce compound II and $\cdot NO_2$ directly. In the latter case, the cross-reaction between $\cdot NO_2$ and the thiolate of Prx has to be considered as described earlier.

Many heme proteins can catalyze peroxidase-like reactions and also use peroxynitrite as an oxidant so it could be tempting to consider them as candidates for determining rate constants of Prx via competition kinetics. This is not advisable, as reactions between heme proteins and peroxynitrite are often complex and yield multiple products, as is the case for hemoglobin (Romero et al., 2003), or do not produce a significant change in absorbance as happens with metmyoglobin (Herold et al., 2004; Martinez et al., 2000).

Table 10.2 Reactions of heme peroxidases with peroxynitrite

| Peroxidase[a] | Rate constant with ONOOH | | Heme product | λ_{obs} | Reference |
	$k\,(M^{-1}s^{-1})$	Condition			
HRP	3.2×10^6	25 °C, pH indep.	Compound I	398 nm	Floris *et al.* (1993)
	1.02×10^6	25 °C, pH 7.4			Ogusucu *et al.* (2007)
PGHS-1	1.7×10^7	8 °C, pH 7	Compound I	408 nm	Trostchansky *et al.* (2007)
MPO	2×10^7	12 °C, pH indep.	Compound II	428, 455 nm	Floris *et al.* (1993)
	6.8×10^6	25 °C, pH 7.4		456 nm	Furtmuller *et al.* (2005)
LPO	3.3×10^5	12 °C, pH 7.4	Compound II	412, 430 nm	Floris *et al.* (1993)
CPO	1.96×10^6	23 °C, pH 5.1	Compound II	400 nm	Gebicka and Didik (2007)

[a] HRP, horseradish peroxidase; PGHS-1, prostaglandin endoperoxide H synthase-1; MPO, myeloperoxidase; LPO, lactoperoxidase; CPO, chloroperoxidase.

3.4. Case 2: One target two peroxides

In this case, the competition is between peroxynitrite and other hydroperoxides for reduced Prx:

$$Prx - S^- + ONOOH \rightarrow Prx - SOH + NO_2^- \qquad (10.13)$$

$$Prx - S^- + ROOH \rightarrow Prx - SOH + RO^- \qquad (10.14)$$

If the rate constant of reaction (10.13) or (10.14) is known, measuring the effect of a given concentration of one oxidizing substrate on the Prx-mediated reduction of the other, and using Eq. (10.10), the unknown rate constant can be calculated easily. Therefore, this methodology allows (a) determining the kinetics of the reaction between peroxynitrite and Prx if the rate constant with other peroxides is known or (b) determining the kinetics of the reaction between other oxidizing substrates and Prx if the rate constant with peroxynitrite has already been determined by an alternative methodology.

Materials

Phosphate buffer (100 mM, pH 7, containing 0.2 mM DTPA)
NaOH approximately 10 mM
Alkaline stock solution of peroxynitrite
Stock solution of reduced Prx (100 μM)
Stock solution of H_2O_2 (or other hydroperoxide). Hydrogen peroxide concentration can be measured spectrophotometrically at 240 nm ($\varepsilon = 43.6$ $M^{-1}cm^{-1}$).

Peroxynitrite (20 μM) in 10 mM NaOH either in the absence or in the presence of increasing concentrations of ROOH is mixed rapidly with 100 mM potassium phosphate buffer at final pH 7.4 and 25 °C with 10 μM reduced enzyme (Fig. 10.5), and peroxynitrite decay is followed in a stopped-flow apparatus at 310 nm. The concentration of peroxynitrite remaining after its reaction with Prx is calculated by using a split time mode. Half experimental data points are obtained during the first 0.1 s of reaction (which is typically enough for the fast reaction between Prx and peroxynitrite to be complete), and the other half during 20 seconds (which is enough to assure that all remaining peroxynitrite is decomposed under our experimental conditions). Data points from 0.1 to 20 s are fitted to an exponential curve, and from the amplitude of peroxynitrite decay of the second phase (which is converted into peroxynitrite concentration as described earlier) the remaining peroxynitrite concentrations after its reaction with Prx in the absence and in the presence of the alternative oxidizing substrate are calculated. Control experiments showing (a) the absence of an effect of the hydroperoxide of choice alone on peroxynitrite decomposition (in the absence of the enzyme) and (b) no change in absorbance by mixing the hydroperoxide and reduced Prx in the absence of peroxynitrite (at 310 nm) should be performed. From Eq. (10.10), and if the rate constant of reaction (10.14) is known, the kinetics of the reaction with peroxynitrite can be calculated. We used this approach to measure the second-order rate constant of oxidation of Prx5 by organic hydroperoxides. This technique has the following advantages: it allows studying the reactivities of many different peroxides with Prx, it is a simple and rapid method, and peroxides are usually not expensive. Drawbacks include the fact that only peroxides reacting at similar rates or slower than peroxynitrite can be studied and since quantifiable peroxynitrite should be >10 μM (we describe the methodology for a 20 μM peroxynitrite concentration), the sensitivity of the methodology is low and requires a large amount of Prx to be used.

This method was originally devised by Peshenko *et al.* (2001) to measure relative rates of reaction between competing peroxides for Prx6 (Peshenko *et al.*, 2001). In the original publication, the authors employed the methodology without using a stopped–flow spectrophotometer; instead of following peroxynitrite, they measured the remaining concentration of H_2O_2. This possibility has the advantage of a higher sensitivity, since the remaining hydrogen peroxide can be measured using fluorescence methods at concentrations below 1 μM (Guilbault *et al.*, 1968). It can be used to study the kinetics of reaction of different peroxides (and not only peroxynitrite) with Prx, provided its rate constant with hydrogen peroxide is known. However, only peroxides reacting at similar rates or slower than H_2O_2 can be studied and, importantly, the remaining H_2O_2 needs to be quantified by a subsequent reaction, which should not be affected by the presence of alternative oxidizing substrates.

4. Reductive Part of the Catalytic Cycle

In order to study the kinetics of the reductive part of the catalytic cycle of Prxs, which are independent of the nature of the oxidizing substrate, as they have a ping-pong mechanism of reaction, direct approaches that take advantage of the changes in thioredoxin fluorescence properties upon interaction with oxidized Prxs have been used.

Materials

Buffer potassium phosphate 100 mM, pH 7.4, plus 0.2 mM DTPA
Reduced homologous thioredoxin
The protein is reduced by incubation with an excess concentration of DTT
during 1 h and excess DTT is removed as described earlier.

Preoxidized Prx: reduced Prx is treated with a 1.1 excess peroxynitrite concentration in the aforementioned buffer immediately before use.

A solution of 1 μM reduced homologous thioredoxin in 100 mM potassium phosphate buffer, pH 7.4, plus 0.2 mM DTPA is mixed with increasing concentrations of preoxidized Prxs in the same buffer, at 25 °C, in a stopped-flow spectrofluorimeter and the total fluorescence increase ($\lambda_{ex} = 280$ nm, $\lambda_{em} > 320$ nm using appropriate filters) is recorded. Data points are fitted to exponential curves from which the observed rate constant of change in fluorescence intensity is obtained. From the slope of the plot of k_{obs} versus Prx concentration, the k_2 value of thioredoxin-mediated Prx reduction is obtained. This method allowed the determination of the reactivity between homologous (human, recombinant, His-tagged) oxidized Prx 5 and reduced thioredoxin 2 (Trujillo et al., 2007b), as well as homologous (from M. tuberculosis, recombinant, His-tagged) oxidized thioredoxin peroxidase and reduced thioredoxin B (Trujillo et al., 2006).

The method is simple and quite sensitive, allowing rate constant determinations in a wide range of possible rate constants (up to 10^7–10^8 M^{-1} s^{-1}), but some disadvantages should be mentioned. (a) Changes in fluorescence intensity in mammalian thioredoxin upon oxidation are lower than in some bacterial thioredoxins, decreasing the sensitivity of the method. (b) Pseudo-first-order conditions, with excess Prx concentrations over thioredoxin, allow rate constants to be calculated without a precise measurement of thioredoxin concentration. However, pseudo-first-order conditions cannot always be achieved, and initial rate approaches require making a calibration curve in order to transform the intensities of fluorescence change into changes in thioredoxin concentration. Moreover, the stable oxidized form of the enzyme is not always necessarily the same that interacts with the reducing substrate during catalysis. Moreover, not all Prxs use thioredoxins or thioredoxin-like proteins as reducing substrates. In most 1-Cys Prx, a physiologically relevant reducing substrate has not yet been determined, with possible candidates being ascorbate (Monteiro et al., 2007) and, in the case of mammalian Prx6, heterodimerization and glutathione-dependent reduction (Ralat et al., 2006).

5. CONCLUSIONS

Peroxiredoxins catalyze peroxynitrite reduction. Rate constants for peroxynitrite-mediated Prxs oxidation, measured by either direct and/or competitive kinetic approaches, are in the 10^6 to 10^8 M^{-1} s^{-1} range (Table 10.3). These rate constants lie far above the expected 10^4 M^{-1} s^{-1} value, considering a peroxidatic thiol pK_a of $\approx$5–6 (Table 10.3), if the Brønsted relationship followed by low-molecular weight thiols is applied

Table 10.3   Reported thiol pK_a and pH-dependent second-order rate constants for peroxynitrite reduction by recombinant Prxs

Peroxiredoxin	Thiol pK_a	Rate constant ($M^{-1}s^{-1}$)	Condition	Reference
Bacterial AhpC	<5		pH 6.8	Bryk *et al.* (2000)
S. typhimurium		1.5×10^6		
M. tuberculosis		1.3×10^6		
H. pylori		1.2×10^6		
T. brucei cytosolic TXNPx	ND	9.0×10^5	pH 7.4, 37 °C	Trujillo *et al.* (2004)
T. cruzi cytosolic TXNPx	ND	7.2×10^5	pH 7.4, 37 °C	Trujillo *et al.* (2004)
M. tuberculosis TPx	ND	1.5×10^7	pH 7.4, 25 °C	Jaeger *et al.* (2004)
Human Prx5		7.0×10^7	pH 7.8[a]	Dubuisson *et al.* (2004);
	5.3	1.2×10^8	pH 7.4, 25 °C	Trujillo *et al.* (2007a)
P. falciparum TPx I	ND	1.0×10^6	pH 7.4, 37 °C	Nickel *et al.* (2005)
S. cerevisiae TPx I	5.7	7.4×10^5	pH 7.4, 25 °C	Ogusucu *et al.* (2007)
S. cerevisiae TPx II	6.3	5.1×10^5	pH 7.4, 25 °C	Ogusucu *et al.* (2007)

[a] Determinations were carried out at room temperature.

(Trujillo *et al.*, 2007a). Although less data regarding other hydroperoxide-mediated thiol oxidation are available, it is very probable that the same considerations apply for those reactions. Thus, data indicate that Prxs are in a way specialized for hydroperoxide (including peroxynitrite) reduction and that not yet identified protein factors, in addition to a low peroxidatic thiol pK_a, are responsible for the mentioned specialization. Moreover, peroxynitrite-oxidized Prxs are generally reduced by thioredoxin or analogous proteins, closing the catalytic cycle. Rate constants for reduction can be measured by a fluorescence direct approach and are in the 10^4 to 10^5 M^{-1} s^{-1} range, similar to reported values for thioredoxin–mediated Prx reduction using steady-state approaches, when other (stable) hydroperoxides were used as oxidizing substrates (Trujillo *et al.*, 2007b), as should be the case for an enzyme catalyzing a bisubstratic reaction by a ping-pong mechanism. Although at first glance one could think that the Prx-catalyzed peroxynitrite reduction would be limited by the reductive part of the catalytic cycle, it should be noted that actual rates of reactions depend not only on rate constants but also on substrate concentrations. It seems very probable that under most physiologically relevant conditions, where steady-state concentrations of peroxynitrite are considered to be in the nanomolar range, whereas the thioredoxin concentration is expected to be in the micromolar range, the rate of peroxynitrite reduction would be limited by the oxidative part of the catalytic cycle. However, *in vivo* situations are even more complicated than that, as, in fact, Prx and thioredoxin are part of a sequential reaction antioxidant system that includes thioredoxin reductase and NADPH (Fig. 10.1), which should also be taken into account. Moreover, peroxynitrite-mediated Prx oxidative inactivation is likely to occur.

One can look the problem from another point of view: would Prxs be important in peroxynitrite detoxification, considering the multiplicity of naturally existing compounds that also react with it, including CO_2 and GSH, to name the more abundant ones in cellular compartments? Kinetics can also help answer this question, as the peroxynitrite fate in biological media would again depend on rate constants and target concentrations. Considering a reduced Prx concentration in the 10 μM range [Prx 2 concentration in red blood cells has been reported to be 250 μM (Moore *et al.*, 1991)] and a mean rate constant for the reaction with peroxynitrite of 5 × $10^6 \, M^{-1} \, s^{-1}$ at pH 7.4 and 25 °C, then the product of rate constant times Prx concentration would be 50 s^{-1}. In the case of CO_2 and GSH, with physiological concentrations near 1.3 and 5 mM, respectively, the aforementioned product would be 46 s^{-1} for CO_2 (Lymar and Hurst, 1995) and 4 s^{-1} for GSH (Koppenol *et al.*, 1992), indicating that Prxs could be considered relevant targets for peroxynitrite reactivity, even in the presence of a physiological concentration of alternative substrates (Trujillo *et al.*, 2007a). Although kinetic data provide a basis for the rationalization of the role of Prxs in peroxynitrite reduction, they should be complemented by cellular studies or animal Prx knockout or overexpressing models, which would allow determining whether this pathway is relevant for peroxynitrite detoxification *in vivo*. Recent data from our laboratory (Piacenza *et al.*, 2008), as well as published by other groups (Barr and Gedamu, 2003; Hattori *et al.*, 2003; Master *et al.*, 2002; Wong *et al.*, 2002), indicate that, at least for some cellular systems, this is indeed the case.

ACKNOWLEDGMENTS

This work was supported by grants from the Howard Hughes Medical Institute and the International Centre of Genetic Engineering and Biotechnology (ICGEB, Trieste) to R. R. and from CONICYT- Fondo Clemente Estable, Uruguay, to M.T.; GFS was partially supported by Grant PDT 63/081 from CONICYT- Fondo Clemente Estable, Uruguay.

REFERENCES

Barr, S. D., and Gedamu, L. (2003). Role of peroxidoxins in *Leishmania chagasi* survival: Evidence of an enzymatic defense against nitrosative stress. *J. Biol. Chem.* **278,** 10816–10823.

Bryk, R., Griffin, P., and Nathan, C. (2000). Peroxynitrite reductase activity of bacterial peroxiredoxins. *Nature* **407,** 211–215.

Denicola, A., Freeman, B. A., Trujillo, M., and Radi, R. (1996). Peroxynitrite reaction with carbon dioxide/bicarbonate: Kinetics and influence on peroxynitrite-mediated oxidations. *Arch. Biochem. Biophys.* **333,** 49–58.

Dubuisson, M., Vander Stricht, D., Clippe, A., Etienne, F., Nauser, T., Kissner, R., Koppenol, W. H., Rees, J. F., and Knoops, B. (2004). Human peroxiredoxin 5 is a peroxynitrite reductase. *FEBS Lett.* **571,** 161–165.

Espenson, J. H. (1995). Chemical kinetics and reaction mechanisms. *In* "Series in Advanced Chemistry" (J.H Espenson, ed.), pp. 46–69. McGraw-Hill, New York.

Ferrer-Sueta, G., Batinic-Haberle, I., Spasojevic, I., Fridovich, I., and Radi, R. (1999). Catalytic scavenging of peroxynitrite by isomeric Mn(III) N-methylpyridylporphyrins in the presence of reductants. *Chem. Res. Toxicol.* **12,** 442–449.

Ferrer-Sueta, G., Quijano, C., Alvarez, B., and Radi, R. (2002). Reactions of manganese porphyrins and manganese-superoxide dismutase with peroxynitrite. *Methods Enzymol.* **349,** 23–37.

Ferrer-Sueta, G., Vitturi, D., Batinic-Haberle, I., Fridovich, I., Goldstein, S., Czapski, G., and Radi, R. (2003). Reactions of manganese porphyrins with peroxynitrite and carbonate radical anion. *J. Biol. Chem.* **278,** 27432–27438.

Floris, R., Piersma, S. R., Yang, G., Jones, P., and Wever, R. (1993). Interaction of myeloperoxidase with peroxynitrite: A comparison with lactoperoxidase, horseradish peroxidase and catalase. *Eur. J. Biochem.* **215,** 767–775.

Ford, E., Hughes, M. N., and Wardman, P. (2002). Kinetics of the reactions of nitrogen dioxide with glutathione, cysteine, and uric acid at physiological pH. *Free Radic. Biol. Med.* **32,** 1314–1323.

Furtmuller, P. G., Jantschko, W., Zederbauer, M., Schwanninger, M., Jakopitsch, C., Herold, S., Koppenol, W. H., and Obinger, C. (2005). Peroxynitrite efficiently mediates the interconversion of redox intermediates of myeloperoxidase. *Biochem. Biophys. Res. Commun.* **337,** 944–954.

Gebicka, L., and Didik, J. (2007). Kinetic studies of the reaction of heme-thiolate enzyme chloroperoxidase with peroxynitrite. *J. Inorg. Biochem.* **101,** 159–164.

Goldstein, S., and Czapski, G. (1995). The reaction of NO˙ with O_2^- and HO_2^-: A pulse radiolysis study. *Free Radic. Biol. Med.* **19,** 505–510.

Guilbault, G. G., Brignac, P. J., Jr., and Juneau, M. (1968). New substrates for the fluorometric determination of oxidative enzymes. *Anal. Chem.* **40,** 1256–1263.

Hattori, F., Murayama, N., Noshita, T., and Oikawa, S. (2003). Mitochondrial peroxiredoxin-3 protects hippocampal neurons from excitotoxic injury *in vivo*. *J. Neurochem.* **86,** 860–868.

Hayashi, Y., and Yamazaki, I. (1979). The oxidation-reduction potentials of compound I/compound II and compound II/ferric couples of horseradish peroxidases A2 and C. *J. Biol. Chem.* **254,** 9101–9106.

Herold, S., Kalinga, S., Matsui, T., and Watanabe, Y. (2004). Mechanistic studies of the isomerization of peroxynitrite to nitrate catalyzed by distal histidine metmyoglobin mutants. *J. Am. Chem. Soc.* **126,** 6945–6955.

Hofmann, B., Hecht, H. J., and Flohé, L. (2002). Peroxiredoxins. *Biol. Chem.* **383,** 347–364.

Jaeger, T., Budde, H., Flohé, L., Menge, U., Singh, M., Trujillo, M., and Radi, R. (2004). Multiple thioredoxin-mediated routes to detoxify hydroperoxides in Mycobacterium tuberculosis. *Arch. Biochem. Biophys.* **423,** 182–191.

Kissner, R., Nauser, T., Bugnon, P., Lye, P. G., and Koppenol, W. H. (1997). Formation and properties of peroxynitrite as studied by laser flash photolysis, high-pressure stopped-flow technique, and pulse radiolysis. *Chem. Res. Toxicol.* **10**, 1285–1292.

Koppenol, W. H., Kissner, R., and Beckman, J. S. (1996). Syntheses of peroxynitrite: To go with the flow or on solid grounds? *Methods Enzymol.* **269**, 296–302.

Koppenol, W. H., Moreno, J. J., Pryor, W. A., Ischiropoulos, H., and Beckman, J. S. (1992). Peroxynitrite, a cloaked oxidant formed by nitric oxide and superoxide. *Chem. Res. Toxicol.* **5**, 834–842.

Lymar, S. V., and Hurst, J. K. (1995). Rapid reaction between peroxynitrite ion and carbon dioxide: Implications for biological activity. *J. Am. Chem. Soc.* **117**, 8867–8868.

Martinez, G. R., Di Mascio, P., Bonini, M. G., Augusto, O., Briviba, K., Sies, H., Maurer, P., Rothlisberger, U., Herold, S., and Koppenol, W. H. (2000). Peroxynitrite does not decompose to singlet oxygen ((1)Delta (g)O(2)) and nitroxyl (NO(-)). *Proc. Natl. Acad. Sci. USA* **97**, 10307–10312.

Master, S. S., Springer, B., Sander, P., Boettger, E. C., Deretic, V., and Timmins, G. S. (2002). Oxidative stress response genes in *Mycobacterium tuberculosis*: Role of ahpC in resistance to peroxynitrite and stage-specific survival in macrophages. *Microbiology* **148**, 3139–31344.

Mendes, P. (1993). GEPASI: A software package for modelling the dynamics, steady states and control of biochemical and other systems. *Comput. Appl. Biosci.* **9**, 563–571.

Mendes, P. (1997). Biochemistry by numbers: Simulation of biochemical pathways with Gepasi 3. *Trends Biochem. Sci.* **22**, 361–363.

Monteiro, G., Horta, B. B., Pimenta, D. C., Augusto, O., and Netto, L. E. (2007). Reduction of 1-Cys peroxiredoxins by ascorbate changes the thiol-specific antioxidant paradigm, revealing another function of vitamin C. *Proc. Natl. Acad. Sci. USA* **104**, 4886–4891.

Moore, R. B., Mankad, M. V., Shriver, S. K., Mankad, V. N., and Plishker, G. A. (1991). Reconstitution of Ca(2+)-dependent K+ transport in erythrocyte membrane vesicles requires a cytoplasmic protein. *J. Biol. Chem.* **266**, 18964–18968.

Nickel, C., Trujillo, M., Rahlfs, S., Deponte, M., Radi, R., and Becker, K. (2005). *Plasmodium falciparum* 2-Cys peroxiredoxin reacts with plasmoredoxin and peroxynitrite. *Biol. Chem.* **386**, 1129–1136.

Ogusucu, R., Rettori, D., Munhoz, D. C., Netto, L. E., and Augusto, O. (2007). Reactions of yeast thioredoxin peroxidases I and II with hydrogen peroxide and peroxynitrite: Rate constants by competitive kinetics. *Free Radic. Biol. Med.* **42**, 326–334.

Peshenko, I. V., Singh, A. K., and Shichi, H. (2001). Bovine eye 1-Cys peroxiredoxin: Expression in *E. coli* and antioxidant properties. *J. Ocul. harmacol. Ther.* **17**, 93–99.

Piacenza, L., Peluffo, G., Alvarez, M. N., Nelly, J. M., Wilkinson, S. R., and Radi, R. (2008). Peroxiredoxins play a major role in protecting *Trypanosoma cruzi* against macrophage- and endogenously-derived peroxynitrite. *Biochem. J.* **410**(2), 359–368.

Quijano, C., Alvarez, B., Gatti, R. M., Augusto, O., and Radi, R. (1997). Pathways of peroxynitrite oxidation of thiol groups. *Biochem. J.* **322**(Pt 1), 167–173.

Radi, R. (1996). Kinetic analysis of reactivity of peroxynitrite with biomolecules. *Methods Enzymol.* **269**, 354–366.

Radi, R. (1998). Peroxynitrite reactions and diffusion in biology. *Chem. Res. Toxicol.* **11**, 720–721.

Radi, R., Beckman, J. S., Bush, K. M., and Freeman, B. A. (1991). Peroxynitrite oxidation of sulfhydryls: The cytotoxic potential of superoxide and nitric oxide. *J. Biol. Chem.* **266**, 4244–4250.

Radi, R., Denicola, A., Alvarez, B., Ferrer-Sueta, G., and Rubbo, H. (2000). The biological chemistry of peroxynitrite. *In* "Nitric Oxide: Biology and Pathobiology" (L. Ignarro, ed.), pp. 57–82. Academic Press.

Ralat, L. A., Manevich, Y., Fisher, A. B., and Colman, R. F. (2006). Direct evidence for the formation of a complex between 1-cysteine peroxiredoxin and glutathione S-trnasferase pi with activity changes in both enzymes. *Biochemistry* **45,** 360–372.

Romero, N., Radi, R., Linares, E., Augusto, O., Detweiler, C. D., Mason, R. P., and Denicola, A. (2003). Reaction of human hemoglobin with peroxynitrite: Isomerization to nitrate and secondary formation of protein radicals. *J. Biol. Chem.* **278,** 44049–44057.

Schonbaum, G. R., and Lo, S. (1972). Interaction of peroxidases with aromatic peracids and alkyl peroxides: Product analysis. *J. Biol. Chem.* **247,** 3353–3360.

Trostchansky, A., O'Donnell, V. B., Goodwin, D. C., Landino, L. M., Marnett, L. J., Radi, R., and Rubbo, H. (2007). Interactions between nitric oxide and peroxynitrite during prostaglandin endoperoxide H synthase-1 catalysis: A free radical mechanism of inactivation. *Free Radic. Biol. Med.* **42,** 1029–1038.

Trujillo, M., Budde, H., Pineyro, M. D., Stehr, M., Robello, C., Flohé, L., and Radi, R. (2004). T*rypanosoma brucei* and *Trypanosoma cruzi* tryparedoxin peroxidases catalytically detoxify peroxynitrite via oxidation of fast reacting thiols. *J. Biol. Chem.* **279,** 34175–34182.

Trujillo, M., Clippe, A., Manta, B., Ferrer-Sueta, G., Smeets, A., Declercq, J. P., Knoops, B., and Radi, R. (2007a). Pre-steady state kinetic characterization of human peroxiredoxin 5: Taking advantage of Trp84 fluorescence increase upon oxidation. *Arch. Biochem. Biophys.* **467,** 95–106.

Trujillo, M., Ferrer-Sueta, G., Thomson, L., Flohé, L., and Radi, R. (2007b). Kinetics of peroxiredoxins and their role in the decomposition of peroxynitrite. *Subcell Biochem.* **44,** 83–113.

Trujillo, M., Mauri, P., Benazzi, L., Comini, M., De Palma, A., Flohé, L., Radi, R., Stehr, M., Singh, M., Ursini, F., and Jaeger, T. (2006). The mycobacterial thioredoxin peroxidase can act as a one-cysteine peroxiredoxin. *J. Biol. Chem.* **281,** 20555–20566.

Trujillo, M., and Radi, R. (2002). Peroxynitrite reaction with the reduced and the oxidized forms of lipoic acid: New insights into the reaction of peroxynitrite with thiols. *Arch. Biochem. Biophys.* **397,** 91–98.

Wong, C. M., Zhou, Y., Ng, R. W., Kung Hf, H. F., and Jin, D. Y. (2002). Cooperation of yeast peroxiredoxins Tsa1p and Tsa2p in the cellular defense against oxidative and nitrosative stress. *J. Biol. Chem.* **277,** 5385–5394.

Wood, Z. A., Schroder, E., Robin Harris, J., and Poole, L. B. (2003). Structure, mechanism and regulation of peroxiredoxins. *Trends Biochem. Sci.* **28,** 32–40.

CHAPTER ELEVEN

NITROCYTOCHROME c: SYNTHESIS, PURIFICATION, AND FUNCTIONAL STUDIES

José M. Souza,* Laura Castro,* Adriana María Cassina,*
Carlos Batthyány,† and Rafael Radi*

Contents

Abstract

Posttranslational protein tyrosine oxidation, to yield 3-nitrotyrosine, is a biologically relevant protein modification related with acute and chronic inflammation and degenerative processes. It is usually associated with a decrease or loss in protein function. However, in some proteins, tyrosine nitration results in an increase or gain in protein function. Nitration of cytochrome c by biological oxidants *in vitro* can be achieved via different mechanisms, which include reactions with peroxynitrite, nitrite plus hydrogen peroxide, and nitric oxide plus hydrogen peroxide, and result in a loss in its electron transport capacity and in a higher peroxidatic activity. This chapter describes the methodology for studying chemical and biological properties of nitrocytochrome c. In particular, we report methods to synthesize tyrosine-nitrated cytochrome c, purify cytochrome c mononitrated species, map the sites of tyrosine nitration, and

* Department of Biochemistry and Center for Free Radical and Biomedical Research, Facultad de Medicina, Universidad de la República, Montevideo, Uruguay
† Institut Pasteur de Montevideo and Center for Free Radical and Biomedical Research, Facultad de Medicina, Montevideo, Uruguay

Methods in Enzymology, Volume 441   © 2008 Elsevier Inc.
ISSN 0076-6879, DOI: 10.1016/S0076-6879(08)01211-1

investigate the functional consequences of nitrated cytochrome *c* on mitochondrial electron transport properties, peroxidatic activity, and apoptosome assembly.

 ## 1. INTRODUCTION

Cytochrome *c* is a mitochondrial peripheral inner membrane protein, with a heme moiety that switches between ferro and ferri forms during electron transfer reactions among complexes III and IV of the respiratory chain (Margoliash, 1996). This 13-kDa water-soluble molecule also participates in the apoptotic pathway of cellular death: once released from mitochondria it forms a complex with other proteins, apoptotic protein activating factor-1 (Apaf-1) and procaspase 9 called apoptosome, leading to the activation of the cascade of proteases executing apoptosis in cells (Jiang and Wang, 2004).

Horse cytochrome *c* is composed of a single polypeptide chain of 104 amino acid residues and a covalently bound heme located in an internal pocket formed by highly conserved amino acid residues. The porphyrin is covalently bound to Cys-14 and Cys-17, with the fifth and sixth coordination positions of the heme–Fe interacting with His-18 and Met-80, respectively (Fig. 11.1). Thus, cytochrome *c* constitutes a hexa-coordinate heme protein with a characteristically high E'^{o} of +260 mV. In addition to these four invariant amino acids, there are others that have been conserved during evolution, for example, four tyrosine residues (Tyr-48, Tyr-67, Tyr-74, and Tyr-97; Fig. 11.1) and nine lysine residues that give the protein its characteristically net positive charge of +8 (Margoliash, 1996). Tyrosine-67 is the closest tyrosine residue to the heme group (7 Å) and is adjacent to Met-80, whereas Tyr-48 is at the opposite side of the heme with respect to Tyr-67, being both internal residues. Tyr-74 and Tyr-97 are solvent accessible and are 12.5 and 14 Å apart from the heme group, respectively. Tyr-74 occludes and separates Tyr-67 from the protein surface (Fig. 11.1).

Early work from 1969 to 1970 with isolated horse heart cytochrome *c* showed that excess amounts of the nitrating and oxidizing compound tetranitromethane led to tyrosine nitration, specifically in Tyr-67. This nitrocytochrome *c* showed spectroscopic and physicochemical differences with native cytochrome *c* and was unable to restore respiration in cytochrome *c*-depleted mitochondria (Skov *et al.*, 1969; Sokolovsky *et al.*, 1970).

Posttranslational nitration of tyrosine to yield 3-nitrotyrosine (NO_2Tyr) has been detected in various organs and cell types both clinically and in animal models of acute and chronic inflammation and degenerative diseases (Ischiropoulos, 1998; Szabo *et al.*, 2007). A number of proteins are modified (either *in vitro* or *in vivo*) by nitration of specific tyrosine residues,

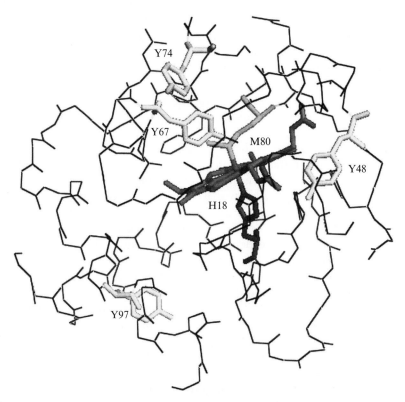

Figure 11.1 Tyrosine residues in cytochrome *c*. The three-dimensional structure of cytochrome *c* was obtained from the Protein Data Bank (1HRC) and downloaded using Pymol (http://pymol.sourceforge.net). The four tyrosine residues are represented in yellow. Additional structures shown are the heme group (red) and its fifth and sixth amino acid ligands, His-18 (blue) and Met-80 (green), respectively. The rest of the protein backbone is drawn with black lines. (See color insert.)

with mitochondrial proteins being preferential targets, including cytochrome *c* (Alonso *et al.*, 2002; Cruthirds *et al.*, 2003; Quijano, 2005). The magnitude of protein tyrosine nitration during inflammatory processes that involve accelerated rates of reactive oxygen species and ·NO production ranges from 10 to 100 μmol 3–nitrotyrosine per mol of tyrosine, quantitatively similar to tyrosine phosphorylation. Incorporation of this bulky moiety into tyrosine can induce alterations to tyrosine-mediated electron transfer reactions, as well as protein structure and function (Radi, 2004).

Substitution of a hydrogen by a nitro group ($-NO_2$) to the *ortho* position of tyrosine can occur via multiple ·NO-dependent-mechanisms (Radi, 2004). Different oxidizing species, such as hydroxyl radical, carbonate radical, oxo-metals, and ·NO_2 derived from either the peroxynitrite ($ONOO^-$) or the hemeperoxidase/nitrite/hydrogen peroxide pathways, abstract one electron from the tyrosine ring to form a tyrosyl radical that then recombines with ·NO_2 to yield NO_2Tyr (Radi, 2004). Alternatively, tyrosyl radical reacts with nitric oxide (·NO) to yield the unstable 3-nitrosotyrosine, which in an appropriate oxidizing environment can evolve via two consecutive one electron oxidation steps to NO_2Tyr (Chen *et al.*, 2004). *In vitro*, cytochrome *c* is nitrated by peroxynitrite and by peroxidase-like mechanisms in the presence of hydrogen peroxide (H_2O_2) and nitrite (NO_2^-) or ·NO (Batthyany *et al.*, 2005; Cassina *et al.*, 2000; Castro *et al.*, 2004; Chen *et al.*, 2004).

Mitochondrial nitration by the peroxidase pathway would require mitochondrial peroxidases. In this sense, cytochrome *c* is a mitochondrial protein with a weak peroxidatic activity that can promote H_2O_2-mediated oxidation of different substrates (Radi *et al.*, 1991a,b). Modifications of cytochrome *c* leading to disruption of the Fe-Met80 bond increase peroxidatic activity. Specifically, peroxynitrite-dependent tyrosine nitration, methionine oxidation by singlet oxygen or hypochlorous acid, and cardiolipin binding increase cytochrome *c* peroxidatic activity significantly (Basova *et al.*, 2007; Cassina *et al.*, 2000; Chen *et al.*, 2002; Estevam *et al.*, 2004).

Based on experimental designs for studies on the chemical and biological properties of nitrocytochrome *c*, this chapter describes protocols to evaluate the effects of cytochrome *c* tyrosine nitration in its structural and functional properties. In particular, we describe procedures for (a) synthesizing tyrosine-nitrated cytochrome *c*, (b) isolating cytochrome *c* mononitrated species, (c) mapping the sites of tyrosine nitration and the spectral characterization of nitrated cytochrome *c*, and (d) examining the functional consequences of nitrated cytochrome *c* on mitochondrial electron transport properties, peroxidatic cytochrome *c* activity and Apaf-1 assembly.

2. SYNTHESIS OF TYROSINE-NITRATED CYTOCHROME *c*

Horse heart cytochrome *c* is from commercial sources; a trichloroacetic acid-free cytochrome *c* preparation (e.g., Sigma) must be used during the described procedures to avoid important amounts of deaminated and polymeric cytochrome *c* forms (Margoliash and Lustgarten, 1962). Different reactions systems can be used to obtain tyrosine-nitrated cytochrome *c*. Early works utilized tetranitromethane (TNM) in a 40-fold excess respect to cytochrome *c* concentration: 1 m*M* cytochrome c^{3+} in 0.1 *M* Tris-HCl, 0.1 *M* KCl, pH 8.0, is reacted with 40 m*M* TNM for 30 min at room temperature (Skov *et al.*, 1969; Sokolovsky *et al.*, 1966). The TNM stock solution is in 95% ethanol and the final ethanol addition is always less than 6%.

The reaction is stopped by passing through a Sephadex-G25 column (PD-10 Amersham Bioscience) in 0.1 M Tris-HCl and 0.1 M KCl, pH 8.0.

Cytochrome c can also be nitrated by biologically relevant nitrating agents such as peroxynitrite, nitrite with hydrogen peroxide, and nitric oxide with hydrogen peroxide. Cytochrome c can be nitrated by the bolus addition of peroxynitrite: 200 μM cytochrome c^{3+} in 200 mM potassium phosphate buffer, 0.1 mM diethylenetriamine pentaacetic acid (DTPA), pH 7.0, is reacted with peroxynitrite, added in six successive bolus of 0.5 mM each, with a quick vortexing during every bolus to obtain a 3 mM final peroxynitrite concentration at 25 °C (Cassina et al., 2000). Because of the alkaline pH of the peroxynitrite stock solution, the pH must be controlled and kept below 7.4. A control sample is generated by decomposing peroxynitrite in the same potassium phosphate buffer (e.g., 2-min incubation) and then cytochrome c is added (reverse order addition) to estimate the potential effect of the by-products NO_2^- and nitrate (NO_3^-). Cytochrome c can be nitrated more efficiently using a continuous flux of peroxynitrite instead of the bolus addition: 1 mM cytochrome c in 200 mM potassium phosphate buffer, 0.1 mM DTPA, pH 7.0, is exposed to 0.13 mM min^{-1} peroxynitrite during 30 min with a motor-driven syringe (Sage Instrument, Boston, MA) under vigorous stirring conditions to eliminate mixing problems. As before, the final pH should be controlled and kept below 7.4 (Batthyany et al., 2005). Finally, the cytochrome c mixture is desalted by passing through a HiTrap column (Amersham Bioscience). In some experiments, NaHCO$_3$ (25 mM) is added to test the effect of the carbon dioxide (in equilibrium with bicarbonate). Under peroxynitrite flux exposure conditions, the yield of tyrosine nitration related with the peroxynitrite infusion is 0.7%, being about 0.34% in the mononitrated species. In terms of total protein ~55% of cytochrome c is nitrated and ~27% represents the mononitrated species (Batthyany et al., 2005).

Cytochrome c tyrosine nitration can be carried out during cytochrome c peroxidatic activity using hydrogen peroxide as the oxidizing agent and sodium nitrite (Castro et al., 2004). Cytochrome c (200 μM) is incubated with 0.5 mM sodium nitrite in 100 mM potassium phosphate, 0.1 mM DTPA, pH 7.0, and hydrogen peroxide is infused at a flux of 16.6 μM min^{-1} during 30 min with a motor-driven syringe pump to reach a 0.5 mM H$_2$O$_2$ accumulated concentration (Castro et al., 2004). Based on a similar assay, cytochrome c can be nitrated by nitric oxide (Chen et al., 2004). Cytochrome c (0.5 mM) in argon-saturated 50 mM sodium phosphate buffer, pH 7.4, 1 mM DTPA is mixed with 5 mM diethylamine nonoate (DEA/NO) plus 2.5 mM H$_2$O$_2$ or under an argon saturation solution of cytochrome c, 0.1 mM •NO (from an •NO-saturated bubble) and 0.2 mM H$_2$O$_2$ during 30 min (Chen et al., 2004). Hydrogen peroxide plus nitrite or •NO induces a lower yield of nitrated cytochrome c species, precluding these methods for structural studies of the modified protein.

3. Purification of the Mononitrated Cytochrome *c* Species

Tyrosine nitration of cytochrome *c* can be followed by mass spectrometry analysis of the intact protein. After cytochrome *c* nitration, the samples are completely desalted by gel filtration or reversed-phase Poros R2 and diluted to 12.3 µg/ml cytochrome *c* in 1% acetic acid, 50% methanol for a direct infusion into the electrospray mass spectrometry (MS) or mixing with a matrix solution [10 mg/ml sinapinic acid in 50% acetonitrile 0.2% trifluoroacetic acid (TFA)] for a matrix-assisted laser desorption ionization time-of-flight (MALDI-TOF) MS analysis. Figure 11.2A shows deconvoluted electrospray MS data from control cytochrome *c* (12,356 Da) corresponding with the predicted cytochrome *c* mass. After exposure to peroxynitrite (0.5 m*M*) (Fig. 11.2B), a new peak of +46 Da appears compatible with a mononitrated cytochrome *c* species (12,402 Da). After the addition of 2 m*M* peroxynitrite (Fig. 11.2C), a third MS peak of +90 Da with respect to control (12,447 Da) is present that is compatible with the addition of two nitro groups into the cytochrome *c*; still, a large amount of control cytochrome *c* and the mononitrated species is present (Fig. 11.2C). These samples are recognized by the affinity-purified polyclonal antibody against NO_2Tyr, in agreement with the mass spectrometry analysis (Brito *et al.*, 1999; Cassina *et al.*, 2000). These data point out that after peroxynitrite cytochrome *c* treatment, a mixture of different nitrated cytochrome *c* species is obtained together with the unmodified cytochrome *c*. Cytochrome *c* is rich in lysine residues, which determine its alkaline isoelectric point (p*I*) of 9.3 (Margoliash, 1996). Tyrosine nitration induces a large change in the tyrosine hydroxyl pK_a from ~10 to about 7.5 (Creighton, 1993). Cytochrome *c* tyrosine nitration also induced a change in the protein p*I*, which can be visualized by native polyacrylamide gel electrophoresis using a 15% polyacrylamide with a 3.5% stacking gel and 0.22 *M* Tris-glycine, pH 8.3, as running buffer. Because of the alkaline p*I* of cytochrome *c*, the sample should be loaded in the anodic side (Fig. 11.3). Upon tyrosine nitration with peroxynitrite or TNM, the appearance of less positively bands is evident (Fig. 11.3), supporting a change in the whole p*I* of the protein.

The change in the p*I* of the nitrated cytochrome *c* allows setting a purification protocol. A strong cationic exchange chromatographic method is utilized with a sulfopropyl-TSK column (7.5 × 75 mm, Tosoh Biosep). The column is equilibrated with 10 m*M* ammonium acetate buffer, pH 9.0, at a 0.7ml/min flow, and the products are separated with a linear gradient of ammonium acetate from 10 to 400 m*M* (pH 9.0) during 50 min. Figure 11.4A shows a time course of cytochrome *c* tyrosine nitration during the incubation with a flow of peroxynitrite (as described earlier). At zero

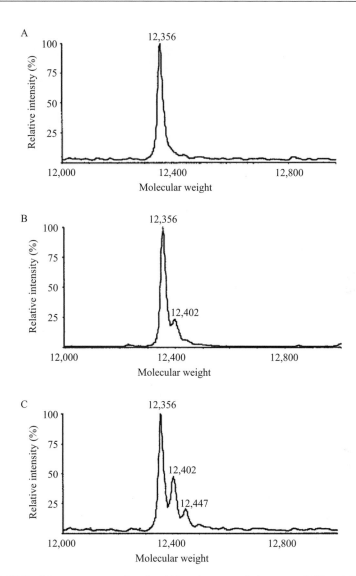

Figure 11.2 Mass spectrum of native (A) and peroxynitrite-treated (B and C) cyto-chrome *c*. Deconvolution masses of cytochrome *c* obtained by electrospray mass spec-trometry: (A) control, (B) cytochrome *c* (200 μM) reacted with 0.5 mM peroxynitrite, and (C) cytochrome *c* reacted with 2 mM peroxynitrite. The nitration reaction was per-formed in 200 mM potassium phosphate buffer and 100 μM DTPA, pH 7.0, in the presence of 25 mM of bicarbonate. Reproduced with permission from Cassina *et al.* (2000).

time cytochrome *c* appears as a major peak (A in Fig. 11.4A) corresponding with the ferricytochrome *c* (3+) form and a small amount of the ferrocyto-chrome *c* (2+) species (A' in Fig. 11.4A) (Table 11.1). Early reaction

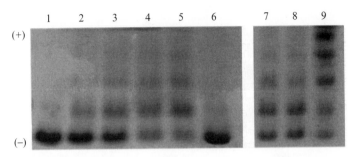

Figure 11.3 Native polyacrylamide electrophoresis. A 15% native polyacrylamide gel loaded with 10 μg of cytochrome *c* and separation preformed toward the cathode. Lane 1, control cytochrome *c*; lanes 2 to 5, cytochrome *c* (200 μM) treated with one, two, four, and six bolus additions of 3 mM peroxynitrite, respectively; lane 6, reverse order addition of peroxynitrite as lane 5; lanes 7 and 8, cytochrome *c* treated with six bolus additions of 3 mM peroxynitrite in the presence or absence of 25 mM bicarbonate, respectively; and lane 9, cytochrome *c* treated with 40 mM TNM. Reproduced with permission from Cassina *et al.* (2000).

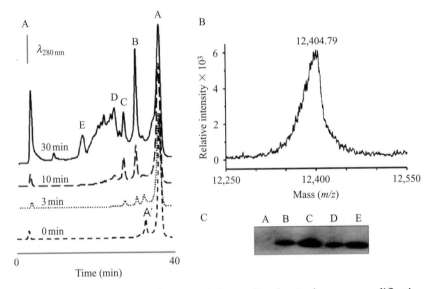

Figure 11.4 Time course of peroxynitrite-mediated cytochrome *c* modification. Cation-exchange HPLC purification and analysis of products by MALDI-TOF and immunochemistry. Cytochrome *c* (1 mM) in 200 mM potassium phosphate buffer and 100 μM DTPA, pH 7.0, was treated with an infusion of peroxynitrite (0.13 m$M \cdot$min^{-1}). (A) The reaction mixture (600 μg protein) was analyzed at different time points by cation-exchange HPLC. (B) MALDI-TOF mass spectrum of an early product (peak B in A) with a mass increased in 45 Da as compared to native protein (same result was obtained for peak C in A, data not shown). (C) Western blot analysis of the main products of peroxynitrite-treated cytochrome *c*. Five micrograms of each fraction in A was ran on a 15% SDS-PAGE and examined by Western blot analysis using an anti-NO$_2$Tyr antibody. Reproduced with permission from Batthyany *et al.* (2005).

Table 11.1 Molecular masses of peroxynitrite-treated cytochrome *c* products obtained from cationic-exchange HPLC (Fig. 11.4)

Peak	Molecular mass (Da)	Assignment	Modified Tyr residue
A	12,361	Native cytochrome c^{3+}	None
A'	12,361	Native cytochrome c^{2+}	None
B	12,406	Mononitrated	Y97
C	12,405	Mononitrated	Y74
D	12,452	Dinitrated	Y97–67 and Y74–67
E	12,495	Trinitrated	Y97–74–67

products, corresponding with peaks B and C in Fig. 11.4A, are two mono-nitrated species (Fig. 11.4B), which show the +45-Da increase in molecular mass (Table 11.1) compatible with the addition of a $-NO_2$ group. Longer incubation times of cytochrome *c* with a flux of peroxynitrite induce the appearance of dinitrated cytochrome *c* species (Fig. 11.4A peak D and Table 11.1) and trinitrated forms (Fig. 11.4A peak E and Table 11.1). All the different nitrated fractions are recognized by the affinity-purified polyclonal antibody against NO_2Tyr (Fig. 11.4C).

4. MAPPING THE SITES OF TYROSINE NITRATION AND SPECTRAL CHARACTERIZATION OF NITRATED CYTOCHROME *C*

Analysis of the site of tyrosine nitration in cytochrome *c* is performed by trypsin digestion followed by HPLC and mass spectrometry examination. Cytochrome *c* samples obtained by sulfopropyl-TSK column (Fig. 11.4A) are lyophilized and resuspended in 100 mM sodium bicarbonate buffer, pH 8.3. Trypsin (sequencing grade modified from Promega) is added at a 1:100 ratio (w:w) of trypsin and cytochrome *c* and incubated at 37 °C for 16 h. Cytochrome *c* peptides are separated and collected by reversed-phase HPLC using an octadecyl-silica column (5 μm, 2.1 × 150 mm, 300 Å, Vydac) coupled with a UV detector set at 220 and 360 nm. The column is equilibrated with 0.1% TFA, and separation is achieved with a linear gradient of TFA–acetonitrile solution (0.07% TFA in acetonitrile) from 0 to 45% in 90 min at a 0.25-ml/min flow. Nitrated peptides show absorbance at 360 nm, as do heme- and tryptophan-containing peptides. Collected peptides are concentrated under N_2 stream and measured by MALDI-TOF/MS using α-cyano-4-hydroxyl-cinnamic acid as the matrix. Peptide masses are analyzed in the MALDI-TOF reflector mode with internal mass calibration standards from cytochrome *c* tryptic peptides. Figure 11.5 shows the nitrated Y97 collected peptide from tryptic digestion cytochrome *c*, with a mass of 1009 Da (964 Da plus 45 Da) and the characteristic pattern of triplet of peaks due to the loss of −16 Da, −16 Da in the mass spectrum of the −NO_2-holding peptides (Sarver *et al.*, 2001). The time course of cytochrome *c* nitration produces the preferential nitration of tyrosine residues 97 and 74 in the mononitrated species (peaks B and C of Fig. 11.4A and Tables 11.1 and 11.2). The dinitrated form of cytochrome *c* shows a mixture of samples containing nitrated Y97 plus Y67 and Y74 plus Y67 (peak D of Fig. 11.4A and Tables 11.1 and 11.2). The trinitrated cytochrome *c* forms (peak E of Fig. 11.4A) are rich in tyrosine-nitrated peptides containing Y97, Y74, and Y67 and a small amount of Y48 (Table 11.2) (Batthyany *et al.*, 2005).

In native cytochrome *c*, the 695-nm band has a pK_a of 9.3 because of the conformational transition between states III and IV, which involves the substitution of Met-80 from the heme, presumably by Lys-79 (Wilson, 1996). Peroxynitrite induced a dose-dependent disappearance of the 695-nm spectral band of cytochrome c^{3+} (Cassina *et al.*, 2000) at pH 7.0, indicating loss of the sixth iron coordination bond with the methionine-80 (Met-80). Also, the characteristic Soret band of cytochrome c^{3+} (410 nm) shows a 2- to 3-nm blue shift (408–407 nm) upon expose to a nitrating agent (Cassina *et al.*, 2000). Disappearance of the 695-nm band can be also induced by changes in the pH (called the alkaline transition), oxidation in the Met-80 by hypochlorous acid, singlet oxygen reaction, or during

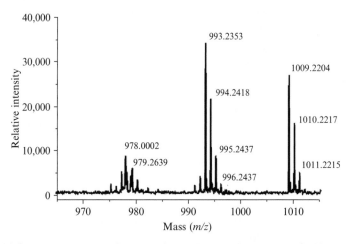

Figure 11.5 Mass spectrum of a nitrated peptide. Peptides were purified by RP-HPLC and analyzed by MALDI-TOF/MS. Herein, as an example, is shown the mass spectrum of peptide 1 (Table 11.2) obtained in positive reflector mode using α-cyano-4-hydroxicinnamic acid as the matrix. The sequence corresponds to E92DLIA-NO_2Y97-LK99. Reproduced with permission from Batthyany *et al.* (2005).

Table 11.2 Molecular masses of nitrated peptides and identification of nitration sites by postsource decay analysis

Peptide	Theoretical mass (Da), $(M+H)^+$	Theoretical mass + 45[a] (Da) $(M+H)^+$	Observed masses (Da)	Assigned sequence	Nitrated tyrosine
1	964.53	1009.53	1009.22	E92DLIAY97LK99	Y97
			993.23		
			978.00		
2	678.37	723.37	723.13	Y74IPGTK79	Y74
			707.14		
			691.15		
3	1623.79	1668.79	1668.77	E61ETLMEY67LENPKK73	Y67
			1652.77		
			1636.77		
4	1598.79	1643.79	1643.93	K39TGQAPGFTY48TDANK53	Y48
			1627.94		
			1611.94		

[a] Addition of a nitro ($-NO_2$) group results in a molecular mass increase of 45 Da.

interaction with cardiolipin (Basova *et al.*, 2007; Chen *et al.*, 2002; Prutz *et al.*, 2001; Rodrigues *et al.*, 2007; Wilson, 1996). The oxidation of Met-80 to methionine sulfoxide induced a irreversible change in cytochrome *c* indicated by the unrecovered 695-nm band at acidic pH and the appearance of a 620-nm band (Prutz *et al.*, 2001). Instead, cytochrome *c* tyrosine nitration, in either Tyr-97 or Tyr-74, showed a decrease in the 695-nm band at pH 7.4, but which can be completely recovered at acidic pH (Fig. 11.6), in agreement with a reversible process and the preservation of M80, which in the mass spectrometry analysis of the modified protein always was intact (Batthyany *et al.*, 2005). Tyrosine nitration of cytochrome *c* induces an early alkaline transition lowering the pK_a of such a conformational change.

5. FUNCTIONAL CONSEQUENCES OF NITRATED CYTOCHROME *c*

Once homogeneous and pure nitrated cytochrome *c* species are obtained, different cytochrome *c* properties related with its electron transfer and redox properties or its participation in the apoptotic pathway can be examined.

5.1. Mitochondrial electron transport and redox properties

To study the effect of nitrocytochrome *c* on mitochondrial respiration, a preparation of mitochondrial particles depleted of cytochrome *c* must be achieved. Rat heart mitochondria are prepared by differential centrifugation (Cassina and Radi, 1996), and the mitochondrial pellet is resuspended at a 25 to 35 mg/ml protein concentration in 0.3 *M* sucrose, 1 m*M* EGTA, 0.1% bovine serum albumin, pH 7.4, 4 °C. Succinate-dependent oxygen consumption should range from 3 to 5 in the respiratory control ratios. The mitochondrial pellet is resuspended in 10 m*M* KCl, 2 m*M* Tris-HCl, pH 7.4, at a protein concentration of 2 mg/ml and incubated for 10 min at 37 °C. After this time, the sample is centrifuged at 10,000*g* at 4 °C and the pellet is divided into two fractions: fraction A, which is resuspended in the same buffer (10 m*M* KCl, 2 m*M* Tris-HCl, pH 7.4), and fraction B, which is resuspended in 150 m*M* KCl, 2 m*M* Tris-HCl, pH 7.4, to obtain the extraction of endogenous cytochrome *c*. Both fractions, A and B, are centrifuged twice using the same condition as described before, and the supernatants of both centrifugations are collected to measure the concentration of cytochrome *c* released following the absorbance at 550 nm of the cytochrome c^{2+} ($\epsilon = 21$ mM^{-1} cm^{-1}) after reduction with sodium dithionite. The amount of extracted cytochrome *c* is about 0.6 nmol cytochrome *c*/mg of mitochondrial protein (Cassina *et al.*, 2000). Oxygen consumption of cytochrome

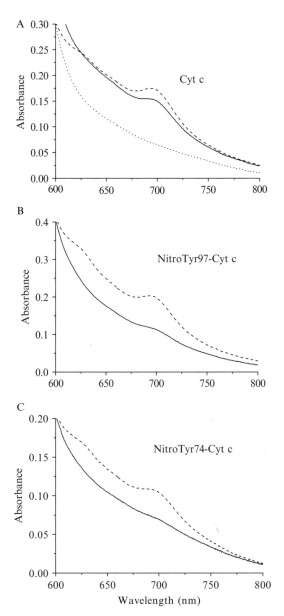

Figure 11.6 Effect of peroxynitrite-mediated tyrosine nitration on the cytochrome *c* band at 695 nm. Spectral of cytochrome *c* control and the mononitrated species (NO$_2$-Y97 and NO$_2$-Y74) reordered between 600 and 800 nm at different pH values with 200 mM potassium phosphate buffer: (A) control, (B) NO$_2$-Y97, and (C) NO$_2$-Y74 at pH 7.4 (—) and pH 5 (---) and for control also at pH 10 ($\cdot\cdot\cdot$). Reproduced with permission from Batthyany *et al.* (2005).

c-depleted mitochondria is studied in a Clark-type electrode (YSI 5300) at 37 °C with a 0.2 to 0.5 mg/ml protein concentration following the addition of 5 m*M* succinate and 0.7 nmol of control or nitrated cytochrome *c*/mg of mitochondrial protein. Supplementation of depleted-mitochondria with cytochrome *c* restores oxygen consumption rates, whereas the mononitrated cytochrome *c* in Y74, the dinitrated cytochrome *c* species, or the cytochrome *c* nitrated in Y67 (obtained by TNM treatment) only restored between 21 and 55% compared with the control (Table 11.3). Nitration in residue 97 showed a lesser effect compared with the other cytochrome *c*-nitrated species (Table 11.3).

Electron transfer properties of cytochrome *c* can be assayed by measuring the reducibility of ferricytochrome *c* by ascorbic acid. Cytochrome c^{3+} or nitrocytochrome *c* (10 μM) in 200 m*M* potassium phosphate buffer, pH 7.2, 0.1 m*M* DPTA is reduced with 50 μM ascorbic acid, and the increase of absorbance ferrocytochrome *c* at 550 nm is followed spectrophotometrically

Table 11.3 Mitochondrial electron transport capacity and peroxidatic activity of nitrated cytochrome *c* species

	Succinate-dependent respiration rates[a]		Peroxidatic activity[b]	
	mM O$_2$/min	% of control	μM/min	% of control
Native	0.071 ± 0.032	100	1.67 ± 0.11	100
NO$_2$Y74	0.039 ± 0.011	55	7.69 ± 0.11	460
NO$_2$Y97	0.064 ± 0.015	90	4.58 ± 0.08	274
NO$_2$Y67	0.015 ± 0.011	21[c]	4.77 ± 0.12	285
Di NO$_2$Y[d]	0.028 ± 0.012	39	11.72 ± 0.15	701
DiNO$_2$Y (Y74–67)[e]	ND[f]	—	6.2 ± 0.2	371
TriNO$_2$Y[g]	ND	—	12.2 ± 0.2	730
Peroxynitrite treated[h]	0.026 ± 0.013	36.6	9.35 ± 0.25	560

[a] Oxygen consumption in cytochrome *c*-depleted rat heart mitochondria (0.5 mg/ml) was measured with the addition of 5 m*M* succinate and 0.7 mmol/mg of cytochrome *c*. Respiration without the addition of cytochrome *c* was 0.004 ± 0.001 mM O$_2$/min.
[b] Native or nitrated cytochrome *c* (0.6 μM) in 100 m*M* potassium phosphate plus 0.1 m*M* DTPA, pH 7.2, was incubated with ABTS and H$_2$O$_2$ at 20 °C. ABTS oxidation was followed at 420 nm. The Tyr67 mononitrated product was formed upon treatment of cytochrome *c* with TNM as indicated in Fig. 12.6. All other species were purified from peroxynitrite infusion experiments, as described in Fig. 12.2.
[c] Data from Sokolovsky *et al.* (1970).
[d] From peroxynitrite treatment, a mix of dinitrated forms, mainly dinitro-Tyr97–67 and dinitro-Tyr74–67 (fraction D in Fig. 11.4).
[e] TNM treatment, dinitro-Tyr74–67 (i.e. fraction D in Fig. 11.4).
[f] Not determined.
[g] TNM treatment, i.e., trinitro-Tyr97–74–67 (i.e. fraction E in Fig. 11.4).
[h] Data from Cassina *et al.* (2000); a mixture of nitrated and native forms.

for 60 min (Cassina *et al.*, 2000). The reducibility of ferricytochrome *c* can also be assayed by superoxide fluxes, using xanthine oxidase and xanthine as a superoxide source (in the presence of catalase to eliminate hydrogen peroxide) (Jang and Han, 2006). Nitrocytochrome *c* is resistant to reduction by both ascorbate and superoxide (Cassina *et al.*, 2000; Jang and Han, 2006).

5.2. Peroxidatic cytochrome *c* activity

Cytochrome *c* shows a weak peroxidatic activity that can induce hydrogen peroxide-mediated oxidation of different substrates such as small organic molecules and unsaturated fatty acids present in model or mitochondrial membranes (Radi *et al.*, 1991a,b, 1993). Cytochrome *c* catalyzes 2,2-azino-bis (3-ethylbenzthiazoline-6-sulfonate) (ABTS) oxidation in the presence of H_2O_2. This assay is performed by incubating cytochrome *c* or nitrocytochrome *c* (0.6 μM) in 100 mM potassium phosphate, 0.1 mM DTPA, pH 7.2, and 1.2 mM ABTS; the reaction is initiated by the addition of 1.2 mM H_2O_2 following the formation of radical $ABTS^{\cdot +}$ at 420 nm ($\epsilon = 36$ mM^{-1} cm^{-1}). A similar assay of peroxidatic activity, which requires less cytochrome *c*, can be performed using luminol instead of ABTS as the substrate and measuring light emission over time. In this assay, 100 nM cytochrome *c* or nitrated cytochrome is incubated in 100 mM potassium phosphate, pH 7.4, 100 μM DTPA, 30 μM luminal, and 1 mM H_2O_2. Reactions are carried out in a plate reader luminometer (Lumistar galaxy, BMG Labtechnologies) analyzing total light emission during 60 min (Fig. 11.7A). This cytochrome *c* peroxidatic activity assay can also be performed measuring the oxygen consumption in liposomes (composed with 30% cardiolipin and 70% phosphatidylcholine) after the addition of hydrogen peroxide (Jang and Han, 2006). Cytochrome *c* tyrosine nitration induces an increase in the peroxidatic activity, showing a relationship between the degree of tyrosine nitration and the increase of the ABTS (Table 11.3) and luminol oxidation velocity (Fig. 11.7A). Both mononitrated cytochrome *c* species (Y97 and Y74) increase the peroxidatic activity, and even a higher increase is observed with the di and trinitrated forms (Table 11.3).

5.3. Preparation of cell-free extracts and caspase-3 assay

Release of cytochrome *c* into the cytoplasm stimulates formation of a multiprotein complex, including Apaf-1 and cytochrome *c* called apoptosome, which activates caspases and triggers apoptotic cell death. The apoptosome assembly can be induced in a cell-free extract made with a high protein concentration of cyotsolic fraction (Liu *et al.*, 1996). Cell-free extracts are generated from Jurkat T lymphocyte cells (ATCC, TIB-152). Cells (5 × 10^6/ml) are pelleted and washed twice with ice-cold phosphate-buffered saline. The cell pellet is suspended in 5 volumes of

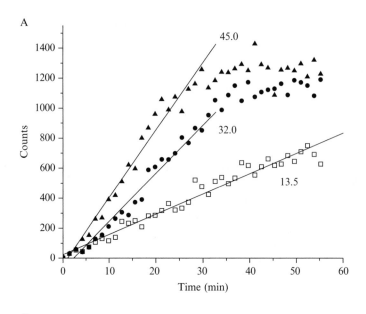

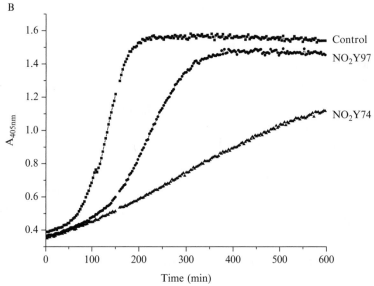

Figure 11.7 Peroxidatic and caspase-inducing activities of nitrocytochrome *c*. (A) Peroxidatic activity of native (□), mono (●)-, or dinitrated (▲) cytochrome *c* (corresponding to peaks A, B, and D, respectively from the cation-exchange HPLC purification, 25 μM each) was assayed in 100 mM potassium phosphate, pH 7.4, plus 100 μM DTPA, 100 μM H_2O_2, and 30 μM luminol. Reactions were carried out in a Lumistar galaxy (BMG Labtechnologies). Numbers are obtained slope values of linear fitting during the first 30 min. (B) Caspase-3 activity. Equal concentrations of control, NO_2-Y74cytochrome *c*, or NO_2-Y97cytochrome *c* (0.4 μM) were added to cytosolic extracts of Jurkat cells in the presence of dATP and ATP. Caspase-3 activity was measured using the chromogenic substrate AcDEVDpNA.

ice-cold buffer [20 m*M* HEPES-KOH, pH 7.5, 10 m*M* KCl, 1.5 m*M* MgCl$_2$, 1 m*M* EDTA, 1 m*M* EGTA, and 1 m*M* dithiothreitol supplemented with protease inhibitors (protease inhibitor cocktail P8340 from Sigma)]. After sitting on ice for 15 min, cells are disrupted with 10 strokes at 500 rpm with a Teflon pestle. Nuclei and undisrupted cells are removed by centrifugation at 1000*g* for 10 min at 4 °C. The supernatant is then centrifuged at 100,000 *g* for 1 h at 4 °C. The resulting supernatant is carefully removed and stored at −80 °C until use. Caspase 3 activity is evaluated following the cleavage of the chromogenic substrate AcDEVDpNA (Upstate, NY) at 405 nm, using a 200-*μ*l final volume of cell-free extract (2–4 mg/ml), 2 m*M* ATP, 2 m*M* dATP, 0.2 m*M* AcDEVDpNA, and different concentrations of control or nitrated cytochrome *c* during a 10-h incubation using a 96-well plate reader. Figure 12.7B shows caspase-3 activity measured in the Jurkat T cell-free extract initiated by the addition of control and mononitrated cytochrome *c* samples (in Tyr-97 and Tyr-74). Both mononitrated cytochrome *c* species show a decrease in initiating apoptosome assembly.

6. CONCLUDING REMARKS

The capacity to purify significant amounts of individual and well-characterized mono- and dinitrated cytochrome *c* species allows for studies on the structural and functional properties of these oxidatively modified forms produced *in vitro* and *in vivo* through the action of nitric oxide-derived oxidants. Incorporation of a nitro group to the structure of cytochrome *c* (i.e., in its tyrosine moieties) causes relevant changes in its physical, chemical and conformational properties, which result in alterations of its normal redox and proapoptotic properties, including the gain of a previously marginal peroxidatic function. Precise understanding of the redox, conformational, and functional alterations in cytochrome *c* secondary to tyrosine nitration at chemical, biochemical, and cellular levels, as well as identification of the specific cytochrome *c*-nitrated forms *in vivo*, requires future research efforts, some of which can be based in methods presented herein for nitrocytochrome *c* synthesis, purification, and characterization.

ACKNOWLEDGMENTS

This work was supported by grants from Howard Hughes Medical Institute (HHMI), International Centre of Genetic Engineering and Biotechnology (ICGEB) to RR, and Comisión Sectorial de Investigación Científica (CSIC), Uruguay to LC. RR is a Howard Hughes International Research Scholar.

REFERENCES

Alonso, D., Encinas, J. M., Uttenthal, L. O., Bosca, L., Serrano, J., Fernandez, A. P., Castro-Blanco, S., Santacana, M., Bentura, M. L., Richart, A., Fernandez-Vizarra, P., and Rodrigo, J. (2002). Coexistence of translocated cytochrome c and nitrated protein in neurons of the rat cerebral cortex after oxygen and glucose deprivation. *Neuroscience* **111,** 47–56.

Basova, L. V., Kurnikov, I. V., Wang, L., Ritov, V. B., Belikova, N. A., Vlasova, I. I., Pacheco, A. A., Winnica, D. E., Peterson, J., Bayir, H., Waldeck, D. H., and Kagan, V. E. (2007). Cardiolipin switch in mitochondria: Shutting off the reduction of cytochrome c and turning on the peroxidase activity. *Biochemistry* **46,** 3423–3434.

Batthyany, C., Souza, J. M., Duran, R., Cassina, A., Cervenansky, C., and Radi, R. (2005). Time course and site(s) of cytochrome c tyrosine nitration by peroxynitrite. *Biochemistry* **44,** 8038–8046.

Brito, C., Naviliat, M., Tiscornia, A. C., Vuillier, F., Gualco, G., Dighiero, G., Radi, R., and Cayota, A. M. (1999). Peroxynitrite inhibits T lymphocyte activation and proliferation by promoting impairment of tyrosine phosphorylation and peroxynitrite-driven apoptotic death. *J. Immunol.* **162,** 3356–3366.

Cassina, A., and Radi, R. (1996). Differential inhibitory action of nitric oxide and peroxynitrite on mitochondrial electron transport. *Arch. Biochem. Biophys.* **328,** 309–316.

Cassina, A. M., Hodara, R., Souza, J. M., Thomson, L., Castro, L., Ischiropoulos, H., Freeman, B. A., and Radi, R. (2000). Cytochrome c nitration by peroxynitrite. *J. Biol. Chem.* **275,** 21409–21415.

Castro, L., Eiserich, J. P., Sweeney, S., Radi, R., and Freeman, B. A. (2004). Cytochrome c: A catalyst and target of nitrite-hydrogen peroxide-dependent protein nitration. *Arch. Biochem. Biophys.* **421,** 99–107.

Chen, Y. R., Chen, C. L., Chen, W., Zweier, J. L., Augusto, O., Radi, R., and Mason, R. P. (2004). Formation of protein tyrosine ortho-semiquinone radical and nitrotyrosine from cytochrome c-derived tyrosyl radical. *J. Biol. Chem.* **279,** 18054–18062.

Chen, Y. R., Deterding, L. J., Sturgeon, B. E., Tomer, K. B., and Mason, R. P. (2002). Protein oxidation of cytochrome C by reactive halogen species enhances its peroxidase activity. *J. Biol. Chem.* **277,** 29781–29791.

Creighton, T. E. (1993). "Proteins: Structures and Molecular Properties." Freeman, New York.

Cruthirds, D. L., Novak, L., Akhi, K. M., Sanders, P. W., Thompson, J. A., and MacMillan-Crow, L. A. (2003). Mitochondrial targets of oxidative stress during renal ischemia/reperfusion. *Arch. Biochem. Biophys.* **412,** 27–33.

Estevam, M. L., Nascimento, O. R., Baptista, M. S., Di Mascio, P., Prado, F. M., Faljoni-Alario, A., Zucchi Mdo, R., and Nantes, I. L. (2004). Changes in the spin state and reactivity of cytochrome C induced by photochemically generated singlet oxygen and free radicals. *J. Biol. Chem.* **279,** 39214–39222.

Ischiropoulos, H. (1998). Biological tyrosine nitration: A pathophysiological function of nitric oxide and reactive oxygen species. *Arch. Biophys. Biochem.* **356,** 1–11.

Jang, B., and Han, S. (2006). Biochemical properties of cytochrome c nitrated by peroxynitrite. *Biochimie* **88,** 53–58.

Jiang, X., and Wang, X. (2004). Cytochrome C-mediated apoptosis. *Annu. Rev. Biochem.* **73,** 87–106.

Liu, X., Kim, C. N., Yang, J., Jemmerson, R., and Wang, X. (1996). Induction of apoptotic program in cell-free extracts: Requirement for dATP and cytochrome c. *Cell* **86,** 147–157.

Margoliash, E., and Lustgarten, J. (1962). Interconversion of horse heart cytochrome C monomer and polymers. *J. Biol. Chem.* **237,** 3397–3405.

Margoliash, E., and Schejter, A. (1996). How does a small protein become so popular? *In* "Cytochrome c: A Multidisciplinary Approach" (A. Mauk, ed.), pp. 3–31. University Science Books, Sausalito, CA.

Prutz, W. A., Kissner, R., Nauser, T., and Koppenol, W. H. (2001). On the oxidation of cytochrome c by hypohalous acids. *Arch. Biochem. Biophys.* **389,** 110–122.

Quijano, C., Cassina, A. M., Castro, L., Rodríguez, M., and Radi, R. (2005). Peroxynitrite: A mediator of nitric-oxide-dependent mitochondrial dysfunction in pathology. *In* "Nitric Oxide Cell Signaling and Gene Expression" (E. Cadenas and S. Lamas, eds.), pp. 99–143. Marcel Dekker, New York.

Radi, R. (2004). Nitric oxide, oxidants, and protein tyrosine nitration. *Proc. Natl. Acad. Sci. USA* **101,** 4003–4008.

Radi, R., Sims, S., Cassina, A., and Turrens, J. F. (1993). Roles of catalase and cytochrome c in hydroperoxide-dependent lipid peroxidation and chemiluminescence in rat heart and kidney mitochondria. *Free Radic. Biol. Med.* **15,** 653–659.

Radi, R., Thomson, L., Rubbo, H., and Prodanov, E. (1991a). Cytochrome c-catalyzed oxidation of organic molecules by hydrogen peroxide. *Arch. Biochem. Biophys.* **288,** 112–117.

Radi, R., Turrens, J. F., and Freeman, B. A. (1991b). Cytochrome c-catalyzed membrane lipid peroxidation by hydrogen peroxide. *Arch. Biochem. Biophys.* **288,** 118–125.

Rodrigues, T., de Franca, L. P., Kawai, C., de Faria, P. A., Mugnol, K. C., Braga, F. M., Tersariol, I. L., Smaili, S. S., and Nantes, I. L. (2007). Protective role of mitochondrial unsaturated lipids on the preservation of the apoptotic ability of cytochrome C exposed to singlet oxygen. *J. Biol. Chem.* **282,** 25577–25587.

Sarver, A., Scheffler, N. K., Shetlar, M. D., and Gibson, B. W. (2001). Analysis of peptides and proteins containing nitrotyrosine by matrix-assisted laser desorption/ionization mass spectrometry. *J. Am. Soc. Mass Spectrom.* **12,** 439–448.

Skov, K., Hofmann, T., and Williams, G. R. (1969). The nitration of cytochrome c. *Can. J. Biochem.* **47,** 750–752.

Sokolovsky, M., Aviram, I., and Schejter, A. (1970). Nitrocytochrome c. I. Structure and enzymic properties. *Biochemistry* **9,** 5113–5118.

Sokolovsky, M., Riordan, J. F., and Vallee, B. L. (1966). Tetranitromethane: A reagent for the nitration of tyrosyl residues in proteins. *Biochemistry* **5,** 3582–3589.

Szabo, C., Ischiropoulos, H., and Radi, R. (2007). Peroxynitrite: Biochemistry, pathophysiology and development of therapeutics. *Nat. Rev. Drug Discov.* **6,** 662–680.

Wilson, M. T., and Greenwood, C. (1996). The alkaline transition in ferricytochrome c. *In* "Cytochrome c: A Multidisciplinary Approach" (A. Mauk, ed.), pp. 611–634. University Science Books, Sausalito, CA.

CHAPTER TWELVE

TYROSINE NITRATION, DIMERIZATION, AND HYDROXYLATION BY PEROXYNITRITE IN MEMBRANES AS STUDIED BY THE HYDROPHOBIC PROBE N-T-BOC-L-TYROSINE TERT-BUTYL ESTER

Silvina Bartesaghi,*,[†] Gonzalo Peluffo,* Hao Zhang,[‡] Joy Joseph,[‡] Balaraman Kalyanaraman,[‡] and Rafael Radi*

Contents

* Department of Biochemistry and Center for Free Radical and Biomedical Research, Facultad de Medicina, Universidad de la República, Montevideo, Uruguay
[†] Department of Histology and Embryology, Facultad de Medicina, Universidad de la República, Montevideo, Uruguay
[‡] Department of Biophysics and Free Radical Research Center, Medical College of Wisconsin, Milwaukee, Wisconsin

Methods in Enzymology, Volume 441
ISSN 0076-6879, DOI: 10.1016/S0076-6879(08)01212-3

Abstract

Protein tyrosine oxidation mechanisms in hydrophobic biocompartments (i.e., biomembranes, lipoproteins) leading to nitrated, dimerized, and hydroxylated products are just starting to be appreciated. This chapter reports on the use of the hydrophobic tyrosine analog *N-t*-BOC-L-tyrosine *tert*-butyl ester (BTBE) incorporated to phosphatidyl choline liposomes to study peroxynitrite-dependent tyrosine oxidation processes in model biomembranes. The probe proved to be valuable in defining the role of biologically relevant variables in the oxidation process, including the action of hydrophilic and hydrophobic peroxynitrite and peroxynitrite-derived free radical scavengers, transition metal catalysts, carbon dioxide, molecular oxygen, pH, and fatty acid unsaturation degree. Moreover, detection of the BTBE phenoxyl radical and relative product distribution yields of 3-nitro-, 3,3′-di-, and 3-hydroxy-BTBE in the membrane fully accommodate with a free radical mechanism of tyrosine oxidation, with physical chemical and biochemical determinants that in several respects differ of those participating in aqueous environments. The methods presented herein can be extended to explore the reaction mechanisms of tyrosine oxidation by other biologically relevant oxidants and in other hydrophobic biocompartments.

1. Introduction

Protein tyrosine nitration has been associated with several pathologies (Beckman *et al.*, 1994), such as inflammation, cardiovascular disease, neurodegeneration, and diabetic complications, and has been revealed as a biomarker of oxidative stress *in vivo* and a predictor of disease progression and severity (Shishehbor *et al.*, 2003; Zhang *et al.*, 2001b; Zheng *et al.*, 2005).

The nitration of protein tyrosine residues constitutes the substitution of hydrogen by a nitro group ($-NO_2$;$+45$ Da) in the three position of the phenolic ring and represents a posttranslational modification produced by nitric oxide ($\cdot NO$)-derived oxidants, such as peroxynitrite (ONOOH, ONOO$^-$),[1] and nitrogen dioxide radical ($\cdot NO_2$).

---

[1] IUPAC-recommended names for peroxynitrite anion (ONOO$^-$) and peroxynitrous acid (ONOOH, pK_a= 6.8) are oxoperoxynitrate (1-) and hydrogen oxoperoxynitrate, respectively. The term peroxynitrite is used to refer to the sum of ONOO$^-$ and ONOOH.

Early work showed that nitration could result in dramatic changes in protein structure and function (Sokolovsky et al., 1966), resulting in either a gain or a loss of function (Radi, 2004); however, it was not until the early nineties (Beckman et al., 1990; Ischiropoulos et al., 1992) when the biological significance of protein tyrosine nitration was really appreciated, after recognizing the formation of strong oxidizing and nitrating intermediates during the biological oxidation of $\cdot$NO (Beckman et al., 1990; Koppenol et al., 1992; Radi et al., 1991a,b). Since then, protein tyrosine nitration has been well established to occur both in vivo and in vitro (Ischiropoulos, 1998; Radi, 2004).

Tyrosine nitration takes place biologically by a variety of routes, all of them based in free radical chemistry. The oxidation of tyrosine[2] by different oxidants (such as hydroxyl radical ($\cdot$OH) carbonate radical (CO_3^{-}), $\cdot$NO$_2$, and oxo-metal complexes) yields tyrosyl radical ($\cdot$Tyr), which reacts with nitrogen dioxide ($\cdot$NO$_2$) or another tyrosyl radical to yield 3-nitrotyrosine (3-NO$_2$-Tyr) and di-tyrosine (3,3'-di-Tyr), respectively. Alternatively, 3-nitrotyrosine can be formed secondary to the reaction of $\cdot$NO with tyrosyl radical to yield the transient 3-nitrosotyrosine, which must be further oxidized by two consecutive one-electron steps with the intermediate formation of an iminoxyl radical (Radi, 2004; Sturgeon et al., 2001); in conjunction with these radicals processes, the addition of $\cdot$OH to the tyrosyl radical leads to the formation of a hydroxylated derivative, 3,4-dihydroxyphenylalanine (DOPA) (Bartesaghi et al., 2007; Radi, 2004; Santos et al., 2000).

It has became evident that nitration of protein tyrosine can occur in both hydrophilic or hydrophobic biocompartments (e.g., in proteins associated to biomembranes and lipoproteins; reviewed in Bartesaghi et al., 2007) and that various factors controlling the nitration pathways may differ in both compartments. For instance, in the hydrophobic phase, several factors can promote tyrosine nitration, whereas others may inhibit this process. Enhanced nitration can be due to different factors, such as a higher concentration of $\cdot$NO$_2$/$\cdot$NO in this phase, membrane-associated transition metal centers (e.g., hemin), or the exclusion of antioxidants such as glutathione, which have a relevant role in aqueous phases. However, the high concentration of unsaturated fatty acids in membranes, which may outcompete for the radical species and therefore inhibit tyrosine nitration, as well as the lack of permeability of some of the oxidants to the hydrophobic compartment [e.g., CO_3^{-} (Bartesaghi et al., 2006), compound I myeloperoxidase (Zhang et al., 2003)] may lead to lower nitration yields relative to the aqueous phase.

Peroxynitrite does not react directly with tyrosine (Alvarez et al., 1999), but peroxynitrite-derived radicals (e.g., $\cdot$OH, CO_3^{-}, NO$_2$) do. Because the peroxynitrite-derived radicals are short-lived, the nitration of tyrosine in membranes by ONOOH can occur mainly by two mechanisms: peroxynitrite

---

[2] The term oxidation refers to the sum of nitration, dimerization, and hydroxylation processes.

can decompose in the aqueous phase and its derived radicals diffuse into the hydrophobic phase or peroxynitrite can enter the hydrophobic compartment and the homolysis takes place in the interior of the membrane or lipoprotein. Radical species have a specific diffusion behavior towards the membrane and a different probability of formation in its interior. For instance, $\cdot NO_2$ can both readily permeate toward the membrane and be formed in its interior; however, $\cdot OH$ has a very short life and is diffusion is minimal (three to four molecular diameters), in which case its action would depend exclusively on ONOOH homolysis inside the hydrophobic compartment. We have previously described the formation of 3-hydroxy-BTBE (BTBE-OH) in model membranes that must be because of a site-specific homolysis in the bilayer (Bartesaghi *et al.*, 2006). In turn, $CO_3^{\cdot-}$ should not be either formed (by the reaction of $ONOO^-$ with CO_2) in the membrane or permeate to it, as anions will be excluded from the hydrophobic phase.

The formation of 3-nitrotyrosine and other oxidation products in the membrane will depend on the distribution of radical intermediates, including the concentration of $\cdot NO_2$ and the competing pathway involving the recombination of two tyrosyl radicals to yield 3,3'-di-tyrosine; in hydrophobic environments, there are diffusional restrictions for the lateral movement of tyrosine residues that would not be present in aqueous phases. In this context, work has been directed to define mechanisms and product yields of tyrosine oxidation in hydrophobic compartments. Two main types of probes have been developed to study tyrosine nitration in membranes: hydrophobic tyrosine analogs such as N-*t*-BOC-L-tyrosine *tert*-butyl ester (BTBE) (Zhang *et al.*, 2001a) and tyrosine-containing transmembrane peptides (Zhang *et al.*, 2003). A description of BTBE and the methods for the detection of its nitration and dimerization products have been reported elsewhere in this series (Zhang *et al.*, 2005).

This chapter focuses on the utilization of BTBE incorporated to phosphatidyl choline (PC) liposomes to study the mechanisms of peroxynitrite-dependent tyrosine oxidation in model biomembranes. Variables that affect tyrosine oxidation yields can be tested, including the effect of transition metal catalysts, carbon dioxide, molecular oxygen, pH, hydrophilic and hydrophobic antioxidants, and fatty acid composition of the membrane and peroxynitrite-derived free radical scavengers (Bartesaghi *et al.*, 2006).

2. Methods

2.1. Synthesis of BTBE and its oxidation products

The synthesis of BTBE and the standards 3-nitro-BTBE and 3,3'-di-BTBE has been described previously (Zhang *et al.*, 2005). 3-Hydroxy-BTBE is generated by a hydroxylation reaction mediated by a Fenton system (Bartesaghi *et al.*,

2006). The Fenton reagent is prepared by diluting 10 mM ferrous ammonium sulfate in 2.5 mM sulfuric acid. The reaction is started by adding 0.3 mM H_2O_2 and 0.3 mM Fe^{II} to BTBE (0.3 mM)-containing liposomes.

2.2. Preparation of BTBE-containing liposomes and peroxynitrite addition

BTBE incorporation to liposomes is carried out as described previously (Bartesaghi et al., 2006; Zhang et al., 2001a). Briefly, a methanolic solution of BTBE (0.35 mM) is added to 35 mM PC lipids dissolved in chloroform. The mixture is then dried under a stream of nitrogen gas. Multilamelar liposomes are formed by thoroughly mixing the dried lipid with 100 mM potassium phosphate buffer, pH 7.4, plus 0.1 mM DTPA. Under these conditions, the BTBE incorporation degree is >98%.

Phosphatidyl choline liposomes of variable fatty acid composition can be prepared containing saturated 1,2-dimyristoyl-sn-glycero-3-phosphocholine (DMPC), 1,2-dilauroyl-sn-glycero-3-phosphocholine (DLPC), or unsaturated (egg PC, soybean PC) fatty acids. In addition, the unsaturation degree can be modulated by working with mixtures of DLPC and 1-palmitoyl-2-linoleoyl-sn-glycero-3-phosphocholine. It is important to note that experiments with liposomes must be performed in each case above the transition phase temperature (i.e., 23, −1, −3, and <0 °C for DMPC, DLPC, egg PC, and soybean PC, respectively).

Peroxynitrite synthesis is performed in a quenched-flow reactor from sodium nitrite (NaNO$_2$) and H_2O_2 under acidic conditions as described previously (Radi et al., 1991b). The H_2O_2 remaining from the synthesis is eliminated by treating the stock solution with manganese dioxide. The peroxynitrite concentration is determined spectrophotometrically at 302 nm ($\varepsilon = 1670$ M^{-1} cm^{-1}). The nitrite concentration in preparations is typically lower than 30%. Nitrite levels must be strictly controlled and are critical for obtaining reproducible data.

Peroxynitrite is added either as a single bolus or as multiple successive boluses under vigorous vortexing or by infusion with a motor-driven syringe under continuous stirring (Trostchansky et al., 2001). In some control experiments, peroxynitrite is allowed to decompose to nitrate and nitrite in 100 mM phosphate buffer, pH 7.4, for 2 min before use, i.e., "reverse order addition" of peroxynitrite (RA).

2.3. Sample preparation

A 200-μl reaction mixture containing liposomes is mixed with 200 μl methanol and vortexed for 1 min to dissolve liposomes. The resulting solution is mixed with 400 μl chloroform and 80 μl 5 M NaCl and is vortexed for 2 min.

The final solution is centrifuged at 5000 rpm for 10 min, and the aqueous phase (supernatant) is removed. The chloroform layer is dried under a stream of nitrogen and kept at −20 °C until use (Bartesaghi *et al.*, 2006; Zhang *et al.*, 2001a). Recovery efficiencies for all compounds are >95% (Zhang *et al.*, 2001a). Immediately before HPLC analysis, samples are resuspended in 100 μl 85% methanol:15% 15 mM KPi, pH 3, sonicated for 10 min, and centrifuged before injection.

Because chloroform is used during sample preparation, and because of the affinity of lipids and BTBE for the hydrophobic components of plastic tubes, experiments should be performed preferentially in glass tubes.

2.4. HPLC analysis

BTBE, 3-nitro-BTBE, and 3,3′-di-BTBE are separated on a Gilson HPLC system equipped with UV–VIS and fluorescence detectors by reversed-phase HPLC using a Partisil ODS-3 10-μm column (250 mm length, 4.6 mm i.d.).

Mobile phase A consists of 15 mM phosphate potassium buffer, pH 3, and mobile phase B consists of methanol. Chromatographic conditions are as follows: flow, 1 ml/min; 75% mobile phase B for 25 min, followed by a linear increase to 100% mobile phase B for 10 min, which is essential for column reconstitution and elution of higher oxidation states of BTBE polymerization products and phospholipids. UV–VIS settings are 280 nm, $\varepsilon = 1200\ M^{-1}$ cm^{-1} for BTBE and 360 nm, $\varepsilon = 1500\ M^{-1}$ cm^{-1} for 3-nitro-BTBE. 3,3′-di-BTBE is detected fluorimetrically at λ_{ex}=294 nm and λ_{em}= 401 nm. Authentic 3-nitro-BTBE and 3,3′-di-BTBE are used as standards.

For the detection of 3-hydroxy-BTBE (i.e., 3,4-dihydroxy-*N-t*-BOC-L-phenylalanine *tert* butyl ester), the HLPC protocol is modified slightly to seek for more hydrophilic compounds derived from peroxynitrite-treated BTBE-containing liposomes. Mobile phase A consists of water, and the gradient is started at 50% methanol to 100% for 35 min.

2.5. Spectrophotometric analysis

In saturated fatty acid-containing PC liposomes (i.e., DLPC and DMPC), 3-nitro-BTBE can be quantitated by direct UV–VIS measurement. Briefly, liposomes are solubilized with 1.2% deoxycholate (Bartesaghi *et al.*, 2006; Buege and Aust, 1978), followed by alkalinization to pH 10 with 5 M NaOH, and 3-nitro-BTBE is measured at 424 nm at pH 10, corresponding to the phenolate anion absorbance ($\varepsilon = 4000\ M^{-1}$ cm^{-1}).

Direct spectrophotometric measurement of 3-nitro-BTBE after deoxycholate solubilization turn out to be practical and reproducible for saturated fatty acid-containing liposomes and less time-consuming than the HPLC experiments. However, this method should not be applied to unsaturated fatty acid-containing liposomes, as peroxynitrite leads to the formation of other absorbing species in the same region of the spectrum, such as nitrated and peroxidized lipids (Radi *et al.*, 1991b; Schopfer *et al.*, 2005).

2.6. Mass spectrometry analysis of 3-hydroxy-BTBE

3-Hydroxy-BTBE is analyzed using an Applied Biosystems QTRAP, triple quadrupole-linear ion trap (LIT) mass spectrometer equipped with a turbo ion spray ionization source (ESI). The mass spectrometer is operated in a positive mode, and the ESI settings are optimized as follows: ion spray voltage, 2500 V; temperature, 375 °C; declustering potential, 50 V; entrance potential, 10 V; nebulizer gas, 40 psi; heater gas, 25 psi. Samples collected from the HPLC are diluted in acidified methanol (0.1% formic acid) and infused continuously (10 μM/min) at an estimated concentration of 10 nM. The molecular ion is identified at m/z 352.2. Fragmentation analysis of hydroxy-BTBE is conducted using the LIT in the enhanced product ion mode of the instrument. Fragmentation experiments of the molecular ion at 352.2 are conducted at different collision-assisted dissociation energies identifying fragments from the parent ion.

2.7. Electron spin resonance (ESR) measurements

Electron spin resonance spin-trapping experiments are performed to detect the one-electron oxidation product of BTBE (i.e., BTBE phenoxyl radical). DLPC liposomes are incubated with 20 mM 2-methyl-nitroso propane (MNP) and peroxynitrite for 1 min. The liposomes are diluted with water, spinned down to remove MNP, and resuspended in phosphate buffer. Samples are subsequently transferred to a 100-μl capillary tube for ESR measurements. ESR spectra are recorded at room temperature on a Bruker EMX spectrometer operating at 9.8 GHz. Typical spectrometer parameters are as follow: scan range, 100 G; field set, 3510 G; time constant, 0.64 ms; scan time, 20 s; modulation amplitude, 5.0 G; modulation frequency, 100 kHz; receiver gain, 2 × 10^5; microwave power, 20 mW. The obtained signals are the result of 200 scans.

3. RESULTS

3.1. Electron spin resonance spin trapping of BTBE phenoxyl radical

Electron spin resonance spin trapping was used to detect the BTBE phenoxyl radical formed during peroxynitrite-mediated BTBE oxidation with the spin trap MNP as described previously (Bartesaghi *et al.*, 2006). BTBE-containing liposomes treated with ONOO$^-$ result in an anisotropic three line signal, suggesting the formation of a partially immobilized phenoxyl radical in the

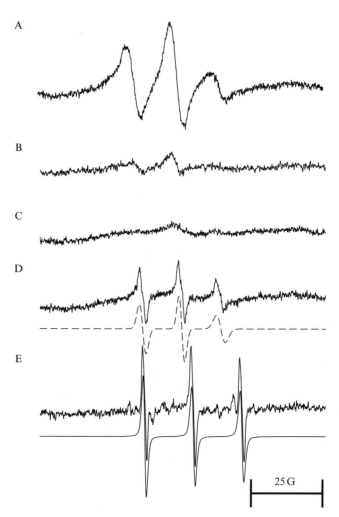

Figure 12.1 ESR spin-trapping measurements. Reaction mixtures consisting of BTBE (2.25 mM) incorporated into 45 mM DLPC liposomes in a phosphate buffer (100 mM, pH 7.4) containing DTPA (0.1 mM) were treated with a 20 mM 2-methyl-2-nitrosopropane (MNP) spin trap and mixed rapidly with 5 mM peroxynitrite. Samples were subsequently transferred to a 100-μl capillary tube for ESR measurements. (A) BTBE-containing liposomes plus peroxynitrite. (B) BTBE-containing liposomes plus decomposed peroxynitrite (RA). (C) Liposomes plus peroxynitrite plus MNP. (D) Sample A was centrifuged and BTBE-containing liposomes were redissolved in ethanol $a_N = 13.8$ G. (E) Product of MNP photolysis, $a_N = 17.1$ G.

interior of the membrane (Fig. 12.1A). No ESR spectrum was obtained when liposomes were treated with decomposed peroxynitrite (Fig. 12.1B) or in the absence of BTBE (Fig. 12.1C). When peroxynitrite-treated BTBE-containing liposomes were dissolved in ethanol, a clear three line signal was obtained (Fig. 12.1D). Despite the low signal-to-noise ratio, we can estimate a hyperfine constant of 13.8 G, which is consistent with the one-electron oxidation of BTBE by peroxynitrite-derived radicals (Mossoba *et al.*, 1982).[3]

MNP photolysis may occur during the experiments and leads to the formation of a three line signal corresponding to the di-*tert*-butyl nitroxide radical, which has a different a_N value, as shown in Fig. 12.1E.

3.2. HPLC and spectroscopic analysis of BTBE, 3-nitro-BTBE, and 3,3'-di-BTBE

BTBE oxidation products were quantitated by UV–VIS and/or fluorimetric measurements after either (a) reverse-phase–HPLC separation of organic extraction material or (b) deoxycholate solubilization. Figure 12.2A shows a typical HPLC chromatogram obtained from peroxynitrite-treated samples, and the elution of BTBE and its oxidation products 3-nitro-BTBE and 3,3'-di-BTBE is shown at 7, 9, and 19 min, respectively.

Alternatively, spectral analysis of 3-nitro-BTBE after liposome solubilization with 1.2% deoxycholate allowed carrying out direct measurements at the peak absorbance of 424 nm at pH 10 (Fig. 12.2B). In DLPC liposomes at pH 7.4, peroxynitrite (0–2 mM) caused a dose-dependent increase in BTBE nitration yields, with similar results obtained for both methods (Fig. 12.1C). While the 3% yield obtained for 3-nitro-BTBE with respect to added peroxynitrite compares well for that obtained for free tyrosine (6–10%), values for the corresponding dimer 3,3'-di-BTBE are considerably lower (0.11 and 0.02% for egg PC and DLPC, respectively) because of the lateral restriction of BTBE within the bilayer (Bartesaghi *et al.*, 2006). The direct spectroscopic measurement of 3-nitro-BTBE should be applied for saturated fatty acid-containing liposomes only, as mentioned in the description of the method.

3.3. Peroxynitrite-mediated BTBE hydroxylation in DLPC liposomes

In order to explore if the ONOOH-derived ·OH in lipophilic environments could form the hydroxylated derivative of BTBE, chromatographic conditions were changed, as this product is less hydrophobic than the parent compound and therefore elutes at shorter times in our HPLC system. As a positive control for reactions of the ·OH radical, BTBE-containing liposomes were incubated with a Fenton system. Indeed a peak eluting at 11 min was identified both in the Fenton and in the peroxynitrite addition experiments, collected, and characterized by mass spectrometry.

---

[3] MNP adducts with free or peptide-bound tyrosyl radicals were reported to have a_N values of 16.5 (Mossoba *et al.*, 1982) and 15.6 G (Zhang *et al.*, 2003), respectively.

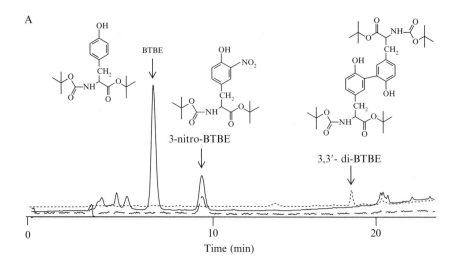

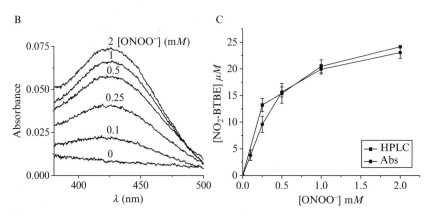

Figure 12.2 Analysis of 3-nitro-BTBE and 3,3'-di-BTBE after peroxynitrite addition. BTBE (0.3 mM) in DLPC liposomes (30 mM) was exposed to peroxynitrite in phosphate buffer (100 mM), pH 7.4, plus 0.1 mM DTPA. (A) After an organic extraction, products were separated by RP-HPLC as described in the text. The HPLC chromatogram shows the elution of BTBE, 3-nitro-BTBE, and 3,3'-di-BTBE after treatment with peroxynitrite (1 mM); structures have been drawn above the peaks. UV-VIS detection was done for BTBE and 3-nitro-BTBE at 280 nm (solid line) and 360 nm (dashed line). 3,3'-Di-BTBE was measured fluorimetrically at 294- and 401-nm excitation and emission wavelengths, respectively (dotted line). (B) Peroxynitrite-treated, BTBE-containing liposomes were solubilized with 1.2% deoxycholate, the pH was adjusted to 10 with NaOH, and UV-VIS spectra of 3-nitro-BTBE were recorded at different peroxynitrite concentrations. (C) Quantitation of 3-nitro-BTBE as a function of peroxynitrite concentration after HPLC separation (■) or deoxycholate solubilization (●). Reproduced from Bartesaghi *et al.* (2006).

    3-Hydroxy-BTBE yields were extremely low, which is consistent with the fact that the competition reaction of $\cdot$OH with the saturated fatty acids [60 mM lauric acid, $k = 6.4 \times 10^8\ M^{-1}\ s^{-1}$ (Barber, 1978)] will predominate over the $\cdot$OH oxidation reaction with 0.3 mM BTBE.

3.4. Mass spectrometry characterization of 3-hydroxy-BTBE

An aliquot ($100\ \mu l$) of the peak eluting at 11 min was diluted in 80% methanol 0.1 % formic acid and infused continuously to the mass spectrometer. Figure 12.3 shows the enhanced resolution mass spectrum of the collected peak after the Fenton reagent and peroxynitrite addition to BTBE-containing liposomes where an ion of m/z 353.2 was identified in both experimental settings. A turbo electrospray ionization source was used,

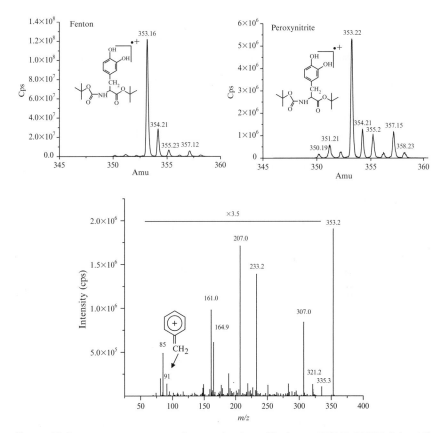

Figure 12.3 Mass spectrometry characterization of hydroxy-BTBE. BTBE (0.3 mM) in DLPC liposomes (30 mM) was exposed to FeSO$_4$ (0.3 mM) + H$_2$O$_2$ (0.3 mM) or peroxynitrite in phosphate buffer (20 mM), pH 6, supplemented with 0.4 mM DTPA and injected to HPLC. A peak eluting with a retention time of 11 min was collected and resuspended in 80% methanol plus 0.1% formic acid prior to injection into the mass spectrometer. (Top) Enhanced resolution mass scan of the peak eluting at 11 min shows an ion with an m/z of 353.2, which corresponds to the cation radical of hydroxy-BTBE for the Fenton (left) and peroxynitrite (right) reactions. (Bottom) MS/MS fragmentation pattern of the m/z 353.2 ion showing the typical fragment ion of m/z 91 (arrow) arising from phenolic cation radicals. Reproduced with modifications from Bartesaghi *et al.* (2006).

which produces mild ionization of molecules after vaporization of the liquid phase, minimizing in–source fragmentation. As opposed to atmospheric pressure chemical ionization (APCI) sources, ESI ionization typically produces protonation (positive mode under acidic conditions) of Bronsted–Lowry acids such as an amino group (e.g., $-NH_3^+$) giving rise to $(M + H^+)$ ions. The calculated monoisotopic molecular mass of BTBE is 337.2, whereas that of hydroxy-BTBE is 353.2. Mass spectrometry analysis of both compounds gives rise in our experimental settings to the molecular ion with a m/z of 337 and 353, respectively. This is indicative of formation of the cation radical ion of these molecules, most probably due to in–source electrochemistry as described previously for ESI ionization. To further confirm this observation, we conducted mass analysis of tyrosine and N-acetyl tyrosine, as both are phenolic compounds as BTBE; in the former molecule, protonation of the amino group is blocked by the acetyl moiety. We detected the well-characterized $[M + H^+]$ ion of tyrosine (m/z 182.2) but in agreement to what was observed for BTBE, the ion corresponding to N-acetyl tyrosine was M^+ with a m/z of 223.1. Furthermore, analysis of the peak corresponding to 3–hydroxy-BTBE carried out without acidification of the sample and in aprotic solvent (100% acetonitrile) to avoid the gas-phase proton transfer reaction still gave rise to the m/z 353.2 species, strongly suggesting formation of the cation radical ion (not shown). The fragmentation pattern observed after Fenton and peroxynitrite addition experiments were identical (Fig. 12.3).

3.5. Enhancers and inhibitors of peroxynitrite-mediated tyrosine oxidation in membranes

In order to study the mechanism of peroxynitrite-mediated tyrosine oxidation in hydrophobic environments, we studied the effect of different molecules known to react with peroxynitrite or its derived radicals and that can up- or downmodulate product formation. Tested scavengers may include polar compounds that will mainly react in the aqueous phase or hydrophobic, which may undergo reactions either in the lipophilic phase or in the aqueous/lipid interphase.

For instance, BTBE nitration and dimerization were inhibited by glutathione, lipoic acid (Trujillo et al., 2005), pHPA, tyrosine, dimethyl sulfoxide, mannitol, desferrioxamine (Bartesaghi et al., 2004), and uric acid in extents that are compatible with their different reactivities with peroxynitrite (e.g., GSH and lipoic acid) and peroxynitrite-derived radicals[4] (Fig. 12.4). It is important to note that the presence of nitrite (NO_2^-) either remaining from

---

[4] Specific rate constants of scavengers and metal centers with peroxynitrite, $\cdot NO_2$, $\cdot OH$, and $CO_3^{\cdot-}$, as well as reactant concentrations, critically influence tyrosine-oxidation yields. Exhaustive kinetic analyses of these processes have been performed elsewhere (Bartesaghi et al., 2004, 2006; Trujillo et al., 2005).

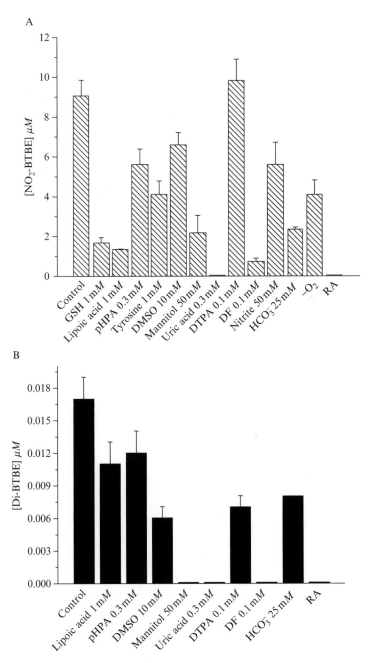

Figure 12.4 Effect of scavengers, carbon dioxide, and oxygen on BTBE nitration and dimerization. BTBE (0.3 mM) in DLPC liposomes (30 mM) was exposed to peroxynitrite (0.5 mM) in the presence of different compounds and concentrations, and 3-nitro-BTBE (A) and 3,3'-di-BTBE (B) were analyzed by RP-HPLC. The condition under low oxygen tensions (-O$_2$) was obtained by saturation of the samples under an argon atmosphere previous to peroxynitrite addition.

peroxynitrite synthesis or added may affect product distribution and yields due to its fast reaction with $\cdot OH$ to yield $\cdot NO_2$ ($k = 6 \times 10^9\ M^{-1}\ s^{-1}$).

It is well known that some transition metal complexes enhance peroxynitrite-mediated tyrosine nitration of phenolic compounds in aqueous media via a catalytic redox cycle mechanism (Beckman *et al.*, 1992; Radi, 2004). The effect of different transition metal complexes on BTBE nitration and dimerization in either saturated (DLPC) or unsaturated (egg PC) BTBE-containing liposomes can be studied. In DLPC liposomes, nitration yields were enhanced in the presence of hemin, Fe-EDTA, and the metal porphyrins Mn (III) *meso*-tetrakis (4-carboxylatophenyl) porphyrin (Mn-tccp) and Fe (III) *meso*-tetrakis (4-carboxylatophenyl) porphyrin (Fe-tccp), while ferrioxamine had no effect. In egg PC liposomes, hemin clearly enhanced BTBE nitration fivefold, whereas Fe-EDTA, Mn-tccp and Fe-tccp did not (Fig. 12.5). It is clear that transition metal complexes act as nitration catalysts in simple saturated fatty acid–containing systems. Indeed, in DLPC liposomes, BTBE nitration is enhanced by hemin and Mn-tccp in a dose-dependent manner (Bartesaghi *et al.*, 2006). The effect of transition metal catalysts may be different in more complex systems, such as unsaturated fatty acid–containing liposomes where lipid peroxidation may play a

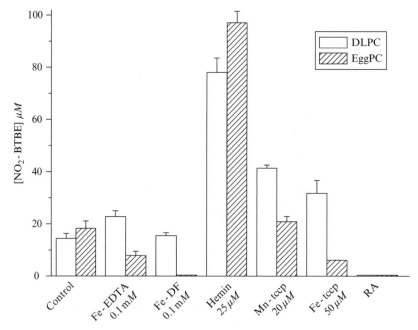

Figure 12.5   Effect of transition metal complexes on BTBE nitration. BTBE (0.3 m*M*) in DLPC and egg PC liposomes (30 m*M*) were incubated with the indicated concentration of different transition metal complexes and treated with peroxynitrite (0.5 m*M*). NO$_2$-BTBE yields were analyzed by RP-HPLC. RA, reverse addition of peroxynitrite.

relevant role in the nitration process (Bartesaghi *et al.*, 2007). Importantly, oxygen depletion known to inhibit lipoperoxidation chain reactions was also strongly inhibitory of BTBE nitration (Fig. 12.4A).

3.6. Effect of pH on nitration, dimerization, and hydroxylation yields

The effect of pH on tyrosine oxidation has been established previously with nitration, dimerization, and hydroxylation yields as a function of pH having distinctive profiles (Beckman *et al.*, 1992; Santos *et al.*, 2000; van der Vliet *et al.*, 1995). The pH profiles are dictated by different reactions, some of which largely depend on the ionization state of the phenolic -OH group of tyrosine. Indeed, the deprotonated form of the tyrosine ring (i.e., phenolate) is the molecular species that reacts readily with $\cdot NO_2$. The effect of pH on BTBE oxidation aids in determining to what extent its incorporation to hydrophobic environments affects the dependency observed for tyrosine and therefore the reaction mechanism. Changes in pH will alter proton concentration in the aqueous phase and may affect the chemistry of BTBE in the bilayer directly or indirectly. 3–Nitro-BTBE formation as a function of pH resulted in a bell-shaped curve with a maximum yield at pH 7.5 (Fig. 12.6), similar to what is observed for nitro-tyrosine. 3,3′-Di-BTBE formation was very low at pH <8, but increased significantly towards

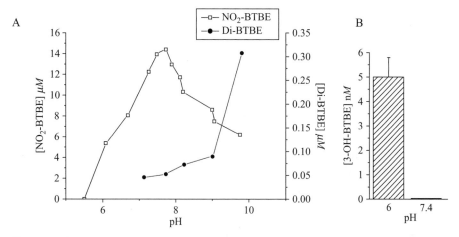

Figure 12.6 BTBE oxidation as a function of pH. BTBE-containing DLPC liposomes were prepared after lipid resuspension in phosphate buffer (100 mM) plus 0.1 mM DTPA at different pH values. Then BTBE (0.3 mM) was treated with peroxynitrite (0.5 mM) at each pH. (A) 3-NO$_2$-BTBE and 3,3′-di-BTBE were analyzed after organic extraction by RP-HPLC. (B) The 3-hydroxy-BTBE concentration at two pH values, estimated by using (3,4-dihydroxy-phenylalanine) DOPA as a standard. Reproduced with modifications from Bartesaghi *et al.* (2006).

alkaline pH. In addition, 3-hydroxy-BTBE was detected at pH 6 but not at pH 7.4.

The pH dependency of the three BTBE oxidation products can be fully rationalized kinetically by free radical–dependent mechanisms and has been reported elsewhere (Bartesaghi et al., 2006). These data are also in agreement with BTBE being partially immersed in the bilayer with the −OH being exposed towards the aqueous phase and therefore capable of ionization; thus, kinetic data are in agreement with structural information indicating that BTBE is located all through the bilayer with the highest concentration near the glycerol backbone of the phospholipids (Zhang et al., 2001a).

4. DISCUSSION

BTBE incorporated to PC liposomes has been revealed to be a useful probe to study tyrosine oxidation processes in membranes (Fig. 12.7), particularly peroxynitrite-mediated nitration, dimerization, and hydroxylation; these oxidation processes require the intermediacy of BTBE phenoxyl radicals as evidenced by ESR spin-trapping studies.

In hydrophobic environments, tyrosine nitration is the predominant pathway, partly because of the physical chemical properties of nitrogen dioxide, that is, it partitions and diffuses favorably in hydrophobic environments and reacts with tyrosyl radicals close to diffusion-controlled rates. In contrast, tyrosine dimerization is hindered because of the limited lateral diffusion of tyrosyl radicals within the lipid bilayer structure, expected to occur at least 100 times slower than in the aqueous phase (Bartesaghi et al., 2006). Small amounts of the hydroxylated tyrosine analog derivative were found, supporting the homolysis of ONOOH in the interior or the immediate proximity of the liposomal membrane. The role of hydrophilic and hydrophobic compounds that either enhance or inhibit tyrosine nitration processes in the membrane can be studied in detail through the use of membrane-containing BTBE. Additionally, the incidence of membrane fatty acid unsaturation and the role of lipid peroxidation processes in tyrosine oxidation product yields and distribution are just starting to be appreciated. While kinetic data on the reactions of $\cdot NO_2$, CO_3^-, and $\cdot OH$ radicals with BTBE or, in general, membrane-associated tyrosine residues are still lacking, the use of known rate constants with tyrosine in computer-assisted simulation studies recapitulates experimental data and fully supports a free radical mechanism of BTBE oxidation by peroxynitrite in biomembranes (Bartesaghi et al., 2006). The use of BTBE can help to understand distinctive factors that affect tyrosine oxidation in hydrophobic environments, which clearly differ in several respects to those existing in aqueous phases. Importantly, BTBE incorporation and oxidation in red blood cell

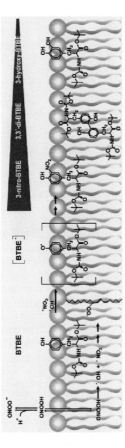

Figure 12.7 Tyrosine oxidation products in membranes induced by peroxynitrite. The structure of the hydrophobic probe BTBE, which undergoes one-electron oxidation to the corresponding BTBE phenoxyl radical either by peroxynitrite-derived radicals (·OH, ·NO₂) or by membrane-derived lipid peroxyl radicals (ROO·), is shown. The transient BTBE phenoxyl radical either reacts at diffusion-controlled rates with ·NO₂ to yield 3-nitro–BTBE or recombines with another phenoxyl radical to yield 3,3'di-BTBE; nitration yields are significantly larger than dimerization yields. The figure also indicates the formation of small amounts of the 3–hydroxy-BTBE from the addition reaction with hydroxyl radical and supports the diffusion and homolysis of ONOOH within the lipid bilayer. (See color insert.)

membranes have been just reported (Romero *et al.*, 2007), opening the use of BTBE as a probe to study free radical-dependent processes in hydrophobic biocompartments. While tyrosine-containing transmembrane peptides (Zhang *et al.*, 2003) reflect the biochemical behavior of a membrane-associated protein more closely, the relative ease of BTBE synthesis, incorporation to model and biological membranes, extraction, and quantitation of reaction products offers a unique possibility to study oxidation mechanisms mediated by peroxynitrite and other reactive oxygen and nitrogen species and their modulation by biomolecules, xenobiotics, and drugs, as well as carbon dioxide and molecular oxygen.

ACKNOWLEDGMENTS

We thank Valeria Valez for her contribution to the artwork. This work was supported by grants from the Howard Hughes Medical Institute and the International Centre of Genetic Engineering and Biotechnology to R.R. and the National Institutes of Health to B.K. and R.R. (2 R01H1063119-05). A donation for research support from Laboratorios Gramón-Bagó Uruguay to R.R. and G.P. through Universidad de la República is gratefully acknowledged. S.B. is supported by a fellowship from Programa de Desarrollo de Ciencias Básicas (PEDECIBA-Química), Universidad de la República, Uruguay. R.R. is a Howard Hughes International Research Scholar.

REFERENCES

Alvarez, B., Ferrer-Sueta, G., Freeman, B. A., and Radi, R. (1999). Kinetics of peroxynitrite reaction with amino acids and human serum albumin. *J. Biol. Chem.* **274,** 842–848.

Barber, D. J. W., and Thomas, J. K. (1978). Reactions of radicals with lecithin bilayers. *Radiat. Res.* **74,** 51–65.

Bartesaghi, S., Ferrer-Sueta, G., Peluffo, G., Valez, V., Zhang, H., Kalyanaraman, B., and Radi, R. (2007). Protein tyrosine nitration in hydrophilic and hydrophobic environments. *Amino. Acids.* **32,** 501–515.

Bartesaghi, S., Trujillo, M., Denicola, A., Folkes, L., Wardman, P., and Radi, R. (2004). Reactions of desferrioxamine with peroxynitrite-derived carbonate and nitrogen dioxide radicals. *Free. Radic. Biol. Med.* **36,** 471–483.

Bartesaghi, S., Valez, V., Trujillo, M., Peluffo, G., Romero, N., Zhang, H., Kalyanaraman, B., and Radi, R. (2006). Mechanistic studies of peroxynitrite-mediated tyrosine nitration in membranes using the hydrophobic probe N-t-BOC-L-tyrosine tert-butyl ester. *Biochemistry* **45,** 6813–6825.

Beckman, J. S., Beckman, T. W., Chen, J., Marshall, P. A., and Freeman, B. A. (1990). Apparent hydroxyl radical production by peroxynitrite: Implications for endothelial injury from nitric oxide and superoxide. *Proc. Natl. Acad. Sci. USA* **87,** 1620–1624.

Beckman, J. S., Ischiropoulos, H., Zhu, L., van der Woerd, M., Smith, C., Chen, J., Harrison, J., Martin, J. C., and Tsai, M. (1992). Kinetics of superoxide dismutase- and iron-catalyzed nitration of phenolics by peroxynitrite. *Arch. Biochem. Biophys.* **298,** 438–445.

Beckman, J. S., Zu Ye, Y., Anderson, P. G., Chen, J., Accavitti, M. A., Tarpey, M. M., and White, C. R. (1994). Extensive nitration of protein tyrosines in human atherosclerosis detected by immunohistochemistry. *Biol. Chem. Hoppe-Seyler.* **375,** 81–88.

Buege, J. A., and Aust, S. D. (1978). Microsomal lipid peroxidation. *Methods Enzymol.* **52,** 302–310.

Ischiropoulos, H. (1998). Biological tyrosine nitration: A pathophysiological function of nitric oxide and reactive oxygen species. *Arch. Biochem. Biophys.* **356**, 1–11.

Ischiropoulos, H., Zhu, L., Chen, J., Tsai, M., Martin, J. C., Smith, C. D., and Beckman, J. S. (1992). Peroxynitrite-mediated tyrosine nitration catalyzed by superoxide dismutase. *Arch. Biochem. Biophys.* **298**, 431–437.

Koppenol, W. H., Moreno, J. J., Pryor, W. A., Ischiropoulos, H., and Beckman, J. S. (1992). Peroxynitrite, a cloaked oxidant formed by nitric oxide and superoxide. *Chem. Res. Toxicol.* **5**, 834–842.

Mossoba, M., Makino, K., and Riesz, P. (1982). Photoionization of aromatic amino acids in aqueous solutions: A spin-trapping and electron spin resonance study. *J. Phys. Chem.* **86**, 3478–3483.

Radi, R. (2004). Nitric oxide, oxidants, and protein tyrosine nitration. *Proc. Natl. Acad. Sci. USA* **101**, 4003–4008.

Radi, R., Beckman, J. S., Bush, K. M., and Freeman, B. A. (1991a). Peroxynitrite oxidation of sulfhydryls: The cytotoxic potential of superoxide and nitric oxide. *J. Biol. Chem.* **266**, 4244–4250.

Radi, R., Beckman, J. S., Bush, K. M., and Freeman, B. A. (1991b). Peroxynitrite-induced membrane lipid peroxidation: The cytotoxic potential of superoxide and nitric oxide. *Arch. Biochem. Biophys.* **288**, 481–487.

Romero, N., Peluffo, G., Bartesaghi, S., Zhang, H., Joseph, J., Kalyanaraman, B., and Radi, R. (2007). Incorporation of the hydrophobic probe N-t-BOC-L-tyrosine tert-butyl ester (BTBE) to red blood cell membranes to study peroxynitrite-dependent reactions. *Chem. Res. Toxicol.* **20**, 1638–1648.

Santos, C. X., Bonini, M. G., and Augusto, O. (2000). Role of the carbonate radical anion in tyrosine nitration and hydroxylation by peroxynitrite. *Arch. Biochem. Biophys.* **377**, 146–152.

Schopfer, F. J., Baker, P. R., Giles, G., Chumley, P., Batthyany, C., Crawford, J., Patel, R. P., Hogg, N., Branchaud, B. P., Lancaster, J. R., and Freeman, B. A. (2005). Fatty acid transduction of nitric oxide signaling: Nitrolinoleic acid is a hydrophobically stabilized nitric oxide donor. *J. Biol. Chem.* **280**, 19289–19297.

Shishehbor, M. H., Aviles, R. J., Brennan, M. L., Fu, X., Goormastic, M., Pearce, G. L., Gokce, N., Keaney, J. F., Penn, M. S., Sprecher, D. L., Vita, J. A., and Hazen, S. L. (2003). Association of nitrotyrosine levels with cardiovascular disease and modulation by statin therapy. *JAMA* **289**, 1675–1680.

Sokolovsky, M., Riordan, J. F., and Vallee, B. L. (1966). Tetranitromethane: A reagent for the nitration of tyrosyl residues in proteins. *Biochemistry* **5**, 3582–3589.

Sturgeon, B. E., Glover, R. E., Chen, Y. R., Burka, L. T., and Mason, R. P. (2001). Tyrosine iminoxyl radical formation from tyrosyl radical/nitric oxide and nitrosotyrosine. *J. Biol. Chem.* **276**, 45516–45521.

Trostchansky, A., Batthyany, C., Botti, H., Radi, R., Denicola, A., and Rubbo, H. (2001). Formation of lipid-protein adducts in low-density lipoprotein by fluxes of peroxynitrite and its inhibition by nitric oxide. *Arch. Biochem. Biophys.* **395**, 225–232.

Trujillo, M., Folkes, L., Bartesaghi, S., Kalyanaraman, B., Wardman, P., and Radi, R. (2005). Peroxynitrite-derived carbonate and nitrogen dioxide radicals readily react with lipoic and dihydrolipoic acid. *Free Radic. Biol. Med.* **39**, 279–288.

van der Vliet, A., Eiserich, J. P., O'Neill, C. A., Halliwell, B., and Cross, C. E. (1995). Tyrosine modification by reactive nitrogen species: A closer look. *Arch. Biochem. Biophys.* **319**, 341–349.

Zhang, H., Bhargava, K., Keszler, A., Feix, J., Hogg, N., Joseph, J., and Kalyanaraman, B. (2003). Transmembrane nitration of hydrophobic tyrosyl peptides: Localization, characterization, mechanism of nitration, and biological implications. *J. Biol. Chem.* **278**, 8969–8978.

Zhang, H., Joseph, J., Feix, J., Hogg, N., and Kalyanaraman, B. (2001a). Nitration and oxidation of a hydrophobic tyrosine probe by peroxynitrite in membranes: Comparison with nitration and oxidation of tyrosine by peroxynitrite in aqueous solution. *Biochemistry* **40,** 7675–7686.

Zhang, H., Joseph, J., and Kalyanaraman, B. (2005). Hydrophobic tyrosyl probes for monitoring nitration reactions in membranes. *Methods Enzymol.* **396,** 182–204.

Zhang, R., Brennan, M. L., Fu, X., Aviles, R. J., Pearce, G. L., Penn, M. S., Topol, E. J., Sprecher, D. L., and Hazen, S. L. (2001b). Association between myeloperoxidase levels and risk of coronary artery disease. *JAMA* **286,** 2136–2142.

Zheng, L., Settle, M., Brubaker, G., Schmitt, D., Hazen, S. L., Smith, J. D., and Kinter, M. (2005). Localization of nitration and chlorination sites on apolipoprotein A-I catalyzed by myeloperoxidase in human atheroma and associated oxidative impairment in ABCA1-dependent cholesterol efflux from macrophages. *J. Biol. Chem.* **280,** 38–47.

CHAPTER THIRTEEN

ASSESSMENT OF SUPEROXIDE PRODUCTION AND NADPH OXIDASE ACTIVITY BY HPLC ANALYSIS OF DIHYDROETHIDIUM OXIDATION PRODUCTS

Francisco R. M. Laurindo, Denise C. Fernandes, *and* Célio X. C. Santos

Contents

Vascular Biology Laboratory, Heart Institute (InCor), University of São Paulo School of Medicine, São Paulo, Brazil

Methods in Enzymology, Volume 441
ISSN 0076-6879, DOI: 10.1016/S0076-6879(08)01213-5

Abstract

Assessment of low-level superoxide in nonphagocytic cells is crucial for assessing redox-dependent signaling pathways and the role of enzymes such as the
NADPH oxidase complex. However, most superoxide probes present inherent
limitations. Particularly, assessment of dihydroethidium (DHE) fluorescence is
limited regarding a lack of possible quantification and simultaneous detection of
its two main products: 2-hydroxyethidium, more specific for superoxide, and
ethidium, which reflects H_2O_2-dependent pathways involving metal proteins.
HPLC separation and analysis of those two main products have been described.
This chapter reports procedures used for the validation of superoxide measurements in vascular system. Superoxide assessment was performed for cultured
cells and tissue fragments incubated with DHE, followed by acetonitrile extraction
and HPLC run, with simultaneous fluorescence detection of 2-hydroxyethidium
and ethidium and ultraviolet detection of remaining DHE. It also describes procedures for DHE-based NADPH oxidase activity assays using HPLC or fluorometry.
Such methods can enhance accuracy and allow better quantitation of vascular
superoxide measurements.

1. Introduction

Knowledge about the pathobiology of superoxide radical in nonphagocytic cells has grown substantially over the last decades, thanks in part to its
well-known direct reaction with nitric oxide (NO)(Beckmann *et al.*, 1994;
Kissner *et al.*, 1997). Multilevel evidence supporting a regulatory role of
superoxide in NO bioactivity is overwhelming (Beckmann *et al.*, 1994).
Furthermore, emergence of the redox signal transduction concept brought
about renewed attention to superoxide as a signaling intermediate on its own
and as the precursor of hydrogen peroxide, a second messenger involved,
among other effects, in the regulation of intracellular and cell surface thiol
proteins or thiol redox buffers (Chen *et al.*, 2003; Forman *et al.*, 2004). While
superoxide reactivity is generally low, some known effects of this radical
include inactivation of Fe-S clusters, particularly from aconitase (Gardner and
Fridovich, 1992), iron mobilization from ferritin (Biemond *et al.*, 1986) and
quinone reduction or diphenol oxidation to semiquinones (Cadenas *et al.*,
1988). A somewhat underappreciated aspect is the increased reactivity of the
protonated uncharged membrane-permeable form of superoxide, the hydro-
peroxyl radical (OOH·), which is generated in higher amounts at acidic pH
values, given the $pK_a \approx 4.8$ of the reaction $OOH· \rightleftharpoons H^+ + O_2^{-·}$. This can

become particularly relevant at the range of pH values found in lysosomes (Thompson *et al.*, 2006) and the endosomal/vesicular secretory system, the latter a relevant site of nonphagocytic NADPH oxidase activation (Li *et al.*, 2006; Miller *et al.*, 2007). In fact, the spontaneous dismutation of superoxide in aqueous media necessarily requires protonation of one or both superoxide species, yielding respective rate constants $k_2 = 9.7 \times 10^7$ and $k_2 = 8.3 \times 10^5$ $M^{-1} \cdot s^{-1}$ at physiological pH values (Halliwell and Gutteridge, 1998).

A corollary and requisite for efficient redox signaling is that reactive oxygen species (ROS) generation is mainly an enzymatic-controlled process rather than an accidental event. In a time-amount basis, the highest-level cellular source of superoxide is probably the mitochondrial electron transport chain, although it is yet unclear to what extent mitochondrial leakage can account for extramitochondrial superoxide-dependent signaling (Nemoto *et al.*, 2000). However, the NADPH oxidase complex is the most prominent and well-studied enzymatic source of signaling ROS (reviewed in Clempus and Griendling, 2006). In the particular case of the vascular system, physiologically relevant NADPH oxidase-associated superoxide generation has been documented in conditions such as shear stress changes (Hwang *et al.*, 2003; Laurindo *et al.*, 1994) and in pathological conditions such as atherosclerosis, hypertension, diabetes melittus, and vascular response to injury (Cave *et al.*, 2006). Vascular ROS signaling involves assumptions common to general molecular signaling, such as (1) reversibility of target end-point reactions; (2) controlled, low-level, sizable-range, stimulus-associated production of the signaling intermediate; (3) accessory pathways allowing fine-tuning; and (4) temporal and/or spatial subcompartmentalization. Specifically, vascular cell redox signal transduction involves low nanomolar levels of superoxide (Lassegue and Clempus, 2003; Souza *et al.*, 2002). This composes a much different picture from that of the phagocyte, in which micromolar-level, burst-like superoxide production is produced inside the phagosome and extracellularly. Therefore, measuring superoxide output in vascular and other nonphagocytic cells, while a fundamental step for understanding redox-dependent signaling, poses a much more difficult challenge.

2. Probes for Assessment of Superoxide

Most methods of assessing superoxide output involve the use of exogenous probes, with the exception of endogenous aconitase inhibition (Tarpey *et al.*, 2004). Whereas spin traps react with superoxide via nucleophilic addition reactions, most probes explore redox reactions of superoxide (Table 13.1). Given the $E^{o\prime}$ of −330 mV for the O_2/O_2^{-} pair and $E^{o\prime}$ of +940 mV for the $O_2^{\cdot}/H_2O_2$ pair, exogenous probes have a thermodynamic window that allows exploring either reductive or oxidative superoxide

Table 13.1 Probes for superoxide detection

	Probes exploring the reductive chemistry of superoxide	Probes exploring the oxidative chemistry of superoxide
Enzymatic	Cytochrome *c*	Aconitase
Spectrophotometric	Nitro blue tetrazolium	—
Chemiluminescent	Lucigenin	Coelenterazine, CLA, luminol
Fluorescent	-	dihydroethidium
Probes exploring nucleophilic addition reactions ESR spin traps		

chemistry (Table 13.1). Whatever the underlying chemistry, however, all probes react with superoxide at rate constants ranging from as low as 5 or up to 10^8 $M^{-1} \cdot s^{-1}$, which are uniformly much lower than rate constants of reactions with superoxide dismutase (SOD) ($k_2 = 2.6 \times 10^9$ $M^{-1} \cdot s^{-1}$) or nitric oxide ($k_2 = 1.9 \times 10^{10}$ $M^{-1} \cdot s^{-1}$) and not substantially higher than the contants of some endogenous nonradical reactants or spontaneous superoxide dismutation. This competitive kinetic drawback can only be overcome by increasing concentrations of probes, usually to micromolar levels. This may introduce significant background in the system and exacerbate artifacts, decreasing specificity and accuracy. Probes exploring reductive superoxide chemistry such as lucigenin tend to be more specific for superoxide, but are prone to redox cycling caused by the reaction of molecular oxygen with the partially reduced probe, generating superoxide (Janiszewski *et al.*, 2002; Spasojevic *et al.*, 2000; Vasquez-Vivar *et al.*, 1998). Redox cycling is particularly troublesome with lucigenin-dependent analysis of NADPH oxidase activity in particulate fractions, particularly when NADH instead of NADPH is used as a substrate (Janiszewski *et al.*, 2002). Probes exploring the oxidative superoxide chemistry tend to present less or no redox cycling, but suffer from low specificity, being oxidizable by other species such as hydrogen peroxide, peroxynitrite, thiyl radical, and heme compounds, among others (Dikalov *et al.*, 2007; Tarpey *et al.*, 2004). Pitfalls and limitations of all such probes have been discussed extensively elsewhere (Dikalov *et al.*, 2007; Tarpey *et al.*, 2004; Zhao *et al.*, 2005).

3. DIHYDROETHIDIUM AS A SUPEROXIDE PROBE

Dihydroethidium (5-ethyl-5,6-dihydro-6-phenyl-3,8-diaminophenanthridine, hydroethidine, DHE) has been used increasingly as a probe for superoxide in biological systems. DHE is a hydrophobic uncharged compound that is able to cross extra- and intracellular membranes and, upon oxidation, becomes positively charged and accumulates in cells by intercalating into DNA, primarily by electrostatic interactions with DNA phosphate groups and further via hydrophobic interactions (Garbett *et al.*, 2004). Its oxidation by different oxidizing systems has been used increasingly for fluorescent analysis of ROS output in cells and tissues, although mechanistic aspects of these reactions have been explored only recently. Initially, DHE-derived red fluorescence observed with rhodamine filter (excitation 490; emission 590 nm) was attributed to ethidium compound formation, a two-electron oxidation product (Fig. 13.1), and *in vitro* experiments showed that the red fluorescence was obtained more specifically with superoxide-generating systems (xanthine or glucose oxidase) rather than with oxidants such as hydrogen peroxide, peroxynitrite, or hydroxyl radical (generated by the Fenton reaction) (Benov *et al.*, 1998; Bindokas *et al.*, 1996). Such initial studies were consistent with the possibility of other DHE products triggered by superoxide in parallel to ethidium and, in addition, suggested a possible superoxide dismutase-mimetic activity for DHE (Benov *et al.*, 1998; Bindokas *et al.*, 1996; Papapostolou *et al.*, 2004).

More recently, important advances have been obtained in the understanding of DHE chemistry. In 2003, Zhao and colleagues demonstrated by HPLC analysis of fluorescent products obtained from DHE oxidation by superoxide that, in addition to ethidium, another similar compound was formed, further characterized as 2-hydroxyethidium. Structurally, this compound presents, in relation to ethidium, an additional hydroxyl group in the two position of aromatic rings, whereas both 2-hydroxyethidium and ethidium bind to DNA due to its positive charge. Most importantly, 2-hydroxyethidium (2-EOH) is generated specifically by superoxide oxidation of DHE, whereas ethidium is associated mainly with pathways involving hydrogen peroxide and metal-based oxidizing systems, including heme proteins and peroxidases, although to an unknown extent the ethidium product can represent the decay of 2-EOH and potentially of other fluorescent intermediate products (Fink *et al.*, 2004).

Determination of total DHE fluorescence in cells has been performed extensively in the literature for assessment of ROS and, more specifically, of superoxide (Fig. 13.2). The techniques for this determination have been well described and are not discussed here. As with other analogous

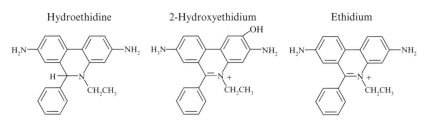

Figure 13.1 Chemical structures of dihydroethidium, 2-hydroxyethidium, and ethidium.

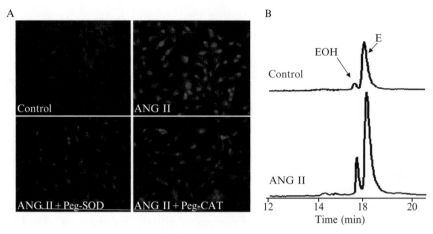

Figure 13.2 (A) Fluorescence microscopic images of cultured vascular smooth muscle cells after incubation with DHE (2 μM) for 10 min, observed under a rhodamine filter in a common fluorescence microscope (Zeiss Axiovert 200). Cells were analyzed in control conditions or after incubation with the NADPH oxidase agonist angiotensin II (ANG II, 100 nM, 4 h) and, in the latter case, also after the addition of inhibitors Peg-superoxide dismutase (Peg-SOD, 50 U/ml) or Peg-catalase (Peg-CAT, 200 U/ml). (B) HPLC chromatogram of control and ANG II-stimulated cell extracts showing 2-EOH and ethidium peaks. After acetonitrile extraction, the supernatant was injected into the HPLC system, as described in text, and identification was performed with fluorescence detector. (See color insert.)

fluorescence techniques, the method is sensitive, but its main drawback is that the total fluorescence of DHE is a sum of the composite spectra of all different products and thus will likely reflect preferentially a measure of total cell redox state rather than production of a specific intermediate. The use of controls with polyethylene glycol (PEG)-SOD or PEG-catalase (similar to those described later) allows to some extent discrimination of the respective contributions of superoxide or hydrogen peroxide to the fluorescence. Moreover, the method is inherently nonquantitative, both from the stoichiometric standpoint (Benov *et al.*, 1998) and for methodological reasons, the latter related to the fact that fluorescence of the target cell is usually determined as a comparative function of control(s) for which fluorescence parameters are arbitrarily adjusted to low levels. Similar considerations apply to the *in situ* microfluorotopography of DHE products in tissue slices, added to the fact that tissues are usually frozen before cutting (Miller *et al.*, 1998). Because this makes the assay quite distant from physiological conditions, these measurements should be interpreted mainly as an *in situ* bioassay of potential ROS sources, particularly enzymatic ones. Nevertheless, with adequate PEG-SOD and PEG-catalase controls, this test can be useful for revealing the main sites of ROS production in a given tissue and, particularly, for allowing spatial correlation with specific enzyme sources being studied.

Both 2-EOH and ethidium are fluorescent products difficult to discriminate between each other by conventional fluorescence microscopy or fluorometry. Thus, HPLC analysis of DHE-derived fluorescent compounds (2-EOH and ethidium) has been developed (Fink *et al.*, 2004; Zhao *et al.*, 2005) and validated in the vascular system (Fernandes *et al.*, 2007) in order to achieve separation and individual analysis of such products. This technique provides a significant increase in the accurary of superoxide output determinations and is a meaningful advance toward the precise quantification of this species in cells and tissue.

4. HPLC Analysis of Dihydroethidium Oxidation in Cells and Tissues: General Considerations

Dihydroethidium oxidation in cells and tissues is basically analyzed after a 30-min incubation of the sample with DHE in specific buffers and *not* in culture medium. Culture media are complex solutions that contain high levels of serum, amino acids, and redox metals, which interfere with DHE oxidation, and can strongly decrease the detection efficiency of DHE-derived compounds. Buffers such as Hanks or Krebs maintain cell homeostasis during sufficient time, thanks to the contents of glucose, bicarbonate, and essential ions such as calcium. In addition, adherent cells will remain attached throughout DHE incubation. The use of phosphate-buffered saline (PBS) is inadequate because it promotes detachment of most primary cells, although some cells lines are resistant (Fernandes *et al.*, 2007). During DHE incubation of cells or tissues, it is essential to prevent artificial secondary oxidizing reactions due to adventitial metals present in buffers and plastic/glassware. For this purpose, we routinely add the chelator diethylenetriaminepenta-acetic acid (DTPA, 100 μM) to all buffers. Furthermore, all solutions are treated routinely with Chelex-100 resin (typically, one spatula of resin is added to 500 ml of buffer, and the solution is maintained under agitation overnight).

Considering the hydrophobic nature of the binding between charged DHE products and DNA, extraction of DHE-derived products was optimized with organic solvents capable of breaking such interactions, such as butanol, chloroform, methanol, or acetonitrile (Fernandes *et al.*, 2007; Papapostolou *et al.*, 2004; Zhao *et al.*, 2005; Zielonka *et al.*, 2005). We compared chloroform versus acetonitrile extraction of 2-EOH and ethidium from DNA *in vitro*, and acetonitrile showed the best recovery of 2-EOH. Therefore, we have used acetonitrile for cell or tissue extraction. Importantly, the remaining DHE is recovered during acetonitrile extraction, and we routinely perform its quantification by HPLC-UV detection (Fig. 13.3) in parallel to fluorescent detection of its derived compounds (Fig. 13.2). This procedure is important to improve the accuracy of analysis

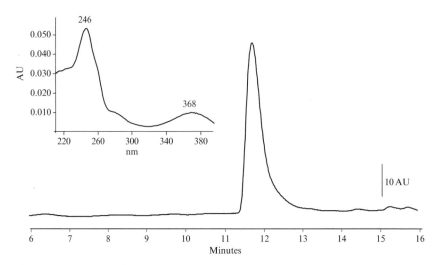

Figure 13.3 HPLC chromatogram showing DHE peak obtained by UV detection (245 nm) under chromatographic conditions described in the text. (Insert) UV spectrum of DHE peak (200–400 nm) exhibiting maximum absorbance at 245 and 370 nm.

for at least two reasons: (i) the presence of remaining DHE confirms the optimal experimental condition that DHE levels should be in excess; of note, the initial DHE concentration may require adjustments for each particular system and type of sample; and (ii) the quantification of remaining DHE levels should be inversely correlated to the total cell or tissue redox state, that is, low remaining DHE levels are expected in more oxidizing systems, with the opposite happening in more reducing systems. In this regard, it is important to remember that DHE is not only oxidized to 2-EOH and ethidium (both detected at "red fluorescent" filter), but also to other noncharacterized fluorescent compounds and maybe to nonfluorescent compounds as well (Papapostolou *et al.*, 2004). Thus, DHE quantification becomes more important as the complexity of the sample and experimental conditions increases.

4.1. Chromatographic separation of fluorescent products derived from DHE oxidation

Because the only structural difference between 2-EOH and ethidium molecules is one hydroxyl group (Fig. 13.1), chromatographic separation is not so easy and requires gradient condition in a reversed-phase column. Columns employed for this separation until now have been C_{18} or ether-linked phenyl Polar-RP (Table 13.2), with the latter described to develop a better separation than the C_{18} column (Zielonka *et al.*, 2006a). So far, we have typically used the C_{18} column (Fernandes *et al.*, 2007). Irrespective of the column, all chromatographic runs take approximately 30 to 40 min at a flow rate of 0.5 ml/min, with the gradient condition based on increasing

Table 13.2 Columns employed in 2-EOH and ethidium separation

Column	Particle size	Dimensions (length × diameter)	Ref.
NovaPack C_{18}	5 μm	150 × 3.9 mm	Fernandes *et al.* (2007)
Kromasil C_{18}	5 μm	250 × 4.6 mm	Zielonka *et al.* (2006b)
Synergil Polar-RP	4 μm	250 × 4.6 mm	Zielonka *et al.* (2006a)

acetonitrile levels (generally from 10 to 40%) (Fernandes *et al.*, 2007; Fink *et al.*, 2004; Robinson *et al.*, 2006; Zhao *et al.*, 2005).

We perform identification of 2-EOH and ethidium with the originally described fluorescence detection (Zhao *et al.*, 2003). Excitation/emission wavelengths for fluorescence detection are usually 510/595 nm, corresponding to the "red fluorescence" of the rhodamine filter at microscopy analysis. The main reason for using such a wavelength pair is the possibility of simultaneously analyzing not only 2-EOH, the more specific superoxide-derived DHE product, but also ethidium, which, although less specific for superoxide, is also an important by-product formed during DHE oxidation by different sources of reactive oxygen species, as well as heme proteins (Fernandes *et al.*, 2007; Papapostolou *et al.*, 2004; Patsoukis *et al.*, 2005). In addition, other reports have described the pair 480/580 nm, which is more sensitive for the detection of 2-EOH rather than ethidium (Fink *et al.*, 2004), and 396/579 nm, which detects 2-EOH specifically (Robinson *et al.*, 2007). Comparison among wavelength pairs for 2-EOH and ethidium fluorescent detection in each specific chromatographic condition is one of the initial steps for optimization of HPLC analysis (see Fig. 13.4). For example, we found that detection at 480/580 nm provided better identification of 2-EOH without loss in the detection of ethidium (Fig. 13.4). Electrochemical detection has been reported for the detection of DHE and its oxidation products (Zielonka *et al.*, 2006b). HPLC conditions are the same as those described for fluorescent detection, with exception of the mobile phase, which is adjusted at pH 2.6 in 50 mM phosphate buffer.

4.2. HPLC system and mobile phase

Chromatographic separation is carried out with a NovaPak C_{18} column (3.9 × 150 mm, 5-μm particle size) in a HPLC system (Waters) equipped with a rheodyne injector (loop volume of 200 μl) and photodiode array (W2996) and fluorescence (W2475) detectors. Addition of a C_{18} precolumn is routinely employed and strongly recommended for maintaining the column in good condition, especially for tissue extract injections. Ethidium and 2-EOH separation is performed with a gradient of solutions A (pure

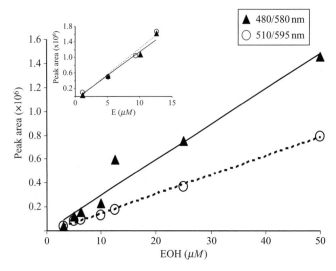

Figure 13.4 Comparison between two different wavelength pairs employed for 2-EOH and ethidium fluorescent detection: 480/580 nm (Fink *et al.*, 2004) and 510/595 nm (Fernandes *et al.*, 2007; Zhao *et al.*, 2003). Using the chromatographic conditions described in the text, the 480/580-nm pair was more sensitive for 2-EOH detection, with no significant differences for ethidium quantification (insert) when compared to the 510/595-nm pair.

acetonitrile) and B (water/10% acetonitrile/0.1% trifluoroacetic acid, v/v/v) at a flow rate of 0.4 ml/min. Runs are started with 0% solution A, increased linearly to 40% solution A during the initial 10 min, kept at this proportion for another 10 min, changed to 100% solution A for an additional 5 min, and to 0% solution A for the final 10 min. DHE is monitored by UV absorption at 245 nm or 370 nm (Fig. 13.3). Ethidium and 2-EOH are monitored by fluorescence detection with excitation at 510 nm and emission at 595 nm. Under such chromatographic conditions, the usual approximate retention times for DHE, 2-EOH, and ethidium are, respectively, 11–12, 18–19, and 19–20 min (Figs. 13.2 and 13.3).

4.3. HPLC routine

In the beginning of each working day, we wash the column with pure acetonitrile for 30 min (at flow rate of 1.0 ml/min) and then the flow of solution A is started at a rate equal to the gradient condition (0.4 ml/min) for 10 min. The first injection should be ultrapure water (100 μl) in order to allow column equilibration with gradient condition and also to observe possible artifactual peaks caused by column impurities or debris. The second injection should be a standard solution of known concentration of DHE (e.g., 50 μM) containing low levels of ethidium and 2-EOH as contaminants (<1 μM; ideally, this should be performed with the pure standards, but the 2-EOH standard is difficult to obtain at present—see later). This chromatogram will show whether ethidium and 2-EOH separation is good and the correlation between actual DHE concentration/area ratios versus a standard DHE calibration curve. Finally, samples can be injected in sequence, unless the column pressure starts to increase, which will probably be because of the apolar lipid content of biological samples. In these cases, it is necessary to clean the column again with water, followed by solution A (pure acetonitrile, as described earlier).

4.4. Column maintenance and preparation

As chromatographic separation is a critical step in DHE oxidation analysis, we perform and recommend the following important (although eventually tedious) procedures for preparing, cleaning, and keeping the column in order to get reliable results. All HPLC solutions (mobile phases, cleaning solutions, and sample ressuspension buffer) must be prepared with ultrapure water (MilliQ) degassed in a sonicator for at least 10 min. *For column storage*, wash the column with low flow (0.5 ml/min) first with ultrapure water for 4 to 5 h and then with an increasing methanol concentration (from 10 to 50%) for 1 h and store the column. *For using the column*, wash the column with ultrapure water for at least 30 min to remove the 50% methanol (from the storage condition) and then proceed as described earlier in "HPLC routine." *For overnight column cleaning*, it is important to maintain a column cleaning routine, that is, if the column is used daily, wash the column with ultrapure water overnight in recycling mode at a low flow rate of 0.1 ml/min. The latter step is performed with pure acetonitrile in case the column pressure is increased at the end of the working day.

4.5. Stock and working solutions

Dihydroethidium and ethidium bromide are acquired from Invitrogen. DHE, 2-EOH, and ethidium solutions should be stored at −20 °C in the dark and sealed. To avoid freeze–thaw of stock solutions, it is recommended to aliquot stock solution in volumes consistent with one working day. Fresh working solutions are made each day and kept on ice under protection from light.

4.5.1. DHE stock solution

The 10 mM DHE stock solution is prepared by diluting 3 mg DHE in 1 ml deoxygenated dimethyl sulfoxide (DMSO) and its concentration should be confirmed at pH 7.4 at 265 nm ($\varepsilon = 1.8 \times 10^4 \ M^{-1} \cdot cm^{-1}$) or 345 nm ($\varepsilon = 9.75 \times 10^3 \ M^{-1} \cdot cm^{-1}$) (Zielonka *et al.*, 2005). Care with ambient light is necessary because it can interfere strongly with rates of formation and/or metabolism of ethidium and 2-EOH during DHE oxidation *in vitro* (Zielonka *et al.*, 2006b) and *in vivo* (Fernandes *et al.*, 2007). It is also recommended that fresh DHE solutions be kept in 1 mM HCl to minimize autooxidation (Zielonka *et al.*, 2005).

4.5.2. 2-Hydroxyethidium stock solution

A 50 μM 2-EOH stock solution can be prepared from incubation of DHE (50 μM) with xanthine/xanthine oxidase (0.5 mM/0.05 U/ml) in aerated PBS (7.78 mM Na$_2$HPO$_4$, 2.20 mM KH$_2$PO$_4$, 140 mM NaCl, 2.73 mM KCl, pH 7.4) containing DTPA (100 μM) (PBS/DTPA) at 37 °C for 30 min in the dark. In order to obtain a pure 2-EOH solution, ≈50 ml of 2-EOH solution is prepared as described earlier, injected in sequential aliquots into HPLC, and collected at the corresponding 2-EOH peak (taking care not to collect the ethidium compound). After dried, the obtained reddish pellet is ressuspended in DMSO, and the concentration of the 2-EOH solution is determined at pH 7.4 at 470 nm ($\varepsilon = 9.4 \times 10^3$ $M^{-1} \cdot cm^{-1}$) (Fernandes *et al.*, 2007; Zielonka *et al.*, 2005). A better alternative to this extremely time-consuming method is to prepare the 2-EOH compound by reacting DHE with Fremy's salt (Zielonka *et al.*, 2005). In addition, 2-EOH can be obtained commercially as "oxyethidium" (Noxygen Science Transfer & Diagnostics GmbH, Germany; http://www.noxygen.de).

4.5.3. Ethidium stock solution

The 50 μM ethidium stock solution is prepared by diluting 2 μl of commercially obtained ethidium bromide (10 mg/ml) in 998 μl of PBS containing DTPA (100 μM), pH 7.4. The ethidium concentration is determined in pH 7.4 at 480 nm ($\varepsilon = 5.8 \times 10^3$ $M^{-1} \cdot cm^{-1}$) (Zielonka *et al.*, 2005).

5. Extraction of Fluorescent DHE-Derived Products in Biological Samples

5.1. Cultured cells

5.1.1. Adherent cells

Cells are spread in six-well plates to gain confluence of 80 to 95% after 18 to 24 h. In the case of vascular smooth muscle or endothelial cells, the cell number is ≈3 × 10^5/well. After specific treatments, cells are washed with PBS/100 μM DTPA two to three times. *All subsequent steps must be done under dim light.* Hanks buffer (1.3 mM CaCl$_2$, 0.8 mM MgSO$_4$, 5.4 mM KCl, 0.4 mM KH$_2$PO4, 4.3 mM NaHCO$_3$, 137 mM NaCl, 0.3 mM Na$_2$HPO$_4$, 5.6 mM glucose, pH 7.4; 500 μl) containing DTPA (100 μM) is added to each well, followed by DHE (50 μM; i.e., 2.5 μl of DHE stock solution per well). Plates are kept in an incubator (at 37 °C/5% CO_2) for 30 min. Thereafter, each well is washed two to three times with PBS/DTPA. Acetonitrile (500 μl) is added to each well, cells are immediately harvested with a scraper, and the lysate is pipetted into an Eppendorf tube, which must be kept over ice. Because acetonitrile is highly volatile, we strongly recommend extraction of only two wells at a time. The lysates are centrifuged (12,000g, 10 min, 4 °C), and supernatants are transferred to an Eppendorf tube and acetonitrile dried under vacuum for 2 to 3 h(Speed-Vac Plus SC-110A, Thermo Savant). Of note, the remaining pellet should be white, denoting that reddish compounds derived from DHE oxidation are in the acetonitrile supernatant. It is important to dry lysates at a low rate in order to avoid turning on the Speed-Vac lamp, which will happen at medium or high rates and may interfere with oxidation of the remaining DHE. Pellets can be maintained at −20 °C in the dark until analysis for up to a few weeks. For HPLC analysis, samples are resuspended in 120 μl PBS/DTPA, and a volume of 100 μl is injected.

5.1.2. Cells in suspension

The following procedures apply to cells that are cultured in suspension or adhered cells that, after the specific experimental protocol, are detached from culture flasks (with trypsin/EDTA or pancreatin). Initially, cells are washed twice with PBS/DTPA by centrifugation (1000g for 5 min at 4 °C). *All subsequent steps must be done under dim light.* Hanks buffer (500 μl) containing DTPA (100 μM) and DHE (50 μM) is added and cell suspensions are kept for 30 min at 37 °C/5% CO_2. Excess DHE is eliminated by washing twice with PBS/DTPA and centrifugation at 1000g for 5 min at 4 °C. The cell pellet is resuspended in 500 μl acetonitrile, sonicated (10 s, two cycles at 8 W), and the lysates are centrifuged at 12,000g, 10 min, 4 °C. Subsequent analysis of supernatants is performed in the same way as described earlier for adherent cells.

5.1.3. Particularities during cell extraction of DHE-derived products

5.1.3.1. Confounding extrinsic factors Because of some confounding effects of light and sonication observed during DHE oxidation *in vitro* (Zielonka *et al.*, 2006b), we compared 2-EOH and ethidium generation in vascular smooth muscle cells submitted or not to sonication or visible light exposure during and/or following acetonitrile extraction. In our system, the sonication procedure had no detectable effect in 2-EOH and ethidium levels. However, after 10 min of exposure to ambient light (≥ 2 fluorescent 40-W tubular lamps, 1 to 1.5 m distance), extracts exhibited an approximately fourfold increase in 2-EOH generation as compared to extracts maintained in the dark or dim light (data not shown). Therefore, we strongly recommend keeping all samples under dim light throughout all steps of manipulation.

5.1.3.2. Interference of peguilated compounds

One important experimental control with cells (or tissues–see later) is to add SOD conjugated to polyethylene glycol (PEG-SOD; 25 U/ml) before DHE incubation. Because PEG penetration into cells promotes disturbances in membrane fluidity by itself, as described by Beckman *et al.* (1998), a 30- to 60-min recovery time should be allowed before further measurements. Indeed, cell exposure to PEG by itself for time periods equivalent to those for PEG-SOD (0.5–4 h) results in a moderate nonspecific decrease in 2-EOH levels (data not shown). In addition, concerning the effects of PEG-catalase, indirect agonistic effects of hydrogen peroxide on superoxide-producing systems such as NADPH oxidase can cause a misleading impression that hydrogen peroxide contributes to 2-EOH fluorescence (Fernandes *et al.*, 2007).

5.1.3.3. Interference of cell transfection reagents

Analysis of DHE-derived fluorescent compounds in cells transfected with cDNA(s) or siRNA, in which transfection procedures employ lipid reagents (p. eg. Lipofectamine, from Invitrogen), may show high basal levels of 2-EOH and ethidium compared to nontransfected cells, which may be sufficient to obscure differences among samples. This is probably a consequence of interference of these lipid reagents with membrane homeostasis. This should be controlled by assessing cells treated only with transfection reagents, in parallel to usual controls (e.g., empty vector or scrambled siRNA).

5.1.3.4. Measurement of neutrophil oxidative burst

In systems producing high amounts of superoxide such as the neutrophil, the DHE concentration may have to be adjusted to higher concentrations ($>50 \, \mu M$) and analysis performed with caution and appropriate controls, considering the large amounts of distinct oxidizing species formed during the respiratory burst.

5.2. Tissues

Contrary to most cells, not all tissues can provide successful results of HPLC analysis of DHE oxidation products. Heme proteins, as well as peroxidases, interfere with DHE oxidation by competing with superoxide for DHE. In this way, tissues that are highly vascularized or contain high levels of heme proteins (blood, liver, spleen) may provide innacurate results. We strongly recommend prior *in situ* or *ex vivo* PBS perfusion of organs in order to remove as much of the remaining blood as possible and careful cleaning of tissues in order to remove clots, thrombi, or adhered blood-trapping connective tissue (e.g., adventitial tissue in vessels). Organs are cut into small fragments (1–5 mg), whereas arteries should be divided into 3- to 10-mm segments (depending on caliber), which are weighed (typical weights of 2–10 mg). *All subsequent steps should be performed under dim light.* Organ/artery segments are incubated in 500 μl PBS/DTPA containing DHE (50 μM; i.e., 2.5 μl of DHE stock solution) in a 1.5-ml Eppendorf for 30 min at 37 °C in the dark. Then, organ/artery segments are washed with PBS twice, transfered to liquid nitrogen, and homogenized with a mortar and pestle. The artery powder is resuspended in 500 μl acetonitrile, sonicated (three cycles at 8 W for 10 s), and lysates are centrifuged at 12,000g for 10 min at 4 °C. Supernatants are transferred to a tube and processed similarly to those of cell extracts as described for intact cells.

5.2.1. Particularities during HPLC analysis of tissues extracts

5.2.1.1. Interference of high lipid content Most tissues, including vessels, have a high content of apolar compounds such as lipids that are co-extracted with acetonitrile. Consequently, after even a few repeated injections of tissue extract supernatants into HPLC, column pressure will likely be increased. In order to minimize this problem, we clean the HPLC system with pure acetonitrile after every five to six tissue supernatant injections.

5.2.1.2. Peguilated compounds as controls Similar to analysis of cell extracts, control samples incubated with PEG-SOD should be analyzed in parallel during protocol validation. One tissue segment is preincubated with PEG-SOD (25 U/ml) for at least 15 min at 37 °C, followed by DHE (50 μM) incubation and further procedures as described previously. Another relevant control is PEG-catalase (200 U/ml) incubation, especially in samples with a high heme protein content.

6. QUANTIFICATION OF FLUORESCENT DHE-DERIVED PRODUCTS

Quantification of fluorescent DHE products is performed by comparison of integrated peak areas between the supernatants of cell or tissue extracts and a standard curve obtained for DHE or each of its products under identical chromatographic conditions. One of the critical steps for the quantification of superoxide from DHE products is estimating the efficiency of organic extraction for each particular experiment, which is challenging. In this regard, the simultaneous detection of DHE and its derived oxidation products (2-EOH and ethidium), respectively, through UV or fluorescence detection (Figs. 13.2 and 13.3) allows the use of DHE as an internal control also submitted to similar organic extraction of each sample. Thus, we often express the results of DHE-derived product analysis as ratios of 2-EOH or ethidium generated per DHE consumed (which is the initial minus remaining DHE concentration) (EOH/DHE and E/DHE). The overall baseline values obtained under optimal conditions with 50 μM DHE at 80% vascular smooth muscle cell confluence are typically 180 ± 40 nmol EOH/μmol DHE and 200 ± 50 nmol E/μmol DHE (i.e., $\approx$20% of initial DHE concentration). For all of the aforementioned protocols, data can also be normalized for the amount of protein or number of cells; we found that results will likely be analogous, although variability tends to be higher. In specific cases where cell numbers vary more substantially among the different conditions (e.g., with a reagent that induces apoptosis), it is necessary to normalize data for the amount of protein or cell number.

6.1. Particularities in quantification of fluorescent DHE-derived products

6.1.1. Coelution of peaks

It is important to note among different types of samples, especially in tissue samples, whether there is any compound that might coelute which DHE (by UV detection). For example, DHE coincubations with FAD, but not NADH or NADPH, yield a double peak at the corresponding DHE retention time (data not shown) under the chromatographic conditions described here. Such a coelution makes it necessary to optimize other HPLC chromatographic conditions.

6.1.2. Possible influence of peroxynitrite

We previously reported *in vitro* that peroxynitrite, in the presence of physiological concentrations of bicarbonate (25 mM), was able to oxidize DHE to 2-EOH with small yields (Fernandes *et al.*, 2007). Although the *in vivo* relevance of this reaction is unknown, in systems potentially generating high amounts of nitric oxide and superoxide, 2-EOH might be influenced by peroxynitrite production.

6.2. What is the meaning of 2-EOH and ethidium quantification in cells and tissues?

Basically, the goal of performing 2-EOH quantification in cells or tissues is to obtain a quantitative assessment of overall superoxide output production in such specific systems during the period of DHE incubation. Ethidium quantification will add specificity to the measurement by separating fluorescence due to other sources, such as H_2O_2-dependent pathways involving metal proteins (Fernandes *et al.*, 2007; Fink *et al.*, 2004; Zhao *et al.*, 2003). In this regard, the methods described here certainly allow, on a quantitative basis, an improved and more specific comparison among different experimental situations of similar samples and even among different samples analyzed simultaneously. Furthermore, these techniques are a major step toward establishing the ideal goal of an absolute quantification of superoxide output in cells and tissues. However, we believe that, at present, there are still some drawbacks that prevent such a goal; consequently, expressing results, for example, as *nmol superoxide*/mg/min, provides an inaccurate picture of real superoxide production. Such drawbacks include (a) the yet unclear stoichiometry of 2-EOH formation versus superoxide produced; (b) the yet unclear stoichometry of 2-EOH formation from DHE, in relation to possible additional unknown intermediates formed; (c) the variable and mostly unclear efficiency rates of organic extraction; and (d) the possible and yet unclear variability in subcellular factors affecting the binding of DHE and its products to DNA. Comments can be made regarding some of these issues. Total fluorescence analysis of DHE oxidation by superoxide-generating systems showed a ratio of 4.6:1 ($O_2^{\cdot-}$:DHE) (Benov *et al.*, 1998), similar to the ratio of 4.6 to 4.8:1 obtained with the association of HPLC and EPR techniques (Fink *et al.*, 2004). Requirement of many superoxide molecules to yield one 2-EOH molecule further indicates the complexity of DHE oxidation pathways, which involve at least one radical intermediate. In fact, pulse radiolysis studies showed that DHE oxidation to 2-EOH requires two steps, the first involving an electron or a hydrogen transfer to form the corresponding one-electron oxidized form of DHE, that is, an aromatic aminyl radical of DHE (Zielonka *et al.*, 2006a). This radical intermediate may decay or react with superoxide to form 2-EOH (Fig. 13.5).

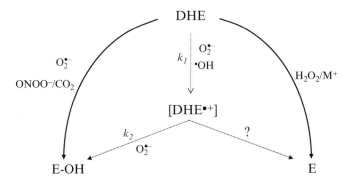

Figure 13.5 Chemical pathways postulated at present to account for DHE-derived red fluorescence. Outer curved arrows summarize the main pathways involving 2-hydroxyethidium formation (EOH), basically from superoxide and to a lesser extent peroxynitrite/CO_2, and ethidium (E), from hydrogen peroxide pathways involving metal proteins. Inner dotted arrows indicate possible intermediate pathways involved in the formation of these products. Reaction between superoxide and hydroxyl radical with DHE has a $k_1 = (0.5–4) \times 10^6\ M^{-1}\,s^{-1}$ and generates the DHE monocation radical intermediate, which can react with superoxide with a $k_2 = 2 \times 10^9\ M^{-1}\,s^{-1}$, which is fivefold faster than DHE radical decay $k = 6 \times 10^8\ M^{-1}\,s^{-1}$. Rate constants were based on pulse radiolysis studies (Zielonka *et al.*, 2006a). It is yet unclear whether ethidium formation can occur from the DHE monocation radical intermediate or whether it results from two-electron DHE oxidation, induced, e.g., by metal-centered proteins, such as heme proteins and citochrome c^{3+}.

All such considerations also apply to the case of NADPH oxidase activity assays, described in the following section, although the lower complexity of this system allows simplification of several procedures.

 ## 7. Assessment of NADPH Oxidase Activity in Cell Membrane Fraction by DHE Oxidation

Assessment of NADPH oxidase activity in subcellular fractions is a valuable tool to help understand the functional aspects of this enzyme complex activity in perspective with structural studies of the distinct oxidase isoforms and regulatory subunits. Until now, the only methods available utilized lucigenin chemiluminescence, which is sensitive, but may be associated with redox-cycling artifacts (Janiszewski *et al.*, 2002) or EPR spin trapping (Laurindo *et al.*, 2002), which, however, is less sensitive and not widely available. This section describes procedures used to assess NADPH oxidase activity in vascular smooth muscle cell membrane fractions using DHE-derived fluorescence oxidation products detected by either HPLC or fluorometry.

7.1. Membrane-enriched fraction separation

Cells are spread in 100-mm dishes to achieve 80 to 95% confluence after 18 to 24 h of growth. In the case of vascular smooth muscle or endothelial cells, the cell number is $\approx 1.5 \times 10^6$/dish. After experimental procedures, cells are washed with cold PBS twice, harvested, homogenized with a scraper in 500 μl of lysis buffer (50 mM Tris, pH 7.4, containing 0.1 mM EDTA, 0.1 mM EGTA, 10 μg/ml aprotinin, 10 μg/ml leupeptin, and 1 mM phenylmethyl-sulfonyl fluoride), sonicated (10 s of three cycles at 8 W), and centrifuged (18,000g for 15 min at 4 °C) to remove intact cells, mitochondria, and nuclei. Supernatants are further centrifuged at 100,000g for 1 h at 4 °C and the transparent pellet is resuspended in 30 to 50 μl of lysis buffer to obtain a membrane-enriched fraction, which represents a sample of plasma membrane and some endomembranes (e.g., endosomal vesicles or endoplasmic reticulum-derived microsomes). The membrane fraction can be used in NADPH oxidase activity measurement by HPLC or in a microplate reader. Analogous methods for tissue samples are described in detail elsewhere in this series (Laurindo *et al.*, 2002).

7.2. Measurement of NADPH oxidase activity by HPLC

The membrane fraction (≈ 20 μg protein) is incubated in PBS/DTPA with DHE (50 μM) and NADPH (300 μM) in a final volume of 120 μl at 37 °C in the dark for 30 min. The reaction is stopped on ice until HPLC injection (volume of 100 μl), which we usually perform over a total period of 4 to 7 h, in the same way as described earlier. We usually perform a quick centrifugation (5 min, 1000g) in order to sediment the homogenate fragments, yielding a clear supernatant for HPLC injection.

7.3. Measurement of NADPH oxidase activity by fluorometry

Since the major DHE-derived product in this assay is 2-EOH, it becomes possible to quantify NADPH oxidase activity with a microplate fluorometric reader (Fernandes *et al.*, 2007), as long as there is no interference of a parallel ethidium increase. The major difference between such a fluorometric assay and HPLC analysis is the use of a lower DHE concentration, combined with DNA addition, which greatly amplifies DHE-derived fluorescence. In each well of 96-well plaques for fluorescence analysis, the membrane fraction ($\approx$10 μg protein) is incubated with DHE (10 μM) in PBS/DTPA in the presence of NADPH (50 μM) and calf thymus DNA (1.25 μg/ml) in a final volume of 120 μl for 30 min at 37 °C in the dark. Total fluorescence is followed in a microplate reader using two different filters: a rhodamine filter (excitation 490 nm and emission 590 nm) and an acridine filter (excitation 490 nm and emission 570 nm) in a spectrofluorometer (Wallac Victor^2 1420-Multilabel Counter, Perkin-Elmer). The acridine filter is more specific for 2-EOH measurements (Zhao *et al.*, 2005).

7.4. Particularities of measurement of NADPH oxidase activity

7.4.1. Controls

In HPLC or fluorometric analysis, it is important to compare incubation of the same membrane fraction with or without superoxide scavengers, such as SOD (300–500 U/ml) or tiron (20 μM). In addition, it is important to validate the assay using compounds known to inhibit NADPH oxidase activity, such as the flavoprotein inhibitor diphenyleneiodonium chloride (10–20 μM), despite its lack of specificity (Laurindo *et al.*, 2002).

7.4.2. Validation of fluorometric assay

The fluorometric assay has been validated for vascular smooth muscle cells under the conditions of our experiments. Since the use of this assay in a broad range of conditions has not been validated, we recommend that each cell type and condition be validated first with HPLC in order to check whether there is interference of the ethidium product, and only then should the fluorometric assay be used.

7.4.3. Influence of contaminant enzymes

NADPH oxidase activity is a relatively nonspecific activity displayed by several electron transfer enzymes and dehydrogenases as well. Because many of these enzymes are flavoproteins, a blocking effect of diphenylene iodonium is not

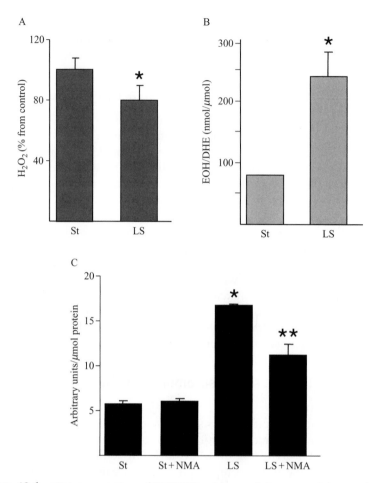

Figure 13.6 Misinterpretation of NADPH oxidase activity assessed in membrane-enriched endothelial cell fractions, possibly as a consequence of endothelial nitric oxide synthase uncoupling. (A) NADPH-triggered H_2O_2 production in membrane fraction of cultured endothelial cells submitted to laminar shear stress (LS, 15 dynes/cm²) in a cone-plate system for 18 h. In agreement with prior studies with EPR, there was a decrease ($\approx 20\%$) in NADPH-driven peroxide levels after shear (Hwang *et al.*, 2003). (B) A similar preparation showed increased superoxide production (approximately sixfold) in NADPH-triggered membrane fractions incubated with DHE and analyzed by HPLC compared to static controls (St). (C) Same as B, plus endothelial NO synthase inhibition with *N*-methyl-l-arginine (NMA, 100 μM), which prevented the increase in NADPH-driven superoxide significantly (although partially). (C) Assays were performed with a microplate reader (acridine filter). Values are mean $\pm$ SE ($n = 3$ experiments). *$P < 0.05$ versus St; **$P < 0.05$ versus LS (Li *et al.*, 2006; Oak and Cai, 2007).

discriminative of the NADPH oxidase complex (Laurindo *et al.*, 2002). Therefore, assessing the NADPH oxidase activity of whole cell homogenates has no meaning, since it will be heavily contaminated by all mitochondrial and cytosolic dehydrogenases. In addition, even membrane-enriched fractions variably congregate other superoxide sources different from NADPH oxidase, which may cause misinterpretation of such enzyme assays. An important example in endothelial cells is endothelial NO synthase (eNOS), which is bound to membrane caveolae and, in the conditions of the NADPH oxidase assay, that is, absence of cofactors (e.g., tetrahydrobiopterin) or substrate (L-arginine), generates superoxide (Vasquez-Vivar et al., 1998). This artifact can become particularly relevant under conditions in which eNOS is strongly recruited to membranes such as after sustained laminar shear stress (Fig. 13.6). Thus, in these conditions, other approaches should be used to assess the NADPH oxidase contribution to ROS generation, such as siRNA for Nox isoforms.

8. Summary and Conclusions

Quantitative assessment of reactive oxygen species production is an essential goal regarding any desired progress toward more comprehensive and integrative understanding of redox signaling at a cellular and organ level. This chapter reported procedures for the validation of superoxide measurements in the vascular system based on HPLC separation and analysis of the two main products of DHE oxidation—2-hydroxyethidium and ethidium. Superoxide assessment was performed for cultured cells and tissue fragments incubated with DHE, followed by acetonitrile extraction and HPLC run, with simultaneous fluorescence detection of the two main products plus UV detection of remaining DHE. It also described procedures for DHE-based NADPH oxidase activity assays using HPLC or fluorometry. Such techniques provide an important advance toward improving the accuracy and precision of quantitative superoxide measurements in the low scale expected for allowing assessment of redox-mediated signal transduction. In addition, the theoretical basis underlying DHE oxidation provides a platform for further improvements.

REFERENCES

Beckman, J. S., Minor, R. L., Jr., White, C. W., Repine, J. E., Rosen, G. M., and Freeman, B. A. (1988). Superoxide dismutase and catalase conjugated to polyethylene glycol increases endothelial enzyme activity and oxidant resistance. J. Biol. Chem. **263**, 6884–6892.

Beckmann, J. S., Ye, Y. Z., Anderson, P. G., Chen, J., Accavitti, M. A., Tarpey, M. M., and White, C. R. (1994). Extensive nitration of protein tyrosines in human atherosclerosis detected by immunohistochemistry. Biol. Chem. Hoppe Seyler **375**, 81–88.

Benov, L., Sztejnberg, L., and Fridovich, I. (1998). Critical evaluation of the use of hydroethidine as a measure of superoxide anion radical. Free Radic. Biol. Med. **25**, 826–831.

Biemond, P., Swaak, A. J., Beindorff, C. M., and Koster, J. F. (1986). Superoxide-dependent and -independent mechanisms of iron mobilization from ferritin by xanthine oxidase: Implications for oxygen-free-radical-induced tissue destruction during ischaemia and inflammation. *Biochem. J.* **239,** 169–173.

Bindokas, V. P., Jordan, J., Lee, C. C., and Miller, R. J. (1996). Superoxide production in rat hippocampal neurons: Selective imaging with hydroethidine. *J. Neurosci.* **16,** 1324–1336.

Cadenas, E., Mira, D., Brunmark, A., Lind, C., Segura-Aguilar, J., and Ernster, L. (1988). Effect of superoxide dismutase on the autoxidation of various hydroquinones: A possible role of superoxide dismutase as a superoxide: Semiquinone oxidoreductase. *Free. Radic. Biol. Med.* **5,** 71–79.

Cave, A. C., Brewer, A. C., Narayanapanicker, A., Ray, R., Grieve, D. J., Walker, S., and Shah, A. M. (2006). NADPH oxidases in cardiovascular health and disease. *Antioxid. Redox. Signal.* **8,** 691–728.

Chen, K., Thomas, S. R., and Keaney, J. F., Jr. (2003). Beyond LDL oxidation: ROS in vascular signal transduction. *Free Radic. Biol. Med.* **35,** 117–132.

Clempus, R. E., and Griendling, K. K. (2006). Reactive oxygen species signaling in vascular smooth muscle cells. *Cardiovasc. Res.* **71,** 216–225.

Dikalov, S., Griendling, K. K., and Harrison, D. G. (2007). Measurement of reactive oxygen species in cardiovascular studies. *Hypertension* **49,** 717–727.

Fernandes, D. C., Wosniak, J. J., Pescatore, L. A., Bertoline, M. A., Liberman, M., Laurindo, F. R., and Santos, C. X. C. (2007). Analysis of dihydroethidium-derived oxidation products by HPLC in the assessment of superoxide production and NADPH oxidase activity in vascular systems. *Am. J. Physiol. Cell. Physiol.* **292,** C413–C422.

Fink, B., Laude, K., McCann, L., Doughan, A., Harrison, D. G., and Dikalov, S. (2004). Detection of intracellular superoxide formation in endothelial cells and intact tissues using dihydroethidium and an HPLC-based assay. *Am. J. Physiol. Cell. Physiol.* **287,** C895–C902.

Forman, H. J., Fukuto, J. M., and Torres, M. (2004). Redox signaling: Thiol chemistry defines which reactive oxygen and nitrogen species can act as second messengers. *Am. J. Physiol. Cell. Physiol.* **287,** C246–C256.

Garbett, N. C., Hammond, N. B., and Graves, D. E. (2004). Influence of the amino substituents in the interaction of ethidium bromide with DNA. *Biophys. J.* **87,** 3974–3981.

Gardner, P. R., and Fridovich, I. (1992). Inactivation-reactivation of aconitase in *Escherichia coli*: A sensitive measure of superoxide radical. *J. Biol. Chem.* **267,** 8757–8763.

Kissner, R., Nauser, T., Bugnon, P., Lye, P. G., and Koppenol, W. H. (1997). Formation and properties of peroxynitrite as studied by laser flash photolysis, high-pressure stopped-flow technique, and pulse radiolysis. *Chem. Res. Toxicol.* **10,** 1285–1292.

Halliwell, B., and Gutteridge, J. M. C. (1998). The chemistry of free radicals and related "reactive species" *In* "Free Radicals in Biology and Medicine" 3rd Ed. p. 62. Oxford University Press, New York.

Hwang, J., Saha, A., Boo, Y. C., Sorescu, G. P., McNally, J. S., Holland, S. M., Dikalov, S., Giddens, D. P., Griendling, K. K., Harrison, D. G., and Jo, H. (2003). Oscillatory shear stress stimulates endothelial production of O2- from p47phox-dependent NAD(P)H oxidases, leading to monocyte adhesion. *J. Biol. Chem.* **278,** 47291–47298.

Janiszewski, M., de Sousa, H. P., Liu, X., Pedro, M. A., Zweier, J. L., and Laurindo, F. R. (2002). Overestimation of NADH-driven vascular oxidase activity due to lucigenin artifacts. *Free Radic. Biol. Med.* **32,** 446–453.

Lassegue, B., and Clempus, R. E. (2003). Vascular NAD(P)H oxidases: Specific features, expression, and regulation. *Am. J. Physiol. Regul. Integr. Comp. Physiol.* **285,** R277–R297.

Laurindo, F. R. M., Souza, H. P., Pedro, M. P., and Janiszewski, M. (2002). Redox aspects of vascular response to injury. *Methods Enzymol.* **352,** 432–454.

Li, H., Witte, K., August, M., Brausch, I., Godtel-Armbrust, U., Habermeier, A., Closs, E. I., Oelze, M., Munzel, T., and Forstermann, U. (2006). Reversal of endothelial nitric oxide synthase uncoupling and up-regulation of endothelial nitric oxide synthase expression lowers blood pressure in hypertensive rats. *J. Am. Coll. Cardiol.* **47,** 2536–2544.

Li, J., Stouffs, M., Serrander, L., Banfi, B., Bettiol, E., Charnay, Y., Steger, K., Krause, K. H., and Jaconi, M. E. (2006). The NADPH oxidase NOX4 drives cardiac differentiation: Role in regulating cardiactranscription factors and MAP kinase activation. *Mol. Biol. Cell.* **17,** 3978–3988.

Miller, F. J., Jr., Filali, M., Huss, G. J., Stanic, B., Chamseddine, A., Barna, T. J., and Lamb, F. S. (2007). Cytokine activation of nuclear factor kappa B in vascular smooth muscle cells requires signaling endosomes containing Nox1 and ClC-3. *Circ. Res.* **101,** 663–671.

Miller, F. J., Jr, Gutterman, D. D., Rios, C. D., Heistad, D. D., and Davidson, B. L. (1998). Superoxide production in vascular smooth muscle contributes to oxidative stress and impaired relaxation in atherosclerosis. *Circ. Res.* **82,** 1298–1305.

Nemoto, S., Takeda, K., Yu, Z. X., Ferrans, V. J., and Finkel, T. (2000). Role for mitochondrial oxidants as regulators of cellular metabolism. *Mol. Cell. Biol.* **20,** 7311–7318.

Oak, J. H., and Cai, H. (2007). Attenuation of angiotensin II signaling recouples eNOS and inhibits nonendothelial NOX activity in diabetic mice. *Diabetes* **56,** 118–126.

Papapostolou, I., Patsoukis, N., and Georgiou, C. D. (2004). The fluorescence detection of superoxide radical using hydroethidine could be complicated by the presence of heme proteins. *Anal. Biochem.* **332,** 290–298.

Patsoukis, N., Papapostolou, I., and Georgiou, C. D. (2005). Interference of non-specific peroxidases in the fluorescence detection of superoxide radical by hydroethidine oxidation: A new assay for H_2O_2. *Anal. Bioanal. Chem.* **381,** 1065–1072.

Robinson, K. M., Janes, M. S., Pehar, M., Monette, J. S., Ross, M. F., Hagen, T. M., Murphy, M. P., and Beckman, J. S. (2006). Selective fluorescent imaging of superoxide *in vivo* using ethidium-based probes. *Proc. Natl. Acad. Sci. USA* **103,** 15038–15043.

Souza, H. P., Liu, X., Samouilov, A., Kuppusamy, P., Laurindo, F. R. M., and Zweier, J. L. (2002). Quantitation of superoxide generation and substrate utilization by vascular NAD (P)H oxidase. *Am. J. Physiol. (Heart Circ. Physiol).* **282,** H466–H474.

Spasojevic, I., Liochev, S. I., and Fridovich, I. (2000). Lucigenin: Redox potential in aqueous media and redox cycling with O-(2) production. *Arch. Biochem. Biophys.* **373,** 447–450.

Tarpey, M. M., Wink, D. A., and Grisham, M. B. (2004). Methods for detection of reactive metabolites of oxygen and nitrogen: *In vitro* and *in vivo* considerations. *Am. J. Physiol. Regul. Integr. Comp. Physiol.* **286,** 431–444.

Thompson, R. J., Akana, H. C., Finnigan, C., Howell, K. E., and Caldwell, J. H. (2006). Anion channels transport ATP into the Golgi lumen. *Am. J. Physiol. Cell Physiol.* **290,** C499–C514.

Vasquez-Vivar, J., Kalyanaraman, B., Martasek, P., Hogg, N., Masters, B. S., Karoui, H., Tordo, P., and Pritchard, K. A., Jr. (1998). Superoxide generation by endothelial nitric oxide synthase: The influence of cofactors. *Proc. Natl. Acad. Sci. USA* **95,** 9220–9225.

Zhao, H., Joseph, J., Fales, H. M., Sokoloski, E. A., Levine, R. L., Vasquez-Vivar, J., and Kalyanaraman, B. (2005). Detection and characterization of the product of hydroethidine and intracellular superoxide by HPLC and limitations of fluorescence. *Proc. Natl. Acad. Sci. USA* **102,** 5727–5732.

Zhao, H., Kalivendi, S., Zhang, H., Joseph, J., Nithipatikom, K., Vasquez-Vivar, J., and Kalyanaraman, B. (2003). Superoxide reacts with hydroethidine but forms a fluorescent product that is distinctly different from ethidium: Potential implications in intracellular fluorescence detection of superoxide. *Free Radic. Biol. Med.* **34,** 1359–1368.

Zielonka, J., Sarna, T., Roberts, J. E., Wishart, J. F., and Kalyanaraman, B. (2006a). Pulse radiolysis and steady-state analyses of the reaction between hydroethidine and superoxide and other oxidants. *Arch. Biochem. Biophys.* **456,** 39–47.

Zielonka, J., Vasquez-Vivar, J., and Kalyanaraman, B. (2006b). The confounding effects of light, sonication, and Mn(III)TBAP on quantitation of superoxide using hydroethidine. *Free Radic. Biol. Med.* **41,** 1050–1057.

Zielonka, J., Zhao, H., Xu, Y., and Kalyanaraman, B. (2005). Mechanistic similarities between oxidation of hydroethidine by Fremy's salt and superoxide: Stopped-flow optical and EPR studies. *Free Radic. Biol. Med.* **39,** 853–863.

CHAPTER FOURTEEN

METHODS TO MEASURE THE REACTIVITY OF PEROXYNITRITE-DERIVED OXIDANTS TOWARD REDUCED FLUORESCEINS AND RHODAMINES

Peter Wardman

Contents

Abstract

The commonest probes for "reactive oxygen and nitrogen species" are reduced fluorescein and rhodamine dyes that fluoresce when oxidized. The reduced dyes are reactive toward peroxynitrite, although probably not directly but via free radical oxidants derived from it: hydroxyl, carbonate, and nitrogen dioxide free radicals. The reaction with peroxynitrite can be monitored by rapid mixing and stopped-flow spectrophotometry, but reliable measurement of reactivity of the peroxynitrite-derived radicals requires specialized techniques such as flash photolysis or pulse radiolysis to monitor the fast reactions in real time. A key feature of oxidation by radicals is that the reaction produces an intermediate fluorescein or rhodamine radical, which normally is oxidized further by oxygen

University of Oxford, Gray Cancer Institute, Mount Vernon Hospital, Northwood, Middlesex, United Kingdom

Methods in Enzymology, Volume 441
ISSN 0076-6879, DOI: 10.1016/S0076-6879(08)01214-7

to yield the fluorescent, stable product. Susceptibility of the yield of fluorescence to interference by antioxidants can be assessed from kinetic parameters, which reflect reactivity. This chapter outlines methods for estimation of key rate constants involving peroxynitrite-derived oxidants.

1. INTRODUCTION

There is a continuing need in chemistry and biology to measure the production and spatial distribution of reactive species derived from superoxide and nitric oxide free radicals. Fluorescent probes (more accurately, *pro*-fluorescent probes) offer the potential of high sensitivity for detection and measurement, but quantitative measures of reactivity toward the species of interest are needed in order to assess probe response and susceptibility toward interfering substances (Wardman, 2007). The relative rates of reaction can be characterized as the product of rate constant and reactant concentration; while the intracellular uptake of probes will be heterogeneous and estimating concentrations may be problematical, rate constants measured in physiological buffer offer a guide to reactivity. This chapter outlines methods used to assess the reactivity of peroxynitrite, and the radical products of its decomposition in physiological media, toward the commonest probes that are oxidized to fluorescent products. These are based on fluoresceins and rhodamines; methodology for the detection of "reactive nitrogen species" using both types of dye has been discussed elsewhere (Ischiropoulos *et al.*, 1999), as has the kinetic analysis of the reactivity of peroxynitrite with biomolecules (Radi, 1996).

While the examples here focus on the dyes used most commonly, the methodology is generally applicable to related probes. These include the newer derivatives such as conjugates of phenols or anilines with oxidized fluorescein or rhodamine, oxidation of the phenol or aniline substituent resulting in cleavage to release free fluorescein (Setsukinai *et al.*, 2003; Wardman, 2007). The approaches used are also applicable to the study of the first reaction step involved in the production of fluorescent products from probes for nitric oxide based on vicinal diamines, such as "DAF-2," in which the first step is nonspecific oxidation to an anilinyl radical (Chatton and Broillet, 2002; Itoh *et al.*, 2000; Wardman, 2007).

2. REACTANTS

2.1. Oxidants derived from peroxynitrite

Peroxynitrite can be formed in biology via nitric oxide and superoxide:

$$NO^{\bullet} + O_2^{\bullet-} \leftrightarrows ONOO^-, \tag{14.1}$$

with its biological chemistry (Radi et al., 2000) reflecting the dominant role of the equilibrium with peroxynitrous acid, which has a $pK_{14.2} \approx 6.8$:

$$ONOOH \rightleftharpoons ONOO^- + H^+. \qquad (14.2)$$

Peroxynitrous acid decomposes in ≈ 1 s but peroxynitrite is much more stable; thus solutions of peroxynitrite for kinetic experiments can be prepared conveniently in dilute alkali (e.g., 1 mM NaOH) and the reactant solution prepared in more concentrated buffer near the desired final pH. Decomposition of peroxynitrous acid involves at least two pathways, generating nitrogen dioxide and hydroxyl free radicals [Eq. (14.3a)] or nitrate and H^+ [Eq. (14.3b)], probably in the ratio $\approx 1:2$:

$$ONOOH \rightarrow NO_2^{\cdot} + {}^{\cdot}OH \qquad (14.3a)$$

$$ONOOH \rightarrow NO_3^- + H^+. \qquad (14.3b)$$

A key variable in experimental samples, and in physiology, is the concentration of bicarbonate/carbon dioxide (and therefore pH):

$$H_2O + CO_2 \rightleftharpoons H_2CO_3 \rightleftharpoons H^+ + HCO_3^-, \qquad (14.4)$$

as peroxynitrite is highly reactive toward CO_2 (Denicola et al., 1996a; Lymar and Hurst, 1995):

$$ONOO^- + CO_2 \rightarrow ONOOCO_2^-, \qquad (14.5)$$

and nitrosoperoxocarbonate formed in reaction (14.5) decomposes very rapidly (Goldstein et al., 2001) to yield either nitrogen dioxide and the carbonate radical anion [Eq. (14.6a)] or nitrate and CO_2 [Eq. (14.6b)], probably in the ratio $\approx 1:2$:

$$ONOOCO_2^- \rightarrow NO_2^{\cdot} + CO_3^{\cdot -} \qquad (14.6a)$$

$$ONOOCO_2^- \rightarrow NO_3^- + CO_2. \qquad (14.6b)$$

The proportion of ${}^{\cdot}OH$ compared to $CO_3^{\cdot -}$ resulting from peroxynitrite formation is controlled by the presence of HCO_3^-/CO_2 buffer, its concentration, and pH. The rate constant for reaction (14.3) is ≈ 1 s^{-1} and the fraction of the total peroxynitrous acid/peroxynitrite present as the undissociated acid is $(1 + K_{14.2}/[H^+])^{-1}$ (≈ 0.2 at pH 7.4); the rate constant for reaction (14.5) is $\sim 3 \times 10^4$ M^{-1} s^{-1}. Hence in the absence of other reactants, reactions (14.6a) and (14.6b) will predominate as the fate of peroxynitrite over the alternative decomposition pathways, reactions (14.3a) and (14.3b) if $k_{14.5}[CO_2] > (k_{14.3} \times (1 + K_{14.2}/[H^+])^{-1})$, that is, if $[CO_2] > 10$ μM at pH 7.4.

Higher levels of CO_2 are required to favor reaction (14.5) if alternative reactants for peroxynitrite are present, but in most physiological environments the high levels of CO_2 serve to drive peroxynitrite to be a source of $NO_2^{\cdot}/CO_3^{\cdot-}$ rather than $NO_2^{\cdot}/{}^{\cdot}OH$ (Denicola et al., 1996b).

Dinitrogen trioxide is often included as a peroxynitrite-derived reagent, formed in equilibrium (14.7):

$$NO_2^{\cdot} + NO^{\cdot} \rightleftharpoons N_2O_3. \tag{14.7}$$

It has been pointed out that rapid removal of $NO_2^{\cdot}$ (in microseconds) by reaction with antioxidants—principally thiols, urate, and ascorbate—will maintain $NO_2^{\cdot}$ at very low steady-state concentrations in the cytosol or plasma (Ford et al., 2002); nitric oxide is also maintained in submicromolar concentrations by reaction with hemes. Even by considering only the competing reaction of $NO_2^{\cdot}$ with glutathione (GSH):

$$NO_2^{\cdot} + GSH \rightarrow NO_2^{-} + GS^{\cdot} + H^{+}, \tag{14.8}$$

as well as neglecting the dissociation of N_2O_3 in equilibrium (14.7), we can see that significant formation of N_2O_3 via (14.7) is unlikely in hydrophilic compartments. Thus the relative rates of (14.7) (considering forward reaction only) vs (14.8) are given by $k_{14.7}[NO^{\cdot}]/k_{14.8}[GSH]$; for reaction (14.7) to be favored requires $[NO^{\cdot}]$ to be $> (k_{14.8}/k_{14.7})[GSH]$. If $k_{14.7f} \approx 1 \times 10^9$ $M^{-1} s^{-1}$ (Ross et al., 1998), $k_{14.8} \approx 2 \times 10^7 M^{-1} s^{-1}$ (Ford et al., 2002), and $[GSH] \approx 5$ mM, then $[NO^{\cdot}]$ must be >100 μM for formation of N_2O_3 to be favored over reaction with GSH. Additional reactions of $NO_2^{\cdot}$ with protein thiols and other antioxidants increase this limit to obviously unrealistically high concentrations of $NO^{\cdot}$, even under pathological conditions, so reaction (14.7) is likely to be a very minor fate of $NO_2^{\cdot}$ in most hydrophilic environments in biology.

2.2. Reduced probes, oxidized products, and intermediate radicals

The two commonest "leuco" dyes used as probes in free radical biology are $2',7'$-dichlorodihydro fluorescein (DCFH$_2$) and dihydro rhodamine 123 (RhH$_2$); the structures are shown in Fig. 14.1, along with the oxidized, stable products (DCF and Rh, respectively). As noted later, all three forms (reduced, radical intermediate, and oxidized) are involved in prototropic equilibria, which are important in influencing pH-dependent reactivity, as well as other properties, such as absorption spectra and quantum yields of

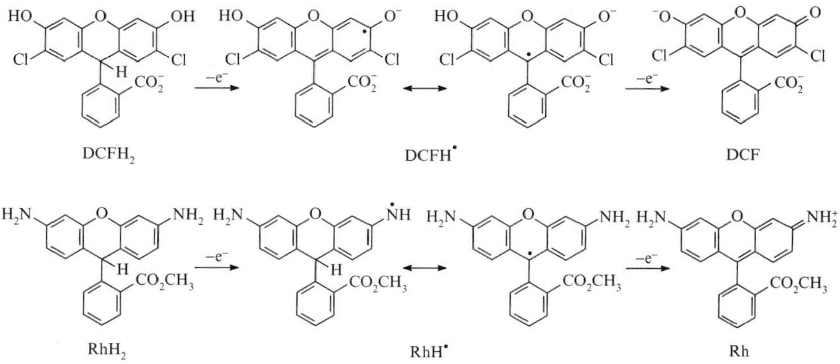

Figure 14.1 Structures of DCFH$_2$; the fluorescent (oxidized) product, DCF; the radical intermediate DCFH˙ (two resonance forms shown); and the corresponding structures for rhodamine 123.

fluorescence. Controlling pH is thus important in measuring reactivity and monitoring oxidized products; Fig. 14.1 shows the prototropic forms thought to be most abundant at pH 7.4, although all the prototropic equilibria have been reasonably well established only for the fluorescein (Wrona and Wardman, 2006). Free radical oxidants such as ˙OH, NO$_2^˙$, and CO$_3^{˙-}$ must necessarily oxidize the reduced dyes to produce free radical intermediates, shown in Fig. 14.1 as DCFH˙ and RhH˙, respectively. Note that only two of the possible resonance structures of each of the latter radicals are given.

The prototropic equilibria associated with DCFH$_2$, DCFH˙, and DCF have been discussed elsewhere (Wrona and Wardman, 2006). DCFH$_2$ has three prototropic equilibria characterizing dissociation of the two phenolic moieties ($pK_a \approx 7.9$ and 9.2) and the carboxylic acid substituent [pK_a probably intermediate between that in DCF (≈ 3.5) and benzoic acid (≈ 4.2)]. The two higher pK_a values for DCFH˙ are probably ≈ 7.3 and 8.8, while the phenol substituent in DCF dissociates with a $pK_a \approx 5.2$.

DCFH$_2$ and similar fluoresceins are commonly loaded into cells by treating cells with diesters of the phenolic moieties, usually the diacetate; cellular esterases cleave the diester to release the free probe. It should be stressed that the reactivity of such probes toward peroxynitrite-derived oxidants reflects the presence of phenolic moieties, especially if ionized [dissociated phenols are much more reactive toward free-radical oxidants than when undissociated, as illustrated with tyrosine (Adams et al., 1972) and a variety of other phenols (Ross et al., 1998)]. Thus the reactivity of the esters will be quite different from the hydrolyzed probes—probably much lower.

3. METHODS

3.1. Direct reaction with peroxynitrite

Methods for measuring the reactivity of peroxynitrite with biomolecules have been described in detail (Radi, 1996) and are only outlined here. Because of possible pH-dependent reactivity, it is preferable to follow reactions at the pH(s) of interest (e.g., 7.4); the lifetime of peroxynitrite around this pH is $\approx$1 s and so mixing the reagents rapidly and having the capacity to follow the reaction on a millisecond timescale (the timescale required for homogeneous mixing of solutions without very specialized techniques) are important. Stopped-flow spectrophotometry, following rapid mixing facilitated by pneumatically driven syringes, with absorbance or fluorescence detection, is the method of choice. Peroxynitrite has a strong absorbance at 302 nm ($\varepsilon \approx 1.7 \times 10^4 \ M^{-1} \ cm^{-1}$) and its decay can be thus monitored, although it may be possible to follow product formation instead of (or preferably as well as) peroxynitrite decay. Dependence of the reaction rate on pH is to be expected not only from equilibrium (14.2) but also from prototropic equilibria in the other reactant, as found, for example, for cysteine (Radi, 1996; Radi et al., 1991).

Oxidation of both $DCFH_2$ and RhH_2 by peroxynitrite/peroxynitrous acid was studied using stopped-flow spectrophotometry (Glebska and Koppenol, 2003); an earlier study used the same method to monitor the reaction between peroxynitrite and RhH_2 (Jourd'heuil et al., 2001). As with the case of applications of these probes in biology, an excess of oxidizable dye over peroxynitrite was used: 10- to 25-fold in these examples. Rather than monitoring peroxynitrite absorbance, which would overlap the absorbance of the reduced dyes, detection of DCF and Rh by absorbance at 500 nm or fluorescence at 530 to 536 nm (with 500 nm excitation) was used. The key observations, made by both groups of authors, were that the rates of formation of DCF/Rh were independent of the concentration of $DCFH_2/RhH_2$ and were similar to the rates of homolysis of peroxynitrite/peroxynitrous acid in the absence of other reactants. The rates varied with pH; bicarbonate/CO_2 accelerated the kinetics of oxidation of $DCFH_2$ or RhH_2 and the decomposition of peroxynitrite, with rates similarly proportional to the HCO_3^- concentration (Glebska and Koppenol, 2003). [Glebska and Koppenol (2003) also noted that at low pH ($\approx$3–5), nitrite contamination in the peroxynitrite caused a slow oxidation of RhH_2, over >5 s, when peroxynitrous acid would have decayed completely even spontaneously; this was ascribed to nitrous acid formation and reaction.]

At first sight these observations are consistent with homolysis, reaction (14.3), or reaction with CO_2 (14.5), being rate limiting. Formation of DCF from as low as $\approx$1 μM $DCFH_2$ by lower concentrations of NO_2^-, the least-reactive radical product of homolysis of peroxynitrite or reaction with CO_2, can be calculated from the rate constant of reaction (14.9) [$1.3 \times 10^7 \ M^{-1} \ s^{-1}$ (Wrona et al., 2005)] to have a half-life of ≈ 50 ms at pH ≈ 7.5, much shorter than the half-life of reaction (14.3):

$$NO_2^{\cdot} + DCFH_2 \rightarrow NO_2^- + DCFH^{\cdot} + H^+. \qquad (14.9)$$

Formation of DCF results from the reaction of DCFH$^{\cdot}$ formed in (14.9) with oxygen; in air-equilibrated buffer reaction (14.10) has a half-life of ≈ 5 μs, as $k_{14.10} \approx 5 \times 10^8 \ M^{-1} \ s^{-1}$ at pH ≈ 7.4 (Wrona and Wardman, 2006):

$$DCFH^{\cdot} + O_2 \rightarrow DCF + O_2^{\cdot -} + H^+. \qquad (14.10)$$

Reaction (14.11) has a half-life of ≈ 3 ms with $[DCFH_2] \approx 1 \ \mu M$ and $[DCFH_2] \gg [CO_3^{\cdot -}]$, since $k_{14.11} \approx 2.6 \times 10^8 \ M^{-1} \ s^{-1}$ at pH ≈ 8 (Wrona et al., 2005):

$$CO_3^{\cdot -} + DCFH_2 \rightarrow HCO_3^- + DCFH^{\cdot}, \qquad (14.11)$$

a time considerably shorter than the shortest half-life (≈ 20 ms) of reaction (14.5) under the conditions used in the study: up to 25 mM HCO$_3^-$ at pH 7 (Glebska and Koppenol, 2003). [Methods for the determination of rate constants for reactions (14.9), (14.11), and (14.12) (see later) are discussed in the following section.]

Again at first sight, then, these kinetic observations are consistent with peroxynitrite oxidizing the reduced dyes directly, if at all, at rates slower than homolysis or reaction with CO_2 *under the conditions used.* Oxidation appears to be indirect, from the decomposition products: $^{\cdot}$OH, NO$_2^{\cdot}$, and CO$_3^{\cdot -}$. However, the efficiency of oxidation varied with pH, being much lower at pH <7 than at higher pH (Glebska and Koppenol, 2003; Kooy et al., 1997). The yields of oxidized dye were maximal at pH 8 to 9 in both studies, at $\approx 35\%$ of the peroxynitrite consumed with 0.2 μM peroxynitrite and 20 μM DCFH$_2$ (Kooy et al., 1997), or ~ 50 to 60% of the peroxynitrite consumed with 1μM peroxynitrite and ~ 13 to 15 μM DCFH$_2$ or RhH$_2$ (Glebska and Koppenol, 2003). A free radical based mechanism, with oxidation of the dyes being achieved in reactions such as (14.9), (14.11) or (14.12):

$$^{\cdot}OH + DCFH_2 \rightarrow \rightarrow H_2O + DCFH^{\cdot} \qquad (14.12)$$

followed by (14.10), might be expected to result in up to about two-thirds the yield of DCF compared to peroxynitrite consumed [i.e., about double the yields of reactions (14.3a) or (14.6a), as peroxynitrite yields two oxidizing radicals and O_2 completes the oxidation of DCFH$^{\cdot}$ to DCF], depending on the extent of any side reactions, such as the addition of ˙OH to the benzoic acid substituent rather than addition/water elimination at the phenolic moiety. Thus the oxidation yield at pH 8 to 9 was somewhat lower than probably expected with radical based oxidation; the relative yields of reactions (14.3a) and (14.6a) have been controversial, but discussion of this issue is outside the scope of this chapter.

However, bicarbonate inhibited oxidation of $DCFH_2$ or RhH_2 by peroxynitrite at pH 7 (Glebska and Koppenol, 2003; Kooy et al., 1994), not immediately explained by a free radical mechanism since $CO_3^{\cdot-}$ certainly leads to dye oxidation with high efficiency (Wrona et al., 2005). A complex alternative reaction scheme has been suggested (Glebska and Koppenol, 2003), involving adduct formation between peroxynitrite and the reduced dyes, followed by protonation and dissociation of the protonated adduct to form DCF or Rh. The earlier study of the RhH_2/peroxynitrite system (Kooy et al., 1997) showed much less inhibition of Rh formation by bicarbonate, at fourfold higher concentrations of bicarbonate, than found by Glebska and Koppenol (2003). Indeed, in another study (Jourd'heuil et al., 1999), adding 25 mM bicarbonate *increased* both the rate and the amount of Rh formation on adding 20 μM peroxynitrite to 50 μM RhH_2 at pH 10.

There are three points worth making about these somewhat conflicting studies. First, the change in oxidation yield with pH between pH $\approx$6 to 8 is over a pH range where $DCFH_2$ has two dissociations of the phenolic moieties, with pK_a values suggested to be $\approx$7.9 and 9.2 (Wrona and Wardman, 2006). It is expected that oxidation of $DCFH_2$ by both $NO_2^{\cdot}$ and $CO_3^{\cdot-}$ would be much slower as the pH is decreased below $\approx$6; thus glycyltyrosine is oxidized >200-fold more rapidly by $NO_2^{\cdot}$ at pH 11.3 than at pH <7, and $CO_3^{\cdot-}$ oxidizes phenol at least an order of magnitude faster at pH $\approx$11 than $\approx$7 (Ross et al., 1998). Hence invoking prototropic dissociations of a complex between $DCFH_2$ or RhH_2 and peroxynitrite to explain pH-dependent yields (Glebska and Koppenol, 2003) is perhaps not necessary. The earlier study with RhH_2 (Kooy et al., 1994) found the yield of Rh to be $\approx$40% of the peroxynitrite added with 5 μM RhH_2 and almost pH independent between pH $\approx$4.2 and 8.3. A second, but related, factor is that the oxidizing radicals are formed in a solvent "cage" in reactions (14.3a) and (14.6a): separation/diffusion of the radical pairs to homogeneous solution and reaction with scavengers, in competition with recombination or "neutralization" such as reaction of $CO_3^{\cdot-}$ with $NO_2^{\cdot}$ to yield CO_2 and NO_3^-, will be more efficient the higher the scavenger concentration— or of course the higher the scavenger reactivity, as influenced, for example, by pH. Indeed, it was shown that both DCF and Rh production decreased significantly when $DCFH_2$ or RhH_2 was reduced below a few micromolar (Kooy et al., 1994, 1997). A third factor not considered is the consequence of superoxide radical formation in reaction (14.10) between $DCFH^{\cdot}$ and oxygen, or the corresponding reaction of $RhH^{\cdot}$. Any involvement of $O_2^{\cdot-}$

can be assessed by comparing results in air-equilibrated and/or O_2-saturated solutions with solutions that have been deaerated, although the accompanying change in DCF/Rh yields on switching formation from (14.10) to (14.14) should be noted. Superoxide is highly reactive toward both NO_2^- and CO_3^{--} (Ross et al., 1998), and its lifetime (with respect to disproportionation) is extended at high pH; it is not impossible that, especially with very low concentrations of $DCFH_2$ (for example) and at high pH, reactions of NO_2^- and/or CO_3^{--} with O_2^{--} might interfere and reduce the oxidation yield. This might be much less a factor at the concentrations of reduced probes used in applications in biology.

Further work is justified, especially with rather higher concentrations of reduced dyes, to resolve these issues. [While few measurements are available, it is quite possible that in many biological experiments in which cells are exposed to $DCFH_2$ diacetate, intracellular concentrations of the hydrolyzed probe are a few hundred micromolar (Wrona et al., 2005).] It may be necessary to avoid artifacts such as inner-filter effects using fluorescence detection, but the simple expedient of exciting and monitoring fluorescence at wavelengths higher than peak absorbance and emission wavelength might suffice. Dye solubility might be a limiting factor, especially at low pH, with mixed aqueous/organic solvents perhaps necessary.

3.2. Reaction of hydroxyl radicals

Reaction (14.12), of $\cdot$OH with $DCFH_2$, very probably occurs by a two-step mechanism in the main. Hydroxyl radicals seldom, if ever, react with organic compounds by direct electron transfer, as might CO_3^{--}, for example. Instead, the reaction seems likely to involve, as with phenol (Land and Ebert, 1967), the addition of $\cdot$OH (mainly ortho) followed by elimination of water to form a phenoxyl radical:

$$\cdot OH + DFCH_2 \rightarrow DCFH_2(OH)^{\cdot} \qquad (14.12a)$$

$$DCFH_2(OH)^{\cdot} \rightarrow DCFH^{\cdot} + H_2O. \qquad (14.12b)$$

However, the individual steps have not been characterized to date with any of these dyes. Indeed, the addition of $\cdot$OH to the benzene ring, as well as phenol/arylamine moieties, is a likely possibility, leading to products other than/additional to DCF or Rh. While a complete analysis of products of reaction of $\cdot$OH with reduced dyes has not been reported, measurements of relative fluorescence using conditions suitable for detecting DCF or Rh show $\cdot$OH radicals with $DCFH_2$ or RhH_2 to generate $\approx$60 to 70% of the fluorescent products DCF or Rh—or at least products fluorescing similarly—compared to CO_3^{--}/NO_2^- (Wrona et al., 2005).

Measurement of the kinetics of reaction of ˙OH with reduced dyes yields a single rate constant reflecting all possible reactions. Two main approaches are possible and both have been applied in this context. The first is to generate ˙OH specifically in a short time and monitor directly the loss of reduced dye, or the formation of the optical absorption characteristic of any of the intermediate/initial products, the majority of which is thought to be the radical DCFH˙ in the case of DCFH$_2$ (Fig. 14.1). Even if two or more products are formed on reaction with ˙OH, monitoring the formation of one yields a rate constant characterizing *the sum* of all possible reactions (Greenstock and Dunlop, 1973). (It is a common misconception that selecting a wavelength where only one product absorbs reflects the kinetics or partial rate constant for the formation of that product alone.)

One of the best ways to generate ˙OH rapidly for the study of fast reactions is by exposing aqueous solutions to a short pulse (e.g., <1 μs) of ionizing radiation, usually high-energy (typically 1–10 MeV) electrons from a linear accelerator or Van de Graaff generator. The general method-ology of this "pulse radiolysis" technique (Asmus, 1984; von Sonntag and Schuchmann, 1994) and some key applications have been outlined in a number of chapters in this series, and only brief details are given here. Applications of pulse radiolysis reviewed in this series include the study of sulfur radicals (Wardman and von Sonntag, 1995), nitric oxide-related reactions (Saran and Bors, 1994), oxygen radicals (Simic, 1990), electron transport in proteins (Salmon and Sykes, 1993), and characterizing redox properties of radicals (Forni and Willson, 1984). While radiolysis of aqueous solutions results in ionization to yield hydrated electrons (e_{aq}^-), ˙OH radicals, and H˙ atoms, saturation of solutions with nitrous oxide ($\approx$25 mM) rapidly scavenges e_{aq}^- [half-life $\approx$3 ns (Ross *et al.*, 1998)] to yield $\approx$90% ˙OH and only $\approx$10% H˙:

$$e_{aq}^- + N_2O + H^+ \rightarrow N_2 + ˙OH. \qquad (14.13)$$

With sensitive spectrophotometric detection and digitally averaging just a few signals, it is easy to monitor accurately absorbance changes of $\approx$0.005 over a few tens of microseconds in an optical cell of a 2-cm path length, and so if the extinction coefficient of the product being measured (or reagent being consumed) is, for example, $\approx$5 $\times$ 10^3 M^{-1} cm^{-1}, only $\sim$ 0.5 μM of radicals need be generated. If the concentration of reactive solute is (say) 10 to 50 μM, then two desirable conditions are met. First, the reactive solute is in quite large excess ($>$20-fold) over the initial concentration of ˙OH, so the kinetics of the reaction are close to exponential [sometimes called "pseudo" first order (Buxton, 1999; Capellos and Bielski, 1972)]. Second, the concentration of N$_2$O is much greater than ($>$50-fold) the reactive solute concentration, so that reactions of e_{aq}^- other than (14.13) can be ignored, as $k_{14.13}$ is diffusion controlled and not much smaller than the highest conceivable rate constant for the reaction of e_{aq}^- with $DCFH_2$ or RhH_2 (Ross et al., 1998). The $\approx 10\%$ of H' compared to 'OH generated in the system is usually ignored, but possible contributions from this minor product of water radiolysis should still be evaluated using available kinetic data.

There are some practical challenges to be considered in monitoring 'OH reactivity directly in this way, for example, by observing formation of the radical product DCFH' or RhH'. The first is that both radicals decay fairly rapidly by disproportionation to form the oxidized dye, and separation in time of formation and decay of, for example, DCFH' is desirable:

$$2\ DCFH' \rightarrow DCF + DCFH_2 \tag{14.14}$$

$$2\ RhH' \rightarrow Rh + RhH_2. \tag{14.15}$$

Reaction (14.14) has been studied in detail (Wrona and Wardman, 2006); the rate of disproportionation was pH dependent with decay faster at higher pH since, at least at ionic strength ≈ 0.1, the dissociated radical reacted faster than the protonated form. Defining the second-order rate constant $2k$ by $-d[R']/dt = 2k_{obs}[R']^2$, where $[R']$ is the sum of all prototropic forms of DCFH', gave a value of $2k_{14.14obs} \sim 3 \times 10^8\ M^{-1}\ s^{-1}$ at pH 7.4. If the initial concentration of 'OH and $DCFH_2$ is, for example, 0.5 and 10 μM, respectively, then the half-life of reaction (14.12) [actually, (14.12a)] is $\approx 7\ \mu s$ if $k_{14.12} \sim 1 \times 10^{10}\ M^{-1}\ s^{-1}$. The first half-life of reaction (14.14) ($= 1/(2k_{14.14}[DCFH'])$) is ≈ 7 ms if $[DCFH'] \approx 0.5\ \mu M$, a factor of ≈ 1000 longer than the timescale of formation in typical experiments. Hence, at least in the case of 'OH, direct observation of initial product formation seems facile.

The caveat is that elimination of water from the intermediate adduct radical, reaction (14.12b), might well complicate the issue. Such reactions have not been studied in the case of $DCFH_2$ or RhH_2, but some reliance may be placed on analogy with phenols (Getoff and Solar, 1986; Land and Ebert, 1967) or anilines (Qin et al., 1985; Singh et al., 2001), respectively. Taking 2-chlorophenol as a model for $DCFH_2$, it might be expected that water elimination (14.12b) might occur over tens of microseconds at pH 7, at least in the absence of catalytic buffer (Getoff and Solar, 1986). Reaction of 'OH with anilines also involves predominantly the addition of 'OH to the aromatic ring followed by elimination of water to form the anilino (N-centered) radical, the latter occurring on the microsecond timescale (Qin et al., 1985). The possibility of such complicating reactions can be examined by evaluating the required proportionality between the observed rate constant and the concentration of $DCFH_2$ or RhH_2. In the case of RhH_2, exponential formation of an absorbance at 380 nm ascribed to RhH' was used to derive an estimate of $k_{14.16} = 1.8 \times 10^{10}\ M^{-1}\ s^{-1}$ (Wrona et al., 2005):

$$^{\bullet}OH + RhH_2 \rightarrow \rightarrow H_2O + RhH^{\bullet}. \qquad (14.16)$$

The widely used competition method for measuring rate constants for radical reactions, although less direct, has the advantage that high concentrations of solutes can be used so that initial attack of the radical occurs in rather less than a microsecond and subsequent reactions of the product radical may be immaterial. In the case of $^{\bullet}OH$ radicals, thiocyanate is widely used as a reference solute, as reaction with $^{\bullet}OH$ yields a fairly intense chromophore [$(SCN)_2^{\bullet-}$, extinction coefficient $\varepsilon \approx 8 \times 10^3$ M^{-1} cm^{-1} at ≈ 475 nm]:

$$^{\bullet}OH + 2\ SCN^- + H^+ \rightarrow H_2O + (SCN)_2^{\bullet-}. \qquad (14.17)$$

It is easily shown, in the case of reaction (14.12) competing with (14.17), that if A_0 is the absorbance of $(SCN)_2^{\bullet-}$ in the absence of $DCFH_2$ and A that in the presence of DCFH, that

$$A_0/A = 1 + (k_{14.12}/k_{14.17})[DCFH_2]/[SCN^-],$$

and so a plot of A_0/A vs $[DCFH_2]/[SCN^-]$ should be linear and have intercept 1 and yields a slope equal to $k_{14.12}/k_{14.17}$. Measurements of absorbance can be made a few microseconds after generating $^{\bullet}OH$ with suitable solute concentrations, before complications arise from the formation of DCF or Rh via disproportionation reactions (14.14) or (14.15), with both oxidized dyes absorbing strongly at 475 nm.

There are three potential problems with this approach. One, which is minor, is that while the "reference" rate constant appears well established [a value of $k_{14.17} = 1.1 \times 10^{10}$ M^{-1} s^{-1}, from several independent studies, is used in the rate constant database (Ross et al., 1998)], another study suggests a slightly higher value of 1.4×10^{10} M^{-1} s^{-1} (Milosavljevic and Laverne, 2005). The second, potentially more serious, is that reaction (14.17) actually occurs in three steps:

$$^{\bullet}OH + SCN^- \rightarrow (HOSCN)^{\bullet-} \qquad (14.17a)$$

$$(HOSCN)^{\bullet-} \rightarrow {}^{\bullet}SCN + OH^- \qquad (14.17b)$$

$$^{\bullet}SCN + SCN^- \rightleftarrows (SCN)_2^{\bullet-}. \qquad (14.17c)$$

The possibility exists that the intermediates formed in reactions (14.17a) and (14.17b) could react with $DCFH_2$ or RhH_2. It is preferable to use a reference compound that produces a measurable chromophore in a single step to reduce the possibility of such confounding reactions (Buxton, 1999).

In practice, rate constants obtained using competition kinetics with thiocyanate usually agree fairly well with estimates using alternative reference compounds, or direct monitoring of products or loss of substrate (Ross et al., 1998). The third potential problem with the thiocyanate method, widely used because of the convenient chromophore $(SCN)_2^{\cdot-}$, is that this latter radical is itself oxidizing and will doubtless be reactive toward reductants such as $DCFH_2$ and RhH_2. However, such secondary reactions are usually much slower than reaction of $\cdot OH$ and easily separated in time-resolved observations.

Using the thiocyanate competition method, an estimate of $k_{14.12} = 1.3 \times 10^{10}$ M^{-1} s^{-1} was reported (Wrona et al., 2005).

In principle, time-resolved measurements of transient intermediates are not essential to apply the competition method to measure rate constants of $\cdot OH$ (and indeed other radicals). One might, for example, measure salicylate as the final, stable product of hydroxylation of benzoate by $\cdot OH$. However, to arrive at the final product the initial radical adduct has to be further oxidized, for example, by O_2 in quite complex reactions (Mvula et al., 2001), and there is the possibility of interference by the competing reagent in these secondary steps. Essentially the same objections apply to using the effects of competing substances to suggest which oxidant is responsible for the final product being measured. Hence the analysis of the reduction in, for example, fluorescence from DCF or Rh by competing antioxidants cannot be used alone to suggest which oxidant is responsible for fluorescence, as antioxidants can compete at both primary and secondary (radical intermediate) stages (Wardman, 2007).

3.3. Reaction of nitrogen dioxide radical

Nitrogen dioxide is easily produced for monitoring its reactions most directly using pulse radiolysis of N_2O-saturated aqueous solutions to generate $\cdot OH$, which yields $NO_2^{\cdot}$ on reaction with nitrite:

$$\cdot OH + NO_2^- \rightarrow OH^- + NO_2^{\cdot}. \qquad (14.18)$$

In this case, reactions of the dyes with $\cdot OH$ [(14.12) and (14.16)] are unwanted but this is accomplished easily by ensuring $[NO_2^-] \gg [DCFH_2]$, for example: $k_{14.18} \approx 9 \times 10^9$ M^{-1} s^{-1} (Ross et al., 1998), not much lower than $k_{14.12}$ or $k_{14.16}$. An upper limit for $[NO_2^-]$ is set by the need to avoid reaction of e_{aq}^- with NO_2^- rather than with N_2O [reaction (14.13)]; in practice $[NO_2^-] \approx 2$ to 5 mM is a reasonable compromise and 10 mM an upper limit, since N_2O is ~ 2.5-fold more reactive toward e_{aq}^- than NO_2^- (Ross et al., 1998) and $[N_2O] \approx 25$ mM.

The simplest method, in principle, to measure the rate constants of reaction (14.9) between $NO_2^{\bullet}$ and $DCFH_2$ and the analogous reaction with dihydrorhodamine,

$$NO_2^{\bullet} + RhH_2 \rightarrow NO_2^{-} + RhH^{\bullet} + H^{+} \tag{14.19}$$

is to monitor the formation of $DCFH^{\bullet}$ or $RhH^{\bullet}$ directly, for example, by the absorbance at ≈ 370 to 405 nm (depending on pH) or ≈ 380 nm, respectively (Wrona and Wardman, 2006; Wrona et al., 2005). The extinction coefficient of $NO_2^{\bullet}$ is too low to be of use [$\varepsilon \approx 200$ M^{-1} cm^{-1} at 400 nm (Schwartz and White, 1981)] and the absorbance of the radical products of oxidation much higher. In practice, the reactivity of $NO_2^{\bullet}$ is about three orders of magnitude lower than that of $^{\bullet}OH$, and rather lower than that of $CO_3^{\bullet-}$, so that direct monitoring formation of $DCFH^{\bullet}$ or $RhH^{\bullet}$ in (14.9) or (14.19) is complicated by the overlapping disproportionating reactions (14.14) or (14.15). Thus $k_{14.9} \approx 1.3 \times 10^7$ M^{-1} s^{-1} (Wrona et al., 2005) (see later) and if $[DCFH_2] = 100$ μM, then the half-life of formation of $DCFH^{\bullet}$ is ≈ 0.5 ms. Solubility restrictions and the desire to show proportionality of the rate constant to concentration mean that, in practice, lower $DCFH_2$ concentrations, and longer timescales of reaction, are desirable. As discussed earlier, the timescale of radical disproportionation (14.14) can then interfere with radical formation. This problem can be minimized by using as low a radical concentration (radiation dose) as possible. This also avoids potential competing reactions such as the dimerization of $NO_2^{\bullet}$ to N_2O_4 and subsequent reactions (see later).

Adding a competing scavenger to react with $NO_2^{\bullet}$ in the system has some advantages. First, the timescale of reaction is shortened by adding the second scavenger so competing radical disproportionation is reduced. Second, some scavengers have attractively intense chromophores extending beyond (extending to higher wavelengths than) the absorbances of DCF or Rh. A good choice is 2,2'-azino-bis-(3-ethylbenzothiazoline-6-sulfonate (ABTS^{2-}), usually as the diammonium salt:

$$NO_2^{\bullet} + ABTS^{2-} \rightarrow NO_2^{-} + ABTS^{\bullet-}, \tag{14.20}$$

where $k_{14.20} = 2.2 \times 10^7$ M^{-1} s^{-1} (Forni et al., 1986). The radical $ABTS^{\bullet-}$ is rather stable in the absence of reductants (Wolfenden and Willson, 1982) and has an intense absorption at 414 nm (Childs and Bardsley, 1975), but the less intense, secondary maximum at 728 nm has $\varepsilon = 1.5 \times 10^4$ M^{-1} s^{-1} (Scott et al., 1993). The choice of the longer wavelength for monitoring $ABTS^{\bullet-}$, while involving a lower extinction coefficient, offers the practical advantage of a complete absence of interference from absorptions of the reduced or oxidized dyes as well as the routine use of long wavelength band-pass optical filters (e.g., passing >650 nm), thus avoiding any possible complications of intense fluorescence influencing absorbance measurements. (It is usual in pulse radiolysis to place the monochromator after the spectrophotometric cell because of technical factors concerning Čerenkov light emission from samples irradiated with electrons initially near the speed of light *in vacuo*; band-pass filters, as well as a synchronized shutter between the light source and the cell, minimize photolytic effects of the analyzing light.)

Standard methods can be used to analyze data resulting from kinetic competition for $NO_2^{\cdot}$ reacting with, for example, $DCFH_2$, reaction (14.9), and $ABTS^{2-}$, reaction (14.20). Thus, as with hydroxyl radical reactions (see earlier discussion), at any time t (> 0) the absorbance (A) at a wavelength where only $ABTS^{\cdot-}$ absorbs is related to the absorbance of $ABTS^{\cdot-}$ in the absence of competing $DCFH_2$, A_0 by

$$A_0/A = 1 + (k_{14.9}/k_{14.20})[DCFH_2]/[ABTS^{2-}],$$

with the ratio $k_{14.9}/k_{14.20}$ being obtained from the linear plot of A_0/A and $k_{14.9}$ thus obtained using the value $k_{14.20} = 2.2 \times 10^7\ M^{-1}\ s^{-1}$ (Forni *et al.*, 1986). While in the case of reactions of $^{\cdot}OH$ and competition with SCN^-, solute concentrations and rate constants are such that the kinetics of formation of the chromophore are too fast for analysis except with equipment with nanosecond time resolution and correspondingly short radiation pulses; in the case of reactions of $NO_2^{\cdot}$, the timescales are such that the formation of $ABTS^{\cdot-}$ is readily observable over tens of microseconds or longer. The kinetics can then be analyzed to estimate $k_{14.9}$ independently. It is shown easily (e.g., Greenstock and Dunlop, 1973) that providing the initial concentrations are such that $[NO_2^{\cdot}]_0 \ll [DCFH_2]_0$ and $[ABTS^{2-}]_0$, the first-order rate constant k_{obs} characterizing the near-exponential formation of $ABTS^{\cdot-}$ is related to $k_{14.9}$ and $k_{14.20}$ by

$$k_{obs} \simeq k_{14.9}[DCFH_2] + k_{14.20}[ABTS^{2-}],$$

and a plot of k_{obs} vs $[ABTS^{2-}]/[DCFH_2]$ should be linear with slope $(k_{14.20}/k_{14.9})$ and intercept $k_{14.9}$.

In experiments of this type to determine $k_{14.9}$ (Wrona *et al.*, 2005), it was observed that the absorption of $ABTS^{\cdot-}$ decayed over milliseconds and interfered with extraction of k_{obs} from absorbance/time data. Providing this concomitant decay is somewhat slower than formation, then the slow decay of $ABTS^{\cdot-}$ can be approximated to an exponential, particularly if only a small fraction of the radicals decay on the timescale of interest:

$$ABTS^{\cdot-} \rightarrow \text{nonabsorbing product(s).} \qquad (14.21)$$

Integration of the appropriate differential equation characterizing the formation/decay of ABTS$^{\cdot-}$ yields the approximation:

$$A_t \simeq A_0 k'_{14.20}[\exp(-(k'_{14.20} + k'_{14.9})t)$$
$$-\exp(-k_{14.21}t)]/(k_{14.21} - k'_{14.20} - k'_{14.9}),$$

where A_t or A_0 is the absorbance of ABTS$^{\cdot-}$ at time t or in the absence of competing reactant, respectively, $k'_{14.9} = k_{14.9}[DCFH_2]_0$ and $k'_{14.20} = k_{14.20}[ABTS^{2-}]_0$. Estimates of $k'_{14.9}$ can be extracted by nonlinear least-squares fit as implemented in several software packages such as Origin (OriginLab Corp.) or SigmaPlot (Systat Software Inc.); the plot of $k'_{14.9}$ vs [DCFH$_2$] should be linear with slope $= k_{14.9}$. This method was used to estimate $k_{14.9} \approx 1.3 \times 10^7 \ M^{-1} \ s^{-1}$ at pH 7.5 (Wrona et al., 2005).

The algebraic treatment of this system is similar to the well-known sequential first-order formation and decay (Capellos and Bielski, 1972), and it is similarly easily shown that the maximum absorbance A_{max} reached in solutions containing ABTS$^{\cdot-}$ and DCFH$_2$ is given by

$$A_{max} \simeq A_0(k'_{14.20}/k_{14.21})\exp(-[k'_{14.20} + k'_{14.9}]t_{max}),$$

where t_{max} is the time at which the absorbance of ABTS$^{\cdot-}$ reaches a maximum. [This method was used in analysis of the oxidation of desfer-rioxamine by NO$_2^{\cdot}$ (Bartesaghi et al., 2004).]

3.4. Reaction of carbonate radical

The carbonate radical can be generated by radiolysis of N$_2$O-saturated solutions containing bicarbonate/carbonate but there are some constraints of pH since $k_{14.22a} \approx 8.5 \times 10^6 \ M^{-1} \ s^{-1}$ while $k_{14.22b} \approx 3.9 \times 10^8 \ M^{-1} \ s^{-1}$ (Ross et al., 1998):

$$^{\cdot}OH + HCO_3^- \rightarrow H_2O + CO_3^{\cdot-} \tag{14.22a}$$

$$^{\cdot}OH + CO_3^{2-} \rightarrow OH^- + CO_3^{\cdot-} \tag{14.22b}$$

While a high pH is an advantage in ensuring greater yields of CO$_3^{\cdot-}$ from $^{\cdot}$OH and avoiding an unwanted reaction of $^{\cdot}$OH with DCFH$_2$ or RhH$_2$, reactivity at pH ≈ 7.4 is often of interest and, as discussed earlier, reactivity of DCFH$_2$ is a function of pH around the physiological range because of prototropic equilibria (dissociation of the phenolic protons). Mixtures of N$_2$O and CO$_2$ as the saturating gas can be used to bring the pH near physiological values, but consideration of the concentration of CO$_2$ at equilibrium must be made since it reacts with e_{aq}^- with a rate constant $\approx 8 \times 10^9$ M^{-1} s^{-1} (Ross et al., 1998), forming the quite powerful reductant, CO_2^-.

An alternative method of generating CO_3^-, useful at pH values near neutral and not requiring radiolysis facilities, is by photolysis of the cobalt complex $Co(NH_3)_4CO_3^+$ (Cope and Hoffman, 1972). In early work (Chen and Hoffman, 1973), a flash photolysis apparatus with the then conventional long path length (22 cm) sample cell with a xenon flash tube was used. Laser sources are now standard, but specialized flash photolysis equipment often uses expensive, high-powered lasers, giving intense, subnanosecond pulses used in a repetitive "pump-probe" mode. However, in the present context, ultrafast processes are not a priority, and in the author's institute, Professor Borivoj Vojnovic and Dr. Simon Ameer-Beg have constructed a relatively low-cost apparatus dedicated to carbonate radical studies. The apparatus uses a frequency-trebled Nd-YAG laser delivering ≈ 8 mJ at 355 nm, with repetition rates up to 20 Hz for averaging signals, in conjunction with a detection laser (HeNe, 633 nm) and silicon photodiode, appropriate for monitoring CO_3^- from its absorption in this region (see later). The photolysis and monitoring laser beams follow the same in-line path through the sample, separated by a dichromatic filter.

Carbonate radical can be monitored directly from its absorbance at ≈ 600 nm ($\varepsilon \approx 2.2 \times 10^3$ M^{-1} cm^{-1}). Radical disproportionation:

$$2\ CO_3^{\cdot -} \rightarrow CO_2 + CO_4^{2-} \tag{14.23}$$

is conveniently slower than that of many other radicals [$2k_{14.23}$ around 4×10^7 M^{-1} s^{-1} depending on ionic strength (Ross et al., 1998)]. In a typical study (Wrona et al., 2005), radical concentrations of ≈ 2 μM were generated by pulse radiolysis using a dose of ≈ 3 Gy. These concentrations yield an absorbance of ≈ 0.01 in a 2-cm path cell, with a first half-life of decay of CO_3^- of ≈ 10 to 25 ms in the absence of other reactants. With ≈ 20 to 80 μM DCFH$_2$, an approximate exponential decay of CO_3^- over ≈ 0.2 to 0.6 ms was first order in [DCFH$_2$] and yielded an estimate of $k_{14.11} \approx 2.6 \times 10^8$ M^{-1} s^{-1} at pH ≈ 8.2. This is likely to be within a factor of two, or probably rather less, of the rate constant at pH ≈ 7.4 since DCFH$_2$ dissociates with pK_a values ≈ 7.9 and 9.2 (see earlier discussion) and the rate constant might decrease between pH ≈ 8.2 and 7.4 with less dissociated substrate at equilibrium. Monitoring CO_3^- radical decay at 600 nm, rather than DCFH$^\cdot$ production at ≈ 400 nm, was preferable even though disproportionation of DCFH$^\cdot$, reaction (14.14), occurred on a somewhat longer timescale than formation ($2k_{14.14} \approx 3.5 \times 10^8$ M^{-1} s^{-1} at pH ≈ 8.2, first half-life of reaction (14.14) ≈ 3 ms at a radical concentration of ≈ 1 μM). Similar experiments with RhH$_2$ yielded an estimate of the rate constant for reaction of CO_3^- of $\approx 7 \times 10^8$ M^{-1} s^{-1} (Wrona et al., 2005).

4. Sources of Error and Scope for Improvements in Methodology

From the aforementioned discussion, it is apparent that a significant complication in the determination of rate constants is frequently the occurrence in parallel of unwanted reactions. These are mainly radical–radical reactions, for example, disproportionation of radicals produced on reaction with the oxidizing radicals, but others should not be ignored. Kinetic and equilibrium data for the reactive dissolution of $NO_2^{\cdot}$:

$$2\,NO_2^{\cdot} + H_2O \rightarrow NO_2^- + NO_3^- + 2\,H^+ \qquad (14.24)$$

and the associated equilibrium:

$$2\,NO_2^{\cdot} \rightleftharpoons N_2O_4 \qquad (14.25)$$

are available (Ross *et al.*, 1998; Schwartz and White, 1981). Such unwanted, competing reactions are minimized by using as low radical concentrations as possible, and their contributions in part allowed for the usual plots of first-order rate constant vs concentration, with the intercept (apparent rate constant at zero concentration of $DCFH_2$, etc.) reflecting hopefully minor contributions of reactions (14.24) and (14.25), or carbonate radical disproportionation, etc. If the intercept is significant compared to the observed first-order rate constant at the lowest substrate concentration, then it implies the standard exponential fit is in any case inappropriate, and fitting to a more complex reaction scheme is justified. This is often facile, with nonlinear least-squares (NLLS) fitting readily available in several commercially available plotting programs. In more complex reaction scenarios, extraction of rate constants is often possible using, for example, Gear's algorithm as implemented in FACSIMILE (ESM Software), particularly appropriate for "stiff" differential equations (where some reactions involve much more rapid change than others); a good alternative is KINFITSIM (Svir *et al.*, 2002). Freely available programs such as Gepasi (http://www.gepasi.org/) are quite widely used for kinetic simulations; the simple Euler integration is practical for simulations and a suitable program is also freely available (Lancaster, 2006).

Often the contributions of unwanted side reactions can be approximated without recourse to complex fitting routines. Examples of approximations to algebraically simple parallel/sequential reaction schemes involving exponential steps were described earlier, requiring only high school calculus and easily implemented using NLLS in Origin or Sigmaplot. More complicated differential equations are sometimes amenable to iterative routines such as

"Solver" in Microsoft Excel. In one such example (Ford *et al.*, 2002), application of this standard plug-in in Excel yielded iteratively the identical solution to the single real root of the relevant equation derived using the more powerful Mathcad software (Parametric Technology Corporation).

While corrections to estimates of rate constants to allow for side reactions can thus often be made quite easily, it seems likely that the uncertainties quoted in publications, usually reflecting merely the standard errors of the linear fits of estimated first-order rate constants vs concentration, are frequently underestimates of the true uncertainties and mask systematic errors. An effort to improve absolute detection sensitivity and stability of light sources in pulse radiolysis and stopped-flow instrumentation is clearly needed. The performance of most "microsecond" pulse radiolysis facilities has hardly changed since the late 1960s, aside from the benefits of digital capture and signal averaging. The application of currently available light sources and detectors should yield dividends in the ability to work with much lower radical concentrations and thus extend downward the range of rate constants that can be determined, as well as eliminating the need for more complex kinetic fits.

5. CONCLUSIONS

The challenges of understanding peroxynitrite reactions have stimulated a great deal of controversy among scientists interested in mechanistic aspects, including the kinetics, of its reactions with biomolecules. There is a need for further kinetic studies to improve understanding. The methodology appropriate to such studies has been outlined in this chapter, and attention has been drawn to areas of particular uncertainty or meriting further development. The study of reaction kinetics is a powerful tool to help distinguish mechanistic hypotheses; it merits greater awareness.

ACKNOWLEDGMENTS

This work was supported by Cancer Research UK. I thank Professor B. Vojnovic and Dr. M. Wrona for helpful discussions.

REFERENCES

Adams, G. E., Aldrich, J. E., Bisby, R. H., Cundall, R. B., Redpath, J. L., and Willson, R. L. (1972). Selective free radical reactions with proteins and enzymes: Reactions of inorganic radical anions with amino acids. *Radiat. Res.* **49,** 278–289.

Asmus, K.-D (1984). Pulse radiolysis methodology. *Methods Enzymol.* **105,** 167–178.

Bartesaghi, S., Trujillo, M., Denicola, A., Folkes, L., Wardman, P., and Radi, R. (2004). Reactions of desferrioxamine with peroxynitrite-derived carbonate and nitrogen dioxide radicals. *Free Radic. Biol. Med.* **36,** 471–483.

Buxton, G. V. (1999). Measurement of rate constants for radical reactions in the liquid phase. *In* "General Aspects of the Chemistry of Radicals" (Z. B Alfassi, ed.), pp. 51–77. Wiley, New York.

Capellos, C., and Bielski, B. H. J. (1972). "Kinetic Systems: Mathematical Description of Chemical Kinetics in Solution." Wiley, New York.

Chatton, J.-Y., and Broillet, M.-C. (2002). Detection of nitric oxide production by fluorescent indicators. *Methods Enzymol.* **359,** 134–148.

Chen, S. N., and Hoffman, M. Z. (1973). Rate constants for the reaction of the carbonate radical with compounds of biochemical interest in neutral aqueous solution. *Radiat. Res.* **56,** 40–47.

Childs, R. E., and Bardsley, W. G. (1975). The steady-state kinetics of peroxidase with 2,2′-azino-di-(3-ethylbenzthiazoline-6-sulphonic acid) as chromogen. *Biochem. J.* **145,** 93–103.

Cope, V. W., and Hoffman, M. Z. (1972). Photochemical generation of CO_3^- radicals in neutral aqueous solution. *J. Chem. Soc. Chem. Commun.* 227–228.

Denicola, A., Freeman, B. A., Trujillo, M., and Radi, R. (1996a). Peroxynitrite reaction with carbon dioxide/bicarbonate: Kinetics and influence on peroxynitrite-mediated oxidations. *Arch. Biochem. Biophys.* **333,** 49–58.

Denicola, A., Souza, J. M., Radi, R., and Lissi, E. (1996b). Nitric oxide diffusion in membranes determined by fluorescence quenching. *Arch. Biochem. Biophys.* **328,** 208–212.

Ford, E., Hughes, M. N., and Wardman, P. (2002). Kinetics of the reactions of nitrogen dioxide with glutathione, cysteine, and uric acid at physiological pH. *Free Radic. Biol. Med.* **32,** 1314–1323.

Forni, L. G., Mora-Arellano, V. O., Packer, J. E., and Willson, R. L. (1986). Nitrogen dioxide and related free radicals: Electron-transfer reactions with organic compounds in solutions containing nitrite or nitrate. *J. Chem. Soc. Perkin Trans.* **2,** 1–6.

Forni, L. G., and Willson, R. L. (1984). Electron and hydrogen atom transfer reactions: Determination of free radical redox potentials by pulse radiolysis. *Methods Enzymol.* **105,** 179–188.

Getoff, N., and Solar, S. (1986). Radiolysis and pulse radiolysis of chlorinated phenols in aqueous solution. *Radiat. Phys. Chem.* **28,** 443–450.

Glebska, J., and Koppenol, W. H. (2003). Peroxynitrite-mediated oxidation of dichlorodihydrofluorescein and dihydrorhodamine. *Free Radic. Biol. Med.* **35,** 676–682.

Goldstein, S., Czapski, G., Lind, J., and Merényi, G. (2001). Carbonate radical ion is the only observable intermediate in the reaction of peroxynitrite with CO_2. *Chem. Res. Toxicol.* **14,** 1273–1276.

Greenstock, C. L., and Dunlop, I. (1973). Kinetics of competing free radical reactions with nitroaromatic compounds. *J. Am. Chem. Soc.* **95,** 6917–6919.

Ischiropoulos, H., Gow, A., Thom, S. R., Kooy, N. E., Royall, J. A., and Crow, J. P. (1999). Detection of reactive nitrogen species using 2,7-dichlorodihydrofluorescein and dihydrorhodamine 123. *Methods Enzymol.* **301,** 367–373.

Itoh, Y., Ma, F. H., Hoshi, H., Oka, M., Noda, K., Ukai, Y., Kojima, H., Nagano, T., and Toda, N. (2000). Determination and bioimaging method for nitric oxide in biological specimens by diaminofluorescein fluorometry. *Anal. Chem.* **287,** 203–209.

Jourd'heuil, D., Jourd'heuil, F. L., Kutchukian, P. S., Musah, R. A., Wink, D. A., and Grisham, M. B. (2001). Reaction of superoxide and nitric oxide with peroxynitrite: Implications for peroxynitrite-mediated oxidation reactions *in vivo*. *J. Biol. Chem.* **276,** 28799–28805.

Jourd'heuil, D., Miranda, K. M., Kim, S. M., Espey, M. G., Vodovotz, Y., Laroux, S., Mai, C. T., Miles, A. M., Grisham, M. B., and Wink, D. A. (1999). The oxidative and nitrosative chemistry of the nitric oxide/superoxide reaction in the presence of bicarbonate. *Arch. Biochem. Biophys.* **365**.

Kooy, N. W., Royall, J. A., and Ischiropoulos, H. (1997). Oxidation of 2',7'-dichlorofluorescin by peroxynitrite. *Free Radic. Res.* **27**, 245–254.

Kooy, N. W., Royall, J. A., Ischiropoulos, H., and Beckman, J. S. (1994). Peroxynitrite-mediated oxidation of dihydrorhodamine 123. *Free Radic. Biol. Med.* **16**, 149–156.

Lancaster, J. R., Jr. (2006). Nitroxidative, nitrosative, and nitrative stress: Kinetic predictions of reactive nitrogen species chemistry under biological conditions. *Chem. Res. Toxicol.* **19**, 1160–1174.

Land, E. J., and Ebert, M. (1967). Pulse radiolysis studies of aqueous phenol. *Trans. Faraday Soc.* **63**, 1181–1190.

Lymar, S. V., and Hurst, J. K. (1995). Rapid reaction between peroxynitrite ion and carbon dioxide: Implications for biological activity. *J. Am. Chem. Soc.* **117**, 8867–8868.

Milosavljevic, B. H., and Laverne, J. A. (2005). Pulse radiolysis of aqueous thiocyanate solution. *J. Phys. Chem. A* **109**, 165–168.

Mvula, E., Schuchmann, M. N., and von Sonntag, C. (2001). Reactions of phenol-OH-adduct radicals: Phenoxyl radical formation by water elimination vs oxidation by dioxygen. *J. Chem. Soc. Perkin Trans.* **2**, 264–268.

Qin, L., Tripathi, G. N. R., and Schuler, R. H. (1985). Radiation chemical studies of the oxidation of aniline in aqueous solution. *Z. Naturforsch. A* **40**, 1026–1039.

Radi, R. (1996). Kinetic analysis of reactivity of peroxynitrite with biomolecules. *Methods Enzymol.* **269**, 354–366.

Radi, R., Beckman, J. S., Bush, K. M., and Freeman, B. A. (1991). Peroxynitrite oxidation of sulfhydryls: The cystostatic potential of superoxide and nitric oxide. *J. Biol. Chem.* **266**, 4244–4250.

Radi, R., Denicola, A., Alvarez, B., Ferrer-Sueta, G., and Rubbo, H. (2000). The biological chemistry of peroxynitrite. *In* "Nitric Oxide Biology and Pathobiology" (L. J. Ignorro, ed.), pp. 57–82. Academic Press, San Diego.

Ross, A. B., Mallard, W. G., Helman, W. P., Buxton, G. V., Huie, R. E., and Neta, P. (1998). "NDRL-NIST Solution Kinetics Database: Ver. 3." Notre Dame Radiation Laboratory and National Institute of Standards and Technology, Notre Dame, Indiana and Gaithersburg, Maryland (see also) http://www.rcdc.nd.edu/.

Salmon, G. A., and Sykes, A. G. (1993). Pulse radiolysis. *Methods Enzymol.* **227**, 522–534.

Saran, M., and Bors, W. (1994). Pulse radiolysis for investigation of nitric oxide-related reactions. *Methods Enzymol.* **233**, 20–34.

Schwartz, S. E., and White, W. H. (1981). Solubility equilibria of the nitrogen oxides and oxyacids in dilute aqueous solution. *In* "Advances in Environmental Science and Engineering" (J. R. Pfafflin and E. N. Ziegler, eds.), pp. 1–45. Gordon and Breach, New York.

Scott, S. L., Chen, W.-J., Bakac, A., and Espenson, J. H. (1993). Spectroscopic parameters, electrode potentials, acid ionization constants, and electron exchange rates of the 2,2'-azinobis(3-ethylbenzothiazoline-6-sulfonate) radicals and ions. *J. Phys. Chem.* **97**, 6710–6714.

Setsukinai, K.-I., Urano, Y., Kakinuma, K., Majima, H. J., and Nagano, T. (2003). Development of novel fluorescent probes that can reliably detect reactive oxygen species and distinguish specific species. *J. Biol. Chem.* **278**, 2170–3175.

Simic, M. G. (1990). Pulse radiolysis studies of oxygen radicals. *Methods Enzymol.* **186**, 89–100.

Singh, T. S., Gejji, S. P., Rao, B. S. M., Mohan, H., and Mittal, J. P. (2001). Radiation chemical oxidation of aniline derivatives. *J. Chem. Soc. Perkin Trans.* **2**, 1205–1211.

Svir, I. B., Klymenko, O. V., and Platz, M. S. (2002). 'KINFITSIM'—a software to fit kinetic data to a user selected mechanism. *Comput. Chem.* **26,** 379–386.

von Sonntag, C., and Schuchmann, H.-P (1994). Pulse radiolysis. *Methods Enzymol.* **233,** 3–20.

Wardman, P. (2007). Fluorescent and luminescent probes for measurement of oxidative and nitrosative species in cells and tissues: Progress, pitfalls, and prospects. *Free Radic. Biol. Med.* **43,** 995–1022.

Wardman, P., and von Sonntag, C. (1995). Kinetic factors that control the fate of thiyl radicals in cells. *Methods Enzymol.* **251,** 31–45.

Wolfenden, B. S., and Willson, R. L. (1982). Radical cations as reference chromogens in kinetic studies of one-electron transfer reactions: Pulse radiolysis studies of 2,2′-azinobis-(3-ethylbenzthiazoline-6-sulphonate). *J. Chem. Soc. Perkin Trans.* **2,** 805–812.

Wrona, M., Patel, K. B., and Wardman, P. (2005). Reactivity of 2′,7′-dichlorodihydro-fluorescein and dihydrorhodamine 123 and their oxidized forms towards carbonate, nitrogen dioxide, and hydroxyl radicals. *Free Radic. Biol. Med.* **38,** 262–270.

Wrona, M., and Wardman, P. (2006). Properties of the radical intermediate obtained on oxidation of 2′,7′-dichlorodihydrofluorescein, a probe for oxidative stress. *Free Radic. Biol. Med.* **41,** 657–667.

CHAPTER FIFTEEN

DETECTION AND CHARACTERIZATION OF PEROXYNITRITE-INDUCED MODIFICATIONS OF TYROSINE, TRYPTOPHAN, AND METHIONINE RESIDUES BY TANDEM MASS SPECTROMETRY

Igor Rebrin, Catherine Bregere, Timothy K. Gallaher, *and* Rajindar S. Sohal

Contents

Abstract

Nitration and oxidation of tyrosine, tryptophan, and methionine residues in proteins are potential markers of their interaction with peroxynitrite. This chapter describes the procedure for the detection of these nitro-oxidative

Department of Pharmacology and Pharmaceutical Sciences, University of Southern California, Los Angeles, California

Methods in Enzymology, Volume 441
ISSN 0076-6879, DOI: 10.1016/S0076-6879(08)01215-9

modifications by tandem mass spectrometry. The peptide YGDLANWMIPGK, shown to contain a nitrohydroxytryptophan in the mitochondrial enzyme succinyl-CoA:3-ketoacid coenzyme A transferase (SCOT) *in vivo*, was synthesized and exposed to peroxynitrite in order to test whether an identical tryptophan derivative could be generated *in vitro*. Data show that the occurrence of specific fragmented ions corresponding to the oxidation of methionine, nitration of tyrosine, and nitration/oxidation of tryptophan residues can be used to identify the sites of the nitration and oxidation of proteins *in vitro* and *in vivo*. It is also demonstrated that a nitrohydroxy addition to the tryptophan, similar to that present in SCOT *in vivo*, can be produced *in vitro*.

1. INTRODUCTION

Several types of posttranslational modifications, induced by nitric oxide-derived reactive intermediates, can target specific amino acid residues in proteins (Beckman, 1996; Beckman *et al.*, 1990; Yamakura and Ikeda, 2006). *In vitro* exposure of proteins, peptides, or amino acids to nitrating agents has been widely demonstrated to cause additions of a nitro group to the benzyl ring of aromatic amino acids and a hydroxyl group to cysteine, methionine, and tryptophan residues (Alvarez and Radi, 2003; Yamakura *et al.*, 2005). Protein nitration can also occur *in situ* in cells under conditions of oxidative stress and inflammation, ostensibly by interaction with peroxynitrite ($ONOO^-$), a potent nitrating and oxidizing agent produced by a reaction between superoxide anion and nitric oxide (Beckman, 1990). The presence of peroxynitrite in mitochondria is considered plausible due to the activity of a hypothetical isoform of nitric oxide synthase and a high rate of $O_2^{-\cdot}$ generation in this organelle (Elfering *et al.*, 2002; Radi, 2004). The *in vitro* as well as *in vivo* detection of tyrosine nitration in proteins has been facilitated greatly by the availability of specific anti-3NT antibodies (Ye *et al.*, 1996). In contrast, a similar detection of nitrated tryptophan residues remained elusive until the recent development of an appropriate antibody (Ikeda *et al.*, 2007). Accumulation of tyrosine-nitrated proteins has been observed in association with several pathological conditions (Ischiropoulos, 1998; Turko and Murad, 2002), as well as during the normal aging process (Kanski *et al.*, 2003, 2005a,b; Sharov *et al.*, 2006; Viner *et al.*, 1999). Thus, the identification of specific protein targets and the individual residues that are modified is desirable for establishing the relationship between oxidation/nitration and the functional effect on the protein.

The application of mass spectrometric (MS) methods has been a highly useful tool for the identification of structural modifications in proteins due to exposure to natural or artificial reactive agents. In particular, matrix-assisted laser desorption/ionization (MALDI) in combination with time-of-flight (TOF) and electrospray ionization mass spectrometry (ESI-MS) have been increasingly applied to investigate protein nitration and oxidation (Nielson and Pennington, 1995; Petersson *et al.*, 2001; Sarver *et al.*, 2001; Turko and Murad, 2005).

Using a combination of methods (Western blot with anti-3-nitrotyrosine monoclonal antibody, HPLC electrochemical detection, MALDI-TOF, and ESI-MS), we demonstrated a novel *in vivo* posttranslational modification in a mitochondrial protein in rat tissues, involving simultaneous additions of nitro and hydroxy groups to tryptophan 372 in succinyl-CoA:3-ketoacid coenzyme A transferase (SCOT) during aging (Rebrin *et al.*, 2007). Thus the application of MS methods for the detection of distinctive modifications (nitro and hydroxy additions) of methionine, tyrosine, and tryptophan residues in proteins can facilitate the identification of cellular targets of oxidative/nitrative stress. This chapter describes procedures used for the (i) synthesis of a YGDLANWMIPGK peptide (tryptic peptide of SCOT, corresponding to amino acids 366–377, shown previously to contain nitro-hydroxytryptophan 372 residue *in vivo*); (ii) nitro-oxidative modification of this peptide by exposure to peroxynitrite; and (iii) tandem mass spectrometry for determination of the specific fragmentation patterns of this synthetic peptide, which are a characteristic of the peroxynitrite-induced modifications of tyrosine, tryptophan, and methionine residues. Such methods can be adapted easily for other proteins.

2. Experimental Procedures

2.1. Synthesis of SCOT peptide

Peptide synthesis should be carried out on a Rainin Symphony peptide synthesizer or some similar instrumentation capable of automated solid-phase peptide synthesis. Using fast-Fmoc chemistry, and starting from the carboxy-terminal residue, the amino acids will undergo deprotection/coupling cycles. The resultant peptide can be cleaved from the resin and deprotected using trifluoroacetic acid. The molecular weight of the neutral monoisotopic peptide YGDLANWMIPGK (1363.66) should be confirmed by MALDI-TOF spectrometry. The peptide is purified and desalted using a preparative-scale reversed-phase HPLC column and the purity confirmed by analytical HPLC analysis. The lyophilized peptide sample may be stored in a dark vial in an oxygen-free environment at −20 °C.

2.2. Nitration by peroxynitrite

Peroxynitrite can be procured from Upstate Biotechnology (Lake Placid, NY) or synthesized from sodium nitrite and hydrogen peroxide in the presence of an acid, according to the procedure described elsewhere

(Koppenol *et al.*, 1996). Peroxynitrite concentration in dilutions of the stock solution made with 0.3 N NaOH can be determined by measurement of absorbance at 303 nm ($\varepsilon = 1670\ M^{-1}cm^{-1}$). To achieve efficient oxidation and nitration of different amino acid residues in the peptide, use of a 10-fold molar excess of peroxynitrite is recommended. Dissolve 5 mg peptide in 100 μl of 100 mM peroxynitrite while stirring the reaction mixture vigorously and incubate at 4 °C. The time course of reaction can be determined in aliquots of 10 μl every hour of the reaction, diluted 1000-fold in 0.1% formic acid in water, and analyzed immediately or stored at −80 °C in small aliquots in order to avoid repetitive freezing/thawing.

2.3. HPLC of the peptide

Following peroxynitrite treatment of the peptide YGDLANWMIPGK, products of the reaction can be separated by the BioBasic 18 reversed-phase capillary column (ThermoFinnigan, San Jose, CA, 100 × 0.18 mm) using a ThermoFinnigan Surveyor MS pump. Equilibrate the column for 5 min at a flow rate of 1.5 μl/min with 95% solution A and 5% solution B (A, 0.1% formic acid in water; B, 0.1% formic acid in acetonitrile). The sample (1–10 μl) can be injected via an autosample injector (Surveyor, ThermoFinnigan). After injecting the sample, hold solution A at 95% and solution B at 5%, followed by a linear gradient of 5 to 65% solution B over 45 min. Increase solution B to 80% over the subsequent 5 min and hold at 80% for an additional 5 min, after which the column should be reequilibrated back to 5% of solution B. Because the peptide YGDLANWMIPGK has no cysteine residues, there is no need for alkylation; however, if the source of the peptides is tryptic digests of proteins or synthetic peptides containing cysteine residues, an additional alkylation step should be carried out prior to injection of the samples on the column, as described previously (Yarian *et al.*, 2005).

2.4. MS/MS spectrometry

Mass analysis can be performed using a ThermoFinnigan LCQ Deca XP Plus ion trap mass spectrometer equipped with a nanospray ion source (ThermoFinnigan) employing a 4-cm metal emitter (Proxeon, Odense, Denmark). Electrical contact and voltage application to the probe tip take place via the nanoprobe assembly. Spray voltage of the mass spectrometer is set to 2.9 kV and heated capillary temperature at 190°. Acquire mass spectra in the m/z 400 to 1800 range. A data-dependent acquisition mode should be used where each of the top five ions for a given scan is subjected to MS/MS analysis.

2.5. Data analysis

Protein identification may be carried out with the MS/MS search software Mascot (Matrix Science), with confirmatory or complementary analyses by TurboSequest (Bioworks Browser 3.2, build 41, from ThermoFinnigan and Sonar MS/MS from Genomics Solutions). Alkylated modification (carbamidomethyl) should be designated as fixed; in contrast, oxidation of methionine and tryptophan, as well as nitration of tyrosine and tryptophan, should be considered as variable modifications in the TurboSequest search. The NCBI rat genome database server complemented with the NCBI nonredundant protein database can be used for the search. Theoretical m/z values for the peptides and their fragmentation ions can be assessed using the "MS/MS fragment ion calculator" from The Institute for Systems Biology at http://db.systemsbiology.net:8080/proteomicsToolkit/. Acceptable cross-correlation scores (X_{corr}) for positive identification of a modification (doubly charged ions) within the peptide should be greater than two. It is recommended that MS/MS spectra of particular interest should be inspected manually.

3. ANTICIPATED RESULTS

3.1. Synthesis, purification, and MS analysis of YGDLANWMIPGK peptide

A homogeneous product should be obtained after solid-phase peptide synthesis and the molecular weight confirmed using MALDI-TOF analysis. The observed m/z value of the singly protonated peptide $[M+H]^+$ should correspond to the theoretical one, within deviation of <0.2 to 0.4 m/z unit (observed $[M+H]^+$ of the peptide YGDLANWMIPGK at m/z 1365.0, calculated $[M+H]^+ = 1364.67$). The purity of the synthetic peptide, indicated by HPLC analysis, should be $>95\%$. After exposure of the synthetic peptide to peroxynitrite, HPLC separation of the reaction products should yield several peaks: peptides with oxidized methionine and nitrohydroxytryptophan should elute at earlier retention times, whereas nitrotyrosine-containing peptides should be retained longer than the unmodified peptide.

Electrospray ionization mass spectra of the reaction products would indicate synthetic peptide YGDLANWMIPGK to be either unmodified (doubly charged precursor ion 682.8) or modified by +16 (oxidation) at methionine (doubly charged precursor ion 690.8) or +45 (nitration) at tyrosine (doubly charged precursor ion 705.3) or +61 (oxidation and nitration) at tryptophan (doubly charged precursor ion 713.3) and +77 addition, corresponding to the oxidation of methionine combined with nitrohydroxytryptophan (doubly charged precursor ion 721.3). The theoretical m/z values of ions, expected from fragmentation of the synthetic peptide YGDLANWMIPGK, are indicated in Table 15.1. MS/MS spectra of the synthetic peptide YGDLANWMIPGK should indicate the distinctive fragmentation pattern, that is, relative dominance of the y3 and b9, but not of y4 or y5 ion peaks. Such fragmentation should be essentially identical to that obtained previously, using tryptic peptide from SCOT, purified from rat heart mitochondria (Rebrin *et al.*, 2007). The occurrence of the indicated fragmentation pattern may be because of the presence of the proline residue (dominant peaks y3 and b9), which can cause suppression and/or disappearance of some of the fragments (including y4 and y5), as also shown in similar proline-containing peptides (Paizs and Suhai, 2005).

3.2. MS/MS analysis of unmodified peptide

The MS/MS spectrum of the unmodified peptide YGDLANWMIPGK, as displayed in Fig. 15.1, should indicate the presence of 8 y and 9 b out of the potential 11 ions. The y3 and b9 ions at m/z 301.2 and 1064.2, respectively, are predominant, while the intensity of other ions is considerably lower. Such a fragmentation profile is virtually identical to the one reported for the native SCOT peptide (Rebrin *et al.*, 2007).

3.3. MS/MS analysis of methionine oxidation

Exposure of peptide to peroxynitrite leads to the nearly ubiquitous presence of methionine sulfoxide containing peptides. The corresponding MS/MS spectrum (single charged precursor ion with m/z $[M+H+16]^+ = 1380.7$ and

Table 15.1 Theoretical m/z values for fragmented ions produced by high collision-induced dissociation of doubly protonated unmodified peptide (theoretical $[M+2H]^{2+}$ at m/z 682.84)

		B ions	Y ions	
Y	1	164.07	—	
G	2	221.09	1201.6	11
D	3	336.12	1144.58	10
L	4	449.2	1029.55	9
A	5	520.24	916.47	8
N	6	634.28	845.43	7
W	7	820.36	731.39	6
M	8	951.4	545.31	5
I	9	1064.49	414.27	4
P	10	1161.54	301.19	3
G	11	1218.56	204.13	2
K		—	147.11	1

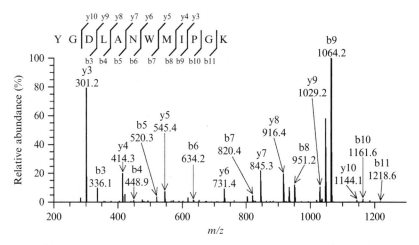

Figure 15.1 Tandem mass spectrum of the untreated synthetic peptide YGDLANW-MIPGK ([M+2H]$^{2+}$ ion at m/z 683.4) displaying a distinctive fragmentation pattern, characterized by the dominance of two ions, y3 and b9. Such a profile was interpreted to be due to the "proline effect," which favors fragmentation on its amino terminal bond and suppresses other ions. For all spectra, ion annotation is based on results presented in the Sequest's "display ion view" window, corroborated by the dta file.

cross-correlation score X_{corr} = 4.3 in TurboSequest analysis), represented in Fig. 15.2, should indicate the presence of 10 y and 9 b out of the potential 11 ions. The presence of methionine sulfoxide dramatically alters the pattern of fragmentation of the synthetic peptide. The ions y5 (m/z 561.4), y6 (m/z 747.3), y7 (m/z 861.4), and y8 (m/z 932.3) become particularly abundant, whereas they are masked in the untreated peptide. The detection of oxidized methionine should also be facilitated by the presence of additional y and/or b ion fragments, which are 64 Da lower in mass than the methionine sulfoxide ion. Such a decrement in mass corresponds to the neutral loss of methane-sulfenic acid (CH3SOH, 64 Da) from the side chain of methionine sulfoxide. The corresponding y5-64 ion at m/z 497.4 is shown in Fig. 15.2. Indeed, this characteristic loss of the 64 Da has been suggested as a valuable "fingerprint" for the detection of methionine oxidation in proteins (Guan *et al.*, 2003).

3.4. MS/MS analysis of tyrosine nitration

Several peptides containing an additional mass of 45 Da, indicative of the presence of a nitro group, can be detected in mass spectra. The program TurboSequest should unambiguously assign the +45 addition on the tyrosine residue of the peptide (singly charged precursor ion with m/z [M+H+45]$^+$=1411.1 and cross-correlation score X_{corr} = 3.83 in Turbo-Sequest analysis). The MS/MS spectrum, shown in Fig. 15.3, indicates the

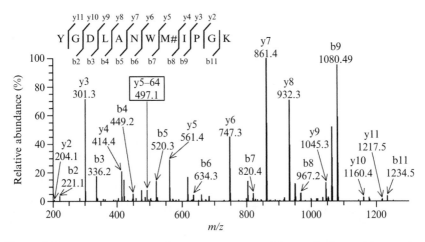

Figure 15.2 Representative tandem mass spectrum of the synthetic peptide containing methionine sulfoxide ([M+2H]$^{2+}$ ion at *m/z* 690.62). Multiple peptides containing methionine sulfoxide are obtained following peroxynitrite exposure. The presence of oxidized methionine alters the fragmentation profile of the peptide: the ions y5, y6, y7, and y8 become quite abundant. The low abundant ion at *m/z* 497.1, corresponding to a neutral mass loss of 64 Da from the y5 ion, constitutes a fingerprint of methionine sulfoxide and is indicated as y5–64. Such a mass decrement is known to be because of the elimination of methanesulfenic acid from the side chain of methionine sulfoxide. M#, methionine sulfoxide.

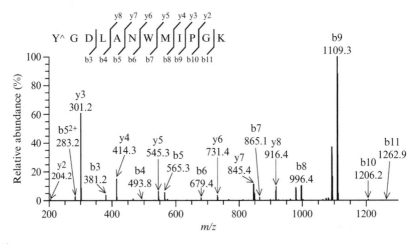

Figure 15.3 Representative tandem mass spectrum of a nitrotyrosine containing peptide ([M+2H]$^{2+}$ ion at *m/z* 705.49). Following exposure to peroxynitrite, a +45-Da mass increment of the peptide is detected and can be attributed to the presence of a nitro group on either tyrosine or tryptophan residues. The fragmented ions unambiguously indicate the addition of the nitro group to the tyrosine residue, as inferred from the presence of the ions b3, b4, b5, and b6 at *m/z* 381.2, 493.8, 565.3, and 679.4, respectively, and the ions y6, y7, and y8 at *m/z* 731.4, 845.4, and 916.4, respectively. Y^, nitrated tyrosine.

presence of 7 y and 9 b out of the potential 11 ions. The ions b3, b4, b5, and b6 at m/z 381.2, 493.8, 565.3, and 679.4, respectively, confirm the addition of a nitro group to the tyrosine residue, whereas the ions y6 ($m/z = 731.4$), y7 ($m/z = 845.4$), and y8 ($m/z = 916.4$) indicate that the tryptophan residue is not modified.

3.5. MS/MS analysis of tryptophan nitration and oxidation

Peptides, containing tryptophan with an additional +61 mass (nitro and hydroxy additions) in association with methionine sulfoxide (+16), are observed with a m/z value of 721.3 (doubly charged precursor ion). In such cases, a relatively low abundance ion y5 corresponding to methionine sulfoxide at $m/z = 561.31$ can be detected. A representative MS/MS spectrum (cross-correlation score $X_{corr} = 3.62$ in TurboSequest analysis) indicates the presence of 9 y and 5 b out of the potential 11 ions (Fig. 15.4). Such a fragmentation pattern is comparable to the one observed for the SCOT peptide *in vivo* (Rebrin *et al.*, 2007). The presence of nitro and hydroxy groups on the tryptophan residue seemingly alters the fragmentation profile of this peptide, as indicated by the predominance of the ions y6 ($m/z = 808.4$), y7 ($m/z = 922.4$), and y9 ($m/z = 1106.4$). In some MS/MS spectra, the y5 ($m/z = 561.2$) ion, corresponding to methionine sulfoxide, is very weak or undetectable. Thus it appears that the double modification on

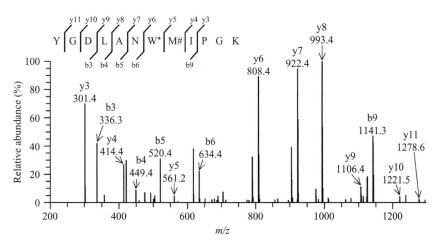

Figure 15.4 Tandem mass spectrum showing the simultaneous presence of nitrohydroxytryptophan and methionine sulfoxide in the peptide ([M+2H]$^{2+}$ ion at m/z 721.0) following peroxynitrite treatment. The fragmentation profile of the peptide is also modified considerably, with y6, y7, and y8 becoming particularly abundant. In contrast, y5 relative intensity is low, suggesting that nitrohydroxytryptophan might repress fragmentation on its C-terminal bond. M#, methionine sulfoxide; W*, nitrohydroxytryptophan.

the tryptophan disfavors the fragmentation of the residue located on its C-terminal side, which makes it difficult to detect the y5 ion corresponding to the methionine residue (or methionine sulfoxide derivative).

3.6. MS/MS analysis of multiple nitration and oxidation products

Prolonged exposure of the synthetic peptide YGDLANWMIPGK to peroxynitrite should also yield peptides containing oxidation on both methionine [methionine sulfoxide (+16) or methionine sulfone (+32)] and tryptophan [hydroxy-tryptophan (+16) or *N*-formylkynurenine (+32) residues]. In addition, multiple peptides containing *N*-formylkynurenine, as well as nitrated tyrosine in combination with methionine sulfoxide or sulfone, can also be observed (data not shown). In contrast, nitration of tryptophan residues alone was not detected. Peptides, containing oxidized derivatives of tryptophan, mainly hydroxy-tryptophan (+16), but also kynurenine (+4) and *N*-formylkynurenine, (+32) are also observed.

4. CONCLUSION

Tandem mass spectrometry is highly informative for the mass determination of fragmented ions originating from peptides. Structural modifications of methionine, tyrosine, and tryptophan residues following exposure of the synthetic peptide YGDLANWMIPGK, derived from SCOT sequence, to peroxynitrite, can be detected by the additions of +16 or +32 (oxidation), +45 (nitration), and +61 (nitration and oxidation) to the mass/charge values of fragmentation ions at a high level of precision. Occurrence of nitrohydroxytryptophan can be demonstrated after *in vitro* exposure of the peptide to peroxynitrite. This novel modification of the tryptophan residue matches the one observed in SCOT *in vivo* during aging. Electrospray ionization MS/MS analysis of nitrative and oxidative modifications provides clear identification of the target amino acid residue(s) as well as estimation of the corresponding mass addition. This analytical approach can be applied to other proteins, which are targets of nitrative and/or oxidative stress.

ACKNOWLEDGMENT

This research was supported by Grant RO1 AG 13563 from National Institute on Aging–National Institutes of Health.

REFERENCES

Alvarez, B., and Radi, R. (2003). Peroxynitrite reactivity with amino acids and proteins. *Amino Acids* **25,** 295–311.

Beckman, J. S. (1990). Ischaemic injury mediator. *Nature* **345,** 27–28.

Beckman, J. S. (1996). Oxidative damage and tyrosine nitration from peroxynitrite. *Chem. Res. Toxicol.* **9,** 836–844.

Beckman, J. S., Beckman, T. W., Chen, J., Marshall, P. A., and Freeman, B. A. (1990). Apparent hydroxyl radical production by peroxynitrite: Implications for endothelial injury from nitric oxide and superoxide. *Proc. Natl. Acad. Sci. USA* **87,** 1620–1624.

Elfering, S. L., Sarkela, T. M., and Giulivi, C. (2002). Biochemistry of mitochondrial nitric-oxide synthase. *J. Biol. Chem.* **277,** 38079–38086.

Guan, Z., Yates, N. A., and Bakhtiar, R. (2003). Detection and characterization of methionine oxidation in peptides by collision-induced dissociation and electron capture dissociation. *J. Am. Soc. Mass Spectrom.* **14,** 605–613.

Ikeda, K., Yukihiro Hiraoka, B., Iwai, H., Matsumoto, T., Mineki, R., Taka, H., Takamori, K., Ogawa, H., and Yamakura, F. (2007). Detection of 6-nitrotryptophan in proteins by Western blot analysis and its application for peroxynitrite-treated PC12 cells. *Nitric Oxide* **16,** 18–28.

Ischiropoulos, H. (1998). Biological tyrosine nitration: A pathophysiological function of nitric oxide and reactive oxygen species. *Arch. Biochem. Biophys.* **356,** 1–11.

Kanski, J., Alterman, M. A., and Schoneich, C. (2003). Proteomic identification of age-dependent protein nitration in rat skeletal muscle. *Free Radic. Biol. Med.* **35,** 1229–1239.

Kanski, J., Behring, A., Pelling, J., and Schoneich, C. (2005a). Proteomic identification of 3-nitrotyrosine-containing rat cardiac proteins: Effects of biological aging. *Am. J. Physiol. Heart Circ. Physiol.* **288,** H371–H381.

Kanski, J., Hong, S. J., and Schoneich, C. (2005b). Proteomic analysis of protein nitration in aging skeletal muscle and identification of nitrotyrosine-containing sequences *in vivo* by nanoelectrospray ionization tandem mass spectrometry. *J. Biol. Chem.* **280,** 24261–24266.

Koppenol, W. H., Kissner, R., and Beckman, J. S. (1996). Syntheses of peroxynitrite: To go with the flow or on solid grounds? *Methods Enzymol.* **269,** 296–302.

Nielson, K. R., and Pennington, M. W. (1995). Mass spectral analysis of peptides containing nitrobenzyl moieties. *Lett. Peptide Sci.* **2,** 301–305.

Paizs, B., and Suhai, S. (2005). Fragmentation pathways of protonated peptides. *Mass Spectrom. Rev.* **24,** 508–548.

Petersson, A. S., Steen, H., Kalume, D. E., Caidahl, K., and Roepstorff, P. (2001). Investigation of tyrosine nitration in proteins by mass spectrometry. *J. Mass Spectrom.* **36,** 616–625.

Radi, R. (2004). Nitric oxide, oxidants, and protein tyrosine nitration. *Proc. Natl. Acad. Sci. USA* **101,** 4003–4008.

Rebrin, I., Bregere, C., Kamzalov, S., Gallaher, T. K., and Sohal, R. S. (2007). Nitration of tryptophan 372 in succinyl-CoA:3-ketoacid CoA transferase during aging in rat heart mitochondria. *Biochemistry* **46,** 10130–10144.

Sarver, A., Scheffler, N. K., Shetlar, M. D., and Gibson, B. W. (2001). Analysis of peptides and proteins containing nitrotyrosine by matrix-assisted laser desorption/ionization mass spectrometry. *J. Am. Soc. Mass Spectrom.* **12,** 439–448.

Sharov, V. S., Galeva, N. A., Kanski, J., Williams, T. D., and Schoneich, C. (2006). Age-associated tyrosine nitration of rat skeletal muscle glycogen phosphorylase b: Characterization by HPLC-nanoelectrospray-tandem mass spectrometry. *Exp. Gerontol.* **41,** 407–416.

Turko, I. V., and Murad, F. (2002). Protein nitration in cardiovascular diseases. *Pharmacol. Rev.* **54,** 619–634.

Turko, I. V., and Murad, F. (2005). Mapping sites of tyrosine nitration by matrix-assisted laser desorption/ionization mass spectrometry. *Methods Enzymol.* **396,** 266–275.

Viner, R. I., Ferrington, D. A., Williams, T. D., Bigelow, D. J., and Schoneich, C. (1999). Protein modification during biological aging: Selective tyrosine nitration of the SER-CA2a isoform of the sarcoplasmic reticulum Ca2+-ATPase in skeletal muscle. *Biochem. J.* **340**(Pt 3), 657–669.

Yamakura, F., and Ikeda, K. (2006). Modification of tryptophan and tryptophan residues in proteins by reactive nitrogen species. *Nitric Oxide* **14,** 152–161.

Yamakura, F., Matsumoto, T., Ikeda, K., Taka, H., Fujimura, T., Murayama, K., Watanabe, E., Tamaki, M., Imai, T., and Takamori, K. (2005). Nitrated and oxidized products of a single tryptophan residue in human Cu,Zn-superoxide dismutase treated with either peroxynitrite-carbon dioxide or myeloperoxidase-hydrogen peroxide-nitrite. *J. Biochem.* (Tokyo) **138,** 57–69.

Yarian, C. S., Rebrin, I., and Sohal, R. S. (2005). Aconitase and ATP synthase are targets of malondialdehyde modification and undergo an age-related decrease in activity in mouse heart mitochondria. *Biochem. Biophys. Res. Commun.* **330,** 151–156.

Ye, Y. Z., Strong, M., Huang, Z. Q., and Beckman, J. S. (1996). Antibodies that recognize nitrotyrosine. *Methods Enzymol.* **269,** 201–209.

CHAPTER SIXTEEN

REDUCTIVE GAS-PHASE CHEMILUMINESCENCE AND FLOW INJECTION ANALYSIS FOR MEASUREMENT OF THE NITRIC OXIDE POOL IN BIOLOGICAL MATRICES

Ulrike Hendgen-Cotta,[1] Marijke Grau,[1] Tienush Rassaf, Putrika Gharini, Malte Kelm, *and* Petra Kleinbongard

Contents

Department of Medicine, Division of Cardiology, Pulmology and Vascular Medicine, CardioBioTech Research Group, University Hospital Aachen, Aachen, Germany
[1] Both authors contributed equally

Methods in Enzymology, Volume 441   © 2008 Elsevier Inc.
ISSN 0076-6879, DOI: 10.1016/S0076-6879(08)01216-0   All rights reserved.

Abstract

There is growing evidence for nitric oxide (NO·) being involved in cell signaling and pathology. Much effort has been made to elucidate and characterize the different biochemical reaction pathways of NO· *in vivo*. However, a major obstacle in assessing the significance of nitrosated species and oxidized meta-bolites often remains: a reliable analytical technique for the detection of NO· in complex biological matrices. This chapter presents refined methodologies, such as chemiluminescence detection and flow injection analysis, compared with adequate sample processing procedures to reliably quantify and assess the circulating and resident NO⁻ pool, consisting of nitrite, nitrate, nitroso, and nitrosylated species.

1. CIRCULATING AND RESIDENT POOL OF NITRIC OXIDE

Nitric oxide (NO·) is a highly diffusible inorganic radical gas that is synthesized by NO synthases (NOS) in the endothelium and other cells. In the vascular system, NO· is produced from the endothelial NOS (eNOS) (Moncada *et al.*, 1991). This heme–containing enzyme catalyzes a five-electron oxidation of one of the basic guanidine nitrogen groups of L-arginine in the presence of multiple cofactors and molecular oxygen (Barbato and Tzeng, 2004). Three biological and interrelated active redox forms of NO· have been described: NO·, nitrosyl (NO^+), and nitroxyl anion (NO^-). NO· serves to regulate important functions in the vessel and in circulating cells. It maintains basal vasodilator tone and participates in the regulation of vascular homeostasis such as modulation of oxidant stress, antiplatelet activity, adhesion molecule expression, and endothelial and smooth muscle cell proliferation (for review, see Moncada and Higgs, 1993). When generated, NO· diffuses radially, representing a paracrine signaling molecule. Diffusion to vicinal smooth muscle leads to vasorelaxation (Palmer *et al.*, 1987). NO· that diffuses luminally into the bloodstream will affect cells, for example, platelets (Alheid *et al.*, 1987; Radomski *et al.*, 1987) and leukocytes (Kubes *et al.*, 1991), and is expected to react with both oxy-and deoxyhemoglobin to form methemoglobin and nitrate ions, and iron–nitrosyl–hemoglobin ($HbFe^{II}NO$), respectively (Doyle and Hoekstra, 1981).

For a long time the signaling actions of NO· in the vasculature have been thought to be short-lived as a result of its rapid irreversible nature of these NO–hemoglobin reactions and the massive concentration of hemoglobin in blood (10,000 μM heme). However, under physiological conditions, the reaction of NO· with hemoglobin is limited, caused on the compartmen-talization of hemoglobin within the erythrocyte (RBCs) (Vaughn *et al.*, 2000). An unstirred layer surrounding the RBC (Liu *et al.*, 1998) and an erythrocyte-free zone along the endothelium in laminar streaming blood

(Butler *et al.*, 1998; Liao *et al.*, 1999) represent diffusional barriers. Intravascular hemolysis leads to a release of hemoglobin into the plasma compartment, resulting in rapid rates of NO• consumption (Olson *et al.*, 2004) and inhibition of vascular relaxation and vasodilation by not activation of the smooth muscle guanylyl cyclase. In addition, the dissociation of cell-free plasma hemoglobin into dimers, which are able to reach the spaces between endothelial and smooth muscle cells, displays further NO• scavenging (Nakai *et al.*, 1998; Olson *et al.*, 2004).

However, a series of studies have appreciated mechanisms by which NO• bioactivity in blood might be sustained, either by the formation of NO•-modified proteins, peptides, and lipids or by oxidation to nitrite. Investigations from our own group have suggested that NO• itself may remain active in the bloodstream for longer than originally assumed and confirmed the earlier notion that the formation of nitroso species may act to conserve and transport NO• (Rassaf *et al.*, 2002a,b,c). A number of intravascular species capable of endocrine vasodilation have been explored, including *S*-nitrosated albumin (Ng *et al.*, 2004; Scharfstein *et al.*, 1994; Stamler *et al.*, 1992) and *S*-nitrosated hemoglobin (Jia *et al.*, 1996), iron-nitrosyl-hemoglobin (Gladwin *et al.*, 2000a), *N*-nitrosamines (Lippton *et al.*, 1982; Rassaf *et al.*, 2002a), nitrated lipids (Lim *et al.*, 2002; Schopfer *et al.*, 2005), and nitrite (Cosby *et al.*, 2003; Rassaf *et al.*, 2006).

The abundance of hemoproteins such as cytochrome P_{450}, cyclooxygenase, and peroxidases in tissues show high affinity for NO•. Therefore heme moieties in tissue appear to be obvious acceptors for endogenous NO• *in vivo*. It has been shown that endothelial NO• production also results in local formation of NO• adducts such as *S*-nitrosothiols (RSNO), iron-nitrosyls, and nitrite that may act as storage forms of NO• in tissues. Biochemical analyses revealed that rat aortic tissue contains equimolar concentrations of *S*- and *N*-nitroso compounds, as well as nitrite and nitrate (Rodriguez *et al.*, 2003). Further, it is suggested that the *S*-nitrosation of myoglobin represents a store of vasoactive NO• in cardiac and skeletal muscle (Rayner *et al.*, 2005; Witting *et al.*, 2001). NO• reacts with free radicals, heme proteins, and thiols, which are abundant at high concentrations in mitochondria. Therefore, interaction with mitochondria is of significance for cell function (Ghafourifar and Colton, 2003). The basal formation of NO• in mitochondria seems to be one of the main regulators of cellular respiration, mitochondrial transmembrane potential, and transmembrane pH gradient (Brown, 1995).

Additionally, NO• plays an important role in myocardial ischemia/reperfusion (I/R) injury. Attenuated eNOS function and reduced NO• generation are critical early events in many cardiovascular diseases (Ignarro *et al.*, 2002). A number of previous studies have demonstrated that treatment with various NO-donating compounds is highly effective in the setting of myocardial I/R injury (Draper and Shah, 1997; Pabla *et al.*, 1996). Results indicate an important capacity for NO•-mediated cardioprotection by regulating mechanisms that might participate in cytotoxicity after severe ischemia: NO• maintains heme proteins in a reduced and liganded state (Fernandez *et al.*, 2003; Herold and Rehmann, 2001; 2003), limits free iron- and heme-mediated oxidative chemistry (Kanner and Harel, 1985; Kanner *et al.*, 1992), regulates mitochondrial function by reversibly inhibiting complexes I and IV consistent with an decreasing reactive oxygen species formation (Brookes and Darley-Usmar, 2002; Poderoso *et al.*, 1996; Shiva *et al.*, 2007b), initiating biogenesis (Nisoli *et al.*, 2003), and limiting apoptotic cytochrome *c* release (Brookes *et al.*, 2000; Kim *et al.*, 1998).

1.1. Nitrite

Nitrite is an endogenous inorganic anion produced by the oxidation of NO• under aerobic conditions (Kelm, 1999). It is present at concentrations of 300–600 nM in plasma (Gladwin *et al.*, 2000b; Grau *et al.*, 2007; Kleinbongard *et al.*, 2003; Rassaf *et al.*, 2003), 200 nM in whole blood (Dejam *et al.*, 2005), 300–700 nM in RBCs (Bryan *et al.*, 2004; Dejam *et al.*, 2005), and 450 nM – 22.5 μM in tissue (Bryan *et al.*, 2004; Rodriguez *et al.*, 2003).

We determined the serum nitrite concentration in blood samples depending on the endothelial NO• pathway. A vasodilator significantly increased nitrite and the forearm blood flow, respectively, whereas the eNOS inhibitor reduced both forearm blood flow and endothelium-dependent vasodilation (Kelm *et al.*, 1999; Lauer *et al.*, 2001). In humans and other mammals, about 90% of the circulating plasma nitrite is derived directly from the L-arginine/NO• pathway (Kleinbongard *et al.*, 2003; Rhodes *et al.*, 1995). Plasma nitrite levels are reduced up to 70% in eNOS knockout mice and upon acute NOS inhibition in wild-type mice (Kleinbongard *et al.*, 2003). Therefore, plasma nitrite seems to derive to a large portion from eNOS, and circulating nitrite could be a specific indicator of eNOS activity *in vivo* in humans (Bode-Böger *et al.*, 1999; Boucher *et al.*, 1999; Kleinbongard *et al.*, 2003; Lauer *et al.*, 2001; Rhodes *et al.*, 1995). In addition, the plasma nitrite concentration reflects numbers of cardiovascular risk factors. In healthy volunteers without cardiovascular risk factors, values were determined around 300 nM (Dejam *et al.*, 2005; Kleinbongard *et al.*, 2006a), but in patients with four risk factors the concentration decreases to 171 nM (Kleinbongard *et al.*, 2006a). During reactive hyperemia of the forearm in healthy young subjects, mean plasma concentrations were found to increase by 53% (Allen *et al.*, 2005) and 54% (Rassaf *et al.*, 2006), whereas in subjects with endothelial dysfunction, no significant increase in plasma nitrite concentration was observed (Rassaf *et al.*, 2006). Investigations by Dejam *et al.* (2005) showed that stimulation of the eNOS by acetylcholine and shear stress increase nitrite concentrations significantly in whole blood. New results suggest that measurement of nitrite in whole blood or RBCs would be advantageous over determination in plasma or serum, if nitrite would be indeed distributed nonuniformly in blood (Dejam *et al.*, 2005). One origin of the elevated nitrite level in RBCs might be intrinsic NO• synthesis in blood cells (Kleinbongard *et al.*, 2006b; Ludolph *et al.*, 2007).

Studies have established a role for nitrite as an endocrine bioavailable storage pool of NO• that is bioactivated along the physiological oxygen gradient to regulate a number of vascular and cellular responses (Bryan *et al.*, 2005; Gladwin, 2005). It has been proposed to participate in hypoxic vasodilation and signaling (Cosby *et al.*, 2003; Crawford *et al.*, 2006; Lundberg and Weitzberg, 2005; Nagababu *et al.*, 2003). During ischemia, nitrite limits apoptosis and cytotoxicity at reperfusion in the mammalian heart, liver, and brain (Duranski *et al.*, 2005; Jung *et al.*, 2006; Webb *et al.*, 2004). In blood, deoxygenated hemoglobin seems to be the predominant catalyzer of nitrite reduction, whereas in tissue hypoxia deoxygenated myoglobin becomes important by its reductase activity, which leads to regulation of mitochrondrial respiration, cardiac energetics, and function (Cosby *et al.*, 2003; Rassaf *et al.*, 2007; Shiva *et al.*, 2007a). Further, nitrite may be converted to NO• by acidic disproportionation (Zweier *et al.*, 1995) and enzymatic conversion via xanthine-oxidoreductase (Webb *et al.*, 2004) and mitochondrial enzymes (Castello *et al.*, 2006).

1.2. Nitrate

Nitrate is the predominant product of oxidative NO• chemistry in biological fluids and tissues. Under physiological conditions, intravascular NO• is inactivated by its reaction with oxygenated hemoglobin to form nitrate (Doyle *et al.*, 1981). Further, it is an oxidation product of nitrite in a hemoglobin-dependent manner after taken up rapidly by RBCs (Kosaka *et al.*, 1982). In skeletal, cardiac, and smooth muscles, oxygenated myoglobin oxidizes NO• and nitrite to nitrate (Doyle *et al.*, 1981). Nitrate levels are normally much higher than nitrite and usually in the range of 5.7 (Bryan *et al.*, 2004) to 60 (Green *et al.*, 1982) μM in human plasma, 10 (Bryan *et al.*, 2004) to 23 (detected in 25 human volunteers) μM in RBCs, and 3 to 49 (Bryan *et al.*, 2004; Rassaf *et al.*, 2007) μM in tissues.

Nitrate displays high background concentrations. These are influenced by a variety of NOS-independent factors, including dietary nitrate intake, formation in saliva, bacterial nitrate synthesis within the bowel, denitrifying liver enzymes, inhalation of atmospheric gaseous nitrogen oxides, and renal function (Kelm, 1999; Lundberg and Weitzberg, 2005; Tannenbaum *et al.*, 1979). Additionally, the plasma half-life of nitrate is fairly long ($\approx$6 h) (Tannenbaum *et al.*, 1979). On the strength of this, one cannot rely on nitrate measurement as an adequate index of systemic NO• generation (Lauer *et al.*, 2001), but in isolated systems it represents an important parameter of NO• consumption.

1.3. Nitrosated and nitrosylated species

Nitroso species and iron–nitrosyls are part of the circulating and resident pool of $NO\cdot$. The nitroso species could be subclassified in mercury-labile and mercury-stable species. The mercury-labile *S*-nitrosothiols may be generated by reaction with $\cdot NO_2$ or N_2O_3 produced during the oxidation of $NO\cdot$ with dissolved oxygen (Wink *et al.*, 2000), by reaction with NO^+ formed from dinitrosyl-iron complexes (Boese *et al.*, 1995), or by reaction with peroxynitrite ($ONOO^-$) derived from the reaction of $NO\cdot$ with O_2^- (Moro *et al.*, 1994). Under anaerobic conditions, RSNO may be formed by direct interaction of $NO\cdot$ with thiols in the presence of electron acceptors (Gow *et al.*, 1997). Once formed, circulating and resident RSNOs can release $NO\cdot$ via specific mechanisms (Ramachandran *et al.*, 2001) or directly transfer NO^+ to another thiol via so-called transnitrosation reactions (Jourd'heuil *et al.*, 2000). Levels of RSNOs can be increased severalfold by exposure to $NO\cdot$ (Marley *et al.*, 2001; Rassaf *et al.*, 2002c; Rossi *et al.*, 2001). A number of clinical investigations have indirectly implicated the involvement of plasma RSNOs in disease processes (Gaston *et al.*, 1998; Tyurin *et al.*, 2001). Many attempts have been made to quantify plasma RSNOs in humans. Various studies suggest that basal RSNO levels in plasma are in fact in the low nanomolar range (Cannon *et al.*, 2001; Goldman *et al.*, 1998; Marley *et al.*, 2000; Moriel *et al.*, 2001; Rassaf *et al.*, 2002a,b). *N*-Nitrosamines (RNNOs) are counted among the mercury-stable species. They are generated endogenously during infections and inflammatory processes (Ohshima and Bartsch, 1994). In the acidic environment of the stomach, RNNOs are formed due to the reaction of nitrite with amino groups of food constituents (Lijinsky, 1980). Under physiological conditions, we have shown that human plasma contains an approximately fivefold higher concentration of RNNOs than RSNOs (Rassaf *et al.*, 2002a). Our finding that the basal concentration of RNNOs exceeds that of RSNOs is important and may suggest a novel storage and/or delivery form of $NO\cdot$ that is differentially regulated from RSNOs.

Under hypoxic and ischemic conditions, the hemoproteins hemoglobin and myoglobin are deoxygenated. Generated $NO\cdot$ from nitrite could be captured by the remaining deoxygenated hemoglobin and myoglobin, respectively, as nitrosylated hemoproteins, which therefore can serve as an index of $NO\cdot$ formation (Gladwin *et al.*, 2005).

The ability of $NO\cdot$ and $NO\cdot$-derived species to nitrate biomolecules serves as the molecular basis for how $NO\cdot$ influences the synthesis and reactions of bioactive lipids (Rubbo *et al.*, 1994; Schopfer *et al.*, 2003). $NO\cdot$ acts as potent inhibitor of lipid peroxidation and low-density lipoprotein oxidation (Goss *et al.*, 1997; Hogg *et al.*, 1993; Rubbo *et al.*, 1994).

The circulating and resident $NO\cdot$ pool, consisting of nitrite, nitrate, nitroso, and nitrosylated species, reflects on the one hand the level of the highly unstable free radical $NO\cdot$, which displays paracrine and endocrine

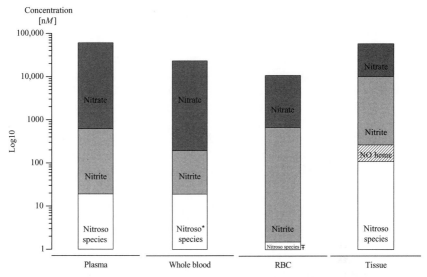

Figure 16.1 Comparison of nitrite, nitrate, and nitroso species levels in human plasma, whole blood, RBCs, and tissue. For human plasma, these are 100 to 600 nM (Grau *et al.*, 2007), 5.7 μM (Bryan *et al.*, 2004), 60 μM (Green *et al.*, 1982), and 19 nM (Kleinbongard *et al.*, 2006b), respectively. For whole blood, 176 nM (Dejam *et al.*, 2005), 23 μM, and estimated 19 nM, and for RBCs 288 nM (Dejam *et al.*, 2005), 680 nM (Bryan *et al.*, 2004), 10.2 μM (Bryan *et al.*, 2004), and <1 nM (Rassaf *et al.*, 2003, 2004), respectively. The tissue nitrite, nitrate, and nitroso species concentrations are 450 nM to 22.5 μM, 3.14 to 48.6 μM, and 10 to 110 nM, respectively, depending on the type of tissue used (Bryan *et al.*, 2004). The NO heme concentration amounts are <1 nM in aorta and lung and between 5.8 (kidney) and 157 nM (brain) (Bryan *et al.*, 2004).

properties, and on the other hand in part reflects the vascular NO· storage pool. An overview of basal levels of nitrite, nitrate, and nitroso species present in human plasma, whole blood, RBCs, and tissue is shown in Fig. 16.1.

2. APPLIED METHODS FOR DETERMINATION OF NITRITE, NITRATE, AND NITROSO SPECIES IN BIOLOGICAL MATRICES AND IRON-NITROSYLS IN TISSUE

The methods described in this chapter—flow injection analysis (FIA) and chemiluminescence detection (CLD)—have emerged in recent years as two of the most commonly used assays due to their specificity, sensitivity, versatility, and practicability (Feelisch *et al.*, 2002; Jourd'heuil *et al.*, 2005; Marley *et al.*, 2000; Wang *et al.*, 2006). Their high validity has been shown compared to other highly sensitive and specific techniques such as electron paramagnetic resonance spectrometry (Tsikas *et al.*, 1997), high-performance liquid chromatography (Gladwin *et al.*, 1999), and gas chromatography–mass spectrometry (Tsikas *et al.*, 1997). Determination of the NO• pool in biological matrices might supply extensive information. It can be employed to determine NOS activity *in vivo* (Andersen *et al.*, 2003; Lauer *et al.*, 2001), diagnose endothelial dysfunction in humans, provide insight into the mechanism of NOx [nitrite, nitrate, and nitroso species (*S*-nitrosothiols and *N*-nitrosamine)] metabolism (Heiss *et al.*, 2005, Kleinbongard *et al.*, 2003; Rassaf *et al.*, 2002a, 2006), and estimate the active NO• pool in blood and tissue. This section describes applied NOx measurement methodologies and sample processing in a variety of matrices such as whole blood, RBCs, plasma, and tissue.

2.1. Quantification of nitrite and nitrate through flow injection analysis in combination with the Griess method

This method can be applied for the detection of nitrite in deproteinized plasma and tissue samples, and plasma and tissue nitrate after enzymatic conversion to nitrite. The principle of the FIA technique is that samples are injected into the system. Within the reaction coil nitrite reacts with the Griess solution containing a 1:1 mixture of sulfanilamide (10 g/liter dissolved in a 2.4% phosphoric acid solution) and *N*-(1naphtyl)ethylenediamine (496 mg/liter dissolved in water). The reaction forms—under acidic conditions—a red-colored azo compound (Kleinbongard *et al.*, 2002). The detector measures the absorbance at the characteristic wavelength of 540 nm. Using an automatic sample injection apparatus, the FIA method is faster (Kleinbongard *et al.*, 2002), but less sensitive than the CLD technique, which is described later (Table 16.1).

Plasma and tissue nitrite can be measured by FIA (Lauer *et al.*, 2003) after deproteination as described in Sections 3.2 and 3.3 through the standard addition procedure (Kleinbongard *et al.*, 2002). As nitrate can only be measured after conversion to nitrite via nitrate reductase, the detected nitrite concentration is the sum of nitrate and nitrite. To evaluate the nitrate concentration it is therefore necessary to subtract the measured nitrite concentration of the respective biological matrix (Kleinbongard *et al.*, 2002).

The FIA technique is not applicable for the measurement of nitrite and nitrate detection in whole blood and RBCs because the sample preparation includes use of a ferricyanide-containing solution. The detection cell of the FIA detector is not stable to the substances in this solution and, moreover, hemoglobin, ethanol, and citrate may interfere with the FIA–Griess

Table 16.1 Comparison of the characteristics of samples and detection units

	CLD (Feelisch *et al.*, 2002)		FIA (Kleinbongard *et al.*, 2002)	
	Nitrite	Nitroso species	Nitrite	Nitrate
Sensitivity (nM)	1	1	5	100
Sample volume (μ)	100	300	20	20
Linearity	1–100,000	1–100,000	10–10,000	100–25,000
Recovery (%)	100 ± 5	100 ± 5	100 ± 10	100 ± 8
Coefficient of variation (%)	5–7	4–6	6–8	3–5

measurement (Kleinbongard *et al.*, 2002). These interactions should be kept in mind when choosing the dilutor and anticoagulant for the sample. Heparin is a noninteracting anticoagulant that can be used. Other substances that may lead to interferences in the Griess reaction include biogenic amines, zinc sulfate, cadmium, manganese, iron, zinc, urea, thiols, proteins, and other plasma constituents (Giovannoni *et al.*, 1997). Nitroso and nitrosyl species cannot be measured through FIA because NO$\cdot$ does not react with the Griess reagent.

2.2. Quantification of nitrite, nitrate, nitrosyl, and nitroso species through reductive gas-phase chemiluminescence detection

Through CLD, the nitrite, nitrate, and nitroso species level in tissue, plasma, whole blood, and RBCs is determined after reductive cleavage by an iodide/triiodide-containing reaction mixture and subsequent determination of the NO$\cdot$ released into the gas phase by its chemiluminescence reaction with ozone (O_3). Tissue NO-heme (iron–nitrosyl) adducts are determined by a parallel injection of replicate aliquots of tissue homogenates into a solution of 0.05 M ferricyanide in phosphate-buffered saline (PBS) at pH 7.5 and 37 °C (Bryan *et al.*, 2004; Rassaf *et al.*, 2007). This modified method employs one-electron oxidation rather than reduction to achieve denitrosylation. Released NO$\cdot$ is quantified by CLD as described for the iodine/iodide assay given earlier. The nitrosylation mechanism presumably involves oxidation of the heme iron "underneath" the ligand, which, because of the weaker NO$\cdot$ affinity of ferric over ferrous heme (Sharma *et al.*, 2007), is associated with a release of NO$\cdot$ into the gas phase (Bryan *et al.*, 2004).

The ozone-based CLD measurement can be used for all samples without a previous deproteination step of the samples (Cox, 1980; Feelisch *et al.*, 2002; Gladwin *et al.*, 2002; Marley *et al.*, 2000).

With larger injection volumes this method can quantify as little as 100 fmol bound NO and has been validated for the use in different biological matrices. To differentiate between compound classes without having to change reaction solutions or conditions, samples can be pretreated with group-specific reagents before analysis. As described in Sections 3.2 and 3.3, prior to nitrate detection the sample is subjected to a conversion step to nitrite. The determination of nitroso species requires nitrite elimination by sulfanilamide and HCl (Kleinbongard *et al.*, 2002).

For the interpretation of data, one has to determine the area under the curve (AUC) by performing a standard curve and dividing the AUC of the samples by the slope of the standard curve, which in turn is obtained by plotting the amounts of nitrite injected into the reaction solution against the AUC of the peak registered. Standard curves should contain the expected concentration range and should be performed daily. As the iodide/triiodide solution is not specific for nitrite, the concentration of other species should be determined and subtracted from the total CLD signal.

3. Sample Processing

3.1. Prior to sample processing

To avoid NOx contamination due to laboratory and plasticware, it is recommended to wash laboratory glass and plasticware and to test non-washable laboratory ware (e.g., pipette tips) for NOx contamination prior to use (Grau *et al.*, 2007; Kleinbongard *et al.*, 2002).

3.2. Whole blood

To measure nitrite, nitrate, and nitroso species in whole blood, plasma, and RBCs via CLD and FIA as described earlier, blood should be drawn into a syringe quickly with a butterfly needle (21 gauge) to avoid hemolysis and to avoid the reaction of nitrite with hemoglobin to form nitrate under oxygenated conditions or nitrosylhemoglobin under deoxygenated conditions (Huang *et al.*, 2005).

3.2.1. Nitroso species in whole blood

Two forms of anticoagulation are used for the determination of NOx. For the determination of nitroso species in whole blood we use a modified protocol (Bryan *et al.*, 2004). Here, blood is diluted with a N-ethylmaleimide (NEM)/EDTA solution (containing 100 mM NEM and 25 mM EDTA in 0.9% NaCl solution) in a 1:10 ratio (v/v; NEM/EDTA solution: whole blood) to prevent coagulation and in order to block SH groups and inhibit transnitrosation reactions, preventing artificial nitrosation, as well as thiolate- and ascorbate-mediated degradation (Rassaf et al., 2003). Nitrite is then eliminated with 0.5% sulfanilamide in 0.1% HCl in a 1:10 ratio (v/v; sulfanilamide solution:whole blood) during a 10-min incubation on ice.

3.2.2. Nitrite, nitrate and nitroso species in plasma and nitrite and nitrate in whole blood and RBC

For the determination of all the following parameters, heparin (1 IU/ml blood) is added to whole blood to prevent coagulation.

To measure NOx in plasma, whole blood needs to be centrifuged immediately at 800g, 4 °C and 15 min to separate plasma from blood cells. Plasma nitrite levels can be measured right after separation via CLD or after dilution with low nitrite containing PBS (e.g., from Serag Wiessner) in a 1:5 ratio and subsequent ultrafiltration (cut-off 10 kDa) for 60 min at 6000g and 4 °C via FIA. For the determination of nitroso species in plasma, nitrite needs to be eliminated by a sulfanilamide solution containing 0.5% sulfanil- amide in 0.1% HCl in a 1:10 ratio (v/v; sulfanilamide solution:plasma/whole blood) during a 15-min incubation on ice prior to measurement. To deter- mine plasma nitrate the untreated sample has to be diluted with methanol at first in a 1:2 ratio to remove proteins. After centrifugation at 21,000g, 4 °C and 15 min, nitrate reductase (0.1 IU/ml), NADPH (2 μM), glucose-6- phosphate (1 mM), and glucose-6-phosphate dehydrogenase (0.3 IU/ml) are added to the supernatant. After incubation for 1 h at room temperature in the dark, nitrate can be measured. Prior to nitrite and nitrate measurements in RBCs and whole blood it is mandatory to add a ferricyanide-based preserva- tion solution in a 1:5 ratio (v/v; preservation solution:RBCs) to the cells to stabilize nitrite (Dejam et al., 2005; Pelletier et al., 2006). The solution consists of 0.8 M ferricyanide, 0.1 M NEM, and Nonidet P-40 (NP-40) substitute (10% of the total volume of preservation solution). Ferricyanide reacts with hemoglobin to form methemoglobin and prohibits the conver- sion of nitrite to nitrate or nitrosylhemoglobin, thus stabilizing nitrite in the presence of millimolar amounts of hemoglobin. NEM is a thiol-alkylating agent, and NP-40, a cytosolic agent, allows the ferricyanide access to the cytosol and red cell contents once the erythrocytic membrane has been broken (Pelletier et al., 2006). The samples are then subjected to a freezing step (−80 °C until the sample is frozen), which yields a better subsequent purification. For nitrite measurement, methanol is added to the frozen samples in a 1:2 ratio to remove proteins. After centrifugation at 21,000g, 4 °C and 15 min the nitrite level of the supernatant can be determined. To evaluate the nitrate content, frozen whole blood and RBCs mixed with preservation solution are first diluted with Millipore water in a 1:5 ratio. After deproteination at 70 °C and 10 min the samples are spun down at

21,000g, 4 °C for 15 min. Nitrate reductase (0.1 IU/ml), glucose-6-phosphate (1 mM), glucose-6-phosphate dehydrogenase (0.3 IU/ml), and NADPH (2 μM) are added to the supernatant. Similar to plasma nitrate the samples are incubated at room temperature in the dark for 1 h before nitrate measurement.

3.2.3. Nitroso species in RBCs

A method for the determination of nitroso species in RBCs is to subject untreated RBCs to hypotonic lysis in water containing NEM (final concentration 10 mM) and EDTA (final concentration 2 mM) (v/v 1:4) but the results revealed that nitroso species are undetectable in RBCs even in the presence of the anion exchange protein-1 and flavoprotein inhibitors, 4,4′ diisothiocyanostilbene-2,2′-disulfonic acid and diphenyleneiodonium, unless traces of nitrite were introduced together with these reagents (Rassaf *et al.*, 2004). The low concentration of NO• adducts in RBCs (<1 nM), despite exposure to comparable levels of plasma nitrite and S-nitrosoalbumin (and presumably NO•), may be a consequence of the lower thiol reactivity of the Cysβ-93 in this hemoglobin (Rassaf *et al.*, 2003).

3.3. Tissue

For determination of the resident NO• pool, tissue should be kept on ice and homogenized in a 1:5 ratio with NEM/EDTA solution (v/v; tissue: NEM/EDTA solution) (100 mM NEM, 2.5 mM EDTA in 0.9% NaCl solution) [modified after Feelisch *et al.* (2002) and Bryan *et al.* (2004) due to the high nitrite content of EDTA] using a glass tissue homogenizer. The homogenate is then split up into four fractions. Prior to measurement these aliquots need additional treatment except the first and second fractions, which are used for the measurement of nitrite and iron-nitrosyl levels. These samples can be injected directly. The third aliquot is used to determine the nitroso species concentration in the tissue homogenate after the elimination of nitrite with 0.5% sulfanilamide in 0.1% HCl in a 1:10 ratio following a 15-min incubation on ice. To determine the nitrate level, the sample is first deproteinized with methanol in a 1:2 ratio. After centrifugation at 14,000g for 15 min, nitrate reductase (0.1 IU/ml), glucose-6-phosphate (1 mM), NADPH (2 μM), and glucose-6-phosphate dehydrogenase (0.3 IU/ml) are added to the supernatant. The suspension needs to incubate for 1 h at 25 °C in the dark prior to measurement.

   An overview of the characteristics of samples, sample processing, and applied analytical techniques is given in Figs. 16.2A and 16.2B and Table 16.1.

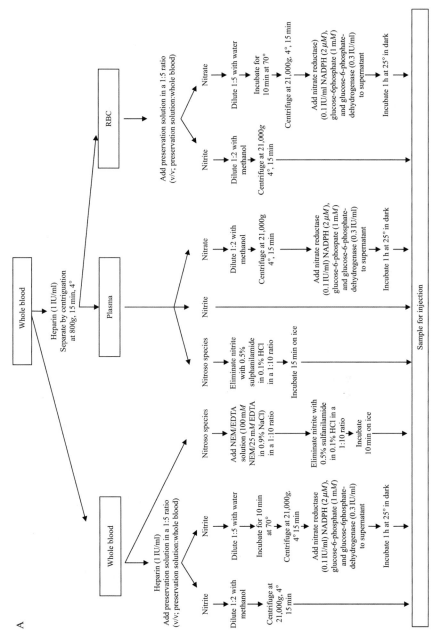

A

Figure 16.2 Continued

B

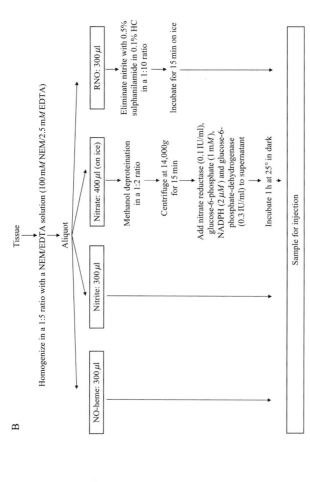

Figure 16.2 Comparison of the characteristics of sample processing and applied analytical techniques to determine nitrite, nitrate, and nitroso species in biological samples. (A) Bedside preparation of samples for nitrite, nitrate, and nitroso species in whole blood, RBCs and plasma; except for nitroso species in RBC because many studies revealed that the concentration is $<1\,nM$ and lies under the detection limit of many techniques (Kleinbongard *et al.*, 2003). Preservation solution: $0.8\,M$ ferricyanide, 10% NP-40, and $100\,mM$ N-ethylmaleimide. (B) Determination of NOx and NO-heme in tissue samples.

4. CONCLUSIONS

With this study we were able to show refined methodologies such as CLD and FIA in comparison with adequate sample processing in order to quantify and assess the circulating and resident NO• pool, consisting of nitrite, nitrate, nitroso, and nitrosylated species.

ACKNOWLEDGMENTS

This work was supported by the Deutsche Forschungsgemeinschaft (Ke405/4-3) (M. Kelm and P. Kleinbongard, University Hospital of Aachen, Germany) and RA 969/4-1 (T. Rassaf, University Hospital of Aachen, Germany). Malte Kelm and part of the research described here were supported by Philip Morris USA Inc. and Philip Morris International (PO#8500020430).

REFERENCES

Alheid, U., Frölich, J. C., and Förstermann, U. (1987). Endothelium-derived relaxing factor from cultured human endothelial cells inhibits aggregation of human platelets. *Thromb. Res.* **47,** 561–571.

Allen, J. D., Cobb, F. R., and Gow, A. J. (2005). Regional and whole-body markers of nitric oxide production following hyperemic stimuli. *Free Radic. Biol. Med.* **38,** 1164–1169.

Andersen, H. R., Nielsen, T. T., Rasmussen, K., Thuesen, L., Kelbaek, H., Thayssen, P., Abildgaard, U., Pedersen, F., Madsen, J. K., Grande, P., Villadsen, A. B., Krusell, L. R., *et al.* (2003). A comparison of coronary angioplasty with fibrinolytic therapy in acute myocardial infarction. *N. Engl. J. Med.* **349,** 733–742.

Barbato, J. E., and Tzeng, E. (2004). Nitric oxide and arterial disease. *J. Vasc. Surg.* **40,** 187–193.

Bode-Böger, S. M., Böger, R. H., Löffler, M., Tsikas, D., Brabant, G., and Frölich, J. C. (1999). L-arginine stimulates NO-dependent vasodilation in healthy humans: Effect of somatostatin pretreatment. *J. Invest. Med.* **47,** 43–50.

Boese, M., Mordvintcev, P. I., Vanin, A. F., Busse, R., and Mulsch, A. (1995). S-nitrosation of serum albumin by dinitrosyl-iron complex. *J. Biol. Chem.* **270,** 29244–29249.

Boucher, J. L., Moali, C., and Tenu, J. P. (1999). Nitric oxide biosynthesis, nitric oxide synthase inhibitors and arginase competition for L-arginine utilization. *Cell. Mol. Life Sci.* **55,** 1015–1028.

Brookes, P., and Darley-Usmar, V. M. (2002). Hypothesis: The mitochondrial NO(★) signaling pathway, and the transduction of nitrosative to oxidative cell signals: An alternative function for cytochrome C oxidase. *Free Radic. Biol. Med.* **32,** 370–374.

Brookes, P. S., Salinas, E. P., Darley-Usmar, K., Eiserich, J. P., Freeman, B. A., Darley-Usmar, V. M., and Anderson, P. G. (2000). Concentration-dependent effects of nitric oxide on mitochondrial permeability transition and cytochrome c release. *J. Biol. Chem.* **275,** 20474–20479.

Brown, G. C. (1995). Nitric oxide regulates mitochondrial respiration and cell functions by inhibiting cytochrome oxidase. *FEBS Lett.* **369,** 136–139.

Bryan, N. S., Fernandez, B. O., Bauer, S. M., Garcia-Saura, M. F., Milsom, A. B., Rassaf, T., Maloney, R. E., Bharti, A., Rodriguez, J., and Feelisch, M. (2005). Nitrite is a signaling molecule and regulator of gene expression in mammalian tissues. *Nat. Chem. Biol.* **1**, 290–297.

Bryan, N. S., Rassaf, T., Maloney, R. E., Rodriguez, C. M., Saijo, F., Rodriguez, J. R., and Feelisch, M. (2004). Cellular targets and mechanisms of nitros(yl)ation: An insight into their nature and kinetics. *In vivo. Proc. Natl. Acad. Sci. USA* **101**, 4308–4313.

Butler, A. R., Megson, I. L., and Wright, P. G. (1998). Diffusion of nitric oxide and scavenging by blood in the vasculature. *Biochim. Biophys. Acta.* **1425**, 168–176.

Cannon, R. O., III, Schechter, A. N., Panza, J. A., Ognibene, F. P., Pease-Fye, M. E., Waclawiw, M. A., Shelhamer, J. H., and Gladwin, M. T. (2001). Effects of inhaled nitric oxide on regional blood flow are consistent with intravascular nitric oxide delivery. *J. Clin. Invest.* **108**, 279–287.

Castello, P. R., David, P. S., McClure, T., Crook, Z., and Poyton, R. O. (2006). Mitochondrial cytochrome oxidase produces nitric oxide under hypoxic conditions: Implications for oxygen sensing and hypoxic signaling in eukaryotes. *Cell Metab.* **3**, 277–287.

Cosby, K., Partovi, K. S., Crawford, J. H., Patel, R. P., Reiter, C. D., Martyr, S., Yang, B. K., Waclawiw, M. A., Zalos, G., Xu, X., Huang, K. T., Shields, H., *et al.* (2003). Nitrite reduction to nitric oxide by deoxyhemoglobin vasodilates the human circulation. *Nat. Med.* **9**, 1498–1505.

Cox, R. D. (1980). Determination of nitrate and nitrite at the parts per billion level by chemiluminescence. *Anal. Chem.* **52**, 332–335.

Crawford, J. H., Isbell, T. S., Huang, Z., Shiva, S., Chacko, B. K., Schechter, A. N., Darley-Usmar, V. M., Kerby, J. D., Lang, J. D., Jr., Kraus, D., Ho, C., Gladwin, M. T., *et al.* (2006). Hypoxia, red blood cells, and nitrite regulate NO-dependent hypoxic vasodilation. *Blood* **107**, 566–574.

Dejam, A., Hunter, C. J., Pelletier, M. M., Hsu, L. L., Machado, R. F., Shiva, S., Power, G. G., Kelm, M., Gladwin, M. T., and Schechter, A. N. (2005). Erythrocytes are the major intravascular storage sites of nitrite in human blood. *Blood* **106**, 734–739.

Doyle, M. P., and Hoekstra, J. W. (1981). Oxidation of nitrogen oxides by bound dioxygen in hemoproteins. *J. Inorg. Biochem.* **14**, 351–358.

Doyle, M. P., Pickering, R. A., DeWeert, T. M., Hoekstra, J. W., and Pater, D. (1981). Kinetics and mechanism of the oxidation of human deoxyhemoglobin by nitrites. *J. Biol. Chem.* **256**, 12393–12398.

Draper, N. J., and Shah, A. M. (1997). Beneficial effects of a nitric oxide donor on recovery of contractile function following brief hypoxia in isolated rat heart. *J. Mol. Cell. Cardiol.* **29**, 1195–1205.

Duranski, M. R., Greer, J. J. M., Dejam, A., Jaganmohan, S., Hogg, N., Langston, W., Patel, R. P., Yet, S.-F., Wang, X., Kevil, C. G., Gladwin, M. T., and Lefer, D. J. (2005). Cytoprotective effects of nitrite during *in vivo* ischemia-reperfusion of the heart and liver. *J. Clin. Invest.* **115**, 1232–1240.

Feelisch, M., Rassaf, T., Mnaimneh, S., Singh, N., Bryan, N. S., Jourd'heuil, D., and Kelm, M. (2002). Concomitant S-, N-, and heme-nitros(yl)ation in biological tissues and fluids: Implications for the fate of NO *in vivo*. *Faseb J.* **16**, 1775–1785.

Fernandez, B. O., Lorkovic, I. M., and Ford, P. C. (2003). Nitrite catalyzes reductive nitrosylation of the water-soluble ferri-heme model FeIII(TPPS) to FeII(TPPS)(NO). *Inorg. Chem.* **13**, 2–4.

Gaston, B., Sears, S., Woods, J., Hunt, J., Ponaman, M., McMahon, T., and Stamler, J. S. (1998). Bronchodilator s-nitrosothiol deficiency in asthmatic respiratory failure. *Lancet* **351**, 1317–1319.

Ghafourifar, P., and Colton, C. A. (2003). Mitochondria and nitric oxide. *Antioxid. Redox Signal.* **5**, 249–250.

Giovannoni, G., Land, J. M., Keir, G., Thompson, E. J., and Heales, S. J. R. (1997). Adaption of the nitrate reductase and Griess reaction methods for the measurement of serum nitrate plus nitrite levels. *Ann. Clin. Biochem.* **34**, 193–198.

Gladwin, M. T. (2005). Nitrite as an intrinsic signaling molecule. *Nat. Chem. Biol.* **1**, 245–246.

Gladwin, M. T., Ognibene, F. P., Pannell, L. K., Nichols, J. S., Pease-Fye, M. E., Shelhamer, J. H., and Schechter, A. N. (2000a). Relative role of heme nitrosylation and β-cysteine 93 nitrosation in the transport and metabolism of nitric oxide by hemoglobin in the human circulation. *Proc. Natl. Acad. Sci. USA* **97**, 9943–9948.

Gladwin, M. T., Schechter, A., Kim-Shapiro, D. B., Patel, R., Hogg, N., Shiva, S., Cannon, R., Kelm, M., Wink, D., Espey, M., Oldfield, E., Pluta, R., *et al.* (2005). The emerging biology of the nitrite anion. *Nat. Chem. Biol.* **1**, 308–314.

Gladwin, M. T., Schechter, A. N., Shelhamer, J. H., Pannell, L. K., Conway, D. A., Hrinczenko, B. W., Nichols, J. S., Pease-Fye, M., Noguchi, C. T., Rodgers, G. P., and Ognibene, F. P. (1999). Inhaled nitric oxide augments nitric oxide transport on sickle cell hemoglobin without affecting oxygen affinity. *J. Clin. Invest.* **104**, 937–945.

Gladwin, M. T., Shelhamer, J. H., Schechter, A. N., Pease-Fye, M. E., Waclawiw, M. A., Panza, J. A., Ognibene, F. P., and Cannon, R. O. (2000b). Role of circulating nitrite and S-nitrosohemoglobin in the regulation of regional blood flow in humans. *Proc. Natl. Acad. Sci. USA* **97**, 11482–11487.

Gladwin, M. T., Wang, X., Reiter, C. D., Yang, B. K., Vivas, E. X., Bonaventura, C., and Schechter, A. N. (2002). S-nitrosohemoglobin is unstable in the reductive erythrocyte environment and lacks O2/NO-linked allosteric function. *J. Biol. Chem.* **277**, 27818–27828.

Goldman, R. K., Vlessis, A. A., and Trunkey, D. D. (1998). Nitrosothiol quantification in human plasma. *Anal. Biochem.* **259**, 98–103.

Goss, S. P., Hogg, N., and Kalyanaraman, B. (1997). The effect of nitric oxide release rates on the oxidation of human low-density lipoprotein. *J. Biol. Chem.* **272**, 21647–21653.

Gow, A. J., Buerk, D. G., and Ischiropoulos, H. (1997). A novel reaction mechanism for the formation of S-nitrosothiol *in vivo*. *J. Biol. Chem.* **272**, 2841–2845.

Grau, M., Hendgen-Cotta, U., Brouzos, P., Drexhage, C., Rassaf, T., Lauer, T., Dejam, A., Kelm, M., and Kleinbongard, P. (2007). Recent methodological advances in the analysis of nitrite in the human circulation: Nitrite as a biochemical parameter of the L-arginine/NO pathway. *J. Chromatogr. B* **851**, 106–123.

Green, L. C., Wagner, D. A., Glogowski, J., Skipper, P. L., Wishnok, J. S., and Tannenbaum, S. R. (1982). Analysis of nitrate, nitrite, and [^{15}N]nitrate in biological fluids. *Anal. Biochem.* **126**, 131–138.

Heiss, C., Kleinbongard, P., Dejam, A., Perré, S., Schroeter, H., Sies, H., and Kelm, M. (2005). Acute consumption of flavanol-rich cocoa and the reversal of endothelial dysfunction in smokers. *J. Am. Coll. Cardiol.* **46**, 1276–1283.

Herold, S., and Rehmann, F. J. (2001). Kinetic and mechanistic studies of the reactions of nitrogen monoxide and nitrite with ferryl myoglobin. *J. Biol. Inorg. Chem.* **6**, 543–555.

Herold, S., and Rehmann, F. J. (2003). Kinetics of the reactions of nitrogen monoxide and nitrite with ferryl hemoglobin. *Free Radic. Biol. Med.* **34**, 531–545.

Hogg, N., Kalyanaaraman, B., Joseph, J., Struck, A., and Parthasarathy, S. (1993). Inhibition of low-density lipoprotein oxidation by nitric oxide: Potential role in atherogenesis. *FEBS Lett.* **334**, 170–174.

Huang, K. T., Keszler, A., Patel, N., Patel, R. P., Gladwin, M. T., Kim-Shapiro, D. B., and Hogg, N. (2005). The reaction between nitrite and deoxyhemoglobin Reassessment of reaction kinetics and stoichiometry. *J. Biol. Chem.* **280**, 31126–31131.

Ignarro, L. J., Napoli, C., and Loscalzo, J. (2002). Nitric oxide donors and cardiovascular agents modulating the bioactivity of nitric oxide: An overview. *Circ. Res.* **90**, 21–28.

Jia, L., Bonaventura, C., Bonaventura, J., and Stamler, J. S. (1996). S-nitrosohaemoglobin: A dynamic activity of blood involved in vascular control. *Nature* **380,** 221–226.

Jourd'heuil, D., Hallen, K., Feelisch, M., and Grisham, M. B. (2000). Dynamic state of S-nitrosothiols in human plasma and whole blood. *Free Radic. Biol. Med.* **28,** 409–417.

Jourd'heuil, D., Jourd'heuil, F. L., Lowery, A. M., Hughes, I., and Grisham, M. B. (2005). Detection of nitrosothiols and other nitroso species *in vitro* and in cells. *Methods Enzmol.* **396,** 118–131.

Jung, K. H., Chu, K., Ko, S. Y., Lee, S. T., Sinn, D. I., Park, D. K., Kim, J. M., Song, E. C., Kim, M., and Roh, J. K. (2006). Early intravenous infusion of sodium nitrite protects brain against *in vivo* ischemia-reperfusion injury. *Stroke* **37,** 2744–2750.

Kanner, J., and Harel, S. (1985). Initiation of membranal lipid peroxidation by activated metmyoglobin and methemoglobin. *Arch. Biochem. Biophys.* **237,** 314–321.

Kanner, J., Harel, S., and Granit, R. (1992). Nitric oxide, an inhibitor of lipid oxidation by lipoxygenase, cyclooxygenase and hemoglobin. *Lipids* **27,** 46–49.

Kelm, M. (1999). Nitric oxide metabolism and breakdown. *Biochim. Biophys. Acta.* **1411,** 273–289.

Kelm, M., Preik-Steinhoff, H., Preik, M., and Strauer, B. E. (1999). Serum nitrite sensitively reflects endothelial NO formation in human forearm vasculature: Evidence for biochemical assessment of the endothelial L-arginine–NO pathway. *Cardiovasc. Res.* **41,** 765–772.

Kim, Y. M., Kim, T. H., Seol, D. W., Talanian, R. V., and Billiar, T. R. (1998). Nitric oxide suppression of apoptosis occurs in association with an inhibition of Bcl-2 cleavage and cytochrome c release. *J. Biol. Chem.* **273,** 31437–31441.

Kleinbongard, P., Dejam, A., Lauer, T., Jax, T., Kerber, S., Gharini, P., Balzer, J., Zotz, R. B., Scharf, R. E., Willers, R., Schechter, A. N., Feelisch, M., *et al.* (2006a). Plasma nitrite concentrations reflect the degree of endothelial dysfunction in humans. *Free Radic. Biol. Med.* **40,** 295–302.

Kleinbongard, P., Dejam, A., Lauer, T., Rassaf, T., Schindler, A., Picker, O., Scheeren, T., Gödecke, A., Schrader, J., Schulz, R., Heusch, G., Schaub, G. A., *et al.* (2003). Plasma nitrite reflects constitutive nitric oxide synthase activity in mammals. *Free Radic. Biol. Med.* **35,** 790–796.

Kleinbongard, P., Rassaf, T., Dejam, A., Kerber, S., and Kelm, M. (2002). Griess method for nitrite measurement of aqueous and protein-containing sample. *Methods Enzymol.* **359,** 158–168.

Kleinbongard, P., Schulz, R., Rassaf, T., Lauer, T., Dejam, A., Jax, T. W., Kumara, I., Gharini, P., Kabanova, S., Özüyaman, B., Schnürch, H.-G., Gödecke, A., *et al.* (2006b). Red blood cells express a functional endothelial nitric oxide synthase. *Blood* **107,** 2943–2951.

Kosaka, H., Imaizumi, K., and Tyuma, I. (1982). Mechanism of autocatalytic oxidation of oxyhemoglobin by nitrite: An intermediate detected by electron spin resonance. *Biochim. Biophys. Acta.* **702,** 237–241.

Kubes, P., Suzuki, M., and Granger, D. N. (1991). Nitric oxide: An endogenous modulator of leukocyte adhesion. *Proc. Natl. Acad. Sci. USA* **88,** 4651–4655.

Lauer, T., Kleinbongard, P., Preik, M., Rauch, B. H., Deussen, A., Feelisch, M., Strauer, B. E., and Kelm, M. (2003). Direct biochemical evidence for eNOS stimulation by bradykinin in the human forearm vasculature. *Basic Res. Cardiol.* **98,** 84–89.

Lauer, T., Preik, M., Rassaf, T., Strauer, B. E., Deussen, A., Feelisch, M., and Kelm, M. (2001). Plasma nitrite rather than nitrate reflects regional endothelial nitric oxide synthase activity but lacks intrinsic vasodilator action. *Proc. Natl. Acad. Sci. USA* **98,** 12814–12819.

Liao, J. C., Hein, T. W., Vaughn, M. W., Huang, K.-T., and Kuo, L. (1999). Intravascular flow decreases erythrocyte consumption of nitric oxide. *Proc. Natl. Acad. Sci. USA* **96,** 8757–8761.

Lijinsky, W. (1980). Significance of *in vivo* formation of N-nitroso compounds. *Oncology* **37**, 223–226.

Lim, D. G., Sweeney, S., Bloodsworth, A., White, C. R., Chumley, P. H., Krishna, N. R., Schopfer, F., ODonnell, V. B., Eiserich, J. P., and Freeman, B. A. (2002). Nitrolinoleate, a nitric oxide-derived mediator of cell function: Synthesis, characterization, and vasomotor activity. *Proc. Natl. Acad. Sci. USA* **99**, 15941–15946.

Lippton, H. L., Gruetter, C. A., Ignarro, L. J., Meyer, R. L., and Kadowitz, P. J. (1982). Vasodilator actions of several N-nitroso compounds. *Can. J. Physiol. Pharmacol.* **60**, 68–75.

Liu, X., Miller, M. J., Joshi, M. S., SadowskaKrowicka, H., Clark, D. A., and Lancaster, J. R., Jr. (1998). Diffusion-limited reaction of free nitric oxide with erythrocytes. *J. Biol. Chem.* **273**, 18709–18713.

Ludolph, B., Bloch, W., Kelm, M., Schulz, R., and Kleinbongard, P. (2007). Short-term effect of the HMG-CoA reductase inhibitor rosuvastatin on erythrocyte nitric oxide synthase activity. *Vasc. Health Risk Manag.* **3**, 1069–1073.

Lundberg, J. O., and Weitzberg, E. (2005). NO generation from nitrite and its role in vascular control. *Arterioscler. Thromb. Vasc. Biol.* **25**, 915–922.

Marley, R., Feelisch, M., Holt, S., and Moore, K. (2000). A chemiluminescense-based assay for s-nitrosoalbumin and other plasma s-nitrosothiols. *Free Radic. Res.* **32**, 1–9.

Marley, R., Patel, R. P., Orie, N., Caeser, E., Darley-Usmar, V., and Moore, K. (2001). Formation of nanomolar concentrations of S-nitroso-albumin in human plasma by nitric oxide. *Free Radic. Biol. Med.* **31**, 688–696.

Moncada, S., and Higgs, A. (1993). The L-arginine-nitric oxide pathway. *N. Engl. J. Med.* **329**, 2002–2012.

Moncada, S., Palmer, R. M. J., and Higgs, E. A. (1991). Nitric oxide, biology pathophysiology and pharmacology. *Pharmacol. Rev.* **43**, 109–142.

Moriel, P., Pereira, I. R. O., Bertolami, M. C., and Abdalla, D. S. P. (2001). Is ceruloplasmin an important catalyst for S-nitrosothiol generation in hypercholesterolemia? *Free Radic. Biol. Med.* **30**, 318–326.

Moro, M. A., Darley-Usmar, V. M., Goodwin, D. A., Read, N. G., Zamora-Pino, R., Feelisch, M., Radomski, M. W., and Moncada, S. (1994). Paradoxical fate and biological action of peroxynitrite on human platelets. *Proc. Natl. Acad. Sci. USA* **91**, 6702–6706.

Nagababu, E., Ramasamy, S., Albernethy, R., and Rifkind, M. (2003). Active nitric oxide produced in the red cell under hypoxic conditions by deoxyhemoglobin-mediated nitrite reduction. *J. Biol. Chem.* **278**, 46349–46356.

Nakai, K., Sakuma, I., Ohta, T., Ando, J., Kitabatake, A., Nakazato, Y., and Takahashi, T. A. (1998). Permeability characteristics of hemoglobin derivatives across cultured endothelial cell monolayers. *J. Lab. Clin. Med.* **132**, 313–319.

Ng, E. S. M., Jourdheuil, D., McCord, J. M., Knight, D., and Kubes, P. (2004). Enhanced S-nitroso-albumin formation from inhaled NO during ischemia/reperfusion. *Circ. Res.* **94**, 559–565.

Nisoli, E., Clementi, E., Paolucci, C., Cozzi, V., Tonello, C., Sciorati, C., Bracale, R., Valerio, A., Francolini, M., Moncada, S., and Carruba, M. O. (2003). Mitochondrial biogenesis in mammals: The role of endogenous nitric oxide. *Science* **299**, 896–899.

Ohshima, H., and Bartsch, H. (1994). Chronic infections and inflammatory processes as cancer risk factors: Possible role of nitric oxide in carcinogenesis. *Mutat. Res.* **305**, 253–264.

Olson, J. S., Foley, E. W., Rogge, C., Tsai, A. L., Doyle, M. P., and Lemon, D. D. (2004). NO scavenging and the hypertensive effect of hemoglobin-based blood substitutes. *Free Radic. Biol. Med.* **36**, 685–697.

Pabla, R., Buda, A. J., Flynn, D. M., Blessé, S. A., Shin, A. M., Curtis, M. J., and Lefer, D. J. (1996). Nitric oxide attenuates neutrophil-mediated contractile dysfunction after ischemia and reperfusion. *Circ. Res.* **78,** 65–72.

Palmer, R. M. J., Ferrige, A. G., and Moncada, S. (1987). Nitric oxide release accounts for the biological activity of endothelium-derived relaxing factor. *Nature* **327,** 524–526.

Pelletier, M. M., Kleinbongard, P., Ringwood, L., Hunter, C. J., Gladwin, M. T., Schechter, A. N., and Dejam, A. (2006). The measurement of blood and plasma nitrite by chemiluminescence: Pitfalls and solutions. *Free Radic. Biol. Med.* **41,** 541–548.

Poderoso, J. J., Carreras, M. C., Lisdero, C., Riobo, N., Schöpfer, F., and Boveris, A. (1996). Nitric oxide inhibits electron transfer and increases superoxide radical production in rat heart mitochondria and submitochondrial particles. *Arch. Biochem. Biophys.* **328,** 85–92.

Radomski, M. W., Palmer, R. M., and Moncada, S. (1987). Endogenous nitric oxide inhibits human platelet adhesion to vascular endothelium. *Lancet* **7,** 1057–1058.

Ramachandran, N., Root, P., Jiang, X. M., Hogg, P. J., and Mutus, B. (2001). Mechanism of transfer of NO from extracellular S-nitrosothiols into the cytosol by cell-surface protein disulfide isomerase. *Proc. Natl. Acad. Sci. USA* **98,** 9539–9544.

Rassaf, T., Bryan, N. S., Kelm, M., and Feelisch, M. (2002a). Concomitant presence of N-nitroso and S-nitroso proteins in human plasma. *Free Radic. Biol. Med.* **33,** 1590–1596.

Rassaf, T., Bryan, N. S., Maloney, R. E., Specian, V., Kelm, M., Kalyanaraman, B., Rodriguez, J., and Feelisch, M. (2003). NO adducts in mammalian red blood cells: Too much or too little? *Nat. Med.* **9,** 481–482.

Rassaf, T., Feelisch, M., and Kelm, M. (2004). Circulating NO pool: Assessment of nitrite and nitroso species in blood and tissues. *Free Radic. Biol. Med.* **36,** 413–422.

Rassaf, T., Flögel, U., Drexhage, C., Hendgen-Cotta, U., Kelm, M., and Schrader, J. (2007). Nitrite reductase function of deoxymyoglobin: Oxygen sensor and regulator of cardiac energetics and function. *Circ. Res.* **100,** 1749–1754.

Rassaf, T., Heiss, C., Hendgen-Cotta, U., Balzer, J., Matern, S., Kleinbongard, P., Lee, A., Lauer, T., and Kelm, M. (2006). Plasma nitrite reserve and endothelial function in the human forearm circulation. *Free Radic. Biol. Med.* **41,** 295–301.

Rassaf, T., Kleinbongard, P., Preik, M., Dejam, A., Gharini, P., Lauer, T., Erckenbrecht, J., Duschin, A., Schulz, R., Heusch, G., Feelisch, M., and Kelm, M. (2002b). Plasma nitrosothiols contribute to the systemic vasodilator effects of intravenously applied NO: Experimental and clinical study on the fate of NO in human blood. *Circ. Res.* **91,** 470–477.

Rassaf, T., Preik, M., Kleinbongard, P., Lauer, T., Heiß, C., Strauer, B. E., Feelisch, M., and Kelm, M. (2002c). Evidence for *in vivo* transport of bioactive nitric oxide in human plasma. *J. Clin. Invest.* **109,** 1241–1248.

Rayner, B. S., Wu, B. J., Raftery, M., Stocker, R., and Witting, P. K. (2005). Human S-nitroso oxymyoglobin is a store of vasoactive nitric oxide. *J. Biol. Chem.* **280,** 9985–9993.

Rhodes, P. M., Leone, A. M., Francis, P. L., Struthers, A. D., and Moncada, S. (1995). The L-arginine: Nitric oxide pathway is the major source of plasma nitrite in fasted humans. *Biochem. Biophys. Res. Commun.* **209,** 590–596.

Rodriguez, J., Maloney, R. E., Rassaf, T., Bryan, N. S., and Feelisch, M. (2003). Chemical nature of nitric oxide storage forms in rat vascular tissue. *Proc. Natl. Acad. Sci. USA* **100,** 336–341.

Rossi, R., Giustarini, D., Milzani, A., Colombo, R., Dalle-Donne, I., and di Simplicio, P. (2001). Physiological levels of s-nitrosothiols in human plasma. *Circ. Res.* **89,** E47.

Rubbo, H., Radi, R., Trujillo, M., Teller, R., Kalyanaraman, B., Barnes, S., Kirk, M., and Freeman, B. A. (1994). Nitric oxide regulation of superoxide and peroxynitrite-dependent lipid peroxydation: Formation of novel nitrogen-containing oxidized lipid derivatives. *J. Biol. Chem.* **269,** 26066–26075.

Scharfstein, J. S., Keaney, J. F., Jr., Slivka, A., Welch, G. N., Vita, J. A., Stamler, J. S., and Loscalzo, J. (1994). *In vivo* transfer of nitric oxide between a plasma protein-bound reservoir and low molecular weight thiols. *J. Clin. Invest.* **94,** 1432–1439.

Schopfer, F. J., Baker, P. R. S., and Freeman, B. A. (2003). NO-dependent protein nitration: A cell signaling event or an oxidative inflammatory response? *Trends Biochem. Sci.* **28,** 646–654.

Schopfer, F. J., Baker, P. R. S., Giles, G., Chumley, P., Batthyany, C., Crawford, J., Patel, R. P., Hogg, N., Branchaud, B. P., Lancaster, J. R., and Freeman, B. A. (2005). Fatty acid transduction of nitric oxide signaling. *J. Biol. Chem.* **280,** 19289–19297.

Sharma, V. S., Traylor, T. G., Gardiner, R., and Mizukami, H. (2007). Reaction of nitric oxide with heme proteins and model compounds of hemoglobin. *Biochemy Mosc.* **26,** 3837–3843.

Shiva, S., Huang, Z., Grubina, R., Sun, J., Ringwood, L. A., MacArthur, P. H., Xu, X., Murphy, E., DarleyUsmar, V. M., and Gladwin, M. T. (2007a). Deoxymyoglobin is a nitrite reductase that generates nitric oxide and regulates mitochondrial respiration. *Circ. Res.* **100,** 654–661.

Shiva, S., Sack, M. N., Greer, J. J., Duranski, M., Ringwood, L. A., Burwell, L., Wang, X., MacArthur, P. H., Shoja, A., Raghavachari, N., Calvert, J. W., Brookes, P. S., *et al.* (2007b). Nitrite augments tolerance to ischemia/reperfusion injury via the modulation of mitochondrial electron transfer. *J. Exp. Med.* **204,** 2089–2102.

Stamler, J. S., Jaraki, O., Osborne, J., Simon, D. I., Keaney, J., Vita, J., Singel, D., Valeri, C. R., and Loscalzo, J. (1992). Nitric oxide circulates in mammalian plasma primarily as an S-nitroso adduct of serum albumin. *Proc. Natl. Acad. Sci. USA* **89,** 7674–7677.

Tannenbaum, S. R., Witter, J. P., Gatley, S. J., and Balish, E. (1979). Nitrate and nitrite: Origin in humans. *Science* **205,** 1333–1337.

Tsikas, D., Gutzki, F.-M., Rossa, S., Bauer, H., Neumann, C., Dockendorff, K., Sandmann, J., and Frölich, J. C. (1997). Measurement of nitrite and nitrate in biological fluids by gas chromatography–mass spectrometry and by the Griess assay: Problems with the Griess assay–solutions by gas chromatography–mass spectrometry. *Anal. Biochem.* **244,** 208–220.

Tyurin, V. A., Liu, S. X., Tyurina, Y. Y., Sussman, N. B., Hubel, C. A., Roberts, J. M., Taylor, R. N., and Kagan, V. E. (2001). Elevated levels of S-nitrosoalbumin in preeclampsia plasma. *Circ. Res.* **88,** 1210–1215.

Vaughn, M. W., Huang, K. T., Kuo, L., and Liao, J. C. (2000). Erythrocytes possess an instrinsic barrier to nitric oxide consumption. *J. Biol. Chem.* **275,** 2342–2348.

Wang, X., Bryan, N. S., MacArthur, P. H., Rodriguez, J., Gladwin, M. T., and Feelisch, M. (2006). Measurement of nitric oxide levels in the red cell: Validation of tri-iodide based chemiluminescence with acid-sulfanilamide pretreatment. *J. Biol. Chem.* **281,** 26994–27002.

Webb, A., Bond, R., McLean, P., Uppal, R., Benjamin, N., and Ahluwalia, A. (2004). Reduction of nitrite to nitric oxide during ischemia protects against myocardial ischemia-reperfusion damage. *Proc. Natl. Acad. Sci. USA* **101,** 13683–13688.

Wink, D. A., Miranda, K. M., Mitchell, J. B., Grisham, M. B., Fukuto, J. M., and Feelisch, M. (2000). The chemical biology of nitric oxide: Balancing nitric oxide with oxidative and nitrosative stress. *In* "Handbook of Experimental Pharmacology" (B Mayer, ed.), pp. 7–29. Berlin, Germany.

Witting, P. K., Douglas, D. J., and Mauk, A. G. (2001). Reaction of human myoglobin and nitric oxide: Heme iron or protein sulfhydryl (s) nitrosation dependence on the absence or presence of oxygen. *J. Biol. Chem.* **276,** 3991–3998.

Zweier, J. L., Wang, P., Samouilov, A., and Kuppusamy, P. (1995). Enzyme-independent formation of nitric oxide in biological tissues. *Nat. Med.* **1,** 804–809.

CHAPTER SEVENTEEN

DETECTION AND MEASUREMENT FOR THE MODIFICATION AND INACTIVATION OF CASPASE BY NITROSATIVE STRESS *IN VITRO* AND *IN VIVO*

Hee-Jun Na,* Hun-Taeg Chung,[†] Kwon-Soo Ha,* Hansoo Lee,*
Young-Guen Kwon,[‡] Timothy R. Billiar,[§] *and* Young-Myeong Kim*

Contents

Abstract

Nitrosative stress, a nitric oxide (NO)-mediated nitrosylation of redox-sensitive thiols, has been linked to the regulation of signal transduction, gene expression, and cell growth and apoptosis and thus may be widely implicated in both physiological and pathological actions of NO. Protein *S*-nitrosylation has been observed to occur *in vitro* and *in vivo* in pathophysiological conditions. Apoptosis can be regulated by *S*-nitrosylation of the redox-sensitive cysteine residue in

* Vascular System Research Center, Kangwon National University, Chunchon, Korea
† Medicinal Resources Research Institute, Wonkwang University, Iksan, Korea
‡ Department of Biochemistry, College of Science, Yonsei University, Seoul, Korea
§ Department of Surgery, Medical School, University of Pittsburgh, Pittsburgh, Pennsylvania

Methods in Enzymology, Volume 441
ISSN 0076-6879, DOI: 10.1016/S0076-6879(08)01217-2

the active site of all caspase family proteases. Detection and measurement for the modification and inactivation of caspases by *S*-nitrosylation remain a new challenge because of the lability of the *S*-nitrosothiol moiety. This chapter describes approaches for assaying and identifying *S*-nitrosylated caspase enzymes *in vitro* and *in vivo*. These methods permit rapid and reproducible assays of *S*-nitrosylated caspases in biological and clinical specimens and should be useful for studies defining a pathophysiological role of NO in several apoptosis-associated human diseases.

1. Nitrosative Stress

Reactive oxygen species such as O_2^-, H_2O_2, and $\cdot OH$, produced during normal aerobic metabolism or when an organism is exposed to a variety of stimuli, may cause widespread damage to biological molecules, particularly redox-sensitive thiol-containing molecules, through direct reactions with free sulfhydryl groups. Such reactions can also be promoted by the sulfhydryl alkylating reagents, *N*-ethylmaleimide and diazenedicarboxylic acid bis-*N*,*N*-dimethylamide (diamide). This oxidative process, called oxidative stress, decreases the biological activity of proteins and the cellular redox potential (GSH/GSSG ratio). Similarly, nitric oxide (NO) and related molecules, synthesized from L-arginine by the catalytic reaction of three isotypes of NO synthase (NOS), can covalently modify the redox-sensitive thiol groups in proteins and small biomolecules through two chemical reactions, *S*-nitrosylation and oxidation (disulfide bond formation). These posttranslational modifications can also cause the inhibition of protein function and the decreased cellular redox potential and thus can serve in the regulation of cellular responses. Modification of biological thiol residues by NO is related to nitrosative stress. The redox-based *S*-nitrosylation of proteins and nonprotein thiols occurs both *in vitro* and *in vivo* (Kim *et al.*, 1997; Saavedra *et al.*, 1997) and regulates gene expression, signal transduction, and cellular homeostasis through alterations in protein function and cellular redox potential (Kim *et al.*, 1995, 1997; Stamler, 1994). Over the past decade, the number of reported proteins for *S*-nitrosylation has increased to over a hundred (Stamler *et al.*, 2001), which is consistent with the ubiquity of regulatory and/or active site thiols across protein classes. Numerous studies have demonstrated that *S*-nitrosylation functions as a prototype of regulatory mechanisms for various signaling pathways in intact cellular systems through redox-based cellular signals and is also implicated in the pathogenesis of several human diseases. This process is largely dependent on the chemical properties and reactivity of NO.

2. BIOLOGICAL REACTIVITY OF NITRIC OXIDE

Nitric oxide, synthesized from L-arginine by the catalytic reaction of three isotypes of NOS, is a small hydrophobic molecule with unique chemical properties that permits it to function as both an autocrine and a paracrine messenger. It is a relatively stable, uncharged radical that readily crosses lipid membranes and interacts with few specific targets such as oxygen-related molecules, metal-containing biomolecules, and protein tyrosine and cysteine residues. NO reacts with molecular oxygen to decay nitrite as a stable oxidized product. However, this reaction is very slow with the physiological concentration of O_2. NO reacts rapidly with superoxide (O_2^-) to produce the strong oxidant peroxynitrite ($ONOO^-$), which interacts with protein tyrosine residues and forms nitrotyrosine. NO also reacts with many heme and nonheme iron-containing proteins. NO activates soluble guanylyl cyclase by binding to its heme iron to generate GTP to cGMP, which plays a key role in vascular tone regulation and neurotransmission. Other examples of NO interactions with heme prosthetic groups include cytochrome p450 (Wink *et al.*, 1993b) and hemoglobin. NO can also bind to nonheme iron moieties of proteins such as aconitase and complexes I and II of the mitochondrial respiratory chain and inhibit their biological activities. However, NO and reactive nitrogen species (RNS, N_2O_3, $ONOO^-$, NO_3^+, NO^-, etc.) can modify lipid radicals and DNA, as well as different protein residues, producing posttranslational modifications that can alter protein functions. Among these posttranslational modifications, NO-mediated protein thiol modifications such as simple oxidation of disulfide bond formation, *S*-nitrosylation, and ADP-ribosylation have been thoroughly studied in the last decade. *S*-Nitrosylation is modification of a cysteine thiol on the protein or peptide by the addition of a NO diatomic group, leading to several cellular signal processes, such as stress response, apoptosis regulation, and gene expression. This modification is a unique cellular process that does not require the aid of enzymatic catalysis, compared to other signaling mechanisms such as phosphorylation. Thus, it has been proposed that it may represent a new paradigm in signal transduction (Martinez-Ruiz and Lamas, 2004).

3. *S*-NITROSYLATING CHEMISTRY

Although many *S*-nitrosylated proteins have been found *in vivo*, the specific enzymatic reaction for *S*-nitrosylation has not yet been identified. All the reactions of nitrosothiol formation are known to occur chemically.

Although NO can react with thiols, this reaction is unfavorable in a biological system. In general, NO interacts with free thiols in the presence of oxygen and tension metals. NO can interact directly with local electron-accepting species such as oxygen, transition metal ions, or superoxide radicals to generate potent-nitrosylating species (e.g., N_2O_3 and NO^+) that engage in nitrosation of sulfur-containing molecules. The reaction of NO with O_2 is known to form the higher nitrogen oxide N_2O_3, which is thought to be the quintessential *S*-nitrosylating species (Wink *et al.*, 1993a). This reaction of NO and O_2 can be favored in hydrophobic membrane environments, where NO and O_2 partition preferentially into the hydrophobic region (Liu *et al.*, 1998), or even in the hydrophobic pockets of proteins (Nedospasov *et al.*, 2000). N_2O_3 can be partially dissociated into $[^+ON \cdot \cdot NO_2]$, which permits the reaction of the nitrosonium (NO^+) moiety with the nucleophile sulfur atom. This indicates that *S*-nitrosylation can be considered the transfer reaction of NO^+, but not the direct addition of NO. NO^+ can be generated by the loss of one electron from NO. Known electron acceptors are transition metal ions such as iron and copper, which react readily with NO *in vivo*. *S*-Nitrosylating species, including NO^+, can be generated by the interaction of NO with iron–sulfur complexes (Kim *et al.*, 2000). Dinitrosyl–iron complexes synthesized by the direct reaction of iron with NO function as a source of NO^+ for *S*-nitrosylation of caspase-3 *in vitro* (Kim *et al.*, 2000). The heme iron–nitrosyl complex (hemoglobin) and nitrosyl–Cu_2^+ (ceruloplasmin) are sources of the *S*-nitrosylating NO donor (Inoue *et al.*, 1999; Joshi *et al.*, 2002). Endogenous low molecular weight *S*-nitrosothiols (RSNOs) such as *S*-nitroso-L-cysteine, *S*-nitroso-L-homocysteine, and *S*-nitrosoglutathione (GSNO) exhibit various functions, including NO storage, transport, and delivery (Hogg, 2002). *S*-Nitrosothiols release NO in the presence of transition metals (i.e., Cu_2^+), ascorbate, or thiols and can undergo *S*-transnitrosation reactions in which the direct transfer of NO^+ equivalent (i.e., $RS^- \cdot \cdot NO^+$) between RSNO and free cysteine residues of proteins takes place (Wang *et al.*, 2002). This transnitrosation reaction is considered to be the predominant mechanism for the biological actions of GSNO. Nonthiol NO donors such as DEA/NO also cause *S*-nitrosylation in a O_2-dependent manner, suggesting that the NO released from DEA/NO is oxidized to nitrosylating NO species such as N_2O_3 (Bauer *et al.*, 2001).

4. *S*-NITROSYLATING SPECIFICITY

The formation of *S*-nitrosylation can be favored in with more ionizable cysteine (free sulfhydryl state), such as those in which the thiolate anion can be stabilized by acid–base interactions with neighboring groups, either belonging to adjacent residues in the primary consensus sequence called the nitrosylating motif (Stamler *et al.*, 1997) or just in the proximity of charged side groups in a tertiary or quaternary structure (Hess *et al.*, 2001; Stamler *et al.*, 1997). Considering such an acid–base motif comprising the flanking acidic (Asp, Glu) and basic (Arg, His, Lys) residues, deprotonation of thiol to the nucleophilic thiolate (RS$^-$) can be easily suppressed and enhanced by neighboring acidic and basic groups, respectively. Thus, it indicates that the pK_a value of different cysteines in a protein depends on the microenvironment of the protein structure and is an important factor in determining the occurrence of *S*-nitrosylation. However, local hydrophobicity surrounding a free cysteine residue might promote specific nitrosylation because NO is a relative hydrophobic molecule and can be sequestered and stabilized in a hydrophobic microenvironment of proteins. This hydrophobic compartmentalization can arise from the surrounding amino acid composition, tertiary protein structure, and protein–protein interactions. Therefore, NO interacts readily with thiolate to promote *S*-nitrosylation in the region of high local hydrophobicity within proteins (Nedospasov *et al.*, 2000).

5. *S*-NITROSYLATION OF CASPASES

Nitrosative stress has been linked to the inhibition of cell growth and apoptosis and may be widely implicated in the pathogenesis of many human diseases. In particular, the caspase proteolytic enzymes consisting of 15 isoforms (Eckhart *et al.*, 2005), which are key mediators in apoptotic cell death, possess a redox-sensitive cysteine residue in the catalytic site. Caspases are the mammalian counterpart of *ced-3*, a protease required for programmed cell death in the nematode *Caenorhabditis elegans* (Ellis and Horvitz, 1986), and are the central initiator and effector molecules of the apoptotic signaling cascade (Wolf and Green, 1999). These evidences suggest that most, or all, of the caspase cysteine proteases can be *S*-nitrosylated at their active-site cysteines with the subsequent inhibition of enzyme activity. Indeed, our previous *in vitro* study directly demonstrated that NO inhibited seven-recombinant enzymes of the caspase family via *S*-nitrosylation (Li *et al.*, 1997). We also showed that NO inhibits the enzymatic activity of caspase-3 and -8 via *S*-nitrosylation of active-site cysteine residues and suppresses apoptosis of hepatocytes *in vitro* and *in vivo* (Kim *et al.*, 1997, 2000; Saavedra *et al.*, 1997). Furthermore, other studies showed that NO prevents cells from apoptotic cell death by the redox-based *S*-nitrosylation of caspases (Dimmeler *et al.*, 1997; Mannick *et al.*, 1997; Mohr *et al.*, 1997; Tenneti *et al.*, 1997). These evidences indicate that endogenous NO generated by NOS exerts an antiapoptotic function by *S*-nitrosylated inhibition of caspase activity. However, it has been shown that the NO donor

S-nitroso-*N*-acetyl-DL-penicillamine (SNAP) inhibited caspase-1 activity in cells as well as the activity of purified recombinant caspase-1 via *S*-nitrosylation and also prevented the proteolytic activation of pro-IL-1β and pro-IL-18 (interferon-γ-inducing factor) by recombinant caspase-1 (Kim *et al.*, 1998). They also showed that lipopolysaccharide (LPS) stimulation elevated the production of both NO and IL-1β in peritoneal macrophages from wild-type mice, whereas macrophages from iNOS-deficient mice released a higher level of IL-1β, without NO production, than normal macrophages. Addition of the iNOS inhibitor N^G-monomethyl-L-arginine (NMA) significantly increased active IL-1β production in normal macrophages, but not in iNOS-deficient cells. In addition, administration of LPS with the iNOS inhibitor aminoguanidine resulted in higher IL-1β and IFN-γ levels in normal mice but not in iNOS-deficient animals. These results provide another role for induced NO in the direct regulation of the inflammatory response via *S*-nitrosylation-dependent caspase-1 inhibition. Thus, *S*-nitrosylation of caspases is a well-established model for the specific redox-based signal pathway and provides a mechanism for pathophysiological regulation via the posttranslational modification of protein residues.

6. Experimental Procedures

Experimental techniques have been developed for the identification and assay of *S*-nitrosylated proteins and enzymes. This chapter details the experimental techniques used for detecting and measuring the modification and inactivation of caspase by *S*-nitrosylation *in vitro* and *in vivo*.

6.1. Assay of *in vitro* caspase inactivation by nitrosative stress

Cultured cells are exposed to the desired apoptogenic stimulants in a NO-generating system such as chemical NO donor or biological NO synthesis (Kim *et al.*, 1997, 2000; Li *et al.*, 1997). The cell pellets are washed with ice-cold phosphate-buffered saline and lysed in 100 m*M* HEPES buffer (pH 7.4) containing the protease inhibitor cocktail (buffer A: 5 mg/ml aprotinin and pepstatin, 10 mg/ml leupeptin, and 0.5 m*M* phenylmethylsulfonyl fluoride) by three cycles of freeze and thawing. The cell lysates are obtained by centrifugation at 13,000*g* for 20 min at 4 °C. Caspase activity is evaluated by measuring the proteolytic cleavage of chromogenic or fluorogenic peptide substrates specific for each caspase (Alexis Biochemicals). To restore the *S*-nitrosylated active-site thiol of caspases to the initial redox state, lysates are preincubated with 5 m*M* dithiothreitol (DTT) for 30 min at room temperature. For colorimetry, the enzyme reaction is initiated by adding cell lysates (100 μg of protein) preincubated with or without DTT into buffer A containing 200 μM chromogenic peptide substrate (final concentration) in a total volume of 150 μl in a 96-well plate. The reaction mixture is incubated at 37 °C for 1 to 2 h. The increase in absorbance of enzymatically released pNA can be measured at 405 nm in a microplate reader every 15 min. For the fluorometric assay, cell lysates (100 μg of protein) are incubated with buffer A containing 50 μM fluorogenic peptide substrate in a total volume of 100 μl in a 96-well plate. Fluorescence intensity is determined in a fluorometer equipped with filters for excitation (400 nm) and detection of emitted light (505 nm) every 15 min at 37 °C for 1 to 2 h. Caspase activity is calculated from the initial velocity by measuring the increased absorbance or fluorescence every 15min. The reaction mixture without enzyme or substrate is used as a control. Finally, inactivation of caspase activity by nitrosative stress or *S*-nitrosylation is determined by comparing enzyme activities between samples preincubated with or without DTT.

6.2. Assay of *in vivo* hepatic caspase inactivation by nitrosative stress

Hepatic apoptosis can also be induced *in vivo* in mice or rats by intraperitoneally injecting with tumor necrosis factor-α plus D-galactosamine, LPS plus D-galactosamine, acetaminophen, or other apoptogenic chemicals (Liu *et al.*, 2003; Mojena *et al.*, 2001; Saavedra *et al.*, 1997). NO generation can be controlled by subcutaneously implanting Alzet osmotic minipumps (Alzet Co., Palo Alto, CA) containing V-PYRRO/NO (liver-specific NO donor), intraperitoneally injecting with a long half-lived NO donor such as glycol-SNAP1 (half-life is 28–30 h), or intraperitoneally administering with LPS. Livers are isolated after perfusion with phosphate-buffered saline and homogenized in 10 volumes of ice-cold butter A with a Dounce homogenizer. Tissue lysates are obtained by centrifugation at 13,000g for 20 min at 4 °C. Lysates are preincubated with or without 5 mM DTT for 30 min at room temperature, and caspase inactivation by nitrosative stress is determined by the same analytic method as described earlier.

6.3. Detection of caspase *S*-nitrosylation by biotin switch-based immunoblot analysis

The biotin switch assay for detecting *S*-nitrosylated caspases is basically performed as described previously (Jaffrey *et al.*, 2001). With proper controls, the method can be used to detect *in vitro* and *in vivo* *S*-nitrosylated caspase. The basic idea is to block the free cysteine residues in a protein with methyl methanethiosulfonate (MMTS) through sulfonylation, and subsequently NO is released from the S-NO group with ascorbate. The newly formed free cysteine group is subsequently labeled with biotin-conjugated hexyl pyridyldithiopropionamide (HPDP). The labeled proteins can then be purified by streptavidin–agarose and identified by Western blot analysis.

6.3.1. Blockade of protein free thiols

Cells or tissues are exposed *in vitro* and *in vivo* to a proper apotogenic stimulant under various NO-generating conditions (Kang *et al.*, 2004; Kim *et al.*, 1997; Lee *et al.*, 2005; Saavedra *et al.*, 1997). Samples are lysed in fresh HENS buffer (250 mM HEPES, pH 7.7, 1 mM EDTA, 0.1 mM neocuproine, and 2.5% SDS) containing fresh 20 mM MMTS (Pierce) to give a final protein concentration of 0.8 to 1.0 mg/ml, incubated with frequent agitation for 30 min at 50 °C, and centrifuged to remove cell debris at 10,000g for 15 min at 4 °C. In this buffer, MMTS acts as a thiol-specific methylthiolating agent blocking free cysteine residues, and SDS acts as a protein denaturant to ensure access of MMTS to buried cysteines. The divalent metal chelator EDTA and the Cu^+ chelator neocuproine prevent metal-catalyzed denitrosylation. Excess MMTS is removed by adding 2 volumes of prechilled acetone (−20 °C), incubating for 10 min at −20 °C, and then centrifuging at 2000g for 10 min at 4 °C. Residual MMTS can be removed by rinsing the tube and pellet with acetone, followed by recentrifugation. Alternatively, MMTS can be eliminated by a spin column. The pellets are resuspended in 0.1 ml HENS buffer per milligram protein. Until this point, all operations are carried out in the dark because RSNO is light sensitive.

6.3.2. Biotinylation of *S*-nitrosylated protein thiols

The protein sample is mixed with 4 mM biotin–HPDP (Pierce, prepared fresh by diluting a 50 mM stock solution in dimethyl sulfoxide) to give a final concentration of 1 mM. The solution is adjusted to 1 mM sodium ascorbate using a 50 mM stock solution prepared in deionized water and incubated for 1 h at room temperature. Ascorbate removes the NO group from *S*-nitrosothiol, but does not reduce disulfides unless divalent metals such as Cu^{2+} are present. EDTA and neocuproine, present in the protein solution, prevent this nonspecific disulfide reduction. Nonreacted biotin–HPDP is removed by dialysis, acetone precipitation, or a spin column. The protein pellet is resuspended in the same volume of HENS buffer. In some reaction mixtures, biotin–HPDP is omitted as a negative control.

6.3.3. Purification of biotinylated proteins and Western blot analysis

Samples of biotinylated proteins are diluted with 2 volumes of neutralization buffer (20 mM HEPES, pH 7.7, 100 mM NaCl, 1 mM EDTA, and 0.5% Triton X-100). To purify the biotinylated proteins, 15 μl of streptavidin–agarose (Pierce) per milligram of protein used in the initial protein sample is added and incubated for 1 h at room temperature with agitation. The resin is washed five times with washing buffer (20 mM HEPES, pH 7.7, 600 mM NaCl, 1 mM EDTA, and 0.5% Triton X-100). The resin is precipitated by centrifugation at 200g for 10 s at room temperature between each washing step. Bound proteins are eluted in elution buffer (20 mM HEPES, pH 7.7, 100 mM NaCl, 1 mM EDTA, and 100 mM of 2-mercaptoethanol) by incubating for 20 min at 37 °C with gentle agitation. Supernatants are separated by SDS-polyacrylamide gel electrophoresis using a 15% gel, transferred onto a nitrocellulose membrane, and blocked in Blocker Blotto (Pierce, Rockford, IL) at room temperature for 2 h. Blots are incubated with caspase antibodies diluted in Blocker Blotto overnight at 4 °C, washed two times for 10 min each with Tris-buffered saline–Tween, and then incubated with a secondary antibody diluted in Blocker Blotto at room temperature for 1 h. Bands are detected with the SuperSignal West Pico chemiluminescent substrate (Pierce) according to the manufacturer's protocol.

7. SUMMARY

Advances in the analytical technology of protein *S*-nitrosylation make it possible to assay and identify the formation of *S*-nitrosylated caspase proteases in living systems under physiological and pathological conditions. The described methods should enable the detection and measurement for modification and inactivation of caspase by *S*-nitrosylation *in vitro* and *in vivo*. Such methods would be capable of revealing the formation of *S*-nitrosylated redox-sensitive cysteine residues in caspase family proteases, which is highly correlated with the inhibition of apoptotic cell death, and could then help quantify the correlations between *S*-nitrosylated caspase levels and health or disease, ultimately leading to important applications in biological and medical research.

ACKNOWLEDGMENT

This work was supported by the Vascular System Research Center Grant from the Korea Science and Engineering Foundation.

REFERENCES

Bauer, P. M., Buga, G. M., Fukuto, J. M., Pegg, A. E., and Ignarro, L. J. (2001). Nitric oxide inhibits ornithine decarboxylase via S-nitrosylation of cysteine 360 in the active site of the enzyme. *J. Biol. Chem.* **276,** 34458–34464.

Dimmeler, S., Haendeler, J., Nehls, M., and Zeiher, A. M. (1997). Suppression of apoptosis by nitric oxide via inhibition of interleukin-1β-converting enzyme (ICE)-like and cysteine protease protein (CPP)-32-like proteases. *J. Exp. Med.* **185,** 601–607.

Eckhart, L., Ballaun, C., Uthman, A., Kittel, C., Stichenwirth, M., Buchberger, M., Fischer, H., Sipos, W., and Tschachler, E. (2005). Identification and characterization of a novel mammalian caspase with proapoptotic activity. *J. Biol. Chem.* **280,** 35077–35080.

Ellis, H. M., and Horvitz, H. R. (1986). Genetic control of programmed cell death in the nematode *C. elegans. Cell* **44,** 817–829.

Hess, D. T., Matsumoto, A., Nudelman, R., and Stamler, J. S. (2001). S-nitrosylation: Spectrum and specificity. *Nat. Cell Biol.* **3,** E46–E49.

Hogg, N. (2002). The biochemistry and physiology of S-nitrosothiols. *Annu. Rev. Pharmacol. Toxicol.* **42,** 585–600.

Inoue, K., Akaike, T., Miyamoto, Y., Okamoto, T., Sawa, T., Otagiri, M., Suzuki, S., Yoshimura, T., and Maeda, H. (1999). Nitrosothiol formation catalyzed by ceruloplasmin: Implication for cytoprotective mechanism *in vivo. J. Biol. Chem.* **274,** 27069–27075.

Jaffrey, S. R., Erdjument-Bromage, H., Ferris, C. D., Tempst, P., and Snyder, S. H. (2001). Protein S-nitrosylation: A physiological signal for neuronal nitric oxide. *Nat. Cell Biol.* **3,** 193–197.

Joshi, M. S., Ferguson, T. B., Jr., Han, T. H., Hyduke, D. R., Liao, J. C., Rassaf, T., Bryan, N., Feelisch, M., and Lancaster, J. R., Jr. (2002). Nitric oxide is consumed, rather than conserved, by reaction with oxyhemoglobin under physiological conditions. *Proc Natl Acad. Sci. USA* **99,** 10341–10346.

Kang, Y. C., Kim, K. M., Lee, K. S., Namkoong, S., Lee, S. J., Han, J. A., Jeoung, D., Ha, K. S., Kwon, Y. G., and Kim, Y. M. (2004). Serum bioactive lysophospholipids prevent TRAIL-induced apoptosis via PI3K/Akt-dependent cFLIP expression and Bad phosphorylation. *Cell Death Differ.* **11,** 1287–1298.

Kim, Y. M., Bergonia, H., and Lancaster, J. R., Jr. (1995). Nitrogen oxide-induced autoprotection in isolated rat hepatocytes. *FEBS Lett.* **374,** 228–232.

Kim, Y. M., Chung, H. T., Simmons, R. L., and Billiar, T. R. (2000). Cellular non-heme iron content is a determinant of nitric oxide-mediated apoptosis, necrosis, and caspase inhibition. *J. Biol. Chem.* **275,** 10954–10961.

Kim, Y. M., Talanian, R. V., and Billiar, T. R. (1997). Nitric oxide inhibits apoptosis by preventing increases in caspase-3-like activity via two distinct mechanisms. *J. Biol. Chem.* **272,** 31138–31148.

Kim, Y. M., Talanian, R. V., Li, J., and Billiar, T. R. (1998). Nitric oxide prevents IL-1β and IFN-γ-inducing factor (IL-18) release from macrophages by inhibiting caspase-1 (IL-1β-converting enzyme). *J. Immunol.* **161,** 4122–4128.

Lee, S. J., Kim, K. M., Namkoong, S., Kim, C. K., Kang, Y. C., Lee, H., Ha, K. S., Han, J. A., Chung, H. T., Kwon, Y. G., and Kim, Y. M. (2005). Nitric oxide inhibition of homocysteine-induced human endothelial cell apoptosis by down-regulation of p53-dependent Noxa expression through the formation of S-nitrosohomocysteine. *J. Biol. Chem.* **280,** 5781–5788.

Li, J., Billiar, T. R., Talanian, R. V., and Kim, Y. M. (1997). Nitric oxide reversibly inhibits seven members of the caspase family via S-nitrosylation. *Biochem. Biophys. Res. Commun.* **240,** 419–424.

Liu, J., Li, C., Waalkes, M. P., Clark, J., Myers, P., Saavedra, J. E., and Keefer, L. K. (2003). The nitric oxide donor, V-PYRRO/NO, protects against acetaminophen-induced hepatotoxicity in mice. *Hepatology* **37,** 324–333.

Liu, X., Miller, M. J., Joshi, M. S., Thomas, D. D., and Lancaster, J. R., Jr. (1998). Accelerated reaction of nitric oxide with O_2 within the hydrophobic interior of biological membranes. *Proc. Natl. Acad. Sci. USA* **95,** 2175–2179.

Mannick, J. B., Miao, X. Q., and Stamler, J. S. (1997). Nitric oxide inhibits Fas-induced apoptosis. *J. Biol. Chem.* **272**, 24125–24128.

Martinez-Ruiz, A., and Lamas, S. (2004). S-nitrosylation: A potential new paradigm in signal transduction. *Cardiovasc. Res.* **62**, 43–52.

Mohr, S., Zech, B., Lapetina, E. G., and Brune, B. (1997). Inhibition of caspase-3 by S-nitrosation and oxidation caused by nitric oxide. *Biochem. Biophys. Res. Commun.* **238**, 387–391.

Mojena, M., Hortelano, S., Castrillo, A., Diaz Guerra, M. J., Garcia-Barchino, M. J., Saez, G. T., and Bosca, L. (2001). Protection by nitric oxide against liver inflammatory injury in animals carrying a nitric oxide synthase-2 transgene. *FASEB J.* **15**, 583–585.

Nedospasov, A., Rafikov, R., Beda, N., and Nudler, E. (2000). An autocatalytic mechanism of protein nitrosylation. *Proc. Natl. Acad. Sci. USA* **97**, 13543–13548.

Saavedra, J. E., Billiar, T. R., Williams, D. L., Kim, Y. M., Watkins, S. C., and Keefer, L. K. (1997). Targeting nitric oxide (NO) delivery *in vivo*: Design of a liver-selective NO donor prodrug that blocks tumor necrosis factor-α-induced apoptosis and toxicity in the liver. *J. Med. Chem.* **40**, 1947–1954.

Stamler, J. S. (1994). Redox signaling: Nitrosylation and related target interactions of nitric oxide. *Cell* **78**, 931–936.

Stamler, J. S., Lamas, S., and Fang, F. C. (2001). Nitrosylation: The prototypic redox-based signaling mechanism. *Cell* **106**, 675–683.

Stamler, J. S., Toone, E. J., Lipton, S. A., and Sucher, N. J. (1997). (S)NO signals: Translocation, regulation, and a consensus motif. *Neuron* **18**, 691–696.

Tenneti, L., D'Emilia, D. M., and Lipton, S. A. (1997). Suppression of neuronal apoptosis by S-nitrosylation of caspases. *Neurosci. Lett.* **236**, 139–142.

Wang, P. G., Xian, M., Tang, X., Wu, X., Wen, Z., Cai, T., and Janczuk, A. J. (2002). Nitric oxide donors: Chemical activities and biological applications. *Chem. Rev.* **102**, 1091–1134.

Wink, D. A., Darbyshire, J. F., Nims, R. W., Saavedra, J. E., and Ford, P. C. (1993a). Reactions of the bioregulatory agent nitric oxide in oxygenated aqueous media: Determination of the kinetics for oxidation and nitrosation by intermediates generated in the NO/O_2 reaction. *Chem. Res. Toxicol.* **6**, 23–27.

Wink, D. A., Osawa, Y., Darbyshire, J. F., Jones, C. R., Eshenaur, S. C., and Nims, R. W. (1993b). Inhibition of cytochromes P450 by nitric oxide and a nitric oxide-releasing agent. *Arch. Biochem. Biophys.* **300**, 115–123.

Wolf, B. B., and Green, D. R. (1999). Suicidal tendencies: Apoptotic cell death by caspase family proteinases. *J. Biol. Chem.* **274**, 20049–20052.

CHAPTER EIGHTEEN

INTERACTIVE RELATIONS BETWEEN NITRIC OXIDE (NO) AND CARBON MONOXIDE (CO): HEME OXYGENASE-1/CO PATHWAY IS A KEY MODULATOR IN NO-MEDIATED ANTIAPOPTOSIS AND ANTI-INFLAMMATION

Hun-Taeg Chung,* Byung-Min Choi,* Young-Guen Kwon,[†] *and* Young-Myeong Kim[‡]

Contents

Abstract

Nitric oxide (NO) and carbon monoxide (CO) are synthesized from L-arginine and heme by the catalytic reaction of NO synthase (NOS) and heme oxygenase (HO). NO, a highly reactive free radical, plays an important role in the regulation of vascular and immune function, antiapoptosis, and neurotransmission by producing cGMP, nitrosyl iron complexes, and *S*-nitrosothiols. CO, a more stable molecule, exerts similar biological activities to those of NO by cGMP production, p38 mitogen-activated protein kinase activation, and nuclear factor-κB activation. NO induces the suppression of apoptosis and inflammation in hepatocytes and macrophages by an elevation in HO-1 and CO production, and these effects

* Medicinal Resources Research Institute, Wonkwang University, Iksan, Korea
† Department of Biochemistry, College of Science, Yonsei University, Seoul, Korea
‡ Vascular System Research Center, Kangwon National University, Chunchon, Korea

Methods in Enzymology, Volume 441
ISSN 0076-6879, DOI: 10.1016/S0076-6879(08)01218-4

were not observed in mice lacking HO-1 as well as in cells treated with a HO-1 inhibitor. These evidences indicate that the HO-1/CO pathway is a key player in NO-mediated cytoprotection and anti-inflammation. This chapter reviews new advances in the interactive relations between iNOS/NO and HO-1/CO pathways in the regulation of apoptosis and inflammation.

1. Introduction

Nitric oxide (NO), synthesized from L-arginine by a family of NO synthase (NOS) isoenzymes, is a small, diffusible, highly reactive molecule with dichotomous regulatory roles under physiological and pathological conditions (Nathan, 1992). Constitutive NOS, such as endothelial NOS (eNOS) and neuronal NOS (nNOS), is activated by a transitory increase generally in cytosolic calcium, which promotes the release of NO over several minutes. Fluid shear stress and vascular endothelial growth factor also activate eNOS by a mechanism that does not require an increase in the intracellular free calcium level and is sensitive to inhibitors of the PI3K/Akt pathway, demonstrating that eNOS can be activated by Akt-dependent phosphorylation at serine 1177 (Dimmeler *et al.*, 1999). Inducible NOS (iNOS) is expressed in many cells, including macrophages and hepatocytes after stimulation with immunological or inflammatory stimulants (Nathan, 1992). When NO was unveiled as the long-sought endothelium–derived relaxing factor in the 1980s, the concept of a gaseous molecule acting as a second messenger responsible for maintaining systemic blood pressure was entirely novel (Palmer *et al.*, 1987). It is now clear that NO possesses numerous vital biological roles, including neurotransmission, host defense, inhibition of platelet aggregation, and regulation of blood flow (Nathan, 1992).

The heme oxygenase-1/carbon monoxide (HO-1/CO) system has shown an explosion of research interest due to its newly discovered physiological effects. This metabolic pathway, first characterized by Tenhunen *et al.* (1968), has only recently revealed its surprising cytoprotective and anti–inflammatory properties (Choi *et al.*, 2003; Otterbein *et al.*, 2000). Research in the HO-1/CO pathway now embraces the entire field of medicine where reactive oxygen/nitrogen species, inflammation, growth control, and apoptosis represent important pathophysiological mechanisms (Li *et al.*, 2007; Ryter *et al.*, 2007). Since Kim *et al.* (1995a,b) first showed that NO can increase HO–1 expression, which is responsible for NO-mediated cytoprotection of hepatocytes from oxidative stress-induced cytotoxicity, interactive relations between NOS/NO and HO/CO pathways have received considerable attention as an emerging field in biomedical research. This chapter highlights the interactive relationship and action of two gaseous molecules, NO and CO, the potential regulatory effects and signals against apoptosis, and the inflammatory process.

2. INDUCTION OF HO-1 BY THE NOS/NO PATHWAY

Since the discovery of NO as a potent physiological regulator of many processes, including vascular tone, neurotransmission, inflammation, and bacterial killing, intensive investigation has focused on the role of NO in the regulation of inducible gene expression such as HO-1 (Kim *et al.*, 1995a) and HSP72 (Kim *et al.*, 1997). NO promotes potent HO-1 induction in many cell types, including hepatocytes and endothelial cells (Kim *et al.*, 1995a; Yee *et al.*, 1996). We initially demonstrated that NO produced by iNOS and the chemical NO donor *S*-nitroso-*N*-acetylpenicillamine (SNAP) increases HO-1 activity and promotes cellular heme degradation in hepatocytes and endothelial cells (Kim *et al.*, 1995a,b; Yee, 1996), leading to the protection of hepatocytes from both oxidative stress- and nitrosative stress-induced cell death. These biochemical events and hepatoprotective effects were effectively suppressed through the inhibition of both iNOS and HO-1 activities by N^G-monomethyl-L-arginine and tin-protoporphyrin. HO-1 induction also follows the application of a number of chemical NO donor compounds, including *S*-nitrosoglutathione, SNAP, and sodium nitroprusside (SNP) (Chen and Maines, 2000; Yee *et al.*, 1996). The most important initial event in the NO-mediated signaling pathway is nitrosylation of iron-containing proteins and redox-sensitive thiols. NO reacts highly with the prosthetic heme group of soluble guanylyl cyclase (sGC) and activates this enzyme, leading to enhanced production of the intracellular signal molecule cyclic $3',5'$-guanosine monophosphate (cGMP) from GTP. The membrane-permeable cGMP analog 8-Br-cGMP has been shown to elevate HO-1 expression in hepatocytes, and ODQ, a selective inhibitor of guanylate cyclase, partially prevented the NO donor SNP-induced increase in HO-1 expression (Choi *et al.*, 2003), suggesting that the cGMP-dependent pathway is involved in the NO action of HO-1 expression. In addition, NO and related molecules can covalently modify the redox-sensitive thiol groups in proteins and small biomolecules through two nitrosative reactions, *S*-nitrosylation and oxidation (disulfide bond formation). This nitrosative stress has been suggested to play an important role in signal transduction and in the regulation of gene expression. Indeed, nitrosative stress by NO-mediated GSH depletion has been shown to be an intercellular signal event underlying the increase in HO-1 expression in response to NO (Yee *et al.*, 1996). One study showed that HO-1 induction required activation of the transcription factor Nrf2, which forms a complex with the redox-sensitive regulatory protein Keap1. We have found that NO can modify the reactive thiol groups of Keap1, leading to substantial conformational changes in Keap1, resulting in dissociation of the Keap1–Nrf2 complex. The liberated Nrf2 migrates to the nucleus where it activates the antioxidant response element of the HO-1 promoter and accelerates its transcription (data not shown). We also showed that CO can stimulate HO-1 expression in HepG2 cells by Nrf2 activation via mitogen-activated protein kinase (MAPK) signaling pathways (Lee *et al.*, 2006), suggesting that CO can stimulate autocrine positive feedback regulation of HO-1 gene expression. These evidences strongly suggest that endogenous and exogenous NO is an important regulator for activation of the HO-1/CO pathway.

3. ANTIAPOPTOTIC ROLES OF THE NOS/HO-1 PATHWAY

Nitric oxide displays antiapoptotic properties in some cells, including hepatocytes. The biochemical mechanism underlying NO-mediated anti-apoptotic effects may be dependent on cell type specificity with multiple signal pathways, such as *S*-nitrosylation of caspases, cGMP-dependent PI3K/Akt pathway, and antiapoptotic gene expression (Kim *et al.*, 1999). As a downstream signal molecule of NO, HO-1 acts as a good antiapoptotic candidate. Fibroblasts and neurons overexpressing HO-1 are resistant to stress-mediated cell death, and fibroblasts deficient in HO-1 are particularly susceptible to stressful or toxic insults (Chen *et al.*, 2000; Ferris *et al.*, 1999). Similarly, human renal epithelial cells resist cisplatin-induced apoptosis and necrosis when HO-1 activity is enhanced by HO-1 gene transfer, and mice lacking HO-1 exhibit increased susceptibility to apoptosis and necrosis after treatment with cisplatin (Shiraishi *et al.*, 2000). Because HO-1 catalyzes the conversion of heme to free iron, CO, and biliverdin, these products may be involved in the protective action of cells from apoptosis or cellular injury, as well as other HO-1-mediated biological activities.

Early studies have provided evidence that the induction of HO-1 by NO inhibits apoptotic cell death of hepatocytes via the regulation of cellular homeostasis of prooxidant iron (Kim *et al.*, 1995a; Yee *et al.*, 1996). This result was reconfirmed by Ferris and co-workers (1999), who showed the correlation of HO-1-mediated cytoprotection with a decrease in intracellular free iron amounts and reproduced the protective effect by iron chelation in HO1-deficient fibroblasts. HO-1 metabolizes heme to free iron, CO, and biliverdin. Liberated iron binds to the unsaturated iron–sulfur cluster (3Fe-4S, inactive form) of the iron response element-binding protein (IRE-BP, i.e., cytosolic aconotase) to form active IRE-BP (4Fe-4S). Active IRE-BP binds to the 5′-untranslated region of ferritin mRNA and increases its stability, leading to an increase in cellular ferritin protein content (Yee *et al.*, 1996). Newly synthesized apoferritins decrease the level of cellular free iron, which is involved in the Fenton reaction to produce a highly cytotoxic hydroxyl radical. Therefore, ferritin induction by the NO/HO-1 pathway protects endothelial cells, hepatocytes, and Jurkat T cells from oxidative stress- and Fas-induced apoptosis (Choi et al., 2004; Kim et al., 1995a; Yee et al., 1996).

Another possible cytoprotective effect of NO-mediated HO-1 expression is associated with the generation of CO, activation of sGS, and subsequent production of cGMP, similar to one of the NO-mediated cytoprotective actions. We showed that (1) the hepatoprotective effect of NO donor-mediated HO-1 induction was significantly suppressed by the biological CO scavenger hemoglobin and (2) the pharmacological CO donor, CO-RM, protected hepatocytes from glucose deprivation-induced apoptosis (Choi et al., 2003). Several protective mechanisms of the HO-1/CO pathway have been suggested (Boruard et al., 2000; Wang et al., 2007; Zuckerbraun et al., 2003). First, the protective effect of CO against hyperoxia-induced endothelial cell apoptosis is likely to occur via inhibition of reactive oxygen species production, Bid activation, and the mitochondrial translocation of Bax (Wang et al., 2007). Second, HO-1-derived CO suppresses the apoptosis of endothelial cells through a mechanism that is dependent on the activation of p38 MAPK (Brouard et al., 2000). Finally, CO protects against tumor necrosis factor (TNF)-α-induced fulminant liver damage through a unique dependence on iNOS expression (Zuckerbraun et al., 2003). Upregulation of iNOS by CO has been demonstrated to operate through an increase in nuclear factor (NF)-κB activation. Thus, hepatocytes lacking iNOS activity were not protected significantly by CO from TNF-α-induced cell death, and the protective effect of CO in wild-type hepatocytes was slightly protected by the inhibition of the iNOS/NO pathway, suggesting the potential existence of an additional protective mechanism by which CO exerts its effects. One possibility is that other protective genes, such as cFLIP, activated by NF-κB can be involved in CO-induced hepatoprotection (Kim et al., 2006). These data demonstrate a functional relationship and essential synergy between the iNOS/NO and HO-1/CO pathways in tandem, resulting in potent hepatoprotection.

Biliverdin is a third generated heme catabolite by HO-1 and is converted to bilirubin by the catalytic reaction of biliverdin reductase. Both compounds are reducing species and hence may play a role in the protective response to vascular injury by oxidative stress (Ryter et al., 2007). However, biliverdin, bilirubin, and iron did not exert any protective effect against oxidative stress-induced hepatic apoptosis (Choi et al., 2003), suggesting that CO may be the key molecule mediating the cytoprotective effect of HO-1. Whether HO-1 exerts its antiapoptotic effect principally through the action of CO, the regulation of cellular iron, or even the generation of bilirubin is a matter for further study. It is, however, possible that two or more of the by-products of HO-1 are required for full protection, or that the requirements differ depending on cell type and apoptotic stimulus. Given the relevance of apoptosis to so many disease processes, the combination effects of heme metabolites will doubtless be an area of active investigation in the future.

4. ANTI-INFLAMMATORY ROLES OF THE NOS/HO-1 PATHWAY

Macrophages stimulated with lipopolysaccharide (LPS) produce NO and CO by concomitant transcriptional activation of iNOS and HO-1, and inhibition of iNOS activity by N-monomethyl-L-arginine significantly suppressed LPS-derived induction of HO-1 without affecting the iNOS protein level (Srisook *et al.*, 2005). As shown in hepatocytes (Kim *et al.*, 1995a), macrophages exposed to an NO donor, spermine NONOate, elevated the expression of HO-1, but not iNOS expression. This indicates that iNOS-derived NO is partially responsible for an increase in HO-1 expression in immune-activated macrophages. These findings raise an important question whether the HO-1/CO pathway can inversely regulate the expression of iNOS or other cytokine genes. Direct transfer of HO-1 to macrophages resulted in LPS-induced NO production (Lin *et al.*, 2003). There are two possible mechanisms for the HO-1-mediated inhibition of NO production. First, HO-1 degrades hemes, acting as a catalytic cofactor of NOS, and subsequently limits the intracellular assembly of dimeric iNOS by preventing heme insertion and decreasing heme availability in macrophages (Albakri *et al.*, 1996; Kim *et al.*, 1995b), leading to the inhibition of NOS activity without affecting iNOS expression. Another possibility is that CO and biliverdin suppress transcriptional iNOS expression and TNF-α production in LPS-stimulated macrophages (Sawle *et al.*, 2005; Wang *et al.*, 2004). HO-1-mediated CO production has also been demonstrated to be critically involved in the anti-inflammatory cytokine interleukin (IL)-10-mediated suppression of TNF-α production in LPS-stimulated macrophages (Kim *et al.*, 2005b; Lee and Chu, 2002). These results provide evidence that authentic CO gas increased LPS-induced expression of IL-10, which then promoted HO-1 expression and CO production, indicating that IL-10 and HO-1 activate a positive feedback circuit amplifying anti-inflammatory capacity via the upregulation of their respective expressions.

Other candidate mechanisms for anti-inflammatory effects of the NO-induced HO-1/CO pathway include the inhibition of trafficking toll-like receptors to lipid rafts (Nakahira *et al.*, 2006), the p38-mediated increase in the expression of HSP70 expression and caveolin-1 (Kim *et al.*, 2005a,b), and the suppression of major histocompatibility complex II expression (Wu *et al.*, 2005). In addition, CO can undergo inhibition of platelet aggregation and thrombosis (Sato *et al.*, 2001), probably by the elevation of sGS-dependent cGMP production, which is the same signal pathway as NO anti-inflammation. Thus, HO-1-derived CO may have a series of regulatory effects on various inflammatory diseases, including rejection endotoxic shock, and atherosclerosis (Ryter *et al.*, 2007).

5. INTERACTIVE RELATIONS BETWEEN NO AND CO

The generation of NO by NOS bears an uncanny resemblance to the production of CO by HO-1. The diatomic gasses, NO and CO, are produced endogenously by both enzymes that have constitutive and inducible forms. The NOS family consists of two constitutive forms and one inducible form as described earlier, whereas there are three isotypes, constitutive HO-2 and HO-3 and inducible HO-1, in the HO family. Both NO and CO share affinity for the heme moiety, and thus both are capable of activating heme-containing sGS. NO stimulates sGC by binding to the heme moiety to form a 5-coordinate complex leading to as much as a 130-fold activation of the purified enzyme. CO can also bind to the heme group of sGC to form a six-coordinate complex with high affinity but leads to a far lower level (4.4-fold) of activation of the enzyme (Stone and Marletta, 1994). This indicates that sGC acts as a common receptor for both NO- and CO-mediated signaling cascades, mediating many of their biological effects via the same signal pathway by elevating cGMP production in select cells or tissues. This common pathway plays an important role in vascular relaxation, immune suppression of platelet aggregation, and neurotransmission (Nathan, 1992). This pathway is also involved in the cytoprotective effect of NO and CO through activation of the PI3K/Akt pathway (Ha *et al.*, 2003; Li *et al.*, 2007). One important difference between NO and CO is that NO is a highly reactive molecule by virtue of its unpaired electron, whereas CO is an inert molecule. NO and related species can interact with various biological molecules, such as molecular oxygen, superoxide, transition metals, protein tyrosine residues, and protein and nonprotein thiol groups, leading to numerous types of biological actions. Although HO-derived CO exerts very similar biological functions, such as immune suppression, cytoprotection, and neurotransmission, to those of the NOS/NO pathway, its interactive target molecules, in addition to sGS, p38 MAPK, and NF-κB, have not yet been elucidated clearly.

6. CONCLUSION

Nitric oxide and its related molecules exert the double-edged effects on cell death and inflammation, depending on its rate of production, the redox state of the cells, and cell types. NO activates the apoptotic signal cascade in some situations, whereas it protects cells against spontaneous or induced apoptosis in other cases. NO directly inhibits the activity of caspases through *S*-nitrosylation of the cysteine thiol at their catalytic site, providing

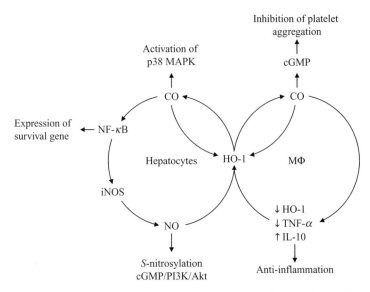

Figure 18.1 Major interactive relationship between iNOS/NO and HO-1/CO pathways for the regulation of apoptotic cell death and inflammation. Typical antiapoptotic and anti-inflammatory pathways are described in hepatocytes and macrophages (MΦ), respectively.

an efficient means to block apoptosis. Other antiapoptotic effects of NO rely on nitrosative stress-mediated transcriptional induction of HO-1. One of the defenses used most widely in nature is the enzyme HO-1. HO-1 has been shown to possess anti-inflammatory, antiapoptotic, and antiproliferative effects, most likely by producing CO, and it is now known to have salutary effects in diseases as diverse as atherosclerosis and sepsis. The protective effects of NO and CO share the common pathway of cGMP-dependent regulation of PI3K/Akt activation, cytochrome *c* release, and bcl-2 expression. This chapter described the interactive relationship between NOS/NO and HO-1/CO pathways conferring cytoprotection and anti-inflammation (Fig. 18.1). The biochemical mechanism and pathophysiological significance of these signal pathways should be further studied to provide new therapeutic strategies for human diseases where alterations of apoptosis and inflammation are involved.

ACKNOWLEDGMENT

This work was supported by the Vascular System Research Center Grant from the Korea Science and Engineering Foundation.

REFERENCES

Albakri, Q. A., and Stuehr, D. J. (1996). Intracellular assembly of inducible NO synthase is limited by nitric oxide-mediated changes in heme insertion and availability. *J. Biol. Chem.* **271,** 5414–5421.

Brouard, S., Otterbein, L. E., Anrather, J., Tobiasch, E., Bach, F. H., Choi, A. M., and Soares, M. P. (2000). Carbon monoxide generated by heme oxygenase 1 suppresses endothelial cell apoptosis. *J. Exp. Med.* **192,** 1015–1026.

Chen, K., Gunter, K., and Maines, M. D. (2000). Neurons overexpressing heme oxygenase-1 resist oxidative stress-mediated cell death. *J. Neurochem.* **75,** 304–313.

Chen, K., and Maines, M. D. (2000). Nitric oxide induces heme oxygenase-1 via mitogen-activated protein kinases ERK and p38. *Cell Mol. Biol.* **46,** 609–617.

Choi, B. M., Pae, H. O., Jeong, Y. R., Oh, G. S., Jun, C. D., Kim, B. R., Kim, Y. M., and Chung, H. T. (2004). Overexpression of heme oxygenase (HO)-1 renders Jurkat T cells resistant to fas-mediated apoptosis: Involvement of iron released by HO-1. *Free Radic. Biol. Med.* **36,** 858–871.

Choi, B. M., Pae, H. O., Kim, Y. M., and Chung, H. T. (2003). Nitric oxide-mediated cytoprotection of hepatocytes from glucose deprivation-induced cytotoxicity: Involvement of heme oxygenase-1. *Hepatology* **37,** 810–823.

Dimmeler, S., Fleming, I., Fisslthaler, B., Hermann, C., Busse, R., and Zeiher, A. M. (1999). Activation of nitric oxide synthase in endothelial cells by Akt-dependent phosphorylation. *Nature* **399,** 601–660.

Ferris, C. D., Jaffrey, S. R., Sawa, A., Takahashi, M., Brady, S. D., Barrow, R. K., Tysoe, S. A., Wolosker, H., Baranano, D. E., Dore, S., Poss, K. D., and Snyder, S. H. (1999). Haem oxygenase-1 prevents cell death by regulating cellular iron. *Nat. Cell Biol.* **1,** 152–157.

Ha, K. S., Kim, K. M., Kwon, Y. G., Bai, S. K., Nam, W. D., Yoo, Y. M., Kim, P. K., Chung, H. T., Billiar, T. R., and Kim, Y. M. (2003). Nitric oxide prevents 6-hydroxydopamine-induced apoptosis in PC12 cells through cGMP-dependent PI3 kinase/Akt activation. *FASEB J.* **17,** 1036–1047.

Kim, H. P., Wang, X., Nakao, A., Kim, S. I., Murase, N., Choi, M. E., Ryter, S. W., and Choi, A. M. (2005a). Caveolin-1 expression by means of p38β mitogen-activated protein kinase mediates the antiproliferative effect of carbon monoxide. *Proc. Natl. Acad. Sci. USA* **102,** 11319–11324.

Kim, H. P., Wang, X., Zhang, J., Suh, G. Y., Benjamin, I. J., Ryter, S. W., and Choi, A. M. (2005b). Heat shock protein-70 mediates the cytoprotective effect of carbon monoxide: Involvement of p38 beta MAPK and heat shock factor-1. *J. Immunol.* **175,** 2622–2629.

Kim, H. S., Loughran, P. A., Kim, P. K., Billiar, T. R., and Zuckerbraun, B. S. (2006). Carbon monoxide protects hepatocytes from TNF-alpha/actinomycin D by inhibition of the caspase-8-mediated apoptotic pathway. *Biochem. Biophys. Res. Commun.* **344,** 1172–1178.

Kim, Y. M., Bergonia, H., and Lancaster, J. R., Jr. (1995a). Nitrogen oxide-induced autoprotection in isolated rat hepatocytes. *FEBS Lett.* **374,** 228–232.

Kim, Y. M., Bergonia, H. A., Muller, C., Pitt, B. R., Watkins, W. D., and Lancaster, J. R., Jr. (1995b). Loss and degradation of enzyme-bound heme induced by cellular nitric oxide synthesis. *J. Biol. Chem.* **270,** 5710–5713.

Kim, Y. M., Bombeck, C. A., and Billiar, T. R. (1999). Nitric oxide as a bifunctional regulator of apoptosis. *Circ. Res.* **84,** 253–256.

Kim, Y. M., de Vera, M. E., Watkins, S. C., and Billiar, T. R. (1997). Nitric oxide protects cultured rat hepatocytes from tumor necrosis factor-alpha-induced apoptosis by inducing heat shock protein 70 expression. *J. Biol. Chem.* **272,** 1402–1411.

Lee, B. S., Heo, J., Kim, Y. M., Shim, S. M., Pae, H. O., Kim, Y. M., and Chung, H. T. (2006). Carbon monoxide mediates heme oxygenase 1 induction via Nrf2 activation in hepatoma cells. *Biochem. Biophys. Res. Commun.* **343,** 965–972.

Lee, T. S., and Chau, L. Y. (2002). Heme oxygenase-1 mediates the anti-inflammatory effect of interleukin-10 in mice. *Nat. Med.* **8,** 240–246.

Li, M. H., Jang, J. H., Na, H. K., Cham, Y. N., and Surh, Y. J. (2007). Carbon monoxide produced by upregulated heme oxygenase-1 in response to nitrosative stress induces expression of glutamate cysteine ligase in PC12 cells via activation of PI3K-Akt and Nrf2-ARE signaling. *J. Biol. Chem.* **282,** 28577–28586.

Lin, H. Y., Juan, S. H., Shen, S. C., Hsu, F. L., and Chen, Y. C. (2003). Inhibition of lipopolysaccharide-induced nitric oxide production by flavonoids in RAW264.7 macrophages involves heme oxygenase-1. *Biochem. Pharmacol.* **66,** 1821–1832.

Nakahira, K., Kim, H. P., Geng, X. H., Nakao, A., Wang, X., Murase, N., Drain, P. F., Wang, X., Sasidhar, M., Nabel, E. G., Takahashi, T., Lukacs, N. W., *et al.* (2006). Carbon monoxide differentially inhibits TLR signaling pathways by regulating ROS-induced trafficking of TLRs to lipid rafts. *J. Exp. Med.* **203,** 2377–2389.

Nathan, C. (1992). Nitric oxide as a secretory product of mammalian cells. *FASEB J.* **6,** 3051–3064.

Otterbein, L. E., Bach, F. H., Alam, J., Soares, M., Tao Lu, H., Wysk, M., Davis, R. J., Flavell, R. A., and Choi, A. M. (2000). Carbon monoxide has anti-inflammatory effects involving the mitogen-activated protein kinase pathway. *Nat. Med.* **6,** 422–428.

Palmer, R. M., Ferrige, A. G., and Moncada, S. (1987). Nitric oxide release accounts for the biological activity of endothelium-derived relaxing factor. *Nature* **327,** 524–526.

Ryter, S. W., Morse, D., and Choi, A. M. (2007). Carbon monoxide and bilirubin: Potential therapies for pulmonary/vascular injury and disease. *Am. J. Respir. Cell Mol. Biol.* **36,** 175–182.

Sato, K., Balla, J., Otterbein, L., Smith, R. N., Brouard, S., Lin, Y., Csizmadia, E., Sevigny, J., Robson, S. C., Vercellotti, G., Choi, A. M., Bach, F. H., *et al.* (2001). Carbon monoxide generated by heme oxygenase-1 suppresses the rejection of mouse-to-rat cardiac transplants. *J. Immunol.* **166,** 4185–4194.

Sawle, P., Foresti, R., Mann, B. E., Johnson, T. R., Green, C. J., and Motterlini, R. (2005). Carbon monoxide-releasing molecules (CO-RMs) attenuate the inflammatory response elicited by lipopolysaccharide in RAW264.7 murine macrophages. *Br. J. Pharmacol.* **145,** 800–810.

Shiraishi, F., Curtis, L. M., Truong, L., Poss, K., Visner, G. A., Madsen, K., Nick, H. S., and Agarwal, A. (2000). Heme oxygenase-1 gene ablation or expression modulates cisplatin-induced renal tubular apoptosis. *Am. J. Physiol. Renal Physiol.* **278,** F726–F736.

Srisook, K., Kim, C., and Cha, Y. N. (2005). Role of NO in enhancing the expression of HO-1 in LPS-stimulated macrophages. *Methods Enzymol.* **396,** 368–377.

Stone, J. R., and Marletta, M. A. (1994). Soluble guanylate cyclase from bovine lung: Activation with nitric oxide and carbon monoxide and spectral characterization of the ferrous and ferric states. *Biochemistry* **33,** 5636–5640.

Tenhunen, R., Marver, H. S., and Schmid, R. (1968). The enzymatic conversion of heme to bilirubin by microsomal heme oxygenase. *Proc. Natl. Acad. Sci. USA* **61,** 748–755.

Wang, W. W., Smith, D. L., and Zucker, S. D. (2004). Bilirubin inhibits iNOS expression and NO production in response to endotoxin in rats. *Hepatology* **40,** 424–433.

Wang, X., Wang, Y., Kim, H. P., Nakahira, K., Ryter, S. W., and Choi, A. M. (2007). Carbon monoxide protects against hyperoxia-induced endothelial cell apoptosis by inhibiting reactive oxygen species formation. *J. Biol. Chem.* **282,** 1718–1726.

Wu, J., Ma, J., Fan, S. T., Schlitt, H. J., and Tsui, T. Y. (2005). Bilirubin derived from heme degradation suppresses MHC class II expression in endothelial cells. *Biochem. Biophys. Res. Commun.* **338,** 890–896.

Yee, E. L., Pitt, B. R., Billiar, T. R., and Kim, Y. M. (1996). Effect of nitric oxide on heme metabolism in pulmonary artery endothelial cells. *Am. J. Physiol.* **271,** L512–L518.

Zuckerbraun, B. S., Billiar, T. R., Otterbein, S. L., Kim, P. K., Liu, F., Choi, A. M., Bach, F. H., and Otterbein, L. E. (2003). Carbon monoxide protects against liver failure through nitric oxide-induced heme oxygenase 1. *J. Exp. Med.* **198,** 1707–1716.

CHAPTER NINETEEN

DETECTION AND CHARACTERIZATION OF *IN VIVO* NITRATION AND OXIDATION OF TRYPTOPHAN RESIDUES IN PROTEINS

Catherine Bregere, Igor Rebrin, *and* Rajindar S. Sohal

Contents

Abstract

Oxygen and nitrogen centered reactive species can cause specific structural modifications in amino acids and proteins, such as the addition of a nitro group onto aromatic residues. Heretofore, studies on protein nitration have mainly focused on the *in vitro* and *in vivo* nitro addition to tyrosine residues (3-nitrotyrosine or 3NT), whereas the formation of nitrotryptophan in proteins *in vivo* and/or its functional significance has remained quite obscure. A novel structural modification, involving the addition of nitro and hydroxy groups to tryptophan, has been detected in the mitochondrial protein succinyl-CoA:3-oxoacid CoA transferase (SCOT) in rat heart. Modified SCOT accumulated progressively with age, which was associated with an elevation of its activity. The specific biochemical properties of this novel amino acid were characterized by a combination of HPLC-electrochemical detection and mass spectrometric analysis. This chapter describes the experimental steps involved in the characterizations

Department of Pharmacology and Pharmaceutical Sciences, University of Southern California, Los Angeles, California

Methods in Enzymology, Volume 441
ISSN 0076-6879, DOI: 10.1016/S0076-6879(08)01219-6

and a procedure for the synthesis of nitrohydroxytryptophan. Similar methodology can be applied to the identification of nitrohydroxytryptophan in other proteins.

1. INTRODUCTION

Certain amino acids within proteins are particularly sensitive to oxidation by reactive oxygen and nitrogen species (ROS and RNS), which, depending on the nature and the location of the modified residue, can cause perturbations in the properties of the protein, such as conformation, catalytic activity, susceptibility to proteolysis, intracellular location, and immunogenicity (Sohal, 2002; Stadtman and Berlett, 1998). One instance of a posttranslational modification induced by the combined action of ROS and RNS is the nitration of proteins, that is, the addition of a nitro group (NO$_2$) onto aromatic residues. Currently, 3–nitrotyrosine (3NT) containing proteins are the main focus of interest, perhaps because of the commercial availability of anti-3NT antibodies (Ye *et al.*, 1996). Thus, a considerable amount of data have accrued, showing, among others, that 3–NT containing proteins accumulate during normal aging (Hong *et al.*, 2007; Kanski *et al.*, 2003, 2005a,b; Sharov *et al.*, 2006; Viner *et al.*, 1999), and in association with pathologies, including neurodegenerative diseases (Beckman *et al.*, 1993; Duda *et al.*, 2000), thereby raising the possibility of a causal association (for review, see Greenacre and Ischiropoulos, 2001).

Despite its relatively lower content in proteins than tyrosine, tryptophan has also been shown to be a specific target of nitration upon *in vitro* exposure to nitrating agents, such as peroxynitrite or the peroxidase/hydrogen peroxide/nitrite system (Herold *et al.*, 2002; Thiagarajan *et al.*, 2004). Using an antibody against 6-nitrotryptophan, Ikeda *et al.* (2007) identified several tryptophan–nitrated proteins following *in vitro* exposure to peroxynitrite. Other authors have reported the presence of nitrotryptophan containing proteins in the liver of mice, treated with acetaminophen, an analgesic that can cause hepatoxicity, in part via RNS-mediated damage. Nevertheless, the nitrated proteins were not individually identified, and no nitrotryptophan containing proteins were observed in the liver of untreated mice (Ishii *et al.*, 2007). Thus far, with the exception of the identification of a bacterial peptide (Kers *et al.*, 2004), nitrotryptophan containing proteins have not been found in mammalian cells *in vivo*. Physiological relevance of such a modification has obviously also remained unknown.

Occurrence of a cross-reaction between a monoclonal anti-3NT antibody and the mitochondrial succinyl-CoA:3-ketoacid transferase (SCOT) protein has been described (Rebrin *et al.*, 2007). Amino acid analysis by HPLC combined with mass spectrometric studies failed to detect the presence of the nitro derivatives, 3-nitrotyrosine and 4- and 5-nitrotryptophan. Instead, a novel amino acid derivative, namely nitrohydroxytryptophan, was detected and identified using HPLC-electrochemical detection (EC) and matrix-assisted laser desorption/ionization (MALDI) and electrospray ionization (ESI) mass spectrometric studies. The amount of the modified SCOT, as well as SCOT catalytic activity, showed an increase with age.

The procedures used for the identification and characterization of nitro-hydroxytryptophan in the SCOT protein in rat heart mitochondria are described. Mitochondria are particularly relevant organelles for the search of nitrated proteins because they are the primary intracellular site of ROS production, and an isoform of nitric oxide synthase may exist in this organelle (Kato and Giulivi, 2006). Thus, peroxynitrite, which arises from a reaction between nitric oxide (NO) and superoxide anion, might be generated in mitochondria.

The example of SCOT is meant here to illustrate the experimental approaches that can be used as a guide to identify and characterize other nitrohydroxytryptophan containing proteins *in vivo*.

2. METHODS

2.1. Isolation of mitochondria and preparation of soluble proteins

Heart tissue, pooled from about 100 rats, is kept in ice-cold antioxidant buffer, consisting of 150 mM potassium phosphate, 2 mM EDTA, and 0.1 mM butylated hydroxytoluene, pH 7.4. Hearts are chopped thinly and homogenized in 5% (w/v) isolation buffer containing 0.3 M sucrose, 0.03 M nicotinamide, and 0.02 M EDTA, pH 7.4, at a low speed using a Polytron homogenizer. To inhibit proteolysis, the homogenization buffer is supplemented with 0.2 mM of freshly prepared phenylmethylsulfonyl fluoride (PMSF) and a complete protease inhibitor cocktail, at the concentration suggested by the manufacturer (Boehringer). Low- and high-speed differential centrifugations for heart mitochondria isolation are, respectively, 700g for 10 min and 10,000g for 5 min. Heart mitochondria are thus isolated within 1 to 2 h after tissue dissection.

Mitochondrial soluble proteins, located in the matrix and intermembrane space compartments, can be obtained as follows: mitochondria ($\approx$5 mg/ml protein) are placed on ice, sonicated for 30 s, and centrifuged at 100,000g for 60 min to separate soluble and membranous proteins. Pellets are resuspended in a buffer containing 50 mM imidazole, pH 7.0, 50 mM sodium chloride, and 5 mM ε-amino-caproic acid, followed by another sonication and an ultracentrifugation, as described earlier. Supernatants from both ultracentrifugation steps are combined and stored in aliquots at −80 °C.

2.2. Purification of the SCOT protein

Mitochondrial soluble proteins (100 mg) are then dialyzed against 25 mM imidazol buffer containing 0.2 mM PMSF, pH 7.85, for 2 h, and applied onto a chromatofocusing column (6 ×100 mm), equilibrated with the same buffer, and eluted with 100 ml of Polybuffer 74 (dilution 1:10), pH 3.9. The flow rate is 6 ml/h and fractions of 2 ml each are collected. Measurements of pH are made in every fifth fraction. Each consecutive fraction is subjected to one-dimensional SDS-PAGE electrophoresis. Gels are electrotransferred onto a PVDF membrane for immunodetection with the polyclonal anti-SCOT antibody (BioSource). SCOT containing fractions are then pooled and concentrated in a volume of 200 μl, using Centricon 30 concentrators. Gel filtration is performed at room temperature using a Shimadzu class VP HPLC system and BioSep-SEC-S 3000 gel permeation column (5 μm, 7.5 × 300 mm) obtained from Phenomenex (Torrance, CA). The column is equilibrated with 25 mM Tris buffer, pH 7.4, containing 75 mM NaCl at a flow rate of 0.5 ml/min. Fifty microliters of concentrated sample containing SCOT is injected onto the column, and absorbance (200–600 nm) is monitored with a diode-array ultraviolet (UV) detector. Fractions (0.5 ml) of the eluate are collected from the gel filtration column between 12 and 20 min. Three consecutive injections are made. The purity of the SCOT protein can be assessed by SDS-PAGE. This purification procedure yields ≈200 μg of electrophoretically pure (>85%) SCOT protein.

2.3. SCOT autolytic fragmentation

The catalytic properties of SCOT allow, under specific conditions, cleavage of the protein at its active site, glutamate 303 (Howard *et al.*, 1986). Two fragments, the amino- and carboxy-terminal fragments, are thus generated, thereby facilitating identification of the nitrated residue. Briefly, the purified SCOT protein is incubated in the absence or presence of 1 mM acetoacetyl-coenzyme A for 5 min at room temperature and then for 1 h at 70 °C. Alternatively, 1 mM succinyl–CoA can also be used. The buffer consists of 50 mM sodium phosphate, pH 7.4. The protein mixtures are separated by one-dimensional SDS-PAGE electrophoresis, following which the gels are either stained with Coomassie blue or electrotransferred to PVDF membranes for immunodetection with the monoclonal anti-3NT antibody (clone 1A6, Upstate) and the polyclonal anti-SCOT antibody. The expected results are shown in Fig. 19.1. The purified full-length SCOT protein (58 kDa), as well as the amino (37 kDa)- and carboxy (21 kDa)-terminal fragments, obtained after substrate-induced cleavage, are visible on the Coomassie-stained gel (Fig. 19.1A). Due to the presence of the peptide recognized by the anti-SCOT serum, the immunoblot using the anti-SCOT serum showed a positive cross-reaction with the amino-terminal

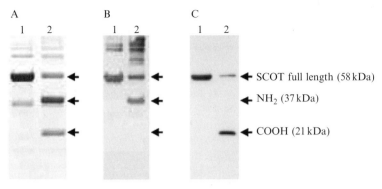

Figure 19.1 (A) Substrate-induced cleavage of the SCOT protein. Lanes 1 and 2 contain the purified SCOT protein incubated in the absence or presence of acetoacetyl-CoA, respectively. The protein bands, namely full-length SCOT and the fragments generated after fragmentation, are shown in a Coomassie-stained gel. (B) An immunoblot using the anti-SCOT serum showing positive reaction with the amino-terminal fragment after cleavage of the protein. (C) An immunoblot with the anti-3NT antibody showing positive cross-reaction with the carboxy-terminal fragment only. The positions of the SCOT full-length protein (58 kDa) and carboxy (COOH, 21 kDa)- and amino (NH₂, 37 kDa)-terminal fragments are indicated by arrows.

fragment (Fig. 19.1B). In contrast, the anti-3NT antibody showed a positive reaction only with the carboxy-terminal fraSgment (Fig. 19.1C), suggesting that the nitrated residue was located exclusively in this particular portion of the protein, which contains only one tyrosine and two tryptophan residues.

2.4. Analysis of nitrated amino acids

The purified SCOT protein can thus be fragmented by the procedure described earlier and the reaction mixtures separated by one-dimensional SDS-PAGE electrophoresis. The gels can then be electrotransferred onto polyvinylidene fluoride (PVDF) membranes to be stained with Coomassie blue. To isolate the protein, bands (10–100 μg) corresponding to the carboxy-terminal fragment are cut and placed in 0.5-ml plastic tubes containing 100 μl of 0.1 M sodium acetate buffer, pH 7.4, mixed with 5% (w/w) pronase from *Staphylococcus griseus* (Boehringer) to hydrolyze the proteins into amino acids. Hydrolysis is carried out overnight at 50 °C. [It can be stopped by the addition of 100 μl 10% (w/v) metaphosphoric acid.] Samples are then centrifuged at 18,000g for 20 min, and supernatants are transferred to autosample microvials for injection onto a HPLC column. Amino acids (tyrosine, 3-nitrotyrosine, tryptophan, 4-nitrotryptophan, 5-nitrotryptophan, 5-hydroxytryptophan, and kynurenine) are separated by HPLC, fitted with a Shimadzu class VP solvent delivery system using a reversed-phase C18 Gemini column (4.6 × 150 mm, 5 μm, Phenomenex). The mobile phase for isocratic elution consists of 25 mM monobasic sodium phosphate, 12.5% methanol, pH 2.7, adjusted with

85% phosphoric acid. The flow rate is 1 ml/min. Under these conditions, separation is completed in 30 min; 5-nitrotryptophan is the last eluted peak, with a retention time of approximately 27 min. For the analysis of tyrosine and 3NT, omit the methanol from the solvent. The elution of 3NT occurs toward the end, with a retention time of approximately 25 min. The amino acids used as calibration standard are prepared in 5% metaphosphoric acid. Amino acids can be detected with a Model 5600 CoulArray electrochemical detector (ESA, Chelmsford, MA), equipped with a four-channel analytical cell, using potentials of +600, +700, +800, and +900 mV. With a signal-to-noise ratio of 4:1, the lower limit for electrochemical detection is 300 fmol for 3-nitrotyrosine and 4- and 5-nitrotryptophan and 200 fmol for tyrosine, tryptophan, 5-hydroxytryptophan, and kynurenine. For quantification, inject each sample twice and average the peak areas. Representative HPLC chromatograms are shown Fig. 19.2. The separation of standards (tyrosine, 3NT, tryptophan, 4-and 5-nitrotryptophan) is shown in trace A (Fig. 19.2). Amino acid analysis of the SCOT carboxy-terminal fragment (trace B, Fig. 19.2) indicates the absence of 3NT and 4- and 5-tryptophan. The amino acid nitrohydroxytryptophan ($W_{OH,N}$ with a retention time of 21 min) occurs exclusively in the full-length protein and the C-terminal fragment. The UV absorption spectrum of

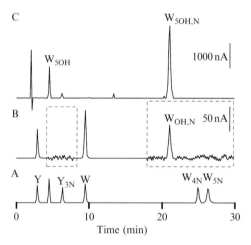

Figure 19.2 Characteristic HPLC–EC chromatographic profile of nitrohydroxytryptophan. The amino acids tyrosine (Y), 3-nitrotyrosine (Y_{3N}), tryptophan (W), and 4–and 5-nitrotryptophan (W_{4N} and W_{5N}) were used as calibration standards (trace A). The elution profile resulting from the amino acid analysis of the SCOT carboxy-terminal fragment shows the absence of nitrotyrosine and nitrotryptophan and the presence of a peak identified as nitrohydroxytryptophan ($W_{OH,N}$) with a retention time of 21 min (trace B). The elution profile obtained after mixing 5-hydroxytryptophan (W_{5OH}, retention time of 5 min) shows an intense peak corresponding to nitrohydroxytryptophan ($W_{5OH,N}$) with an identical retention time to the residue observed in the carboxy-terminal portion of SCOT *in vivo* (trace C). The electrochemical signal intensities represented by the vertical bars are 1000 nA outside and 50 nA inside the dotted box.

nitrohydroxytryptophan, detected with a diode array, is similar to those of nitrated aromatic amino acids, with a characteristic emission peak at 340 to 360 nm. The retention time and UV spectrum of nitrohydroxytryptophan observed during HPLC analysis of the digest of the SCOT protein should be identical to the tryptophan derivative produced *in vitro* (trace C, Fig. 19.2).

2.5. Synthesis of nitrated 5-hydroxytryptophan

5-Hydroxynitrotryptophan can be synthesized by the reaction between 5-hydroxytryptophan and tetranitromethane. The procedure should be carried out on ice in the dark (amber glass vial). Dissolve 10 mg of solid 5-hydroxytryptophan in 50 μl of tetranitromethane, overlay the solution with 450 μl of 70% acetonitrile, and stir gently. It should be noted that, because of the insolubility of tetranitromethane (bottom) in acetonitrile (top), the solution in the vial separates into two phases. In due course, the dark red tetranitromethane phase migrates into the upper acetonitrile containing phase. Progress of the reaction can be monitored by examining a 50-μl aliquot (100-fold dilution of the upper phase of the reaction mixture in 5% metaphosphoric acid) by HPLC, as described previously. The highest yield ($\approx$70%) of 5-hydroxynitrotryptophan is obtained after a prolonged period of reaction (up to 6 h). Further incubation will result in the appearance of additional minor peaks with retention times of more than 30 min, probably due to the occurrence of multiple nitration reactions on the indole ring of tryptophan. In contrast, if the reaction is carried out at pH 8.0, using 50 mM phosphate buffer, the formation of 5-hydroxynitrotryptophan is accompanied by relatively high amounts of multiple nitration by-products, and a rapid decay of 5-hydroxynitrotryptophan is detected within several minutes in phosphate buffer in the pH range of 7.0 to 8.0. Therefore, to obtain high yield and purity, as well as prolonged stability, the buffer should be excluded. No significant decomposition of 5-hydroxynitrotryptophan is observed after storage at 4 °C for up to 5 days if the reaction is acidified with 0.1% trifluoroacetic and 5% metaphosphoric acid.

2.6. Matrix-assisted laser desorption ionization mass spectrometric analysis

Matrix-assisted laser desorption ionization time-of-flight (MALDI-TOF) mass spectra can be acquired on AXIMA CFR (Shimadzu, MD) operated in the positive ion linear mode using α-cyano-4-hydroxycinnamic acid (10 mg/ml in 70% acetonitrile) as a matrix compound. Spectra should represent the average of at least 100 laser shots. To optimize fragmentation spectra, examine various settings of laser power (60, 90, 120, and 150) in order. External mass calibration can be achieved using methionine, tryptophan, 5-hydroxytryptophan, and 5-nitrotryptophan, whose protonated ions $[M+H]^+$ masses are 150.1, 205.1,

221.09, and 250.08, respectively. One milligram of dried sample of amino acids is dissolved in 100 μl of 0.1% trifluoroacetic acid (TFA). One microliter of the sample solution is then mixed with 1 μl of matrix solution, and the resulting mixture is deposited on a stainless-steel sample holder and let dry for several minutes on air. The HPLC fraction containing the putative peak corresponding to nitrohydroxytryptophan is collected between 20 and 22 min during HPLC separation. The eluate is then concentrated using a Speed-Vac (Thermo Savant) and dissolved in 10 μl of 0.1% TFA. Three independent MALDI measurements should be made for each sample to evaluate the reproducibility of the ion peaks.

The theoretical mass of the protonated ion for nitrohydroxytryptophan is 266.08. Under MALDI conditions, the nitro group undergoes photodeoxygenation and yields fragmented ions, which correspond to the loss of one oxygen or two oxygen atoms. The peak abundance of these fragmented ions varies depending on the initial concentration of the compound. Thus, if picomolar amounts of 5-nitrotryptophan are used, the most intense ions observed correspond to 5-nitrosotryptophan. Such a concentration effect has been described previously (Sarver *et al.*, 2001). The expected masses of these ions are indicated in Table 19.1 and can be considered a fingerprint of the presence of a nitrohydroxytryptophan. The presence of such a nitrotryptophan derivative should also be confirmed by treating the sample with dithionite, a reducing agent that converts nitro (NO_2) to amino (NH_2) group. Thus, under MALDI, a new peak corresponding to amino-hydroxytryptophan can be observed at *m/z* 236.38. Altogether, MALDI mass spectrometric studies can ascertain the presence of such a tryptophan derivative. Further characterization of nitrohydroxytryptophan can be performed by ESI tandem mass spectrometry after tryptic digestion of the SCOT full-length protein or carboxy-terminal fragment, using the methodology described in the chapter sixteen.

3. Conclusions

The experimental procedures described in this chapter can be used for the detection of nitrohydroxytryptophan in proteins *in vivo*. Briefly, a preliminary screen for nitrated targets can be made in any organelle of choice using the monoclonal anti-3NT antibody. Purification steps are a prerequisite for identification of the nitrated protein and HPLC-EC amino acid analysis. Identification of the nitrated residues is made using various nitrated amino acids as standards. Nitrohydroxytryptophan is generated by the reaction between 5-hydroxytryptophan and tetranitromethane. The presence of a nitrated amino acid is confirmed using MALDI-TOF mass spectrometric analysis, and the lability of the nitro group is used as a fingerprint of nitration. Treatment with dithionite can provide additional

Table 19.1 Chemical structures, properties, and theoretical m/z values of nitrohydroxytryptophan and its derivatives by MALDI analysis

Structures	Comments	Theoretical m/z of the protonated ions
Nitro-hydroxy-tryptophan	The protonated molecular ion is abundantly formed if the synthetic nitrohydroxytryptophan is used. In contrast, the peak intensity will be much lower if using an HPLC eluate from a protein digest.	266.23
Nitroso-hydroxy-tryptophan	Photodecomposition product of nitrohydroxytryptophan following laser exposure: there is a loss of one oxygen atom from the nitro group. The fragmented ion will be predominant if the initial concentration of nitrohydroxytryptophan is low.	250.23
Nitrene-hydroxy-tryptophan	Photodecompostion product of nitrohydroxytryptophan following laser exposure: note the loss of two oxygen atoms from the nitro group. This fragmented ion is not observed consistently.	234.23
Amino-hydroxy-tryptophan	The compound obtained after treatment with dithionite.	236.25

evidence of the presence of a nitrated aromatic amino acid derivative. Altogether, these experimental approaches should facilitate the identification and characterization of nitrohydroxytryptophan containing proteins *in vivo* as well as *in vitro*.

ACKNOWLEDGMENT

This research was supported by Grant RO1 AG 13563 from National institute on Aging–National Institutes of Health.

REFERENCES

Beckman, J. S., Carson, M., Smith, C. D., and Koppenol, W. H. (1993). ALS, SOD and peroxynitrite. *Nature* **364,** 584.

Duda, J. E., Giasson, B. I., Chen, Q., Gur, T. L., Hurtig, H. I., Stern, M. B., Gollomp, S. M., Ischiropoulos, H., Lee, V. M., and Trojanowski, J. Q. (2000). Widespread nitration of pathological inclusions in neurodegenerative synucleinopathies. *Am. J. Pathol.* **157,** 1439–1445.

Greenacre, S. A., and Ischiropoulos, H. (2001). Tyrosine nitration: Localisation, quantification, consequences for protein function and signal transduction. *Free Radic. Res.* **34,** 541–581.

Herold, S., Shivashankar, K., and Mehl, M. (2002). Myoglobin scavenges peroxynitrite without being significantly nitrated. *Biochemistry* **41,** 13460–13472.

Hong, S. J., Gokulrangan, G., and Schoneich, C. (2007). Proteomic analysis of age dependent nitration of rat cardiac proteins by solution isoelectric focusing coupled to nanoHPLC tandem mass spectrometry. *Exp. Gerontol.* **42,** 639–651.

Howard, J. B., Zieske, L., Clarkson, J., and Rathe, L. (1986). Mechanism-based fragmentation of coenzyme A transferase: Comparison of alpha 2-macroglobulin and coenzyme A transferase thiol ester reactions. *J. Biol. Chem.* **261,** 60–65.

Ikeda, K., Yukihiro Hiraoka, B., Iwai, H., Matsumoto, T., Mineki, R., Taka, H., Takamori, K., Ogawa, H., and Yamakura, F. (2007). Detection of 6-nitrotryptophan in proteins by Western blot analysis and its application for peroxynitrite-treated PC12 cells. *Nitric Oxide* **16,** 18–28.

Ishii, Y., Ogara, A., Katsumata, T., Umemura, T., Nishikawa, A., Iwasaki, Y., Ito, R., Saito, K., Hirose, M., and Nakazawa, H. (2007). Quantification of nitrated tryptophan in proteins and tissues by high-performance liquid chromatography with electrospray ionization tandem mass spectrometry. *J. Pharm. Biomed. Anal.* **44,** 150–159.

Kanski, J., Alterman, M. A., and Schoneich, C. (2003). Proteomic identification of age-dependent protein nitration in rat skeletal muscle. *Free Radic. Biol. Med.* **35,** 1229–1239.

Kanski, J., Behring, A., Pelling, J., and Schoneich, C. (2005a). Proteomic identification of 3-nitrotyrosine-containing rat cardiac proteins: Effects of biological aging. *Am. J. Physiol. Heart Circ. Physiol.* **288,** H371–H381.

Kanski, J., Hong, S. J., and Schoneich, C. (2005b). Proteomic analysis of protein nitration in aging skeletal muscle and identification of nitrotyrosine-containing sequences *in vivo* by nanoelectrospray ionization tandem mass spectrometry. *J. Biol. Chem.* **280,** 24261–24266.

Kato, K., and Giulivi, C. (2006). Critical overview of mitochondrial nitric-oxide synthase. *Front. Biosci.* **11,** 2725–2738.

Kers, J. A., Wach, M. J., Krasnoff, S. B., Widom, J., Cameron, K. D., Bukhalid, R. A., Gibson, D. M., Crane, B. R., and Loria, R. (2004). Nitration of a peptide phytotoxin by bacterial nitric oxide synthase. *Nature* **429,** 79–82.

Rebrin, I., Bregere, C., Kamzalov, S., Gallaher, T. K., and Sohal, R. S. (2007). Nitration of tryptophan 372 in succinyl-CoA:3-ketoacid CoA transferase during aging in rat heart mitochondria. *Biochemistry* **46,** 10130–10144.

Sarver, A., Scheffler, N. K., Shetlar, M. D., and Gibson, B. W. (2001). Analysis of peptides and proteins containing nitrotyrosine by matrix-assisted laser desorption/ionization mass spectrometry. *J. Am. Soc. Mass Spectrom.* **12,** 439–448.

Sharov, V. S., Galeva, N. A., Kanski, J., Williams, T. D., and Schoneich, C. (2006). Age-associated tyrosine nitration of rat skeletal muscle glycogen phosphorylase b: Characterization by HPLC-nanoelectrospray-tandem mass spectrometry. *Exp. Gerontol.* **41,** 407–416.

Sohal, R. S. (2002). Role of oxidative stress and protein oxidation in the aging process. *Free Radic. Biol. Med.* **33,** 37–44.

Stadtman, E. R., and Berlett, B. S. (1998). Reactive oxygen-mediated protein oxidation in aging and disease. *Drug Metab. Rev.* **30,** 225–243.

Thiagarajan, G., Lakshmanan, J., Chalasani, M., and Balasubramanian, D. (2004). Peroxynitrite reaction with eye lens proteins: Alpha-crystallin retains its activity despite modification. *Invest. Ophthalmol. Vis. Sci.* **45,** 2115–2121.

Viner, R. I., Ferrington, D. A., Williams, T. D., Bigelow, D. J., and Schoneich, C. (1999). Protein modification during biological aging: Selective tyrosine nitration of the SERCA2a isoform of the sarcoplasmic reticulum Ca^{2+}-ATPase in skeletal muscle. *Biochem. J.* **340,** 657–669.

Ye, Y. Z., Strong, M., Huang, Z. Q., and Beckman, J. S. (1996). Antibodies that recognize nitrotyrosine. *Methods Enzymol.* **269,** 201–209.

CHAPTER TWENTY

In Vivo Real-Time Measurement of Nitric Oxide in Anesthetized Rat Brain

Rui M. Barbosa,* Cátia F. Lourenço,* Ricardo M. Santos,*
Francois Pomerleau,[†] Peter Huettl,[†] Greg A. Gerhardt,[†] *and*
João Laranjinha*

Contents

Abstract

During the last two decades nitric oxide (·NO) gas has emerged as a novel and ubiquitous intercellular modulator of cell functions. In the brain, ·NO is

* Faculty of Pharmacy and Center for Neurosciences and Cell Biology, University of Coimbra, Coimbra, Portugal
† Department of Anatomy and Neurobiology, Center for Microelectrode Technology, University of Kentucky, Lexington, Kentucky

Methods in Enzymology, Volume 441   © 2008 Elsevier Inc.
ISSN 0076-6879, DOI: 10.1016/S0076-6879(08)01220-2

implicated in mechanisms of synaptic plasticity but it is also involved in cell death pathways underlying several neurological diseases. Because of its hydrophobicity, small size, and rapid diffusion properties, the rate and pattern of ·NO concentration changes are critical determinants for the understanding of its diverse actions in the brain. ·NO measurement *in vivo* has been a challenging task due to its low concentration, short half-life, and high reactivity with other biological molecules, such as superoxide radical, thiols, and heme proteins. Electrochemical methods are versatile approaches for detecting and monitoring various neurotransmitters. When associated with microelectrodes inserted into the brain they provide high temporal and spatial resolution, allowing measurements of neurochemicals in physiological environments in a real-time fashion. To date, electrochemical detection of ·NO is the only available technique that provides a high sensitivity, low detection limit, selectivity, and fast response to measure the concentration dynamics of ·NO *in vivo*. We have used carbon fiber microelectrodes coated with two layers of Nafion and *o*-phenylenediamine to monitor the rate and pattern of ·NO change in the rat brain *in vivo*. The analytical performance of microelectrodes was assessed in terms of sensitivity, detection limit, and selectivity ratios against major interferents: ascorbate, dopamine, noradrenaline, serotonin, and nitrite. For the *in vivo* recording experiments, we used a microelectrode/micropipette array inserted into the brain using a stereotaxic frame. The characterization of *in vivo* signals was assessed by electrochemical and pharmacological verification. Results support our experimental conditions that the measured oxidation current reflects variations in the ·NO concentration in brain extracellular space. We report results from recordings in hippocampus and striatum upon stimulation of *N*-methyl-ᴅ-aspartate-subtype glutamate receptors. Moreover, the kinetics of ·NO disappearance *in vivo* following pressure ejection of a ·NO solution is also addressed.

1. INTRODUCTION

During the last two decades nitric oxide (·NO) has emerged as a novel and ubiquitous intercellular modulator of cell functions (Boehning and Snyder, 2003; Pacher *et al.*, 2007). In the brain, ·NO is implicated in mechanisms of synaptic plasticity but it is also involved in cell death pathways underlying several neurological diseases (Garthwaite *et al.*, 1988; Hara and Snyder, 2007; Moncada and Bolanos, 2006). Because of its small size and rapid diffusion characteristics, the rate and pattern of ·NO concentration changes are critical determinants for the understanding of its diverse actions in the brain. Essentially, a large body of knowledge has been collected on the following: (1) how neural inputs activate the synthesis of ·NO by nitric oxide synthase and (2) what is the impact of ·NO in biological cascades?

However, the qualitative nature of information on the relationship of ·NO production and actions has imparted obvious limitations in the prediction of ·NO bioactivity *in vivo*. This is particularly critical for a diffusible signaling molecule, overcoming storage and selective membrane receptor recognition, for which a transient concentration increase is translated into a biological action. Imaging approaches have provided some clues to the properties of ·NO diffusion in the brain; however, electrochemical microelectrodes inserted into the brain possess the spatial and temporal resolution required to measure real-time ·NO generation and diffusion properties (Ledo *et al.*, 2005; Taha, 2003; Takata *et al.*, 2005). This chapter discusses the use of microelectrode/micropipette array for the measurement of extracellular concentration dynamics of ·NO in the extracellular space of brain tissue. In addition, results from recordings in hippocampus and striatum upon stimulation of *N*-methyl-D-aspartate (NMDA)-subtype glutamate receptors are shown. Moreover, the kinetics of ·NO disappearance *in vivo* following pressure ejection of a ·NO solution is also addressed.

In neurons, the synthesis of ·NO involves the stimulation of NMDA glutamate receptors and the influx of Ca^{2+} to the cytosol, which, upon binding to calmodulin, activates neuronal nitric oxide synthase, nNOS (Garthwaite and Boulton, 1995). While glutamate is responsible for the majority of excitatory transmission in the brain, excessive exposure to this amino acid may trigger toxic pathways associated with neurological disorders (Coyle and Puttfarcken, 1993; Obrenovitch and Urenjak, 1997). ·NO contributes to both sides of this glutamate effect: neurotransmission and excitotoxic insult (Bliss and Collingridge, 1993; Dawson and Dawson, 1998; Dawson *et al.*, 1991); therefore, achievement of real-time measurement *in vivo* following NMDA receptor activation is a critical piece of information in deciphering ·NO roles in brain physiology and pathophysiology. The methodologies shown here contribute to this endeavour.

2. Experimental Considerations

2.1. Microelectrode fabrication

Carbon fiber microelectrodes are fabricated as described previously (Ledo *et al.*, 2002). Briefly, single carbon fibers (30 μm Textron Lowell, MA) are inserted into borosilicate glass capillaries (1.16 mm i.d × 2.0 mm o.d.; Harvard Apparatus Ltd., UK) filled with acetone. After solvent evaporation at room temperature, the capillaries are pulled on a vertical puller (Harvard Apparatus Ltd.). The protruding carbon fibers are cut by tweezers under a microscope to obtain an exposed carbon surface with a tip length of 200 ± 50 μm length. Then, the stem end of the microelectrode is filled with nonviscous epoxy using a syringe for improving the seal between the glass and the carbon fiber. Finally, the electrical contact between the carbon fiber and a copper wire is provided by conductive silver paint (RS, UK).

The microelectrodes are tested for their general recording properties in phosphate-buffered saline (PBS) medium prior to their surface modification by fast cyclic voltammetry (FCV) at a 200 V/s scan rate between −1.0 and +1.0 V for 30 s (Ensman Instruments, USA). The stability of the background current, as well as sharp transients at reversal potentials, gives the indication that the microelectrode behaves in a more capacitive fashion and therefore will perform as a good quality microelectrode. Usually, a microelectrode with a poor glass–carbon fiber seal or a defective electrical contact is rejected by the FCV test.

2.2. Microelectrode surface modification

2.2.1. Nafion

Prior to coatings, the microelectrodes are oven dried for 4 min at 170 °C to remove traces of humidity. The microelectrodes are coated with Nafion by dipping the active surface area of a carbon fiber into a fresh Nafion solution (5% in aliphatic alcohols) at room temperature for 1 to 2 s and drying in an oven for 4 min at 170 °C. Lower temperature curing produces thicker films compared to higher temperature curing (Gerhardt and Hoffman, 2001). The Nafion layer acts as an anionic repellant, preventing molecules such as ascorbic acid to access the surface of the electrode. From our experience and those described in the literature, we have established two layers of Nafion as the optimal number of coatings to achieve selectivity against anions without compromising the sensitivity and response time of the microelectrode.

2.2.2. *o*-Phenylenediamine (*o*-PD)

The organic molecule *o*-PD is used to create an exclusion layer at the carbon fiber surface, acting as a molecular filter that minimizes the access of interfering large molecules such as ascorbate, dopamine, noradrenaline, and serotonin to the carbon surface and also preventing fouling of the carbon surface (Friedemann *et al.*, 1996).

The *o*-PD solution must be prepared fresh in a degassed solution of 0.05 M PBS stored in a brown glass bottle to prevent oxidation from light. If the solution turns yellow, *o*-PD will no longer electropolymerize.

After Nafion coating, the *o*-PD layer is deposited by electropolymerization by placing the electrode tip in a 5 mM *o*-PD solution and applying a constant potential of 0.7 V vs an Ag/AgCl reference electrode for 30 min. The microelectrode is then removed from the *o*-PD solution, rinsed with water, and put into a 50-ml beaker containing 0.05 mM PBS solution for at least 1 h before calibration. Microelectrodes are coated on the day of use and are not reused.

2.3. Recording system

The recording system used is Fast Analytical Sensing Technology (FAST-16), which is a high-speed electrochemistry potentiostat (Quanteon, L.L.C., Nicholasville, KY) comprising hardware elements, including the control box, head stage, and a 16-bit A/D card in the computer and Windows-based FAST-16 software. Data can be processed through the analysis software package or exported to other Windows-based applications for further analysis and plotting of data.

Measurements are carried out using constant voltage amperometry in a two-electrode configuration mode. The potential of the working microelectrode is always applied versus an Ag/AgCl reference electrode. A 10 nA/V final gain is used for all recordings using a combination head stage gain of 20 nA/V and secondary of 2× gain (hardware box).

2.4. Oxidation potential of ·NO at the electrode surface

The electrochemical oxidation of ·NO on solid electrodes such as carbon fiber produces the nitrosonium cation (NO^+), which, upon reaction with OH^- in aqueous solution, yields NO_2^- that, in turn, is further oxidized to nitrate (Malinski and Taha, 1992). The oxidation potential depends on electrode surface composition and electrochemical pretreatment of the carbon fiber, as well as chemical modifications of the electrode surface.

Figure 20.1 shows the effect of varying the applied potential (0.2 to 1.2 V) on the amperometric response of a carbon fiber coated with Nafion

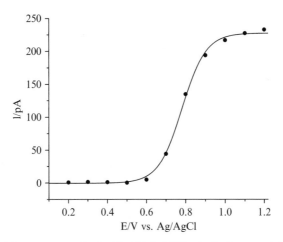

Figure 20.1 Voltammogram for 1 μM ·NO in PBS solution prepared from ·NO gas. Amperometric currents were recorded with a carbon fiber microelectrode coated with Nafion and o-PD. Calculated half-wave potential: 0.78 V vs Ag/AgCl reference electrode.

($2\times$) and *o*-PD in the presence of $\cdot$NO (1 μM). No response is observed for potentials lower than 0.5 V. The amperometric current increases as a function of the applied potential in a sigmoid fashion from which the half-wave potential of 0.78 V vs Ag/AgCl was calculated. The value for $\cdot$NO oxidation potential is in agreement with most of the reports found in the literature (Bedioui and Villeneuve, 2003). Based on our results, we selected 0.9 V as the optimal working potential for amperometric measurements of $\cdot$NO oxidation current.

2.5. *In vitro* testing: Calibration and selectivity

Each microelectrode is calibrated with a fresh saturated solution of $\cdot$NO prepared in a fume hood by bubbling $\cdot$NO gas (Gasin, Portugal), which has been purified by passage through 10 M deoxygenated KOH solution in a custom-designed apparatus using only glass or stainless steel tubing and fittings, in 0.05 M PBS saturated previously with argon in a Vacutainer tube.

The concentration of the $\cdot$NO saturated solution is checked by means of the ISO-NOP 2-mm Pt sensor connected to the amperometer ISO-NO Mark II (World Precision Instruments, Inc., USA) calibrated by chemical generation of $\cdot$NO from the reaction of NO_2^- with excess iodide and sulfuric acid. The concentration of a saturated $\cdot$NO solution is 1.74 $\pm$ 0.03 mM (Mesaros *et al.*, 1997).

Microelectrodes are calibrated by amperometry using the FAST-16 system with an applied potential of 0.9 V vs Ag/AgCl reference electrode (Bioanalytical Systems, RE-5). The microelectrodes are calibrated in a slowly stirred solution of 0.05 M PBS, pH 7.4 (20 ml), using a small magnetic stir bar in conjunction with a battery-operated stir plate (Stuart, Barlworld Scientific Limited, UK) to reduce the AC (50/60 Hz) line interference.

Solutions of $\cdot$NO and ascorbic acid are made fresh daily. Dopamine, noradrenaline, and serotonin are prepared in 1% perchloric acid to prevent oxidation and stored in a freezer at $-18\,^\circ$C.

After stabilization of background current for at least 20 min, the microelectrodes are calibrated. First, 250 μM ascorbic acid (final concentration) is added and then, after allowing time for the current to stabilize, three 10-μl consecutive additions of $\cdot$NO solution are added with a gas–tight syringe in order to achieve a final concentration in the range of 0.4 to 2 μM. After the addition of $\cdot$NO, the other potential interferents are added at a final concentration of 100 μM nitrite, 10 μM dopamine, 10 μM noradrenaline, and 10 μM serotonin.

The calibration parameters calculated for $\cdot$NO are the slope (sensitivity), limit of detection (LOD), and linearity (R^2). The LOD is defined as the analyte concentration that yields an electrode response equivalent to three times the background noise of the recording system.

Table 20.1 Nitric oxide calibration parameters and selectivity ratios for carbon fiber microelectrodes coated with two layers of Nafion and o-PD[a]

Calibration parameter	Carbon fiber microelectrode coated with Nafion and o-PD
Sensitivity ($pA/\mu M$)	317 ± 42 ($n = 12$)
Linearity (R^2)	0.9997 ± 0.0001 ($n = 12$)
Detection limit (nM)	6 ± 1 ($n = 12$)
Selectivity ratio	
Ascorbate	$>10,000{:}1$ ($n = 11$)
Nitrite	$2,318{:}1 \pm 437$ ($n = 11$)
Dopamine	$152{:}1 \pm 28$ ($n = 12$)
Noradrenaline	$211{:}1 \pm 37$ ($n = 12$)
Serotonin	$138{:}1 \pm 30$ ($n = 6$)

[a] Data given as mean $\pm$ SEM.

Selectivity ($\cdot$NO vs interferents) is calculated as the ratio of microelectrode sensitivity for $\cdot$NO over interferents and is calculated by dividing the $\cdot$NO slope by the interferent slope (Table 20.1).

2.6. Micropipette attachments

A micropipette puller (Sutter Instrument Co, Model P30, CA) is used to pull single barrel glass capillaries (0.58 mm i.d $\times$ 1.0 mm o.d.; AM, Systems Inc., WA). The pulled glass micropipettes are bumped to a final inner diameter of 10 to 15 μm. The micropipette is attached to the carbon fiber microelectrode using sticky wax (Kerr Brand Kerr Lab Corporation, Orange, CA) that has been softened by flame. Precise placement is done using a microscope fitted with a reticule to achieve a distance between the tip of the micropipette and the microelectrode of 250 $\pm$ 50 μm between the tips of the micropipette and the $\cdot$NO microelectrode (Fig. 20.2).

2.7. *In vivo* recordings

All animal procedures are approved by the local institutional animal care and use committee and are in accordance with the European Community Council Directive for the Care and Use of Laboratory Animals (86/609/ECC).

In vivo studies are carried out on male Wistar rats (2–3 months old; weight 246–374 g) handled and cared at the animal house facilities of the Center for Neurosciences and Cell Biology (Coimbra) under a 12:12 light: dark cycle with free access to food and water. Rats are anesthetized with urethane (1.25–1.50 g/kg, i.p.) and placed in a stereotaxic frame

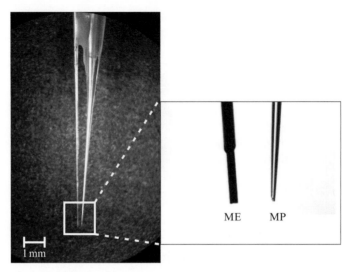

Figure 20.2 The microelectrode/micropipette array. (Inset) Magnification of the array tip. ME, microelectrode; MP, micropipette.

(Stoelting Co., USA) essentially as described previously (Burmeister *et al.*, 2002; Friedemann and Gerhardt, 1996). An incision is made in the scalp, and the skin is reflected to expose the surface of the skull. A Dremel rotary is used to drill a hole in the skull over the brain area of interest. Body temperature is maintained at $37\,^{\circ}C$ using a deltaphase isothermal pad (BrainTree Scientific, MA).

An Ag/AgCl reference electrode (200 μm diameter) is prepared by placing a silver wire (GoodFellow, UK) in a HCl (1 M) solution saturated with NaCl and by applying a voltage of 1.5 V for 10 min. A new reference electrode must be prepared for each experiment. A small hole is drilled in the skull remote from the recording areas. The reference electrode is inserted through the skull opening into the brain. The opening is kept moistened by a physiological saline solution (0.9% NaCl).

In vivo experiments are carried out using local applications of the glutamate receptor agonist NMDA in the CA1 region of the hippocampus using the following coordinates: anterior–posterior (AP) -2.8, medial–lateral (ML) -2.0 dorsoventral (DV) -2.6 and AP -4.1, ML -2.8, DV -2.4. In addition, we measured $\cdot$NO production in the striatum at coordinates AP $+1.0$, ML -2.5, and DV -5.0. The coordinates are calculated from bregma based on the rat brain atlas of Paxinos and Watson (2007).

The single barrel micropipette is filled with the NMDA solution (100 μM in 0.9% saline, pH 7.4) by a syringe fitted with a flexible microfilament (MicroFil, World Precision Instruments, UK). Then, the microelectrode and micropipette array is inserted into the rat brain using a hydraulic micromanipulator (Narishige International Limited, UK). Prior to microelectrode insertion, the dura matter should be removed gently to expose the brain surface.

After allowing a stable baseline current, volumes of NMDA solution (low nanoliter range) are pressure ejected from the micropipette using a Picospritezer III (Parker Hannifin Corp., General Valve Operation, USA). The volume ejected is checked using a stereomicroscope (Meiji EMZ 13, Japan) fitted with an eyepiece reticule in order to allow viewing the decrease of the volume ejected from the glass pipette (Friedemann and Gerhardt, 1992). During the experiments the skull and exposed brain surface are bathed with saline to prevent drying of the brain surface. An overdose of anesthetic is used to euthanize animals following each experiment.

2.8. Verification of recording site

At the end of each experiment the brain is removed rapidly and placed into freshly thawed 4% paraformaldehyde in PBS and soaked for at least 24 h at 4 °C. The brain is then transferred to a solution of 30% sucrose in PBS and stored until it sinks to the bottom of the tube. Tissue is then frozen with a cryo spray (Thermo Electron Corporation, UK), and 40-μm coronal sections are cut in a cryostat (Model CM 1900, Leica, Germany). Brain sections are stained with cresyl violet solution and analyzed under a light microscope to verify the exact location of the recording site.

2.9. Analysis of *in vivo* signals

A number of parameters may be analyzed for characterizing *in vivo* electrochemical signals. The most common are the following: amplitude (A_{max}); decay time (T_{50}), time required for the signal to decay 50% of the maximal amplitude; T_{total} duration of the transient signal; rise time (T_r), time required for the signal to reach maximal amplitude.

Usually, signals follow a first-order decay rate characterized by a constant referred as k (s^{-1}). When k is multiplied by the signal maximal amplitude, the disappearance rate of $\cdot$NO is obtained as $\mu M \cdot s^{-1}$.

3. *IN VIVO* EXPERIMENTS: RESULTS AND DISCUSSION

3.1. Dynamics of nitric oxide signal following ejection of nitric oxide solution in the hippocampus

Characterization of the $\cdot$NO signal *in vivo* is critical when using electrochemical methods. The amperometric recording at a constant potential of 0.9 V shown in the left trace of Fig. 20.3 demonstrates that the

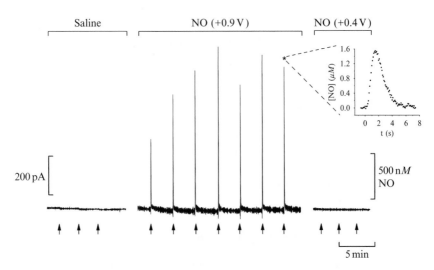

Figure 20.3 Representative tracings of microelectrode responses to pressure ejection of a 0.9% NaCl saline solution versus ·NO ejection at 0.9 and 0.4 V applied potential in the CA1 region of the hippocampus. Arrows along the X axis indicate ejection times (pressure: 10–20 psi; time: 0.5 s; volumes delivered: 6–25 nl). (Inset) Expanded timescale of a representative ·NO signal.

microelectrode inserted in the CA1 region of the hippocampus does not respond to local application by pressure ejection of a saline solution. However, repeated ejections of ·NO solution induce transient signals (300–600 pA amplitude corresponding to approximately 700–1500 nM ·NO). When the applied potential of the microelectrode is changed from 0.9 to 0.4 V, no measurable currents are recorded.

The pressure ejection of ·NO solution at a distance of 300 μm from the microelectrode tip produces a peak (see Fig. 20.3, inset) characterized by a time rise of 1.5 s, a first-order decay rate constant (k) of 0.78 ± 0.01 s^{-1} ($n = 8$ signals). The calculated half-life ($t_{1/2}$) of ·NO is 0.9 s. The averaged k value multiplied by the amplitude of the ·NO signal (for the signal marked with an asterisk in Fig. 20.3) yields a rate of disappearance of 1.2 $\mu M \cdot$s^{-1}.

The first-order rate of ·NO disappearance in intact brain tissue, as shown in Fig. 20.3, is consistent with work by Thomas and colleagues (2001), who observed a cell-dependent first-order rate of ·NO disappearance. Of note, and as pointed out by Thomas *et al.* (2001), the first-order rate of disappearance may result from multiple processes, occurring simultaneously and giving an overall first-order disappearance. The nature of the reactions that lead to ·NO removal is not clear and deciphering such a mechanism will be of great relevance in shaping the ·NO concentration *in vivo* (Hall and Garthwaite, 2006).

The electrochemical methods combined with ·NO microelectrodes could be of valuable importance in studying the underlying mechanism of ·NO disappearance and the agents that affect its decay in intact brain.

3.2. *In vivo* ·NO measurement following NMDA receptor stimulation

3.2.1. Hippocampus

It is currently accepted that stimulation of the glutamate NMDA receptor represents the major pathway by which ·NO is produced in the brain (Garthwaite *et al.*, 1989). Activation of the ionotropic NMDA receptor involves calcium channel-mediated influx of Ca^{2+} through the channel, leading to formation of the Ca^{2+}/calmodulin complex that activates neuronal NO synthase (nNOS)(Boehning and Snyder, 2003; Garthwaite and Boulton, 1995).

Immunohistochemical studies have shown that nNOS is highly expressed in rat hippocampus, namely in the CA1 region(Burette *et al.*, 2002). Thus, it is of great interest to perform *in vivo* recordings of ·NO production in this particular region of the hippocampus to investigate its role as a neuromodulator of important brain functions such as learning and memory.

Figure 20.4 shows a representative *in vivo* amperometric recording (0.9 V vs Ag/AgCl) using a microelectrode coated with Nafion and *o*-PD, with local pressure ejection of 100 μM NMDA to induce the production of ·NO in the CA1 region of the hippocampus at coordinates AP −2.8, ML −2.5, DV −3.5.

Following NMDA stimulus, the oxidation current increased almost instantaneously (<1 s), which corresponds for this particular microelectrode (·NO sensitivity of 257 pA/μM calibrated before experiment) to 394, 211, and 173 nM ·NO for the first, second, and third peaks, respectively. The average $T_{rise}, T_{50,}$ and T_{total} for the three peaks shown is 19, 13, and 48 s, respectively.

The signal amplitude elicited by repeated local applications of NMDA at 10 to 15 min intervals is variable. In most of the experiments (57%), repeated NMDA stimulation diminished the transient signals (Fig. 20.4A). However, occasionally (14% of the experiments) the second NMDA stimulation practically abolished the ·NO signal. In approximately 28% of the experiments, NMDA induced a biphasic signal in the CA1 region resembling a spike/dome response (Fig. 20.4B) documented by other studies of glutamate and dopamine corelease from striatum following potassium stimulation (Burmeister *et al.*, 2002; Walker *et al.*, 2007). Therefore, in the case of the spike/dome type of signals one may consider that, following ·NO production and appearance in the extracellular medium, an additional

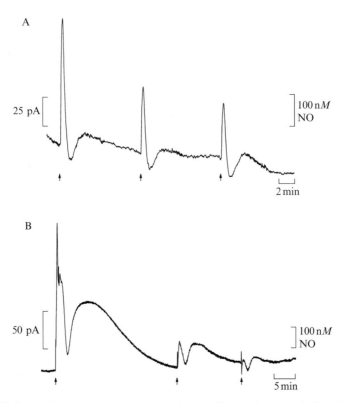

Figure 20.4   Real-time *in vivo* amperometric recordings of NMDA-induced production of ·NO measured with a microelectrode coated with Nafion and *o*-PD following
insertion in the CA1 region of the rat hippocampus. NMDA (100 μM) is applied locally
by pressure ejection (10 psi, 1 s, 30 nl) at times indicated by the arrows. Two types
of responses are consistently observed and categorized into spike (A) and spike/dome
(B) signals.

compound is being oxidized at the electrode surface that, despite the
electrode selectivity for ·NO, is temporarily present in a high concentration. Such a slower putative component could be attributed to the release of
other electroactive neurotransmitters (e.g., noradrenaline, serotonin)
because the CA1 region of the hippocampus has a relatively rich noradren-
ergic innervation originating from the locus coeruleus and serotoninergic
innervation from the dorsal and median raphe nuclei (Vizi and Kiss, 1998).
In fact, ·NO has been implicated in the mechanisms of release of neurotransmitters and may regulate the extracellular concentration of different
transmitters. For instance, it has been shown that local perfusion of NMDA
into the hippocampus significantly decreased the extracellular levels of
serotonin and dopamine (Segieth *et al.*, 2001; Tao and Auerbach, 1996;
Wegener *et al.*, 2000; Whitton *et al.*, 1994). Conversely, other studies

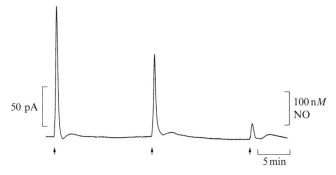

Figure 20.5 Real time *in vivo* amperometric recordings of NMDA-induced production of ·NO measured with a microelectrode coated with Nafion and *o*-PD following insertion in the striatum. NMDA (100 μM) is applied locally repeatedly by pressure ejection (10 psi, 1 s, 30 nl) at times indicated by the arrows.

demonstrate that NMDA-stimulated noradrenaline release in the rat hippocampus is facilitated by ·NO (Lauth 1995; Luo and Vincent, 1994; Stout and Woodward, 1995).

3.2.2. Striatum
The application of 100 μM NMDA in the striatum induces a rapid and transient oxidation signal that decays back to baseline in about 60 s, reaching a maximum amplitude of 400 nM ·NO (Fig. 20.5).

3.3. Validation of nitric oxide signals *in vivo*

Electrochemical methods differ from dialysis in the way that compounds can be measured *in situ* and in a real-time fashion, thus allowing the study of the dynamics affecting *in vivo* concentration. However, despite the high selectivity of ·NO microelectrodes for major brain interferents, it is necessary to verify the identity of the measured species. Critical guidelines have been proposed for validation of the selectivity of *in vivo* microsensors (Phillips and Wightman, 2003). Essentially, the electrochemical verification of the signals is perhaps the most important but should be complemented by anatomical, physiological, and pharmacological verifications. Ultimately, to confirm the identity of the analytical signals, independent chemical verification by other methods (e.g., dialysis, varying applied potential) must be performed.

Figures 20.6 and 20.7 document instrument and pharmacological control of the analytical signal, respectively. In order to assess electrochemical verification, we carried out experiments by stimulating the CA1 region of the hippocampus with NMDA at high (0.9 V) and low (0.4 V) microelectrode holding potentials (Fig. 20.6).

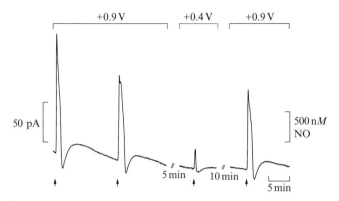

Figure 20.6 Representative amperometric *in vivo* recordings obtained with a micro-electrode coated with Nafion and *o*-PD and recorded at 0.9 V and 0.4 V vs Ag/AgCl reference following NMDA stimulation of the CA1 region of the rat hippocampus.

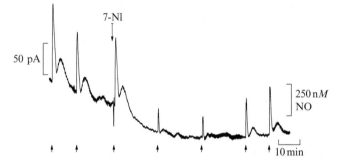

Figure 20.7 Representative amperometric *in vivo* recording showing the effect of the nitric oxide synthase inhibitor 7–nitroindazole administered i.p. (50 mg/kg) at the time indicated by the downward arrow. NMDA (100 μM) is applied locally repeatedly by pressure ejection at times indicated by the upward arrows.

As expected for a compound with the electrochemical behavior documented in Fig. 20.1, at 0.4 V only a residual current is obtained when compared with that measured at 0.9 V. As shown in Fig. 20.1 at potentials lower than 0.5 V, •NO is not oxidized at the Nafion- and *o*-PD-coated carbon fiber surface. The residual current is likely due to the oxidation of other molecules possibly released by neurons that are oxidized at 0.4 V vs Ag/AgCl.

In addition to the electrochemical verification at low and high holding microelectrode potentials, a pharmacological experiment using an inhibitor of nitric oxide synthase, 7–nitroindazole (7-NI), is shown in Fig. 20.7. Following administration of 7-NI (50 mg/kg, i.p.), it takes approximately 20 min for the inhibition of •NO production in the brain to be observed;

as shown in Fig. 20.7, the signal decreases significantly at about 20 min but, noteworthy, the signal returns to almost preapplication levels after about 60 min.

Taken together, results from Figs. 20.6 and 20.7 help validate the accurate detection and measurement of ·NO.

4. CONCLUDING REMARKS

Considering that ·NO may integrate the activity of nonsynaptically connected neurons within a defined volume of tissue, a critical determinant of ·NO bioactivity in the brain is its concentration dynamics in the microenvironment of the brain parenchyma. The challenge of accurate real-time measurement of ·NO at nanomolar concentrations in the brain represents a major impediment to quantitatively studying its role as an intercellular signaling molecule, modulating either physiological (e.g., plasticity) or pathophysiological (cell death) responses. The advantage of electrochemical methods, using modified carbon fiber microelectrodes, over other approaches has been reviewed elsewhere (Ledo *et al.*, 2004), showing it to be the most relevant for an *in vivo* situation, affording a high sensitivity, real-time response with high spatial resolution.

The approach described here for the measurement of ·NO concentration dynamics (rate and pattern of change) *in vivo* may be used to quantitatively answer basic and simple unsolved questions (e.g., what is a physiological and a pathological concentration of ·NO in the brain?) and, from a more holistic stance, to relate ·NO production via activation of NOS with neuronal processes and the impact of ·NO in the flow of information in the brain.

ACKNOWLEDGMENTS

This work was partially supported by Grant PTDC/AGR-ALI/71262/2006 from FCT (Portugal). RMS and CFL acknowledge FCT fellowships SFRH/BD/31051/2006 and SFRH/BD/27333/2006, respectively.

REFERENCES

Bedioui, F., and Villeneuve, N. (2003). Electrochemical nitric oxide sensors for biological samples: Principle, selected examples and applications. *Electroanalysis* **15,** 5–18.
Bliss, T. V. P., and Collingridge, G. L. (1993). A synaptic model of memory: Long-term potentiation in the hippocampus. *Nature* **361,** 31–39.
Boehning, D., and Snyder, S. H. (2003). Novel neural modulators. *Annu. Rev. Neurosci.* **26,** 105–131.

Burette, A., Zabel, U., Weinberg, R. J., Schmidt, H. H. H. W., and Valtschanoff, J. G. (2002). Synaptic localization of nitric oxide synthase and soluble guanylyl cyclase in the hippocampus. *J. Neurosci.* **22,** 8961–8970.

Burmeister, J. J., Pomerleau, F., Palmer, M., Day, B. K., Huettl, P., and Gerhardt, G. A. (2002). Improved ceramic-based multisite microelectrode for rapid measurements of L-glutamate in the CNS. *J. Neurosci. Methods* **119,** 163–171.

Coyle, J. T., and Puttfarcken, P. (1993). Oxidative stress, glutamate, and neurodegenerative disorders. *Science* **262,** 689–695.

Dawson, V. L., and Dawson, T. M. (1998). Nitric oxide in neurodegeneration. *Prog. Brain Res.* **118,** 215–229.

Dawson, V. L., Dawson, T. M., London, E. D., Bredt, D. S., and Snyder, S. H. (1991). Nitric-oxide mediates glutamate neurotoxicity in primary cortical cultures. *Proc. Natl. Acad. Sci. USA* **88,** 6368–6371.

Friedemann, M. N., and Gerhardt, G. A. (1992). Regional effects of aging on dopaminergic function in the Fischer-344 rat. *Neurobiol. Aging* **13,** 325–332.

Friedemann, M. N., and Gerhardt, G. A. (1996). *In vivo* electrochemical studies of the dynamic effects of locally applied excitatory amino acids in the striatum of the anesthetized rat. *Exp. Neurol.* **138,** 53–63.

Friedemann, M. N., Robinson, S. W., and Gerhardt, G. A. (1996). o-phenylenediamine-modified carbon fiber electrodes for the detection of nitric oxide. *Anal. Chem.* **68,** 2621–2628.

Garthwaite, J., and Boulton, C. L. (1995). Nitric oxide signaling in the central nervous system. *Annu. Rev. Physiol.* **57,** 683–706.

Garthwaite, J., Charles, S. L., and Chess-Williams, R. (1988). Endothelium-derived relaxing factor release on activation of NMDA receptors suggests role as intercellular messenger in the brain. *Nature* **336,** 385–388.

Garthwaite, J., Garthwaite, G., Palmer, R. M., and Moncada, S. (1989). NMDA receptor activation induces nitric oxide synthesis from arginine in rat brain slices. *Eur. J. Pharmacol.* **172,** 413–416.

Gerhardt, G. A., and Hoffman, A. F. (2001). Effects of recording media composition on the responses of Nafion-coated carbon fiber microelectrodes measured using high-speed chronoamperometry. *J. Neurosci. Methods* **109,** 13–21.

Hall, C. N., and Garthwaite, J. (2006). Inactivation of nitric oxide by rat cerebellar slices. *J. Physiol.* **577,** 549–567.

Hara, M. R., and Snyder, S. H. (2007). Cell signaling and neuronal death. *Annu. Rev. Pharmacol. Toxicol.* **47,** 117–141.

Lauth, D., Hertting, G., and Jackisch, R. (1995). 3,4-Diaminopyridine-evoked noradrenaline release in rat hippocampal slices: Facilitation by endogenous or exogenous nitric oxide. *Brain Res.* **692,** 174–182.

Ledo, A., Barbosa, R. M., Frade, J., and Laranjinha, J. (2002). Nitric oxide monitoring in hippocampal brain slices using electrochemical methods. *Methods Enzymol.* **359,** 111–125.

Ledo, A., Barbosa, R. M., Gerhardt, G. A., Cadenas, E., and Laranjinha, J. (2005). Concentration dynamics of nitric oxide in rat hippocampal subregions evoked by stimulation of the NMDA glutamate receptor. *Proc. Natl. Acad. Sci. USA* **102,** 17483–17488.

Ledo, A., Frade, J., Barbosa, R. M., and Laranjinha, J. (2004). Nitric oxide in brain: Diffusion, targets and concentration dynamics in hippocampal subregions. *Mol. Aspects Med.* **25,** 75–89.

Luo, D., and Vincent, S. R. (1994). NMDA-dependent nitric oxide release in the hippocampus *in vivo*: Interactions with noradrenaline. *Neuropharmacology* **33,** 1345–1350.

Malinski, T., and Taha, Z. (1992). Nitric oxide release from a single cell measured *in situ* by a porphyrinic-based microsensor. *Nature* **358,** 676–678.

Mesaros, S., Grunfeld, S., Mesarosova, A., Bustin, D., and Malinski, T. (1997). Determination of nitric oxide saturated (stock) solution by chronoamperometry on a porphyrine microelectrode. *Anal. Chim. Acta.* **339,** 265–270.

Moncada, S., and Bolanos, J. P. (2006). Nitric oxide, cell bioenergetics and neurodegeneration. *J. Neurochem.* **97,** 1676–1689.

Obrenovitch, T. P., and Urenjak, J. (1997). Altered glutamatergic transmission in neurological disorders: From high extracellular glutamate to excessive synaptic efficacy. *Prog. Neurobiol.* **51,** 39–87.

Pacher, P., Beckman, J. S., and Liaudet, L. (2007). Nitric oxide and peroxynitrite in health and disease. *Physiol. Rev.* **87,** 315–424.

Paxinos, G., and Watson, C. (2007). "The Rat Brain in Stereotaxic Coordinates." Academic Press, New York.

Phillips, P. E., and Wightman, R. M. (2003). Critical guidelines for validation of the selectivity of *in vivo* chemical microsensors. *Trends Anal. Chem.* **22,** 509–514.

Segieth, J., Pearce, B., Fowler, L., and Whitton, P. S. (2001). Regulatory role of nitric oxide over hippocampal 5-HT release *in vivo. Naunyn Schmiedebergs Arch. Pharmacol.* **363,** 302–306.

Stout, A. K., and Woodward, J. J. (1995). Mechanism for nitric oxide's enhancement of NMDA-stimulated [3H]norepinephrine release from rat hippocampal slices. *Neuropharmacology* **34,** 723–729.

Taha, Z. H. (2003). Nitric oxide measurements in biological samples. *Talanta* **61,** 3–10.

Takata, N., Harada, T., Rose, J. A., and Kawato, S. (2005). Spatiotemporal analysis of NO production upon NMDA and tetanic stimulation of the hippocampus. *Hippocampus* **15,** 427–440.

Tao, R., and Auerbach, S. B. (1996). Differential effect of NMDA on extracellular serotonin in rat midbrain raphe and forebrain sites. *J. Neurochem.* **66,** 1067–1075.

Thomas, D. D., Liu, X., Kantrow, S. P., and Lancaster, J. R., Jr. (2001). The biological lifetime of nitric oxide: Implications for the perivascular dynamics of NO and O2. *Proc. Natl. Acad. Sci. USA* **98,** 355–360.

Vizi, E. S., and Kiss, J. P. (1998). Neurochemistry and pharmacology of the major hippocampal transmitter systems: Synaptic and nonsynaptic interactions. *Hippocampus* **8,** 566–607.

Walker, E., Wang, J., Hamdi, N., Monbouquette, H. G., and Maidment, N. T. (2007). Selective detection of extracellular glutamate in brain tissue using microelectrode arrays coated with over-oxidized polypyrrole. *Analyst* **132,** 1107–1111.

Wegener, G., Volke, V., and Rosenberg, R. (2000). Endogenous nitric oxide decreases hippocampal levels of serotonin and dopamine *in vivo. Br. J. Pharmacol.* **130,** 575–580.

Whitton, P. S., Maione, S., Biggs, C. S., and Fowler, L. J. (1994). N-methyl-d-aspartate receptors modulate extracellular dopamine concentration and metabolism in rat hippocampus and striatum *in vivo. Brain Res.* **635,** 312–316.

CHAPTER TWENTY-ONE

NITRIC OXIDE AND CARDIOBIOLOGY-METHODS FOR INTACT HEARTS AND ISOLATED MYOCYTES

Joshua M. Hare, Farideh Beigi, *and* Konstantinos Tziomalos

Contents

Division of Cardiology and Interdisciplinary Stem Cell Institute, Miller School of Medicine, University of Miami, Miami, Florida

Methods in Enzymology, Volume 441
ISSN 0076-6879, DOI: 10.1016/S0076-6879(08)01221-4

Abstract

The cross talk between reactive oxygen species (ROS) and reactive nitrogen species (RNS) plays a pivotal role in the regulation of myocardial and vascular function. Both nitric oxide and redox-based signaling involve the posttranslational modification of proteins through *S*-nitrosylation and oxidation of specific cysteine residues. Disruption of this cross talk between ROS and RNS contributes to the pathogenesis of heart failure. Therefore, the elucidation of these complex chemical interactions may improve our understanding of cardiovascular pathophysiology. This chapter discusses the significant role of spatial confinement of nitric oxide synthases, NADPH oxidase, and xanthine oxidoreductase in the regulation of myocardial excitation–contraction coupling. This chapter describes techniques for assessing oxidative and nitrosative stress. A variety of assays have been developed that quantify *S*-nitrosylated proteins. Among them, the biotin-switch method directly evaluates endogenously nitrosylated proteins in a reproducible way. Identification of the biotinylated or *S*-nitrosylated proteins subjected to the biotin-switch assay are described and evaluated with a one-dimensional gel (Western blot) or with the newly developed two-dimensional fluorescence difference gel electrophoresis proteomic analysis. Quantifying the number of free thiols with the monobromobimane assay in a protein of interest allows estimation of cysteine oxidation and, in turn, the state of nitroso-redox balance of effector molecules. In summary, this chapter reviews the biochemical methods that assess the impact of nitroso/redox signaling in the cardiovascular system.

1. INTRODUCTION

Normal cardiovascular performance requires delicate balancing of many complex biochemical processes. Disequilibrium of these processes may lead to myocardial dysfunction or result in structural heart disease. A variety of altered signaling systems may contribute to the progression of myocardial dysfunction.

Posttranslational modification of proteins such as phosphorylation and nitrosylation are two signaling systems that are highly conserved throughout evolution (Hess *et al.*, 2005; Stamler *et al.*, 2001). These fundamental chemical modification systems provide dynamic regulation of protein function in health and are often deranged in disease.

The molecular mechanism of nitrosylation-based signaling is less well understood in comparison to phosphorylation, but emerging evidence demonstrates the ubiquitous nature of this signaling pathway. In addition, a consensus motif for thiol nitrosylation on proteins (Stamler *et al.*, 1997) and enzymes that regulate denitrosylation (Liu *et al.*, 2001, 2004; Que *et al.*, 2005) are now described, two key findings supporting the role of *S*-nitrosylation as a second-messenger signaling system.

2. NITROSO-REDOX SIGNAL MEDIATORS

Reactive oxygen (ROS) and nitrogen (RNS) species alter numerous cellular processes. Relevant ROS in biologic systems are superoxide (O_2^-), hydrogen peroxide (H_2O_2), and hydroxyl radical ($OH^{\cdot}$) (Finkel, 1999), and RNS of biological importance include nitric oxide (NO), small and large molecular weight *S*-nitrosothiols (SNOs), and peroxynitrite ($ONOO^-$) (Stamler *et al.*, 1992) (Fig. 21.1). A key concept in understanding nitroso-redox balance in the cardiovascular system is that the effects of ROS and RNS depend mainly on the amount and location of their production (Hare and Stamler, 2005). Increasing evidence shows that ROS play an important role in signal transduction (Finkel, 1999). At low concentrations they act as second messengers through inhibition of phosphatases, acting downstream of effectors such as epidermal growth factor, tumor necrosis factor α, β-adrenergic agonists, and interleukin-1β (Chen and Keaney, 2004; Finkel, 1999). At higher concentrations, however, they take on pathophysiologic roles.

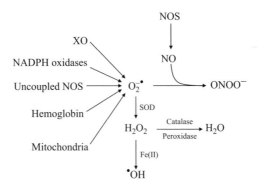

Figure 21.1 Sources of reactive oxygen and nitrogen species. Superoxide is produced by a variety of mechanisms, including the normal functioning of oxidase enzymes, hemoglobin, and mitochondria, as well as by uncoupled NOS. SOD catalyzes the dismutation of $O_2^{-\cdot}$ to hydrogen peroxide and water. H_2O_2 can be converted by catalase or peroxidases to water and molecular oxygen. It may also be converted to the hydroxyl radical through the Fenton reaction, which requires Fe^{2+} as a cofactor. Nitric oxide is produced primarily by one of the three forms of NOS. NO may interact with $O^{2-\cdot}$ to form the highly reactive peroxynitrite. Reproduced with permission from Zimmet and Hare (2006).

ROS can also affect the oxidative modification of DNA, proteins, lipids, and sugars, potentially leading to toxicity. Both reversible and irreversible oxidative modification of proteins might interrupt signaling pathways and lead to organ malfunction (McCord, 1993).

Even though the molecular mechanisms by which ROS and RNS modulate cellular signaling pathways remain incompletely understood, it is clear that cysteine thiol residues and metal centers can both undergo reactions with NO or related species (Forrester *et al.*, 2007; Hess *et al.*, 2005; Stamler *et al.*, 2001). Many investigators showed that covalent modification of cysteine thiol residues by NO (*S*-nitrosylation) represents a ubiquitous second messenger signaling system (Hess *et al.*, 2005). In the past decade, there has been an explosion of data on the role of *S*-nitrosylation as a posttranslational modification and on its signaling specificity in biological systems (Hess *et al.*, 2005; Whalen *et al.*, 2007). A descriptive protocol for measurement of SNOs by biotin–switch assays was originally established by Jaffrey *et al.* (2001) and is now being widely applied in various fields (Forrester *et al.*, 2007; Sun *et al.*, 2007; Whalen *et al.*, 2007). Importantly, the assay has the potential to identify and quantify endogenous SNO proteins in pathways such as neurodegeneration (Chung *et al.*, 2004; Uehara *et al.*, 2006), heart failure (Sun *et al.*, 2007; Whalen *et al.*, 2007) and receptor trafficking (Wang *et al.*, 2006).

3. NITRIC OXIDE AND SIGNALING THROUGH CYSTEINE THIOLS

Nitric oxide and its related ROS, RNS, modify the cysteine thiols and transition metal centers of a large spectrum of proteins, which can be regulated by *S*-nitrosylation of a single cysteine residue within an acid–base or hydrophobic structural motif (Fig. 21.2) (Hess *et al.*, 2005).

Nitric oxide mediates signaling events via two important mechanisms. First, NO binds and affects the activity of enzymes with transition metal centers, leading to activation of soluble guanylyl cyclase and inhibition of cytochrome *c*, which in turn leads to the production of cGMP. Activation of protein kinase G by cGMP initiates a cascade of signaling events with diverse consequences, including vascular smooth muscle relaxation (Schmidt and Walter, 1994). NO also exerts widespread signaling through the *S*-nitrosylation of proteins and small molecules (Gow *et al.*, 1997), which has been demonstrated in more than a hundred proteins in multiple cells and tissues (Gow *et al.*, 2002; Hess *et al.*, 2005; Jaffrey *et al.*, 2001; Stamler *et al.*, 2001). Among them, proteins involved in the regulation of

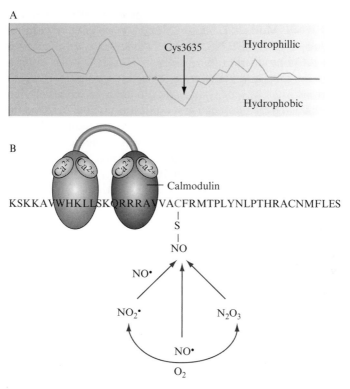

Figure 21.2 Local hydrophobicity promotes *S*-nitrosylation. The ryanodine receptor is specifically *S*-nitrosylated at Cys3635, which is located in a hydrophobic region of a calmodulin-binding domain. A hydrophobic milieu protects NO and SNO from hydrolysis and also stabilizes ROS. Both protein structure and protein–protein interactions can contribute to the generation of these hydrophobic motifs. Reproduced with permission from Hess *et al.* (2005).

myocardial contractility are also modified by *S*-nitrosylation (Hare, 2003; Sun *et al.*, 2006; Xu *et al.*, 1998). Cysteines susceptible to nitrosylation are located between an acidic and a basic amino acid, leading to the proposal of a prototypic consensus acid–base motif for *S*-nitrosylation (Stamler *et al.*, 1997). This motif was shown to be predictive in several examples in which the essential cysteines have been well characterized (Hess *et al.*, 2005).

4. Nitroso-Redox Balance or Imbalance

Nitroso–redox balance may be operationally defined by the idea that RNS and ROS work together in biological systems to achieve optimal signaling (Hare, 2004; Hare and Stamler, 2005). The concept of imbalance arises because this signaling can be disrupted by either increased ROS or decreased RNS. Moreover, because there is cross talk between the enzymes that produce ROS and RNS, NO deficiency can, in some cases, result in increased ROS production (Khan *et al.*, 2004). ROS may compete directly with NO for the same cysteine thiol, leading to reversible or irreversible modifications, such as nitrosylation or oxidation, respectively (Fig. 21.3A). A major characteristic of oxidative modification is the oxidation of thiols that leads to formation of disulfide bonds. These reactions may involve other cysteine moieties or glutathione. In this process, sulfenic acid (S–OH), sulfinic acid (S–O_2H) and sulfonic acid (SO_3H) are formed (Figure 21.3 B).

Thus, the interactions between ROS and RNS are multifaceted and strike a balance that can be disrupted at both cell and organ levels in cardiovascular disease states.

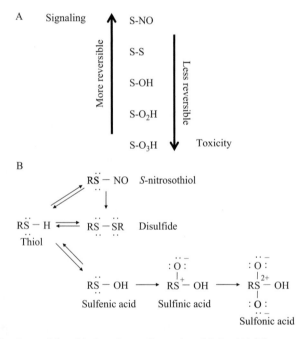

Figure 21.3 Potential oxidation fates of cysteine thiols. (A) The progression from more to less reversible reactions entails a loss of modulatory control and a transition from signaling to toxicity. (B) Thiols represent a ubiquitous target for modification by RNS and ROS. Protein thiols may undergo reversible reactions to form *S*-nitrosothiols or disulfides. S–NO subserves a critical and widespread signaling mechanism and as such can be considered a posttranslational modification system akin to phosphorylation. Thiols are also susceptible to oxidation to sulfenic acids and to sulfinic and sulfonic acids, which are progressively less-reversible reactions and are therefore maladaptive to the extent that they block the more reversible and physiologic regulation mediated by nitrosylation. Reproduced with permission from Zimmet, JM *et al.* (2006).

Oxidative stress is a disturbance in the oxidation–reduction state of the cell in which ROS production exceeds antioxidant defense. Measuring the content of free thiols (number of reduced cysteines) is a marker of the balance between ROS and RNS. The monobromobimane fluorescence technique can be applied for this purpose (Mochizuki *et al.*, 2007; Xu *et al.*, 1998) and is described later. By analogy, *nitrosative stress* is the imbalance in NO signaling caused by increased amounts of RNS, which may be caused by or associated with redox imbalance. In addition, levels of protein *S*-nitrosylation are kept in equilibrium with *S*-nitrosoglutathione (GSNO) reductase (Liu *et al.*, 2004; Que *et al.*, 2005), a newly appreciated enzyme that metabolizes *S*-nitrosothiols and protects against excessive *S*-nitrosylation and against nitrosative stress (Que *et al.*, 2005).

5. Nitric Oxide Synthases, NADPH Oxidase, and Xanthine Oxidoreductase (XOR)

5.1. Nitric oxide synthases

Nitric oxide has a major role in cardiovascular homeostasis as it represents an important messenger in many signal transduction processes. Shear stress and chemical mediators such as acetylcholine, bradykinin, substance P, and β-adrenoceptor agonists stimulate the production of NO from arginine in a complex reaction, which is catalyzed by neuronal nitric oxide synthase (NOS1) or endothelial NOS (NOS3). These NOS enzymes are Ca^{2+} dependent and require NADPH and O_2 as cofactors. A third type of NOS, which is inducible (NOS2) by inflammatory cytokines and by angiotensin II in a Ca^{2+}-independent manner, is found at high levels in the failing heart.

5.2. Spatial localization of NOS

The production of NO by NOS (NOS1, NOS2, and NOS3) occurs through oxidation of the terminal guanidino nitrogen of L-arginine, resulting in the formation of NO and L-citrulline (Michel and Feron, 1997). NO activity in cardiomyocytes is highly dependent on the site of production, which in turn is regulated by the spatial localization of NOS (Barouch *et al.*, 2002). NOS1 has been identified in the sarcoplasmic reticulum, where it directly influences sarcoplasmic reticulum (SR) calcium cycling (Xu *et al.*, 1999), whereas NOS3 is localized in caveolae of the sarcolemma and T-tubules. NOS3 activity is regulated by caveolin-3 and β-adrenergic, muscarinic, and bradykinin receptors (Feron *et al.*, 1998; Hare *et al.*, 2000). The activation of NOS3 signaling via cGMP reduces myocardial contractile force. In contrast, NOS1 activation has primarily positive inotropic effects on the heart (Barouch *et al.*, 2002). Thus, NO does not act as a free diffusible messenger but its actions depend on the specific NOS isoform that produces NO (Hare and Stamler, 2005).

5.3. NADPH oxidase

NADPH oxidase is a flavocytochrome b heterodimer that consists of two protein subunits, p22-phox and either p91-phox in fibroblasts or Nox1 in smooth muscle cells. NAD/NADPH oxidase is membrane bound and is the most powerful source of endogenous $O_2^{\cdot-}$ production. It has been proposed that protein kinase C and phospholipase D are involved in the activation of NADPH oxidase. Upon activation, at least two cytosolic proteins are synthesized, p47-phox and p67-phox, as well as rac2, which is a GTP-binding protein. When p47-phox is phosphorylated extensively, these proteins interact with each other and the flavocytochrome. NADPH oxidase expression varies in different cells, including endothelial, vascular smooth muscle, fibroblast, and inflammatory. Work suggests that the specificity of NADPH oxidases containing different Nox subunits may partially be achieved by precise subcellular localization of the Nox proteins. In this regard, the Nox1- and Nox4-containing NADPH oxidases in vascular smooth muscle have been localized to caveolae and focal adhesions, respectively (Hilenski *et al.*, 2004).

5.4. Xanthine oxidoreductase

Xanthine oxidoreductase is a molybdenum and iron-containing flavoprotein homodimer. Even though it encoded by a single gene, it is expressed in two forms, as xanthine dehydrogenase (XDH) and as xanthine oxidase (XO). XDH can be converted to XO either by reversible thiol oxidation or by irreversible proteolytic cleavage. XOR has a crucial role in the purine degradation pathway, as it oxidizes hypoxanthine to xanthine, leading to the production of uric acid. During these reactions, two molecules of hydrogen peroxide and two molecules of superoxide are also formed (Berry and Hare, 2004).

Xanthine oxidase transfers electrons to molecular oxygen readily and appears to be responsible for the majority of ROS production, whereas XDH prefers NAD^+ as an electron acceptor. XO activity is blocked by the antigout drug allopurinol. However, once NAD^+ is reduced to NADH, XDH is able to act as an NADH oxidase. This process ultimately leads to the generation of superoxide and is not inhibited by allopurinol (Berry and Hare, 2004).

Xanthine oxidoreductase is expressed principally in the liver and small intestine. It is also present in endothelial cells and circulates in the blood (Jarasch *et al.*, 1981). Investigations support XOR expression (Cappola *et al.*, 2001; Linder *et al.*, 1999; Muxfeldt and Schaper, 1987) in the SR (Khan *et al.*, 2004; Hare and Stamler, 2005) of cardiac myocytes. In addition to its well-known activities, XO may also act as a GSNO reductase (Hare and Stamler, 2005; Trujillo *et al.*, 1998), thus speculating that the presence of XO in the SR may regulate *S*-nitrosylation of the ryanodine receptor and other SR proteins.

6. NITRIC OXIDE-REDOX BALANCE IN EXCITATION– CONTRACTION COUPLING

In cardiac myocytes, a cascade of events leading to a rapid increase in cytosolic calcium and muscle contraction is initiated by depolarization of the plasma membrane. The L-type calcium channel, the port of calcium entry, triggers a large release of calcium from the SR through the coupled ryanodine receptor (RyR) channel, a process known as calcium-induced calcium release (Hare, 2003). In the relaxation phase, calcium removal from the cytoplasm is mediated by SR reuptake via the calcium ATPase (SER-CA2a) and by sarcolemmal extrusion via the sodium–calcium exchanger (NCX) protein. Both NO and the redox status affect excitation–contraction coupling through interactions with calcium-handling proteins such as the L-type Ca^{2+} channel and the RyR, the contractile apparatus, and respiratory complexes (Hare, 2003).

The effect of NO in myocardial contractility involves both cGMP-dependent and –independent mechanisms (Campbell *et al.*, 1996; Paolocci *et al.*, 2000). cGMP inhibits L-type Ca^{2+} channel activity, whereas *S*-nitrosylation and oxidation of this channel have a biphasic effect that is either stimulatory at low concentration or inhibitory at high levels (Campbell *et al.*, 1996; Sun *et al.*, 2006).

Similarly, the cardiac ryanodine receptor (RyR2) is regulated by target cysteine nitrosylation or oxidation (Eu *et al.*, 2000; Sun *et al.*, 2001). Additional *S*-nitrosylation of cysteines thiols beyond basal level can activate the channel reversibly (Xu *et al.*, 1998). However, oxidation of cysteine residues may lead to irreversible activation of the channel favoring SR leak and calcium depletion (Xu *et al.*, 1998).

7. TECHNIQUES FOR ASSESSING OXIDATIVE AND NITROSATIVE STRESS

7.1. *In vitro* techniques

A number of assays are being used in order to assess oxidative and nitrosative stress *in vitro* and can be broadly divided into photometric, fluorimetric, and luminescence based. All these techniques are based on the use of a tracer that reacts with the ROS or RNS under consideration and yields measureable products (Tarpey and Fridovich, 2001). The major *in vitro* techniques are described below.

7.1.1. Determination of superoxide production

The most frequently used photometric methods to measure superoxide are the reduction of ferricytochrome C and the reduction of nitro blue tetrazolium (NBT) (Brandes and Janiszewski, 2005). Both techniques are based on the difference in extinction coefficient between ferricytochrome C and NBT and their products, ferrocytochrome C and formazan, respectively (Brandes and Janiszewski, 2005). Because other ROS can also reduce either ferricytochrome C and NBT, superoxide dismutase should be used in both assays in order to discriminate the superoxide-induced reduction (Brandes and Janiszewski, 2005). Limitations of these methods include the relatively long incubation time (up to 1 h) and the small changes in optical density that are frequently observed with the ferricytochrome C assay (Brandes and Janiszewski, 2005; Tarpey and Fridovich, 2001) and the potential of NBT to react with molecular oxygen and produce superoxide (Auclair *et al.*, 1978).

Chemiluminescent methods are based on the use of an enhancer that emits light during the reaction with the ROS of interest. Lucigenin is relatively specific for superoxide and is the most widely used enhancer for its determination (Brandes and Janiszewski, 2005). However, lucigenin can undergo autooxidation and generate superoxide (Janiszewski *et al.*, 2002; Tarpey *et al.*, 1999). Application of low concentrations of lucigenin reduces but does not completely eliminate the production of superoxide (Janiszewski *et al.*, 2002) and also reduces the acquired light signal (Brandes and Janiszewski, 2005). Other enhancers have also been used but whether they offer advantages over lucigenin is a matter of debate (Brandes and Janiszewski, 2005; Janiszewski *et al.*, 2002; Tarpey *et al.*, 1999) (Fig. 21.4).

Fluorimetric methods can also be used to quantify superoxide and determine its production site (Brandes and Janiszewski, 2005), and dihydroethidium (DHE) is the most sensitive fluorescent probe for this purpose (Brandes and Janiszewski, 2005). The reaction of DHE with superoxide yields 2-hydroxyethidium, a product distinct from ethidium, which is generated when DHE interacts with other ROS (Zhao *et al.*, 2003, 2005). Therefore, the determination of 2-hydroxyethidium could potentially improve the specificity of this assay (Zhao *et al.*, 2005).

7.1.2. Determination of other ROS and RNS

Peroxynitrite and hydroxyl radicals are determined primarily with fluorimetric assays, using scavengers to discriminate between them (Brandes and Janiszewski, 2005). However, the lack of specific scavengers limits

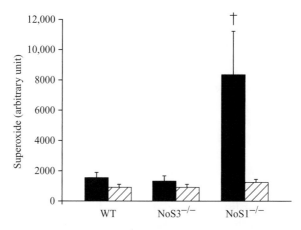

Figure 21.4 Superoxide production in cardiac tissue from NOS knockout mice. Basal lucigenin-enhanced chemiluminescence recordings were elevated in NOS1$^{-/-}$ but not in NOS3$^{-/-}$. Incubation with xanthine (filled bars) produced dramatic increases in lucigenin-detected superoxide. Importantly, this increase was fourfold greater in NOS1$^{-/-}$ compared with wild-type and NOS3$^{-/-}$ hearts ($\dagger P < 0.05$; $n = 4$ mice of each strain), indicating that NOS1 deficiency leads to augmented cardiac XOR superoxide production. Inhibition by allopurinol (striped bars) demonstrates that this increase is caused by XO production of superoxide. These experiments demonstrate ROS/RNS cross talk within the cardiac myocyte, as well as NOS isoform specificity based on subcellular localization of NOS1 and NOS3. Reproduced with permission from Khan *et al.* (2004).

the capability of these assays to differentiate the produced radical (Brandes and Janiszewski, 2005). Hydrogen peroxide is also determined with fluorimetric assays, but because it is not as reactive as other ROS, peroxidase has to be used in combination (Brandes and Janiszewski, 2005). Peroxynitrite, hydroxyl radicals, and hydrogen peroxide can also be measured with chemiluminescent methods using luminol as a probe and specific inhibitors to discriminate the produced species (Brandes and Janiszewski, 2005).

Electron spin resonance (ESR) can also be used to determine ROS *in vitro*. This technique is usually combined with "spin traps," that is, agents that intercept ROS and substantially improve the sensitivity of the method (Halliwell and Whiteman, 2004). A limitation of this method is that the product of the trap–ROS reaction can be either metabolized or reduced by endogenous antioxidants (Halliwell and Whiteman, 2004). Newer probes are resistant to reducing agents and might provide more valid results (Rizzi *et al.*, 2003). ESR has been used to measure ROS in human tissue samples (Valgimigli *et al.*, 2002) and can also be applied in whole animals (Han *et al.*, 2001) but not in humans because of safety considerations (Halliwell and Whiteman, 2004).

7.2. *In vivo* techniques

Both ROS and RNS have a very short half-life, which poses significant difficulties in their direct determination in humans. To overcome this obstacle, a variety of endogenous molecules have been identified that act as targets for ROS or RNS and serve as indirect markers of oxidative and nitrosative stress. In order to be useful in clinical practice, these markers should ideally be stable, accumulate to detectable concentrations, correlate with oxidative/nitrosative stress severity, and be specific for the oxidative/ nitrosative pathway of interest (Dalle-Donne *et al.*, 2006). The major *in vivo* techniques are described below.

7.2.1. Markers of lipid peroxidation

The most frequently used markers of oxidative/nitrosative stress are lipid peroxidation products, including unsaturated reactive aldehydes [principally manoldialdehyde (MDA) as well as 4-hydroxy-2-nonenal (HNE) and 2-propenal (acrolein)] and isoprostanes (Dalle-Donne *et al.*, 2006). MDA is a by-product of arachidonate metabolism (Dalle-Donne *et al.*, 2006). MDA is usually measured with the thiobarbituric acid-reactive substance (TBARS) assay, in which TBA reacts with MDA and generates a stable chromogen (Dalle-Donne *et al.*, 2006). The latter can be quantified easily by spectrophotometry (Dalle-Donne *et al.*, 2006). However, other lipid peroxidation products except MDA can also interact with TBA and produce compounds with similar absorption ranges (Dalle-Donne *et al.*, 2006). Nowadays, separation of the MDA–TBA product is performed with high-performance liquid chromatography (HPLC), which yields more reliable results (Dalle-Donne *et al.*, 2006; Griendling and Fitzgerald, 2003; Halliwell and Whiteman, 2004). Gas chromatography (GC)-negative ion chemical ionization mass spectrometry (MS) can also reliably determine MDA concentrations in plasma (Griendling and Fitzgerald, 2003; Kadiiska *et al.*, 2005). MDA can also be measured in the urine; however, intake of several foods interferes with urine MDA levels (Halliwell and Whiteman, 2004; Richelle *et al.*, 1999).

The most widely used types of isoprostanes for determining oxidative stress are the F_2-isoprostanes, which are produced *in situ* in the cell membrane through ROS-induced oxidation of arachidonic acid (Dalle-Donne *et al.*, 2006; Griendling and Fitzgerald, 2003). Then, they are cleaved by phospholipases, enter the circulation, and are excreted in the urine (Dalle-Donne *et al.*, 2006; Griendling and Fitzgerald, 2003). Among F_2-isoprostanes, 8-iso-prostaglandin F2α has been used extensively for the assessment of oxidative stress (Dalle-Donne *et al.*, 2006; Griendling and Fitzgerald, 2003). F_2-isoprostanes can be found in almost all biological fluids and are considered one of the most reliable markers of oxidative stress (Dalle-Donne *et al.*, 2006; Halliwell and Whiteman, 2004). In contrast to reactive aldehydes, diet does not modify the levels of F_2-isoprostanes in blood or urine (Richelle et al., 1999). Several assays are used for F_2-isoprostane measurements, including GC-MS, GC-tandem MS (GS-MS/MS), liquid chromatography (LC)-MS, LC-MS/MS, enzyme immunoassays, and radio-imunoassays (Dalle-Donne et al., 2006; Kadiiska et al., 2005). MS-based assays are considered the most reliable; immunoassays are less costly, easier to apply, and less time-consuming but further work is required to optimize their performance (Bessard et al., 2001; Dalle-Donne et al., 2006; Halliwell and Whiteman, 2004).

Oxidized low-density lipoprotein (LDL) is also an indicator of lipid peroxidation (Armstrong et al., 2006). Oxidized LDL is not a single compound, as the extent and site of oxidation can vary significantly (Tsimikas and Witztum, 2001). Several different forms of oxidized LDL have been used as markers of lipid oxidation, including minimally modified LDL and MDA-LDL (Tsimikas and Witztum, 2001), and are determined with enzyme-linked immunosorbent assay (Holvoet et al., 1998; Tsimikas et al., 2005).

7.2.2. Markers of protein oxidation

3-Nitrotyrosin represents a marker of nitrosative stress, particularly peroxynitrite production (Dalle-Donne et al., 2006; Halliwell and Whiteman, 2004). However, there is considerable variability in the reported 3-nitrotyrosin levels, which has been attributed to the ex vivo generation of 3-nitrotyrosin and 3-nitrotyrosin degradation during sample preparation and analysis (Daiber et al., 2003; Souza et al., 2000). Both immunoassay- and HPLC-based assays are associated with artifacts (Dalle-Donne et al., 2006; Kaur et al., 1998). In contrast, electron capture-negative chemical ionization GC/MS, LC-MS/MS, and GC-tandem MS are quite sensitive and specific (Gaut et al., 2002). It has been suggested that GC-tandem MS is the most sensitive technique for 3-nitrotyrosin measurement (Dalle-Donne et al., 2006) and can be applied in both plasma and urine samples (Tsikas et al., 2003; Tsikas et al., 2005).

7.2.3. Markers of DNA oxidation

8-Hydroxy-20-deoxyguanosine is the most commonly used marker of DNA oxidation and can be quantified by HPLC, GC-MS, LC-MS, and antibody-based assays (Beckman et al., 2000; Kadiiska et al., 2005; Toyokuni et al., 1997). Following DNA repair, this compound is excreted in the urine where it can be measured (Dalle-Donne et al., 2006; Helbock et al., 1998). Urinary 8-hydroxy-20-deoxyguanosine levels are not affected by diet (Gackowski et al., 2001) and are considered to depict systemic oxidative stress (Dalle-Donne et al., 2006). However, ex vivo oxidation of DNA during sample preparation and analysis introduces artifacts (Halliwell and

Whiteman, 2004; Helbock *et al.*, 1998). The antibody-based assays are free from artifactual DNA oxidation (Toyokuni *et al.*, 1997) and appear to correlate well with the results of more specific methods, such as HPLC (Shimoi *et al.*, 2002; Yoshida *et al.*, 2002). Nitrosative DNA damage can be evaluated by measuring 8-nitroguanine with HPLC (Hsieh *et al.*, 2002). It was suggested that 8-nitroguanine specifically reflects peroxynitrite levels but it has become clear that other RNS can also attack DNA and produce this adduct (Byun *et al.*, 1999).

7.2.4. Other markers of oxidative stress

A decreased glutathione/glutathione disulfide plasma ratio reflects oxidative stress (Dalle-Donne *et al.*, 2006) and can be determined with spectrophotometric and fluorometric assays, GC-MS, and HPLC-electrospray ionization-MS (Dalle-Donne *et al.*, 2006; Steghens *et al.*, 2003). However, several artifacts during sample preparation frequently result in erratic findings (Rossi *et al.*, 2002).

Antioxidant capacity is regarded as an indirect indicator of oxidative stress (Halliwell and Whiteman, 2004). Several methods have been developed in order to measure total antioxidative capacity in erythrocytes, particularly in the serum, but their sensitivity and specificity are questionable (Halliwell and Whiteman, 2004).

8. TECHNIQUES USED FOR MEASURING *S*-NITROSYLATION

8.1. Assays used to quantify SNO proteins

There are several methods available to quantify nitrosylated proteins, including colorimetric, chemiluminescent, fluorescent and antibody-based (Gow *et al.*, 2007). In colorimetric assays (the Saville assay, see later), NO is displaced from nitrosylated proteins by $HgCl_2$, and the increase in NO_2^- concentration (colorimetric determination; Griess reagents) reflects the extent of nitrosylation (Gow *et al.*, 2007). However, this technique is considered rather insensitive for biological samples (Gow *et al.*, 2007). In chemiluminescence-based assays, NO reacts with ozone, NO_2 is produced and then decays to NO_2-releasing light (Gow *et al.*, 2007). NO is released by proteins either by photolysis or by chemical reduction; the latter appears to yield more specific results (Gow *et al.*, 2007). In the latter method, *S*-nitrosocysteine is formed after adding excess cysteine and is then reduced to NO by CuCl (Gow *et al.*, 2007). It should be noted that because heme binds NO, a significant attenuation of the signal may occur in samples containing proteins with a heme moiety (Gow *et al.*, 2007). The addition of CO to the system prevents this phenomenon (copper/cysteine/CO assay or 3C assay) (Gow *et al.*, 2007). Fluorescence-based assays also involve the use of HgCl$_2$ as a SNO bond-reducing agent but are based on the generation of a fluorescent product, triazolofluorescein, from 4,5-diaminofluorescein (Gow *et al.*, 2007). Antibodies directed against the SNO moiety are used primarily for immunohistochemical identification of nitrosylated proteins; use of both positive and negative controls is a sine qua non in these assays (Gow *et al.*, 2007).

8.2. SNO measurements of heart proteins by biotin-switch assay

This assay is based on three steps: "blocking" of free cysteine thiols with *S*-methyl methanethiosulfonate (MMTS), "reduction" of *S*-nitrosothiols to thiols with ascorbate, and "labeling" of the reduced thiols with biotin-HPDP. Complete blocking is an essential step in the evaluation of background biotinylation, as the specificity of the assay is based on the ability of the ascorbate to reduce SNOs into free thiols. Factors that limit ascorbate specificity include the presence of transition metals and ultraviolet (UV) radiation; therefore, adding metal chelators and avoiding direct UV light exposure are essential (Forrester *et al.*, 2007; Jaffrey *et al.*, 2001). Before performing the biotin switch on heart tissue it is advantageous to perfuse the harvested whole heart with a buffer solution (e.g., Krebs–Hensleit). A relatively blood-free heart tissue would decrease the biotinylation background caused by hemoglobin. Detection of endogenous *S*-nitrosothiols in heart tissue is intricate, as factors such as hemoglobin, MMTS (blocking agent), HPDP-biotin (labeling agent), ascorbate (reduction agent), and UV radiation may all affect the labeling for tissue homogenates. In order to minimize background biotinylation and distinguish between true endogenous nitrosylation from artifacts it is crucial to include positive and negative control samples. Treatment of samples with a NO donor such as GSNO or *S*-nitrosocysteine would increase the level of endogenous nitrosylation and is used as a positive control.

It is believed that ascorbate and UV radiation promote protein biotinylation, which is SNO independent and involves reduction of protein disulfides (Forrester *et al.*, 2007; Huang and Chen, 2006). Forrester *et al.* (2007) demonstrated that ascorbate may rapidly reduce protein thiyl radicals by exposure to direct or indirect UV radiation and provide free thiols for biotinylation. These false biotinylation signals in samples can be avoided by eliminating ascorbate in the assay and/or UV photolysis treatment, which are considered negative controls for the assay (Forrester *et al.*, 2007) (Fig. 21.5).

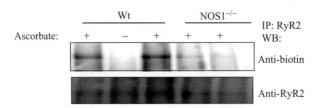

Figure 21.5 *S*-Nitrosothiol identification of ryanodine receptor from wild-type and NOS1$^{-/-}$ mouse hearts as assessed by the biotin-switch assay. A representative Western blot protein gel that demonstrates immunoprecipitation of *S*-nitrosylated/biotinylated RyR2 as detected by antibiotin and anti-RyR2 antibodies on a 3 to 8% Tris–acetate gel in the presence (+) and absence (−) of ascorbate. Results show the significantly lower amount of SNO-RyR2 detected in NOS1$^{-/-}$ than in wild-type mice. Reproduced from Gonzalez *et al.* (2007).

8.3. Biotinylated/SNO protein identification techniques

Identification of the biotinylated or S-nitrosylated proteins subjected to the biotin-switch assay is an essential step in evaluating the amount of SNO proteins in the heart. Methods in use are based on (1) SDS–PAGE one-dimensional (1D) gel (Western blot) or (2) 2D fluorescence difference gel electrophoresis (DIGE) proteomic analysis. 1D gel (Western blot) analysis is the classical technique that primarily separates biotinylated proteins on SDS–PAGE gels in total homogenates followed by detection with the horseradish peroxidase-conjugated antibiotin antibody. For individual protein identification, the samples can be incubated with (a) streptavidin-agarose beads, (b) protein-G Sepharose beads for immunoprecipitation, or (c) subjected to different mass-spectrometry analysis. Completion of (a) and (b) requires a second Western blot gel to be applied. In (a) the proteins cannot be detected by antibiotin, whereas in (b) individual proteins can be identified with antibiotin and reconfirmed with antibody. In (c), the biotinylated proteins are identifiable depending on prior sample treatments. The mass-spectrometry technique is a powerful tool for this purpose and can be combined with the other two methods; however, it should be noted that large amounts of protein are needed in order to obtain accurate results. The 2D fluorescence DIGE proteomics combined with mass spectrometry is a newly developed technique (Sun *et al.*, 2007), which utilizes fluorescent labeling at the site of S–NO on proteins substituting the biotin-HPDP labeling. Development of this sophisticated technique (DIGE) opens new insight for future research as to be able to simultaneously identify SNO proteins in various samples. It should be kept in mind that the fluorescent-labeling technique can also cause background restrictions similar to biotin labeling, thus including positive and negative controls are highly essential.

9. ACTIVITY MEASUREMENTS

9.1. Nitric oxide metabolites

The Saville Griess assay is an established reagent system (since 1879) that can detect nitrite (picomoles) in a variety of biological materials. The Griess assay can detect the amount of NO in samples by measuring the formation of nitrite (NO_2^-), which is one of the primary and stable metabolites of NO. The assay is based on the diazotization reaction of sulfanilamide and N-1-napthylethylenediamine dihydrochloride under acidic conditions. The first step is the conversion of nitrate to nitrite utilizing nitrate reductase. The second step is addition of the Griess reagents, which convert nitrite into a deep purple azo compound. Photometric measurement of the absorbance as a consequence of this azo chromophore determines the nitrite concentration accurately. Several relevant assay kits are available commercially and offer accurate and convenient protocols (e.g., nitrite colorimetric assay kit, Cayman Chemical).

Accordingly, heart homogenates centrifuged in phosphate-buffered saline (pH 7.4) buffer at 10,000g for 20 min. The lysate is centrifuged through 10-kDa Millipore filtration tubes. The combined oxidation products of NO (nitrites and nitrates) in the sample can be reduced with nitrate reductase. After addition of the Griess reagent, nitrite can be quantified colorimetrically at 540 nm. Standards samples are prepared by serial dilutions of sodium nitrite (Saraiva *et al.*, 2005).

It is important to note that this assay cannot be used to analyze nitrate and nitrite from an *in vitro* assay of nitric oxide synthase in which excess NADPH has been added.

9.2. NADPH oxidase activity and protein abundance

NADPH-dependent superoxide ($O_2^{\cdot-}$) production can be measured in homogenates using lucigenin-enhanced chemiluminescence (β-NADPH 300 μM/liter) at room temperature on a microplate luminator (Veritas, Turner Biosystems, Sunnyvale, CA). Measurements of protein abundance of NADPH oxidase can be obtained by electrophoretical separation of its subunits—Gp91-phox, P22-phox, P47-phox, and P67-phox—and then analyzed by Western blot (Saraiva *et al.*, 2005).

9.3. Xanthine oxidoreductase activity

Xanthine oxidoreductase activity can be measured by several assays, including Amplex Red fluorescence and lucigenin-enhanced chemiluminescence for the production of uric acid, superoxide, and peroxynitrite. The horseradish peroxidase-linked Amplex Red fluorescence assay (available commercially at Molecular Probes) is described briefly. In principle, in this assay XO catalyzes the oxidation of purine bases to uric acid and superoxide. In the reaction mixture, the superoxide degrades to H_2O_2, which subsequently reacts with the horseradish peroxidase-linked Amplex Red reagent to generate the red fluorescent oxidation product, resorufin.

Accordingly, heart homogenates are passed through Sephadex G-25 columns (GE healthcare, Piscataway, NJ). The processed effluent is added to a working solution containing Amplex Red reagent (50 μM), xanthine (0.1 mM), and horseradish peroxidase type II (0.1 U/ml). The reaction mixture is incubated at 37 °C for 30 min, and the H_2O_2 production is measured. Fluorescence readings are made in duplicate in 96–well plates at ex/em= 544/590 nm using 100 μl total volume per each well. The obtained values are plotted against a standard curve with known concentrations of XOR (Saraiva *et al.*, 2005).

9.4. Free thiol assessment by monobromobimane assay

Free thiols of proteins can be assessed using the fluorescent probe for cysteine monobromobimane applied in solutions or in suspensions of cells and tissues. Here, the free thiols for the ryanodine receptor (RyR2) from the heart homogenate are determined (Mochizuki *et al.*, 2007) as described previously (Kosower and Kosower, 1987) with minor modifications. Total heart homogenates are incubated with 1 mM monobromobimane for 1 h at room temperature in the dark. The reaction is stopped with 1 mM L-cysteine. Proteins are resolved on Western blot gels and transilluminated with UV light (302 nm). Total RyR2 is identified upon silver staining of the gel and confirmed with the anti-RyR antibody. Free cysteine content is expressed as the ratio of the optical density of the UV signal to the total RyR signal (silver staining) (Fig. 21.6).

10. CONCLUSIONS

Nitric oxide and redox-based signaling mediate numerous signaling pathways. NO, in addition to guanylyl cyclase activation, exerts posttranslational modification of proteins through *S*-nitrosylation. A critical interactivity can occur between RNS and ROS at the site of cysteine thiol residues. Cross talk between ROS- and RNS-generating systems strikes a tightly regulated balance that is disrupted in a wide array of cardiovascular diseases, including hypertension, coronary heart disease, and heart failure. A large number of assays and markers are currently available to evaluate oxidative and nitrosative stress both *in vitro* and *in vivo*. The generally indirect downstream nature of most of these chemical assays represents a significant limitation. Increasingly it is appreciated that S-NO and free thiol

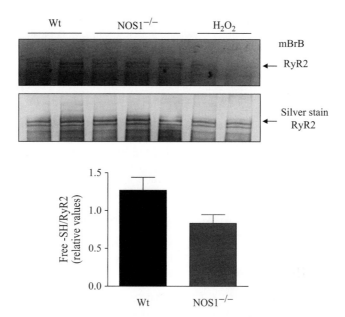

Figure 21.6 Free thiol assessment of ryanodine receptor by monobromobimane assay. Representative Western blot gels demonstrating the RyR2 protein band from wild-type and NOS1 knockout whole mice hearts after monobromobimane assay, silver staining, and anti-RyR2 antibody. Hydrogen peroxide is used as the negative control. The bar graph demonstrates the densitometric values of measured free thiol versus total RyR2. Results demonstrate that RyR2 in NOS1$^{-/-}$ mice are significantly more ($P < 0.05$) oxidized than wild-type mice. Reproduced from Gonzalez *et al.* (2007).

levels should be measured directly, and in this regard the biotin switch assay and modifications thereof are increasingly being used to assess nitroso-redox balance. We continue to recommend assessing oxidative/nitrosative stress coupled to functional assays (for review, see Zimmet and Hare, 2006) to enhance the validity of experimental studies.

ACKNOWLEDGMENTS

Konstantinos Tziomalos is supported by a grant from the Hellenic Antihypertensive Society. Joshua M. Hare is supported by Grants NIH 2RO1 HL-65455-05, NIA RO1 AG025017, NIH U54HL081028-01, and NIH/NHLBI R01HL084275.

REFERENCES

Armstrong, E. J., Morrow, D. A., and Sabatine, M. S. (2006). Inflammatory biomarkers in acute coronary syndromes. III. Biomarkers of oxidative stress and angiogenic growth factors. *Circulation* **113,** e289–e292.

Auclair, C., Torres, M., and Hakim, J. (1978). Superoxide anion involvement in NBT reduction catalyzed by NADPH-cytochrome P-450 reductase: A pitfall. *FEBS Lett.* **89,** 26–28.

Barouch, L. A., Harrison, R. W., Skaf, M. W., Rosas, G. O., Cappola, T. P., Kobeissi, Z. A., Hobai, I. A., Lemmon, C. A., Burnett, A. L., O'Rourke, B., Rodriguez, E. R., Huang, P. L., *et al.* (2002). Nitric oxide regulates the heart by spatial confinement of nitric oxide synthase isoforms. *Nature* **416,** 337–339.

Beckman, K. B., Saljoughi, S., Mashiyama, S. T., and Ames, B. N. (2000). A simpler, more robust method for the analysis of 8-oxoguanine in DNA. *Free Radic. Biol. Med.* **29,** 357–367.

Berry, C. E., and Hare, J. M. (2004). Xanthine oxidoreductase and cardiovascular disease: Molecular mechanisms and pathophysiological implications. *J. Physiol.* **555,** 589–606.

Bessard, J., Cracowski, J. L., Stanke-Labesque, F., and Bessard, G. (2001). Determination of isoprostaglandin F2alpha type III in human urine by gas chromatography-electronic impact mass spectrometry: Comparison with enzyme immunoassay. *J. Chromatogr. B Biomed. Sci. Appl.* **754,** 333–343.

Brandes, R. P., and Janiszewski, M. (2005). Direct detection of reactive oxygen species *ex vivo. Kidney Int.* **67,** 1662–1664.

Byun, J., Henderson, J. P., Mueller, D. M., and Heinecke, J. W. (1999). 8-Nitro-2′-deoxyguanosine, a specific marker of oxidation by reactive nitrogen species, is generated by the myeloperoxidase-hydrogen peroxide-nitrite system of activated human phagocytes. *Biochemistry* **38,** 2590–2600.

Campbell, D. L., Stamler, J. S., and Strauss, H. C. (1996). Redox modulation of L-type calcium channels in ferret ventricular myocytes: Dual mechanism regulation by nitric oxide and *S*-nitrosothiols. *J. Gen. Physiol.* **108,** 277–293.

Cappola, T. P., Kass, D. A., Nelson, G. S., Berger, R. D., Rosas, G. O., Kobeissi, Z. A., Marban, E., and Hare, J. M. (2001). Allopurinol improves myocardial efficiency in patients with idiopathic dilated cardiomyopathy. *Circulation* **104,** 2407–2411.

Chen, K., and Keaney, J. (2004). Reactive oxygen species-mediated signal transduction in the endothelium. *Endothelium* **11,** 109–121.

Chung, K. K., Thomas, B., Li, X., Pletnikova, O., Troncoso, J. C., Marsh, L., Dawson, V. L., and Dawson, T. M. (2004). *S*-nitrosylation of parkin regulates ubiquitination and compromises parkin's protective function. *Science* **304,** 1328–1331.

Daiber, A., Bachschmid, M., Kavakli, C., Frein, D., Wendt, M., Ullrich, V., and Munzel, T. (2003). A new pitfall in detecting biological end products of nitric oxide-nitration, nitros(yl)ation and nitrite/nitrate artefacts during freezing. *Nitric Oxide* **9,** 44–52.

Dalle-Donne, I., Rossi, R., Colombo, R., Giustarini, D., and Milzani, A. (2006). Biomarkers of oxidative damage in human disease. *Clin. Chem.* **52,** 601–623.

Eu, J. P., Sun, J., Xu, L., Stamler, J. S., and Meissner, G. (2000). The skeletal muscle calcium release channel: Coupled O2 sensor and NO signaling functions. *Cell* **102,** 499–509.

Feron, O., Dessy, C., Opel, D. J., Arstall, M. A., Kelly, R. A., and Michel, T. (1998). Modulation of the endothelial nitric-oxide synthase-caveolin interaction in cardiac myocytes: Implications for the autonomic regulation of heart rate. *J. Biol. Chem.* **273,** 30249–30254.

Finkel, T. (1999). Signal transduction by reactive oxygen species in non-phagocytic cells. *J. Leukocyte Biol.* **65,** 337–340.

Forrester, M. T., Foster, M. W., and Stamler, J. S. (2007). Assessment and application of the biotin switch technique for examining protein *S*-nitrosylation under conditions of pharmacologically induced oxidative stress. *J. Biol. Chem.* **282,** 13977–13983.

Gackowski, D., Rozalski, R., Roszkowski, K., Jawien, A., Foksinski, M., and Olinski, R. (2001). 8-Oxo-7,8-dihydroguanine and 8-oxo-7,8-dihydro-2′-deoxyguanosine levels in human urine do not depend on diet. *Free Radic. Res.* **35,** 825–832.

Gaut, J. P., Byun, J., Tran, H. D., and Heinecke, J. W. (2002). Artifact-free quantification of free 3-chlorotyrosine, 3-bromotyrosine, and 3-nitrotyrosine in human plasma by electron capture-negative chemical ionization gas chromatography mass spectrometry and liquid chromatography-electrospray ionization tandem mass spectrometry. *Anal. Biochem.* **300,** 252–259.

Gonzalez, D. R., Beigi, F., Treuer, A. V., and Hare, J. M. (2007). Deficient ryanodine receptor *S*-nitrosylation increases sarcoplasmic reticulum calcium leak and arrhythmogenesis in cardiomyocytes. *Proc. Natl. Acad. Sci. USA* **104,** 20612–20617.

Gow, A., Doctor, A., Mannick, J., and Gaston, B. (2007). *S*-Nitrosothiol measurements in biological systems. *J. Chromatogr. B Analyt. Technol. Biomed. Life Sci.* **851,** 140–151.

Gow, A. J., Buerk, D. G., and Ischiropoulos, H. (1997). A novel reaction mechanism for the formation of *S*-nitrosothiol *in vivo. J. Biol. Chem.* **272,** 2841–2845.

Gow, A. J., Chen, Q., Hess, D. T., Day, B. J., Ischiropoulos, H., and Stamler, J. S. (2002). Basal and stimulated protein *S*-nitrosylation in multiple cell types and tissues. *J. Biol. Chem.* **277,** 9637–9640.

Griendling, K. K., and Fitzgerald, G. A. (2003). Oxidative stress and cardiovascular injury. I. Basic mechanisms and *in vivo* monitoring of ROS. *Circulation* **108,** 1912–1916.

Halliwell, B., and Whiteman, M. (2004). Measuring reactive species and oxidative damage *in vivo* and in cell culture: How should you do it and what do the results mean? *Br. J. Pharmacol.* **142,** 231–255.

Han, J. Y., Takeshita, K., and Utsumi, H. (2001). Noninvasive detection of hydroxyl radical generation in lung by diesel exhaust particles. *Free Radic. Biol. Med.* **30,** 516–525.

Hare, J. M. (2003). Nitric oxide and excitation-contraction coupling. *J. Mol. Cell Cardiol.* **35,** 719–729.

Hare, J. M. (2004). Nitroso-redox balance in the cardiovascular system. *N. Engl. J. Med.* **351,** 2112–2114.

Hare, J. M., Lofthouse, R. A., Juang, G. J., Colman, L., Ricker, K. M., Kim, B., Senzaki, H., Cao, S., Tunin, R. S., and Kass, D. A. (2000). Contribution of caveolin protein abundance to augmented nitric oxide signaling in conscious dogs with pacing-induced heart failure. *Circ. Res.* **86,** 1085–1092.

Hare, J. M., and Stamler, J. S. (2005). NO/redox disequilibrium in the failing heart and cardiovascular system. *J. Clin. Invest.* **115,** 509–517.

Helbock, H. J., Beckman, K. B., Shigenaga, M. K., Walter, P. B., Woodall, A. A., Yeo, H. C., and Ames, B. N. (1998). DNA oxidation matters: The HPLC-electrochemical detection assay of 8-oxo-deoxyguanosine and 8-oxo-guanine. *Proc. Natl. Acad. Sci. USA* **95,** 288–293.

Hess, D. T., Matsumoto, A., Kim, S. O., Marshall, H. E., and Stamler, J. S. (2005). Protein *S*-nitrosylation: Purview and parameters. *Nat. Rev. Mol. Cell Biol.* **6,** 150–166.

Hilenski, L. L., Clempus, R. E., Quinn, M. T., Lambeth, J. D., and Griendling, K. K. (2004). Distinct subcellular localizations of Nox1 and Nox4 in vascular smooth muscle cells. *Arterioscler. Thromb. Vasc. Biol.* **24,** 677–683.

Holvoet, P., Vanhaecke, J., Janssens, S., Van de, W. F., and Collen, D. (1998). Oxidized LDL and malondialdehyde-modified LDL in patients with acute coronary syndromes and stable coronary artery disease. *Circulation* **98,** 1487–1494.

Hsieh, Y. S., Chen, B. C., Shiow, S. J., Wang, H. C., Hsu, J. D., and Wang, C. J. (2002). Formation of 8-nitroguanine in tobacco cigarette smokers and in tobacco smoke-exposed Wistar rats. *Chem. Biol. Interact.* **140,** 67–80.

Huang, B., and Chen, C. (2006). An ascorbate-dependent artifact that interferes with the interpretation of the biotin switch assay. *Free Radic. Biol. Med.* **41,** 562–567.

Jaffrey, S. R., Erdjument-Bromage, H., Ferris, C. D., Tempst, P., and Snyder, S. H. (2001). Protein *S*-nitrosylation: A physiological signal for neuronal nitric oxide. *Nat. Cell Biol.* **3,** 193–197.

Janiszewski, M., Souza, H. P., Liu, X., Pedro, M. A., Zweier, J. L., and Laurindo, F. R. (2002). Overestimation of NADH-driven vascular oxidase activity due to lucigenin artifacts. *Free Radic. Biol. Med.* **32,** 446–453.

Jarasch, E. D., Grund, C., Bruder, G., Heid, H. W., Keenan, T. W., and Franke, W. W. (1981). Localization of xanthine oxidase in mammary-gland epithelium and capillary endothelium. *Cell* **25**, 67–82.

Kadiiska, M. B., Gladen, B. C., Baird, D. D., Germolec, D., Graham, L. B., Parker, C. E., Nyska, A., Wachsman, J. T., Ames, B. N., Basu, S., Brot, N., Fitzgerald, G. A., *et al.* (2005). Biomarkers of oxidative stress study II: Are oxidation products of lipids, proteins, and DNA markers of CCl4 poisoning? *Free Radic. Biol. Med.* **38**, 698–710.

Kaur, H., Lyras, L., Jenner, P., and Halliwell, B. (1998). Artefacts in HPLC detection of 3-nitrotyrosine in human brain tissue. *J. Neurochem.* **70**, 2220–2223.

Khan, S. A., Lee, K., Minhas, K. M., Gonzalez, D. R., Raju, S. V., Tejani, A. D., Li, D., Berkowitz, D. E., and Hare, J. M. (2004). Neuronal nitric oxide synthase negatively regulates xanthine oxidoreductase inhibition of cardiac excitation-contraction coupling. *Proc. Natl. Acad. Sci. USA* **101**, 15944–15948.

Kosower, N. S., and Kosower, E. M. (1987). Thiol labeling with bromobimanes. *Methods Enzymol.* **143**, 76–84.

Linder, N., Rapola, J., and Raivio, K. O. (1999). Cellular expression of xanthine oxidore-ductase protein in normal human tissues. *Lab. Invest.* **79**, 967–974.

Liu, L., Hausladen, A., Zeng, M., Que, L., Heitman, J., and Stamler, J. S. (2001). A metabolic enzyme for S-nitrosothiol conserved from bacteria to humans. *Nature* **410**, 490–494.

Liu, L., Yan, Y., Zeng, M., Zhang, J., Hanes, M. A., Ahearn, G., McMahon, T. J., Dickfeld, T., Marshall, H. E., Que, L. G., and Stamler, J. S. (2004). Essential roles of S-nitrosothiols in vascular homeostasis and endotoxic shock. *Cell* **116**, 617–628.

McCord, J. M. (1993). Oxygen-derived free radicals. *New Horiz.* **1**, 70–76.

Michel, T., and Feron, O. (1997). Nitric oxide synthases: Which, where, how, and why? *J. Clin. Invest.* **100**, 2146–2152.

Mochizuki, M., Yano, M., Oda, T., Tateishi, H., Kobayashi, S., Yamamoto, T., Ikeda, Y., Ohkusa, T., Ikemoto, N., and Matsuzaki, M. (2007). Scavenging free radicals by low-dose carvedilol prevents redox-dependent Ca^{2+} leak via stabilization of ryanodine receptor in heart failure. *J. Am. Coll. Cardiol.* **49**, 1722–1732.

Muxfeldt, M., and Schaper, W. (1987). The activity of xanthine oxidase in heart of pigs, guinea pigs, rabbits, rats, and humans. *Basic Res. Cardiol.* **82**, 486–492.

Paolocci, N., Ekelund, U. E., Isoda, T., Ozaki, M., Vandegaer, K., Georgakopoulos, D., Harrison, R. W., Kass, D. A., and Hare, J. M. (2000). cGMP-independent inotropic effects of nitric oxide and peroxynitrite donors: Potential role for nitrosylation. *Am. J. Physiol. Heart Circ. Physiol.* **279**, H1982–H1988.

Que, L. G., Liu, L., Yan, Y., Whitehead, G. S., Gavett, S. H., Schwartz, D. A., and Stamler, J. S. (2005). Protection from experimental asthma by an endogenous broncho-dilator. *Science* **308**, 1618–1621.

Richelle, M., Turini, M. E., Guidoux, R., Tavazzi, I., Metairon, S., and Fay, L. B. (1999). Urinary isoprostane excretion is not confounded by the lipid content of the diet. *FEBS Lett.* **459**, 259–262.

Rizzi, C., Samouilov, A., Kutala, V. K., Parinandi, N. L., Zweier, J. L., and Kuppusamy, P. (2003). Application of a trityl-based radical probe for measuring superoxide. *Free Radic. Biol. Med.* **35**, 1608–1618.

Rossi, R., Milzani, A., Dalle-Donne, I., Giustarini, D., Lusini, L., Colombo, R., and Di Simplicio, P. (2002). Blood glutathione disulfide: *In vivo* factor or in vitro artifact? *Clin. Chem.* **48**, 742–753.

Saraiva, R. M., Minhas, K. M., Raju, S. V., Barouch, L. A., Pitz, E., Schuleri, K. H., Vandegaer, K., Li, D., and Hare, J. M. (2005). Deficiency of neuronal nitric oxide synthase increases mortality and cardiac remodeling after myocardial infarction: Role of nitroso-redox equilibrium. *Circulation* **112**, 3415–3422.

Schmidt, H. H., and Walter, U. (1994). NO at work. *Cell* **78**, 919–925.

Shimoi, K., Kasai, H., Yokota, N., Toyokuni, S., and Kinae, N. (2002). Comparison between high-performance liquid chromatography and enzyme-linked immunosorbent assay for the determination of 8-hydroxy-2′-deoxyguanosine in human urine. *Cancer Epidemiol. Biomarkers Prev.* **11**, 767–770.

Souza, J. M., Choi, I., Chen, Q., Weisse, M., Daikhin, E., Yudkoff, M., Obin, M., Ara, J., Horwitz, J., and Ischiropoulos, H. (2000). Proteolytic degradation of tyrosine nitrated proteins. *Arch. Biochem. Biophys.* **380**, 360–366.

Stamler, J. S., and Hausladen, A. (1998). Oxidative modifications in nitrosative stress. *Nat. Struct. Biol.* **5**, 247–249.

Stamler, J. S., Lamas, S., and Fang, F. C. (2001). Nitrosylation, the prototypic redox-based signaling mechanism. *Cell* **106**, 675–683.

Stamler, J. S., Singel, D. J., and Loscalzo, J. (1992). Biochemistry of nitric oxide and its redox-activated forms. *Science* **258**, 1898–1902.

Stamler, J. S., Toone, E. J., Lipton, S. A., and Sucher, N. J. (1997). (S)NO signals: Translocation, regulation, and a consensus motif. *Neuron* **18**, 691–696.

Steghens, J. P., Flourie, F., Arab, K., and Collombel, C. (2003). Fast liquid chromatography-mass spectrometry glutathione measurement in whole blood: Micromolar GSSG is a sample preparation artifact. *J. Chromatogr. B Analyt. Technol. Biomed. Life Sci.* **798**, 343–349.

Sun, J., Morgan, M., Shen, R. F., Steenbergen, C., and Murphy, E. (2007). Preconditioning results in *S*-nitrosylation of proteins involved in regulation of mitochondrial energetics and calcium transport. *Circ. Res.* **101**, 1155–1163.

Sun, J., Picht, E., Ginsburg, K. S., Bers, D. M., Steenbergen, C., and Murphy, E. (2006). Hypercontractile female hearts exhibit increased *S*-nitrosylation of the L-type Ca^{2+} channel alpha1 subunit and reduced ischemia/reperfusion injury. *Circ. Res.* **98**, 403–411.

Sun, J., Xin, C., Eu, J. P., Stamler, J. S., and Meissner, G. (2001). Cysteine-3635 is responsible for skeletal muscle ryanodine receptor modulation by NO. *Proc. Natl. Acad. Sci. USA* **98**, 11158–11162.

Tarpey, M. M., and Fridovich, I. (2001). Methods of detection of vascular reactive species: Nitric oxide, superoxide, hydrogen peroxide, and peroxynitrite. *Circ. Res.* **89**, 224–236.

Tarpey, M. M., White, C. R., Suarez, E., Richardson, G., Radi, R., and Freeman, B. A. (1999). Chemiluminescent detection of oxidants in vascular tissue: Lucigenin but not coelenterazine enhances superoxide formation. *Circ. Res.* **84**, 1203–1211.

Toyokuni, S., Tanaka, T., Hattori, Y., Nishiyama, Y., Yoshida, A., Uchida, K., Hiai, H., Ochi, H., and Osawa, T. (1997). Quantitative immunohistochemical determination of 8-hydroxy-2′-deoxyguanosine by a monoclonal antibody N45.1: Its application to ferric nitrilotriacetate-induced renal carcinogenesis model. *Lab. Invest.* **76**, 365–374.

Trujillo, M., Alvarez, M. N., Peluffo, G., Freeman, B. A., and Radi, R. (1998). Xanthine oxidase-mediated decomposition of *S*-nitrosothiols. *J. Biol. Chem.* **273**, 7828–7834.

Tsikas, D., Mitschke, A., Suchy, M. T., Gutzki, F. M., and Stichtenoth, D. O. (2005). Determination of 3-nitrotyrosine in human urine at the basal state by gas chromatography-tandem mass spectrometry and evaluation of the excretion after oral intake. *J. Chromatogr. B Analyt. Technol. Biomed. Life Sci.* **827**, 146–156.

Tsikas, D., Schwedhelm, E., Stutzer, F. K., Gutzki, F. M., Rode, I., Mehls, C., and Frolich, J. C. (2003). Accurate quantification of basal plasma levels of 3-nitrotyrosine and 3-nitrotyrosinoalbumin by gas chromatography-tandem mass spectrometry. *J. Chromatogr. B Analyt. Technol. Biomed. Life Sci.* **784**, 77–90.

Tsimikas, S., Brilakis, E. S., Miller, E. R., McConnell, J. P., Lennon, R. J., Kornman, K. S., Witztum, J. L., and Berger, P. B. (2005). Oxidized phospholipids, Lp(a) lipoprotein, and coronary artery disease. *N. Engl. J. Med.* **353**, 46–57.

Tsimikas, S., and Witztum, J. L. (2001). Measuring circulating oxidized low-density lipo-protein to evaluate coronary risk. *Circulation* **103**, 1930–1932.

Uehara, T., Nakamura, T., Yao, D., Shi, Z. Q., Gu, Z., Ma, Y., Masliah, E., Nomura, Y., and Lipton, S. A. (2006). S-nitrosylated protein-disulphide isomerase links protein misfolding to neurodegeneration. *Nature* **441**, 513–517.

Valgimigli, M., Valgimigli, L., Trere, D., Gaiani, S., Pedulli, G. F., Gramantieri, L., and Bolondi, L. (2002). Oxidative stress EPR measurement in human liver by radical-probe technique: Correlation with etiology, histology and cell proliferation. *Free Radic. Res.* **36**, 939–948.

Wang, G., Moniri, N. H., Ozawa, K., Stamler, J. S., and Daaka, Y. (2006). Nitric oxide regulates endocytosis by S-nitrosylation of dynamin. *Proc. Natl. Acad. Sci. USA* **103**, 1295–1300.

Whalen, E. J., Foster, M. W., Matsumoto, A., Ozawa, K., Violin, J. D., Que, L. G., Nelson, C. D., Nelson, C. D., Benhar, M., Keys, J. R., Rockman, H. A., Koch, W. J., *et al.* Regulation of beta-adrenergic receptor signaling by S-nitrosylation of G-protein-coupled receptor kinase 2. *Cell* **129**, 511–522.

Xu, K. Y., Huso, D. L., Dawson, T. M., Bredt, D. S., and Becker, L. C. (1999). Nitric oxide synthase in cardiac sarcoplasmic reticulum. *Proc. Natl. Acad. Sci. USA* **96**, 657–662.

Xu, L., Eu, J. P., Meissner, G., and Stamler, J. S. (1998). Activation of the cardiac calcium release channel (ryanodine receptor) by poly-S-nitrosylation. *Science* **279**, 234–237.

Yoshida, R., Ogawa, Y., and Kasai, H. (2002). Urinary 8-oxo-7,8-dihydro-2′-deoxygua-nosine values measured by an ELISA correlated well with measurements by high-performance liquid chromatography with electrochemical detection. *Cancer Epidemiol. Biomarkers Prev.* **11**, 1076–1081.

Zhao, H., Joseph, J., Fales, H. M., Sokoloski, E. A., Levine, R. L., Vasquez-Vivar, J., and Kalyanaraman, B. (2005). Detection and characterization of the product of hydroethidine and intracellular superoxide by HPLC and limitations of fluorescence. *Proc. Natl. Acad. Sci. USA* **102**, 5727–5732.

Zhao, H., Kalivendi, S., Zhang, H., Joseph, J., Nithipatikom, K., Vasquez-Vivar, J., and Kalyanaraman, B. (2003). Superoxide reacts with hydroethidine but forms a fluorescent product that is distinctly different from ethidium: Potential implications in intracellular fluorescence detection of superoxide. *Free Radic. Biol. Med.* **34**, 1359–1368.

Zimmet, J. M., and Hare, J. M. (2006). Nitroso-redox interactions in the cardiovascular system. *Circulation* **114**, 1531–1544.

CHAPTER TWENTY-TWO

MICROSCOPIC TECHNIQUE FOR THE DETECTION OF NITRIC OXIDE-DEPENDENT ANGIOGENESIS IN AN ANIMAL MODEL

Seung Namkoong,* Byoung-Hee Chung,* Kwon-Soo Ha,* Hansoo Lee,* Young-Guen Kwon,† *and* Young-Myeong Kim*

Contents

Abstract

Nitric oxide (NO) plays an important role in maintaining vascular homeostasis. The importance of NO in the vasculature is demonstrated by several experimental conditions, such as vascular endothelial growth factor (VEGF)-induced angiogenesis. Thus, the NO metabolic pathway in endothelial cells could be one of the main contributing factors for angiogenesis. Although several methods have

* Vascular System Research Center, Kangwon National University, Chunchon, Korea
† Department of Biochemistry, College of Science, Yonsei University, Seoul, Korea

Methods in Enzymology, Volume 441
ISSN 0076-6879, DOI: 10.1016/S0076-6879(08)01222-6

been used for measuring *in vitro* angiogenesis, a proper technique has not been developed for identifying *in vivo* NO-dependent angiogenesis. This chapter provides a new intravital microscopic method for detecting and measuring NO-dependent angiogenesis in a mouse model. This technique showed strong abdominal neovascularization in wild-type mice, but not eNOS knockout mice, locally injected with VEGF, as well as stimulation of angiogenesis in NO donor-injected mice. This technique also revealed the inhibitory effect of the NOS inhibitor N^{G}-iminoethyl-L-ornithine in VEGF-mediated *in vivo* angiogenesis. This chapter describes intravital microscopy as a new imaging technique for detecting NO-dependent angiogenesis in an animal model.

1. INTRODUCTION

Nitric oxide (NO), first characterized as a major endothelial-derived relaxing factor, synthesized from L-arginine by the catalytic reaction of nitric oxide synthase (NOS), is a gaseous molecule with an astonishingly wide range of physiological and pathophysiological activities, including the regulation of vessel tone and angiogenesis in wound healing, inflammation, ischemic cardiovascular diseases, and malignant diseases.

Angiogenesis is the formation of new blood vessels from preexisting vessels, which enables the delivery of oxygen and nutrients and is strongly regulated in many physiological and pathological conditions. The angiogenic process is complex and involves several discrete steps, such as extracellular matrix degradation, proliferation and migration of endothelial cells, and morphological differentiation of endothelial cells to form tubes (Hanahan and Folkman, 1996). A number of angiogenic factors can trigger the release of NO, synthesized by eNOS. Vascular endothelial growth factor (VEGF), a potent mitogen for endothelial cells, promotes angiogenesis of endothelial cells, as well as the development of the vascular system (Ferrara and Davis-Smyth, 1997).

The first evidence that NO was critically involved in angiogenesis was reported by Ziche *et al.* (1993), who identified the NO donor sodium nitroprusside as an *in vitro* angiogenic indicator that possessed a simulating effect on endothelial cell proliferation. They also showed that the vasoactive agent substance P promoted endothelial cell proliferation and migration, which were abolished by the NOS inhibitor (Ziche *et al.*, 1994). Furthermore, this study revealed that VEGF stimulated NO production in human endothelial cells, which was highly correlated with its angiogenic activity. Both events were blocked by the NOS inhibitor, indicating that NO production contributes to the proliferative and migratory roles of VEGF, which is believed to be responsible for stimulating angiogenesis (van der Zee *et al.*, 1997). In addition, pharmacological inhibitors of NOS effectively inhibited angiogenesis induced by immune-activated monocytes (Leibovich *et al.*, 1994), sphingosine-1-phosphate (Rikitake *et al.*, 2002), angiotensin II (Tamarat *et al.*, 2002), and fractalkine (Lee *et al.*, 2006). More direct evidence for a functional role of NO in angiogenesis was demonstrated by several studies showing that angiogenesis in ischemic and wounded sites was abolished in eNOS-knockout mice and restored by NOS gene transfer (Lee *et al.*, 1999; Murohara *et al.*, 1998). Thus, the NO metabolic pathway in endothelial cells may be a key factor in the control of physiological or pathological angiogenesis.

One of the most critical technical problems in the field of angiogenesis is the correct analysis of varied results from *in vivo* and *in vitro* assays currently in use. Intravital microscopy represents a technique that allows for direct, repetitive, and quantitative measurement of a variety of microcirculatory and morphological parameters *in vivo* (Lehr *et al.*, 1993). Barker and co-workers (1999) improved this technique and investigated microcirculatory changes associated with vascular delay in mice. Currently, this technique is modified for measuring various physiological and pathological phenomena in an animal model, including neovascularization or angiogenesis (Lee *et al.*, 2006; Tozer *et al.*, 2005). This chapter outlines the microscopic technique method used in our laboratories to detect angiogenesis generated by the NO-releasing factor, VEGF with the eNOS inhibitor and eNOS-deficient mice.

2. EXPERIMENTAL PROCEDURES

2.1. Intracellular NO detection

Human umbilical vein endothelial cells (HUVECs) are exposed to VEGF (10 ng/ml) with or without 10 μM N^G-monomethyl-L-arginine (NMA) for 4 h, washed twice with serum-free medium, and then incubated with 5 μM DAF-FM diacetate (Molecular Probes Inc.) for 1 h at 37 °C. After the excess probe is removed, cells are incubated for an additional 20 min to allow for complete deesterification of the intracellular DAF-FM diacetate to the non-permeable and nonfluorescent DAF-FM, which is converted to the highly fluorescent triazol form in the presence of NO. The fluorescence images are captured from at least 10 randomly selected cells per dish using a confocal laser microscope. The relative levels of intracellular NO are determined from the fluorescence intensity of DAF-FM.

2.2. *In vitro* tube formation assay

Twenty-four-well culture plates are coated with growth factor-reduced Matrigel according to the manufacturer's instructions. HUVECs incubated with M199 and 1% fetal bovine serum (FBS) for 6 h are harvested after trypsin treatment, resuspended in M199 with 1% FBS, plated onto a layer of Matrigel at a density of 2×10^5 cells/well, followed by the addition of 10 ng/ml VEGF with or without 10 μM L-NIO. VEGF and L-NIO are provided to the cells for 30 min at room temperature before plating. Matrigel cultures are incubated at 37 °C for 20 h. Tube formation is observed using an inverted phase-contrast microscope and images are captured with a video graphic system.

2.3. Animals

Male BALB/c mice (Orient, Sungnam, Korea) and eNOS knockout mice (Jackson Lab, Bar Harbor, ME) aged 6 to 8 weeks are used in this study. They are maintained at a specific pathogen-free housing facility with free access to tap water and standard laboratory food throughout the experiments.

2.4. Anesthesia

Both surgery and repetitive intravital fluorescence microscopy are performed under anesthesia by inhalation of 1.5% isoflurane and a mixture of O_2/N_2O using a vaporizer (Surgivet, Waukesha, WI).

2.5. Abdominal window preparation

After anesthesia, abdominal wall windows are implanted into the mice. A titanium circular mount with eight holes on the edge (19 mm outer diameter, 14 mm inner diameter, and 0.7 mm thick) is inserted between the skin and abdominal wall (Fig. 22.1A). Growth factor-reduced Matrigel (100 μl, BD Biosciences, Franklin Lakes, NJ) containing 100 ng VEGF (Upstate Biotechnology, Lake Placid, NY) with or without 100 μM of L-NIO/mouse (eNOS inhibitor; Calbiochem, San Diego, CA), as well as 100 μM of glycol-SNAP1, is then applied to the inner window space (Figs. 17.1B and 17.1C). Thereafter, the observation window is sealed with a circular glass coverslip and fixed by a snap ring for subsequent intravital microscopy (Fig. 22.1D). The window is well tolerated by the animals, which show no changes in sleeping or feeding habits.

2.6. Intravital microscopy

For *in vivo* microscopic investigation, mice are anesthetized and receive 50 μl of 25 mg/ml fluorescein isothiocyanate (FITC)-labeled dextran (250,000 mol wt; Sigma, St. Louis, MO) intravenously via the tail vein. The mice are then placed on a Zeiss Axiovert 200M microscope (Zeiss; Oberkocchen, Germany). The fluorescence epi–illumination microscopy setup includes a 100-W mercury lamp and filter set for blue light (440–475 nm, excitation wavelength; 530–550 nm, emission wavelength). Fluorescence images of five random locations of each window are conducted with an electron-multiplying charge-coupled device camera (Photon Max 512; Princeton

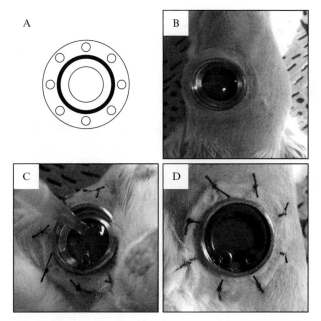

Figure 22.1   (A) Surgical procedure used to prepare the chamber displayed sequentially. Schematic titanium circular mounts (B) for insertion between the skin and the abdominal wall of mice. (C) The Matrigel mixtures were applied to the window inner space. (D) The window was sealed with a cover glass and fixed with a snap ring for subsequent intravital microscopy. (See color insert.)

Instruments, Trenton, NJ) and digitized for subsequent off-line analysis using MetaMorph (Universal Imaging Corp., Downingtown, PA). Repetitive microscopic observations are performed at 24 h as well as at 2 and 4 days. The assay is scored from 0 (least positive) to 6 (most positive) in a double-blinded manner.

2.7. Statistical analysis

Data are presented as means ± standard deviation (SD) of at least six mice per group. Statistical comparisons between groups are performed using one-way ANOVA followed by Student's t test. A $P < 0.05$ is considered statistically significant.

3. RESULTS

3.1. *In vitro* detection of NO-dependent angiogenesis

Because NO has been implicated as a mediator of angiogenesis, NO production was determined during VEGF-induced angiogenesis. HUVECs exposed to VEGF demonstrated elevated NO levels as well as stimulated

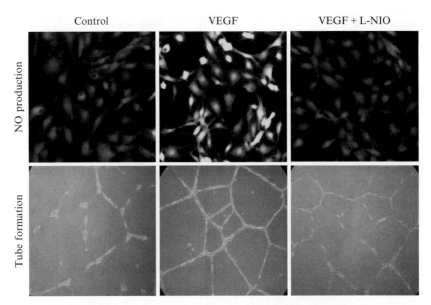

Figure 22.2 VEGF stimulates intracellular NO production and *in vitro* angiogenesis in HUVECs. (Top) Cells were treated with VEGF (10 ng/ml) in the presence or absence of L-NIO (10 μM) for 7 h and further incubated with 5 μM DAF-FM diacetate for another 1 h at 37 °C. Fluorescence images were captured using a confocal laser microscope. (Bottom) HUVECs were cultured on a layer of Matrigel with or without the indicated concentrations of FKN or 10 ng/ml VEGF for 20 h. Tube formation was observed using an inverted phase-contrast microscope with a video graphic system. The area covered by the tube network was quantitated using Image-Pro Plus software. All quantitative data are mean ± SD ($n = 3$). ⋆$P < 0.01$ versus control. (See color insert.)

in vitro angiogenesis compared with control, and these increases were inhibited by cotreatment with L-NIO (Fig. 22.2), indicating that NO plays a critical role in VEGF-induced *in vitro* angiogenesis.

3.2. Intravital microscopic detection of VEGF-induced angiogenesis in mice

For intravital microscopy, implantation of abdominal windows could be performed easily without complications. Moreover, daily microscopic inspection allowed for detection of microcirculation as well as the progression of angiogenesis. The quality of microscopic images was sufficient for the complete observation period of 4 days. In addition, the *in vivo* microscopic images displayed morphological and functional changes of microvascular structures. We examined the functional role of NO in angiogenesis by intravital microscopy using wild-type and eNOS-deficient mice locally administered with VEGF and L–NIO. VEGF injection resulted in an increase in characteristic signs of angiogenesis, such as capillary sprouting, tortuosity, and bud formation in wild-type mice over a 4-day observation period (Fig. 22.3A). The angiogenic response of VEGF was blocked by cotreatment with L–NIO. However, development and remodeling of newly formed microvessels, as well as angiogenic response, were not observed over the entire experimental study period in VEGF-injected eNOS-deficient mice. Statistically, analyzed data also showed that VEGF-induced *in vivo* angiogenesis was suppressed by cotreatment with L–NIO, which mimicked a role of eNOS deficiency in a mouse model (Fig. 22.3B). This technique is useful for directly detecting the functional role of NO in VEGF-induced angiogenesis in an animal model.

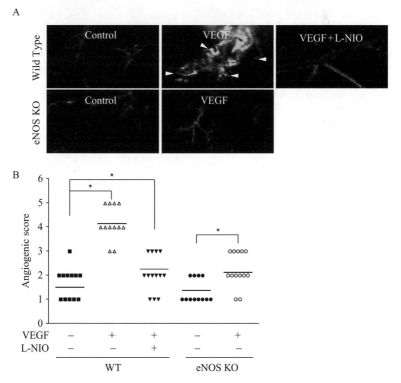

Figure 22.3 VEGF induces angiogenesis in wild-type mice, but not in eNOS-deficient mice. (A) Matrigel containing VEGF (100 ng/mouse) alone or plus L–NIO (100 μM/ mouse) was injected into the abdominal window of wild-type or eNOS-deficient mice, and after 4 days, the *in vivo* angiogenic activity was assessed after intravenous injection with FITC-dextran using an intravital microscope ($\times$50). (B) The *in vivo* angiogenic activity was scored from 0 (least positive) to 6 (most positive) in a double-blind manner. Data shown are mean $\pm$ SD ($n = 6$). *$P < 0.05$ versus control. (See color insert.)

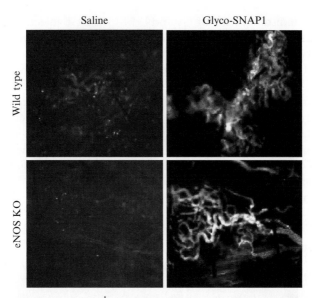

Figure 22.4 The NO donor glyco-SNAP1 stimulates angiogenesis in both wild-type and eNOS-deficient mice. Matrigel containing glycol-SNAP1 (100 μM/mouse) was injected into the abdominal window of wild-type or eNOS-deficient mice, and after 3 days, *in vivo* angiogenic activity was determined after intravenous injection with FITC-dextran using an intravital microscope. (See color insert.)

3.3. Intravital microscopic detection of NO donor-induced angiogenesis in mice

The NO donor glycol-SNAP1 is relatively stable and slowly releases NO because its half-life is 28 to 30 h, indicating that this compound is suitable for *in vivo* experiments. Using the intravital microscopic technique, we showed that activation of glycol-SNAP1-mediated NO release promoted the development of the intricate architecture of the neovascular network (angiogenesis) in both wild-type and eNOS knockout mice (Fig. 22.4). This animal model may serve to understand molecular mechanisms and allow for *in vivo* assessment of NO-mediated angiogenesis in several pathological conditions using a microscopic imaging technique.

4. SUMMARY AND CONCLUSION

Microscopic techniques have been show to allow direct and repetitive analysis of microcirculation in a variety of experimental models. This includes approaches based either on the exterioration or on the *in situ* visualization of organs and tissues (Schuder *et al.*, 1999; Szczesny *et al.*, 2000).

Moreover, using various fluorescent dyes, the intravital microscopic technique enables direct examination of the angiogenic response in vessels, organs, and tissues. This advanced technique has been implemented in different studies to analyze tumor formation and metastasis, as well as to observe the interaction between endothelial cells and monocytes *in vivo* (Hanahan and Folkman, 1996; Tozer *et al.*, 2005). We have developed intravital fluorescence microscopy for imaging new vessel formation *in vivo* following injection with angiogenic factors VEGF and fractalkine (Lee *et al.*, 2006). This chapter demonstrated that this technique significantly extends the capabilities of detecting and measuring NO-dependent angiogenesis in a living animal model.

ACKNOWLEDGMENT

This work was supported by a Vascular System Research Center Grant from the Korea Science and Engineering Foundation.

REFERENCES

Barker, J. H., Frank, J., Bidiwala, S. B., Stengel, C. K., Carroll, S. M., Carroll, C. M., van Aalst, V., and Anderson, G. L. (1999). An animal model to study microcirculatory changes associated with vascular delay. *Br. J. Plast. Surg.* **52,** 133–142.

Ferrara, N., and Davis-Smyth, T. (1997). The biology of vascular endothelial growth factor. *Endocr. Rev.* **18,** 4–25.

Hanahan, D., and Folkman, J. (1996). Patterns and emerging mechanisms of the angiogenic switch during tumorigenesis. *Cell* **86,** 353–364.

Lee, P. C., Salyapongse, A. N., Bragdon, G. A., Shears, L. L., Watkins, S. C., Edington, H. D., and Billiar, T. R. (1999). Impaired wound healing and angiogenesis in eNOS-deficient mice. *Am. J. Physiol.* **277,** H1600–H1608.

Lee, S. J., Namkoong, S., Kim, Y. M., Kim, C. K., Lee, H., Ha, K. S., Chung, H. T., Kwon, Y. G., and Kim, Y. M. (2006). Fractalkine stimulates angiogenesis by activating the Raf-1/MEK/ERK- and PI3K/Akt/eNOS-dependent signal pathways. *Am. J. Physiol. Heart Circ. Physiol.* **291,** H2836–H2846.

Lehr, H. A., Leunig, M., Menger, M. D., Nolte, D., and Messmer, K. (1993). Dorsal skinfold chamber technique for intravital microscopy in nude mice. *Am. J. Pathol.* **143,** 1055–1062.

Leibovich, S. J., Polverini, P. J., Fong, T. W., Harlow, L. A., and Koch, A. E. (1994). Production of angiogenic activity by human monocytes requires an L-arginine/nitric oxide-synthase-dependent effecter mechanism. *Proc. Natl. Acad. Sci. USA* **91,** 4190–4194.

Murohara, T., Asahara, T., Silver, M., Bauters, C., Masuda, H., Kalka, C., Kearney, M., Chen, D., Symes, J. F., Fishman, M. C., Huang, P. L., and Isner, J. M. (1998). Nitric oxide synthase modulates angiogenesis in response to tissue ischemia. *J. Clin. Invest.* **101,** 2567–2578.

Rikitake, Y., Hirata, K., Kawashima, S., Ozaki, M., Takahashi, T., Ogawa, W., Inoue, N., and Yokoyama, M. (2002). Involvement of endothelial nitric oxide in sphingosine-1-phosphate-induced angiogenesis. *Arterioscler. Thromb. Vasc. Biol.* **22,** 108–114.

Schuder, G., Vollmar, B., Richter, S., Pistorius, G., Fehringer, M., Feifel, G., and Menger, M. D. (1999). Epi-illumination fluorescent light microscopy for the *in vivo* study of rat hepatic microvascular response to cryothermia. *Hepatology* **29,** 801–808.

Szczesny, G., Nolte, D., Veihelmann, A., and Messmer, K. (2000). A new chamber technique for intravital microscopic observations in the different soft tissue layers of mouse hindleg. *J. Trauma* **49,** 1108–1115.

Tamarat, R., Silvestre, J. S., Kubis, N., Benessiano, J., Duriez, M., deGasparo, M., Henrion, D., and Levy, B. I. (2002). Endothelial nitric oxide synthase lies downstream from angiotensin II-induced angiogenesis in ischemic hindlimb. *Hypertension* **39,** 830–835.

Tozer, G. M., Ameer-Beg, S. M., Baker, J., Barber, P. R., Hill, S. A., Hodgkiss, R. J., Locke, R., Prise, V. E., Wilson, I., and Vojnovic, B. (2005). Intravital imaging of tumour vascular networks using multi-photon fluorescence microscopy. *Adv. Drug Deliv. Rev.* **57,** 135–152.

van der Zee, R., Murohara, T., Luo, Z., Zollmann, F., Passeri, J., Lekutat, C., and Isner, J. M. (1997). Vascular endothelial growth factor/vascular permeability factor augments nitric oxide release from quiescent rabbit and human vascular endothelium. *Circulation* **95,** 1030–1037.

Ziche, M., Morbidelli, L., Masini, E., Amerini, S., Granger, H. J., Maggi, C. A., Geppetti, P., and Ledda, F. (1994). Nitric oxide mediates angiogenesis *in vivo* and endothelial cell growth and migration *in vitro* promoted by substance P. *J. Clin. Invest.* **94,** 2036–2044.

Ziche, M., Morbidelli, L., Masini, E., Granger, H., Geppetti, P., and Ledda, F. (1993). Nitric oxide promotes DNA synthesis and cyclic GMP formation in endothelial cells from postcapillary venules. *Biochem. Biophys. Res. Commun.* **192,** 1198–1203.

Author Index

Subject Index

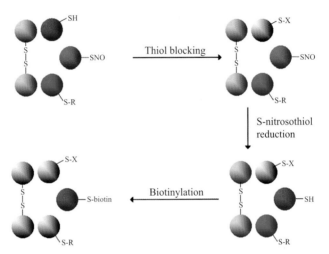

Nicholas J. Kettenhofen *et al.*, Figure 4.1 The biotin-switch method. The biotin-switch assay involves selectively replacing the nitroso group with a biotin label. This is accomplished in three steps. The first step involves blockade of all thiol groups. The second step involves the selective reduction of *S*-nitrosothiols to their parent thiols, avoiding reduction protein–protein disulfides or protein-mixed disulfides. The third step involves labeling of the nascent thiols with a thiol-specific biotinylation agent, thus tagging only the *S*-nitrosated proteins.

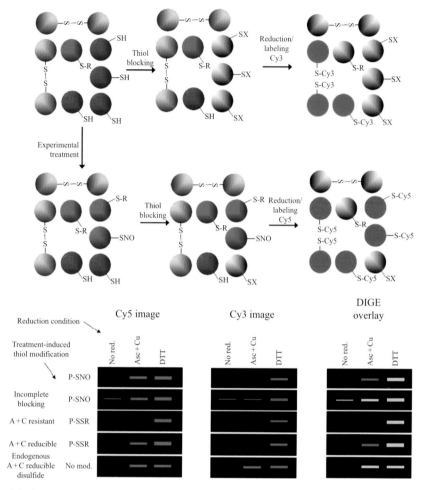

Nicholas J. Kettenhofen *et al.*, Figure 4.2   The CyDye-switch method. (Top) Protein cysteine residues can exist in protein disulfides as mixed disulfides and as free thiols. The experimental treatment (e.g., addition of NO donor or activation of NOS) will convert some of these thiols to *S*-nitrosothiols and some to disulfides. In the CyDye-switch method, the control sample is treated with a thiol blocker, and the resulting proteins are reduced with ascorbate/Cu and labeled with green Cy3. Proteins that show up as green fluorescence will therefore represent unblocked thiols and endogenous disulfides that were reduced (e.g., by the ascorbate/Cu treatment or by DTT). The experimental sample is treated in the same way, but labeled with red Cy5. In this case, red fluorescence will be associated with the same proteins that are labeled in the control experiment, but in addition, both *S*-nitrosothiols, and potentially some mixed disulfides that were formed from the experimental treatment, will also be labeled in red. The labeled protein from both control and experiment are then pooled at equal protein concentration and run on a gel. (Bottom) Typical patterns that may be expected in various scenarios.

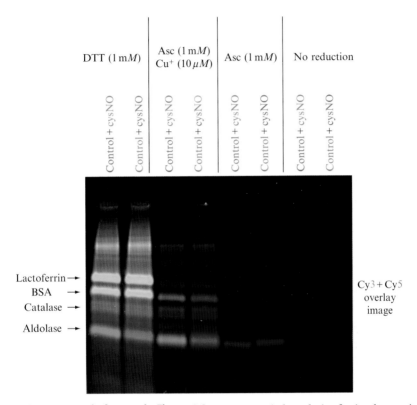

Nicholas J. Kettenhofen *et al.*, Figure 4.3 CyDye-switch analysis of a simple protein mixture. An equimolar mix (1 mg/ml total protein) of aldolase, catalase, BSA, and lactoferrin was treated with either 100 μM S-nitrosocysteine (CysNO) or buffer alone (control) for 30 min. Samples were then labeled with Cy3 (green) or Cy5 (red) by CyDye switch using various reduction conditions as indicated. Equal amounts of experimentally paired samples were then pooled and separated by SDS-PAGE. An overlay image of Cy3 and Cy5 fluorescence is presented.

pH 3   pH 10
–   +

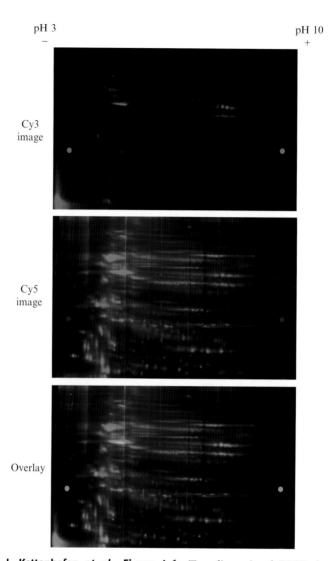

Cy3
image

Cy5
image

Overlay

Nicholas J. Kettenhofen *et al.*, Figure 4.6 Two-dimensional DIGE detection of
S-nitrosated proteins in a complex mixture. NHBE cells were lysed, and cytosolic proteins were treated with either 100 μM *S*-nitrosocysteine (CysNO) or buffer alone (control) for 30 min. Samples were then labeled using the CyDye-switch protocol with
ascorbate (1 mM)/Cu(II) (10 μM) reduction. The control sample was labeled with Cy3
(green) while the treated sample was labeled with Cy5 (red). Equal amounts of protein
from each sample were then pooled and coseparated by two-dimensional electrophoresis. The gel was imaged for both Cy3 and Cy5 fluorescence. Cy3 (top), Cy5 (middle),
and overlay (bottom) images are presented.

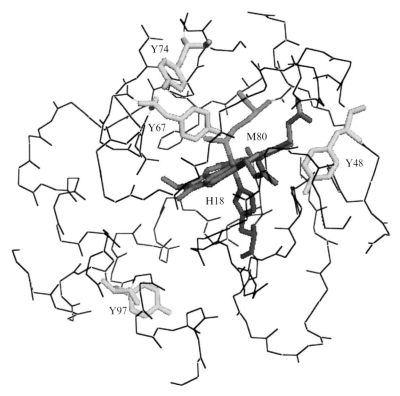

José M. Souza *et al.*, Figure 11.1 Tyrosine residues in cytochrome *c*. The three-dimensional structure of cytochrome *c* was obtained from the Protein Data Bank (1HRC) and downloaded using Pymol (http://pymol.sourceforge.net). The four tyrosine residues are represented in yellow. Additional structures shown are the heme group (red) and its fifth and sixth amino acid ligands, His-18 (blue) and Met-80 (green), respectively. The rest of the protein backbone is drawn with black lines.

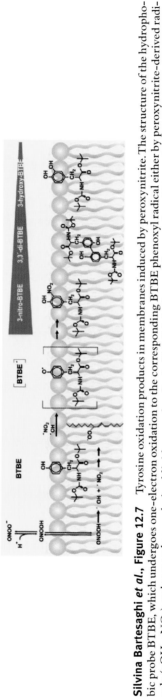

Silvina Bartesaghi *et al.*, Figure 12.7 Tyrosine oxidation products in membranes induced by peroxynitrite. The structure of the hydrophobic probe BTBE, which undergoes one-electron oxidation to the corresponding BTBE phenoxyl radical either by peroxynitrite–derived radicals (·OH, ·NO$_2$) or by membrane–derived lipid peroxyl radicals (ROO·), is shown. The transient BTBE phenoxyl radical either reacts at diffusion–controlled rates with ·NO$_2$ to yield 3–nitro–BTBE or recombines with another phenoxyl radical to yield 3,3′di-BTBE; nitration yields are significantly larger than dimerization yields. The figure also indicates the formation of small amounts of the 3–hydroxy-BTBE from the addition reaction with hydroxyl radical and supports the diffusion and homolysis of ONOOH within the lipid bilayer.

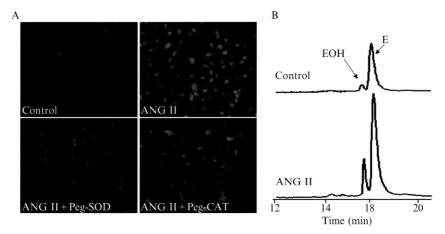

Francisco R. M. Laurindo *et al.*, Figure 13.2   (A) Fluorescence microscopic images of cultured vascular smooth muscle cells after incubation with DHE (2 μM) for 10 min, observed under a rhodamine filter in a common fluorescence microscope (Zeiss Axiovert 200). Cells were analyzed in control conditions or after incubation with the NADPH oxidase agonist angiotensin II (ANG II, 100 nM, 4 h) and, in the latter case, also after the addition of inhibitors Peg-superoxide dismutase (Peg-SOD, 50 U/ml) or Peg-catalase (Peg-CAT, 200 U/ml). (B) HPLC chromatogram of control and ANG II–stimulated cell extracts showing 2-EOH and ethidium peaks. After acetonitrile extraction, the supernatant was injected into the HPLC system, as described in text, and identification was performed with fluorescence detector.

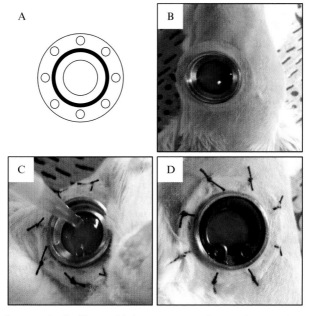

Seung Namkoong *et al.*, Figure 22.1   (A) Surgical procedure used to prepare the chamber displayed sequentially. Schematic titanium circular mounts (B) for insertion between the skin and the abdominal wall of mice. (C) The Matrigel mixtures were applied to the window inner space. (D) The window was sealed with a cover glass and fixed with a snap ring for subsequent intravital microscopy.

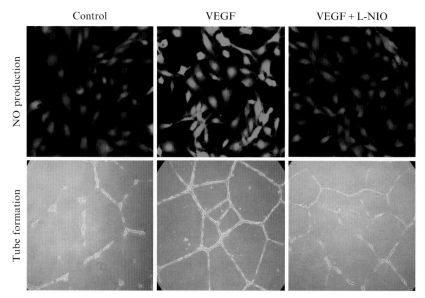

Seung Namkoong *et al.*, Figure 22.2   VEGF stimulates intracellular NO production and *in vitro* angiogenesis in HUVECs. (Top) Cells were treated with VEGF (10 ng/ml) in the presence or absence of L-NIO (10 μM) for 7 h and further incubated with 5 μM DAF-FM diacetate for another 1 h at 37 °C. Fluorescence images were captured using a confocal laser microscope. (Bottom) HUVECs were cultured on a layer of Matrigel with or without the indicated concentrations of FKN or 10 ng/ml VEGF for 20 h. Tube formation was observed using an inverted phase–contrast microscope with a video graphic system. The area covered by the tube network was quantitated using Image-Pro Plus software. All quantitative data are mean $\pm$ SD ($n = 3$). $\star P < 0.01$ versus control.

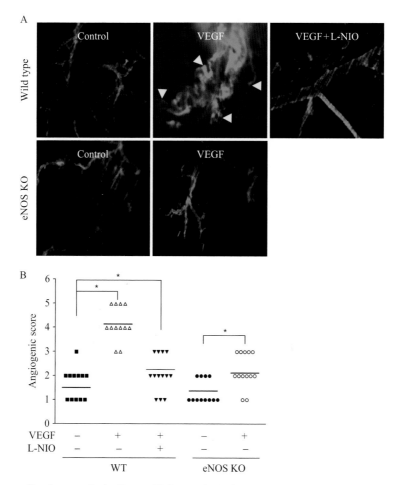

Seung Namkoong *et al.*, Figure 22.3   VEGF induces angiogenesis in wild-type mice, but not in eNOS-deficient mice. (A) Matrigel containing VEGF (100 ng/mouse) alone or plus L-NIO (100 μM/mouse) was injected into the abdominal window of wild-type or eNOS-deficient mice, and after 4 days, the *in vivo* angiogenic activity was assessed after intravenous injection with FITC–dextran using an intravital microscope ($\times$50). (B) The *in vivo* angiogenic activity was scored from 0 (least positive) to 6 (most positive) in a double-blind manner. Data shown are mean $\pm$ SD ($n = 6$). $\star P < 0.05$ versus control.

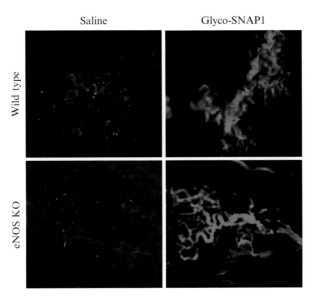

Seung Namkoong *et al.*, Figure 22.4 The NO donor glyco-SNAP1 stimulates angiogenesis in both wild-type and eNOS-deficient mice. Matrigel containing glycol-SNAP1 (100 μM/mouse) was injected into the abdominal window of wild-type or eNOS-deficient mice, and after 3 days, *in vivo* angiogenic activity was determined after intravenous injection with FITC–dextran using an intravital microscope.